Marine Ecology and Biodiversity

Marine Ecology and Biodiversity

Editor: Jonas Bailey

www.callistoreference.com

Callisto Reference,
118-35 Queens Blvd., Suite 400,
Forest Hills, NY 11375, USA

Visit us on the World Wide Web at:
www.callistoreference.com

ISBN: 978-1-64116-086-5 (Hardback)

Cataloging-in-Publication Data

Marine ecology and biodiversity / edited by Jonas Bailey.
    p. cm.
Includes bibliographical references and index.
ISBN 978-1-64116-086-5
1. Marine ecology. 2. Marine biodiversity. 3. Biodiversity. 4. Aquatic ecology.
I. Bailey, Jonas.
QH541.5.S3 M37 2019
577.7--dc23

Table of Contents

Preface

I am honored to present to you this unique book which encompasses the most up-to-date data in the field. I was extremely pleased to get this opportunity of editing the work of experts from across the globe. I have also written papers in this field and researched the various aspects revolving around the progress of the discipline. I have tried to unify my knowledge along with that of stalwarts from every corner of the world, to produce a text which not only benefits the readers but also facilitates the growth of the field.

Marine ecology and biodiversity is the scientific study of marine ecosystems in relation to their biotic and abiotic environment. Marine ecosystems have a large biodiversity and are essential for the sustenance of terrestrial and marine environments. This is because marine organisms are responsible for maintaining carbon, nitrogen, phosphorus and nutrient cycles as well as contributing significantly to the world's total photosynthetic output. Pollution caused by human activities, mostly through the widespread use of fertilizers and pesticides that flow into the oceans, has a drastic effect on the marine ecosystem. The presence of garbage, plastics and oils may result in an increase in ocean acidification and eutrophication. These effects are witnessed through a drop in productivity of marine ecosystems and a reduction in marine biodiversity. This book strives to provide a fair idea about marine ecology and marine biodiversity, as well as helps to develop a better understanding of the diverse aspects of these fields. The various studies that are constantly contributing towards advancing technologies and evolution of these fields are examined in detail. It is an essential guide for both academicians and those who wish to pursue these disciplines further.

Finally, I would like to thank all the contributing authors for their valuable time and contributions. This book would not have been possible without their efforts. I would also like to thank my friends and family for their constant support.

Editor

Colonization record of *Isognomon bicolor* (Mollusca: Bivalvia) on pipeline monobuoys in the Brazilian south coast

Vanessa Ochi Agostini[1*] and Carla Penna Ozorio[2]

Abstract

Background: The introduction of alien species in a system affecting native species due to competition for food and space. The purse oyster *Isognomon bicolor* (Bivalvia, Pteriidae), a potential exotic invader, presents gaps in its distribution and ecology knowledge in Brazilian coast. For this reason, our objective was to investigate the occurrence of this oyster on pipeline monobuoys in Rio Grande do Sul, southern Brazil. Specimens were collected monthly during surveys carried out from March 2010 to June 2011 at the Tramandaí Beach, along two pipeline monobuoys, MN-601 and MN-602, randomly from the depths of three and 22 m.

Results: We found *I. bicolor* dwelling on artificial hard substrates present in the Rio Grande do Sul coast, Brazil. This Caribbean bivalve was accidentally introduced to the Brazilian coast by ballast water from ships and, until now; it had not been recorded on the coast of Rio Grande do Sul state, which is under subtropical conditions.

Conclusions: The occurrence of *I. bicolor* in the faunistic surveys accomplished in the Rio Grande do Sul coast represents the first documented record for this species in this region. As this species is a potential exotic invader, with a high dispersal capacity, control measures must be employed to prevent its spread.

Keywords: Alien species, Artificial substrate, Biofouling, Bivalvia, Exotic species, Hard substrate, Pipeline, Pteriidae, Purse oyster, Tramandaí beach

Introduction

According to Rios' (2009) catalogue of mollusks of the Brazilian coast, Bivalvia is represented by a total of 391 species, with five belonging to family Pteriidae Gray, 1847. Among the South Atlantic coast species, only one, *Pteria hirundo* (Linnaeus, 1758), has been recorded until now from the Rio Grande do Sul (RS) state. The occurrence of purse oyster *Isognomon bicolor* (C. B. Adams, 1845) in the faunistic surveys accomplished in the Rio Grande do Sul coast represents the first documented record for this species in this region, thus increasing the number of Pteriidae in the benthic communities of the South Atlantic Ocean. *Isognomon bicolor* is a bivalve found in intertidal and subtidal artificial and natural hard substrates. They are seen in tide pools and other marine shallow areas and, on rocky shores with high energy, occupying cracks in protected sites. This species is an exotic mollusk to Brazilian marine fauna, native to the Western Central Atlantic Coast living usually in the Caribbean region. They are often found in high densities (Domaneschi & Martins, 2002; Santos et al. 1845) and interfere with the survival capacity of native species (Breves-Ramos et al. 2009). As there are disagreements about the identification of *Isognomon* species (Domaneschi & Martins, 2002), its current distribution on the Brazilian coast is not well established. However *I. bicolor* can be considered an invasive organism based on population size (Martinez, 2012) and has been recorded to North, Northeast and Southwest and more recently to the South of Brazil (Domaneschi & Martins, 2002; Loebmann et al., 2010; Martinez, 2012; Dias et al., 2013, Santos et al., 2015).

The first record of *Isognomon* to Brazil was made by Matthews & Kempf (1970) with the observation of *Isognomon* cf. *alatus* in Atol das Rocas, Rio Grande do

* Correspondence: voagostini@gmail.com
[1]Universidade Federal do Rio Grande (FURG), Programa de Pós-Graduação em Oceanografia Biológica (PPGOB), Instituto de Oceanografia (IO), Avenida Itália, Km 8, CEP 96203-900 Rio Grande, RS, Brazil
Full list of author information is available at the end of the article

Norte. Nevertheless, Domaneschi & Martins (2002) believe that the species found by these authors would likely be *I. bicolor*. According to Santos et al., (2015), the probable vector to introduce *I. bicolor* on the Brazilian coast is the ballast water of international cargo ships. This introduction is affecting native species due to competition for food and space (Kado, 2003) and may have a negative influence on hard substrates communities and consequently in organisms of economic interest.

I. bicolor, a potential exotic invader, with a high dispersal capacity (Martinez, 2012) presents gaps in its distribution and ecology knowledge (Breves et al., 2014), so its records in any coast are vital to promoting an efficient environmental management.

Material and methods

Isognomon bicolor specimens were collected monthly between March 2010 and June 2011, during samplings on two offshore pipeline monobuoys, MN-601 (30°00'40" S; 50°05'42" W) and MN-602 (30°01'52" S; 50°04'36" W) at Tramandaí Beach, RS, Brazil. These petroleum industry artifacts belong to company TRANSPETRO (Petrobras Transportes S.A.) and have been installed for transporting of oil products (MN-601) and crude oil (MN-602). The first is located 4 km offshore, 20 m deep, and the second is 6 km offshore, 24 m deep. The connection between the tanker and monobuoys is made by 16 in. rubber pipeline, and the product transport occurs by piping to the Almirante Soares Dutra Terminal (Tedut) in Osório. In 2009, this monobuoys/pipeline system was used by 224 ships.

The sampling was performed by professional divers of the TRANSPETRO through aleatory scrapings between depths of three and 22 m, considering both superficies, metal (vertical position) and rubber (diagonal position), depending on sea conditions. The organisms were removed of 900 cm^2 for posterior identification. Then, specimens were transported to the laboratory, separated from community and identified using Domaneschi & Martins (2002) description.

Usual measures for this bivalve were taken from each animal and digital images were taken (Fig. 1). The density was estimated, being the individuals fixed in ethyl alcohol 70 % and deposited in the scientific collection of the Centro de Estudos Costeiros Limnológicos e Marinhos (CECLIMAR) at Federal University of Rio Grande do Sul (UFRGS) (30 voucher specimens).

According to Domaneschi & Martins (2002), intraspecific variation and shell deformation due to large individuals' density in restricted areas as cavities or cracks may make it difficult to distinguish between *I. alatus* and *I. bicolor*. However, *I. bicolor* can be identified according to the redescription by the same authors.

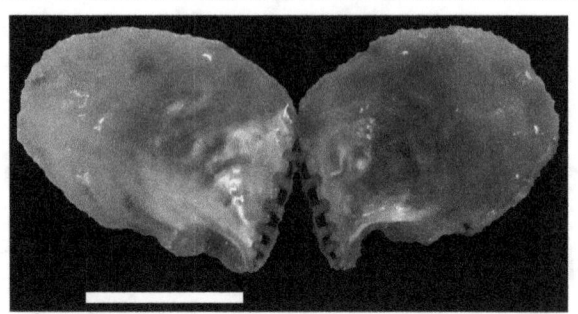

Fig. 1 Exotic species *Isognomon bicolor* found in pipeline monobuoys MN-601 (30°00'40" S; 50°05'42" W) and MN-602 (30°01'52" S; 50°04'36" W) at Tramandaí Beach, RS, Brazil between March 2010 and June 2011 (Scale: 10 mm)

I. bicolor has (*i*) a shell with height greater than the length, the left valve slightly convex to the right; (*ii*) variation in color from blackened brown, slightly red, to cream-yellow tonnes or slightly greenish; (*iii*) outer surface ornamented with irregular, concentric and overlapped lamellae; (*iv*) inside the valves, the area that encloses the body is moderately concave, covered with bright nacre, the color uniformly iridescent; (v) height (mm): 1.3 to 36.0 and hinge length (mm): 0.9 to 17.5.

Results and discussion

Isognomon bicolor (Fig. 1) was observed between three and 22 m depths on both monobuoys, fouling the structures along with the barnacle *Megabalanus coccopoma* (Darwin, 1854) and the mussel *Perna perna* (Linnaeus, 1758). A large population was verified, putatively because the artifacts are offshore, permanently covered and under stable environmental conditions. Usually, *I. bicolor* individuals fixed by byssus were in places where the direct wave impact was absent or minimized by substrate relief (Lopes, 2009).

The *I. bicolor* densities were estimated in 2245 individuals per square meter at 3 m depth and 1129 individuals per square meter at 22 m at summer, being similar along of the months (Agostini, 2011). The densities obtained at 3 m deep support the results recorded in Arraial do Cabo, Rio de Janeiro to this species (2300-2500 ind. m^{-2}) (Lopes, 2009). However, in the mollusks inventory done previously by Rios et al. (1979) in the TRANSPETRO monoboys, *I. bicolor* was not listed. The family Pteriidae Gray, 1847 (1820) has been represented along of Brazil, including the Southern region, with four living genera only: *Isognomon* Solander, 1786, *Pteria* Scopoli, 1777, *Pinctada* Röding, 1798 and *Crenatula* Lamarck, 1804 (Domaneschi & Martins, 2002 Coan et al. 2000; Rios, 2009). According to Rios (2009), the *Isognomon* genus is represented in the Brazilian malacofauna just by *I. alatus* (Gmelin, 1791) and *I. radiatus* (Anton, 1838). However, Domaneschi & Martins (2002) points out the possibility that the *I. alatus* populations of São Sebastião, São Paulo, Brazil, could be *I. bicolor*.

In the other hand, many researchers recorded *I. bicolor* in the coast of the states of Santa Catarina, Paraná, São Paulo, Rio de Janeiro, Espírito Santo, Bahia, Pernambuco, Paraíba, Rio Grande do Norte, Ceará and Piauí (Domaneschi & Martins, 2002; Loebmann et al., 2010, Zamprogno et al., 2010; Rocha-Barreira et al., 2010; Martinez, 2012; Dias et al. 2013; Oliveira & Creed 2008; Gomes and Silva 2013, and Santos et al., 2015). About Brazilian Southern coast, the most established population of *I. bicolor* is at Zimbros Beach in Santa Catarina state (Domaneschi & Martins 2002), approximately 370 Km away straight from Tramandaí Beach in the Rio Grande do Sul. Their works have reported its colonizing in both artificial and natural substrates (rocky shore) and have attributed its adaptive success mainly due to the lack of competitors or specialized predators.

I. bicolor is a native Caribbean mollusk. Artificial materials floating on the sea could be responsible for the *I. bicolor* transport for areas otherwise impossible naturally to the dispersal mechanisms (Thiel & Gutow, 2005), allowing its contact with new rocky substrates similar to its natural environment in the Caribbean region (Tunnell et al., 2010). Another reason could be the ballast water. Breves et al. (2014) recorded *I. bicolor* to Uruguay coast (34°S) and emphasized its extremely rapid and huge expansion in the South Atlantic Ocean. They claim that ship hull fouling or the ballast water was responsible for its introduction. To the Brazilian coast, also, it is believed that this species was accidentally introduced to the ballast water (Breves-Ramos et al., 2009) and ship hulls (Lopes, 2009).

Alien marine species can be introduced into a new geographic area by many ways including ballast water or biofouling. The accumulation of unwanted organisms on hard surfaces occurs on hulls and other submerged parts of vessels (including ducts), shells or carapaces of other species, equipment associated with fishing, mariculture or diving, and even marine debris (GISP, 2008). However, according to Molnar et al. (2008), shipping is the most common pathway (69 %) of species introduction. Maritime transport vessels, through ballast water and fouling on the hulls, are known vectors throughout the history of dispersal of species across oceans (Coutts et al., 2003).

The presence of this mollusk in Tramandaí pipeline monobuoys is probably related to the large influx of Brazilian vessels on site, which can release larvae in the place by encrusted individuals in the hull and by ballast water. The hard substrate closer to the MN-601 and MN-602 monobuoys is the rocky shore in Torres, which is located approximately 100 km away. However,

Agostini (2011) conducted a species inventory of hard substrates in this place and did not register *Isognomon bicolor* at any site.

Any artificial substrate into the sea is susceptible to colonization by organisms. Structures, such as piers, ships hulls, and pipeline monobuoys, offer their surfaces for invertebrates' fixation, turning into feeding areas for many other organisms (Bumbeer & Rocha 2012).

According to López (2003), the snails *Stramonita haemastoma* (Linnaeus, 1767) and *Morula nodulosa* (C. B. Adams, 1845) eat *I. bicolor*. The first snail occurs in both monobuoys (Agostini, 2011), and could be a controlling factor of *I. bicolor* populations. But, López et al. (2010) reported that this gastropod predator prefers the bivalve *Perna perna* (Linnaeus, 1758), which also is present in Tramandaí monobuoys (Rios et al. 1979; Agostini, 2011).

Conclusions

The occurrence of *Isognomon bicolor* in pipeline monobuoys of Tramandaí Beach represents the first documented record for this species in Rio Grande do Sul state and according to Lopes (2009) this species showed an expressive increase in its density along the Brazilian coast from the mid-1990s. Thus, it is necessary the application of management and control measures, mainly related to disposal of ballast water (Regulation Norman 20) and the encrusted community on hulls, to preventing *I. bicolor* spread. Monobuoys and pipelines should also be considered a vector of bioinvasion from offshore to onshore zone, linking the species brought by ships to the natural substrates such as rocky shore.

Acknowledgements

To the company TRANSPETRO (Petrobrás - S.A.) for the collection of charred material from pipeline monobuoys, and to the CECLIMAR for providing the workspace and allowing the deposition of the vouchers in the scientific collection.

Funding

There is no funding.

Authors' contributions

VOA (design, collection, and data analysis). CPO (design, data analysis and intellectual contribution). Both authors read and approved the final manuscript.

Competing interests

The authors declare that they have no competing interests.

Author details

[1]Universidade Federal do Rio Grande (FURG), Programa de Pós-Graduação em Oceanografia Biológica (PPGOB), Instituto de Oceanografia (IO), Avenida Itália, Km 8, CEP 96203-900 Rio Grande, RS, Brazil. [2]Universidade Federal do Rio Grande do Sul (UFRGS), Instituto de Biociências, Departamento de Zoologia, Avenida Bento Gonçalves, 9500, CEP 91501-970 Porto Alegre, RS, Brazil.

References

Agostini VO. Levantamento dos macroinvertebrados de substratos consolidados naturais e artificiais do litoral norte do Rio Grande do Sul, Brasil e caracterização do processo de bioincrustação em substrato metálico sob condições marinhas costeiras subtropicais. Trabalho de Conclusão de Curso (Bacharelado em Biologia Marinha e Costeira) – Universidade Federal do Rio Grande do Sul, Imbé; 2011. 106 f. doi:10.13140/RG.2.2.30219.49440.

Breves-Ramos A, Junqueira AOR, Lavrado HP, Silva SHG, Ferreira-Silva MAG. Population structure of the invasive bivalve *Isognomon bicolor* on rocky shores of Rio de Janeiro State (Brazil). J Mar Biol Assoc UK. 2009;90(3):453–9.

Breves A, Scarabino F, Carranza A, Leoni V. First records of the non-native bivalve *Isognomon bicolor* (C. B. Adams, 1845) rafting to the Uruguayan coast. Check List. 2014;10(3):684–6.

Bumbeer J de A, Rocha RM da. Detection of introduced sessile species on the near shore continental shelf in southern Brazil. Zoologia. 2012;29(2):126–34.

Coan EV, Scott PV, Bernard FR. Bivalve seashells of Western North America. Santa Barbara: Santa Barbara Museum of Natural History; 2000. p. 764.

Coutts ADM, Moore KM, Hewitt CL. Ships' sea-chests: an overlooked transfer mechanism for non-indigenous marine species? Mar Pollut Bull Oxford. 2003; 46:1504–15.

Dias TLP, Mota ELS, Gondim AI, Oliveira JM, Rabelo EF, de Almeida SM, et al. *Isognomon bicolor* (C. B. Adams, 1845) (Mollusca: Bivalvia): First record of this invasive species for the States of Paraíba and Alagoas and new records for other localities of Northeastern Brazil. Check List. 2013;9(1):157–61.

Domaneschi O, Martins CM. *Isognomon bicolor* (C.B. Adams) (Bivalvia, Isognomonidae): primeiro registro para o Brasil, redescrição da espécie e considerações sobre a ocorrência e distribuição de *Isognomon* na costa brasileira. Rev Bras Zool. 2002;19(2):611–27.

Global Invasive Species Programme (GISP). Marine Biofouling: an assessment of risks and management initiatives. Compiled by Lynn Jackson on behalf of the Global Invasive Species Programme and the UNEP Regional Seas Programme. 2008. p. 68.

Gomes LEdeO, Silva EC. New record of Isognomon bicolor (C. B. Adams, 1845) (Bivalvia, Isognomonidae) to Bahia Litoral North. Pan-Am J Aquat Sci. 2013; 8(4):361–3.

Kado R. Invasion of Japanese shores by the NE Pacific barnacle Balanus gleula and its ecological and biogeographical impact. Mar Ecol Prog Ser. 2003;249: 199–206.

Loebmann D, Mai ACG, Lee JT. The invasion of five alien species in the Delta do Parnaíba Environmental Protection Area, Northeastern Brazil. Rev Biol Trop. 2010;58(3):909–23.

López MS. Efecto de la potencial presa exótica *Isognomon bicolor* (Adams, 1845) sobre la ecología trófica de *Stramonita haemastoma* (Kool, 1987) en el intermareal rocoso de Arraial do Cabo, RJ, Brasil. Jaén: MS thesis, Universidad Internacional de Andalucía Sede Antonio Machado de Baeza; 2003.

Lopes MS. Informe sobre as Espécies Exóticas Invasoras Marinhas no Brasil. Ministério do Meio Ambiente; 2009. 441 pp. http://www.mma.gov.br/ estruturas/sbf2008_dcbio/_publicacao/147_publicacao07072011012531.pdf.

López MS, Coutinho R, Ferreira CEL, Rilov G. Predator–prey interactions in a bioinvasion scenario: differential predation by native predators on two exotic rocky intertidal bivalves. Mar Ecol Prog Ser. 2010;403:101–12.

Martinez AS. Spatial distribution of the invasive bivalve *Isognomon bicolor* on rocky shores of Arvoredo Island (Santa Catarina, Brazil). J Mar Biol Assoc UK. 2012;92(3):495–503.

Matthews HR, Kempf M. Moluscos marinhos do norte e nordeste do Brasil. 11 - Moluscosdo Arquipélago de Fernando de Noronha (com algumas referências ao Atol das Rocas). Arq Ciênc Mar. 1970;10(l):l-53.

Molnar JL, Gamboa RL, Revenga C, Spalding M. Assessing the global threat of invasive species to marine biodiversity. Front Ecol Environ. 2008;6(9):485–492. doi:10.1890/070064

Oliveira AES, Creed JC. Mollusca, Bivalvia, Isognomon bicolor (C. B. Adams 1845): Distribution extension. Check List. 2008;4(4):386–8.

Rios EC. Compendiun of Brazil Sea Shells. 2nd ed. Rio Grande: Evangrap; 2009. p. 663.

Rios EC, Lopes-Pitoni VL, Veintenheimer-Mendes IL. Moluscos marinhos em bóias no Rio Grande do Sul, Brasil. In: Encontro dos Malacologistas brasileiros, 5, 1979. Mossoró. Anais. Porto Alegre: FZBRS; 1979. p. 103–7.

Rocha-Barreira CA, Matthews-Cascon H, Mai ACG. Moluscos. In: Mai ACG, Loebmann D, editors. Guia Ilustrado: Biodiversidade do Litoral do Piauí. Sorocaba: Paratodos; 2010. p. 62–93.

Santos HF, Borzone CA, Tavares YAG. Distribuição espacial e temporal de *Isognomon bicolor* (C.B. Adams, 1845) (Bivalvia, Isognomonidae) no litoral paranaense, Brasil. Trabalho Científico;2015. Available at: http://www.mma.gov.br/. Accessed 14 Aug 2015.

Thiel M, Gutow L. The ecology of rafting in the marine environment. II. The rafting organisms and community. Oceanography Mar Biol Ann Rev. 2005;43: 279–419.

Tunnell Jr., JW, Andrews J, Barrera NC, Moretzsohn F. Encyclopedia of Texas: Identification, Ecology, Distribution, and History. Texas A&M University Press; 2010. 512 pp. https://muse.jhu.edu/book/395.

Zamprogno GC, Fernandes LL, Fernandes FC. Spatial variability in the population of *Isognomon bicolor* (C.B. Adams, 1845) (Mollusca, Bivalvia) on rocky shores in Espirito Santo, Brazil. Braz J Oceanography. 2010;58(1):23–9.

2

Tamoya haplonema (Cnidaria: Cubozoa) from Uruguayan and adjacent waters: oceanographic context of new and historical findings

Valentina Leoni[1,2,3*†], Silvana González[1,2,3†], Leonardo Ortega[1†], Fabrizio Scarabino[1,2,4†], Gabriela Failla Siquier[5], Alicia Dutra[6], Luis Rubio[1,2], Martin Abreu[7], Wilson Serra[1,2,3], Ana Gabriella Alonzo Campi[6], Sergio N. Stampar[8†] and André C. Morandini[9]

Abstract

New records of the cubozoan jellyfish *Tamoya haplonema* in Uruguayan waters are reported together with historical records for the region, and associated with the oceanographic conditions at the moment of the finding. Occurrences of the species are mainly associated with positive Sea Surface Temperature Anomalies especially during summer months when the intrusion of warm oceanic waters to the Uruguayan coastline is stronger. This was particularly strong during 2012–2013, when a dry period enhanced this scenario. This species is the only cubozoan present in Uruguay, with a sporadic occurrence and so far has no appreciated negative effects on public health. However, from observed increasing frequency of positive temperature anomalies it would be reasonable to predict a future southward shift in the latitudinal distribution of *T. haplonema*. In this context, occurrence of this toxic species along Uruguayan coastal waters must be considered with particular attention to the potential negative impact on tourism and on general public health.

Keywords: Box jellyfish, *Tamoya haplonema*, Uruguay, South Atlantic

Introduction

The cnidarian class Cubozoa comprises approximately 50 described species (Bentlage et al. 2010), mostly restricted to warm waters of the globe (Marques et al. 2003; Morandini et al. 2005). Cubozoans have ecological and medical importance playing an important role in marine food webs, affecting fishing activities and tourism (Kingsford and Mooney 2014). However, knowledge on their ecology and on how oceanographic conditions affect population dynamics and distribution are rare for cubomedusae when compared to scyphomedusae and hydromedusae. This disparity could be explained by a limited knowledge on cubozoan taxonomy, the rarity of some taxa, as well as their large temporal and spatial variation in abundance (Kingsford and Mooney 2014).

The box jellyfish *Tamoya haplonema* Müller, 1859 is one of the two valid species of the genus, also represented by the recently described *Tamoya ohboya* (Collins et al. 2011). As many other cubozoans, *T. haplonema* causes severe envenomation (Haddad et al. 2009). Morandini and Marques (1997) reported the first case of envenomation in southeastern Brazilian waters, which produced burning sensation followed by a 7-day period of itching and a permanent scar on the affected skin.

Tamoya haplonema is associated to warm neritic waters of both western and eastern Atlantic Ocean (Mianzan and Cornelius 1999), although records from eastern Atlantic might be considered doubtful (Pagès et al. 1992; Gershwin and Gibbons 2009). In the Southwestern Atlantic, the southernmost record is at 38°39′S, 58°40′W, off Puerto Quequén, Buenos Aires Province, Argentina (Mianzan and Cornelius 1999; Pastorino 2001).

* Correspondence: valenleoni64@gmail.com
†Equal contributors
[1]Dirección Nacional de Recursos Acuáticos, Constituyente 1497, C.P. 11200 Montevideo, Uruguay
[2]Museo Nacional de Historia Natural, C.C. 399, C.P 11000 Montevideo, Uruguay
Full list of author information is available at the end of the article

In the present study, we aimed to: 1) present new records of *Tamoya haplonema* in Uruguayan coastal water, 2) review and summarize previous records of this species for Uruguay and adjacent waters, and 3) explore possible association between such records and the oceanographic and climatic context.

Materials and methods

Study area

The most relevant oceanographic feature of the Southwestern Atlantic Ocean is the Brazil-Malvinas Confluence Zone, where the warm poleward-flowing Brazil Current and the cold equatorward-flowing Malvinas/Falklands Current converge (Fig. 1, Gordon 1981; 1989). The location of this zone shifts seasonally reaching higher latitudes during austral summer (i.e., December 21st to March 21st, Maamaatuaiahutapu et al. 1994; Barré et al. 2006). Shallow coastal waters near the borders of Uruguay and Argentina (i.e., depths < 50 m) are highly influenced by the Río de la Plata freshwater discharge that exhibits monthly variations, with higher discharges during the austral fall-winter and the lowest discharge during the

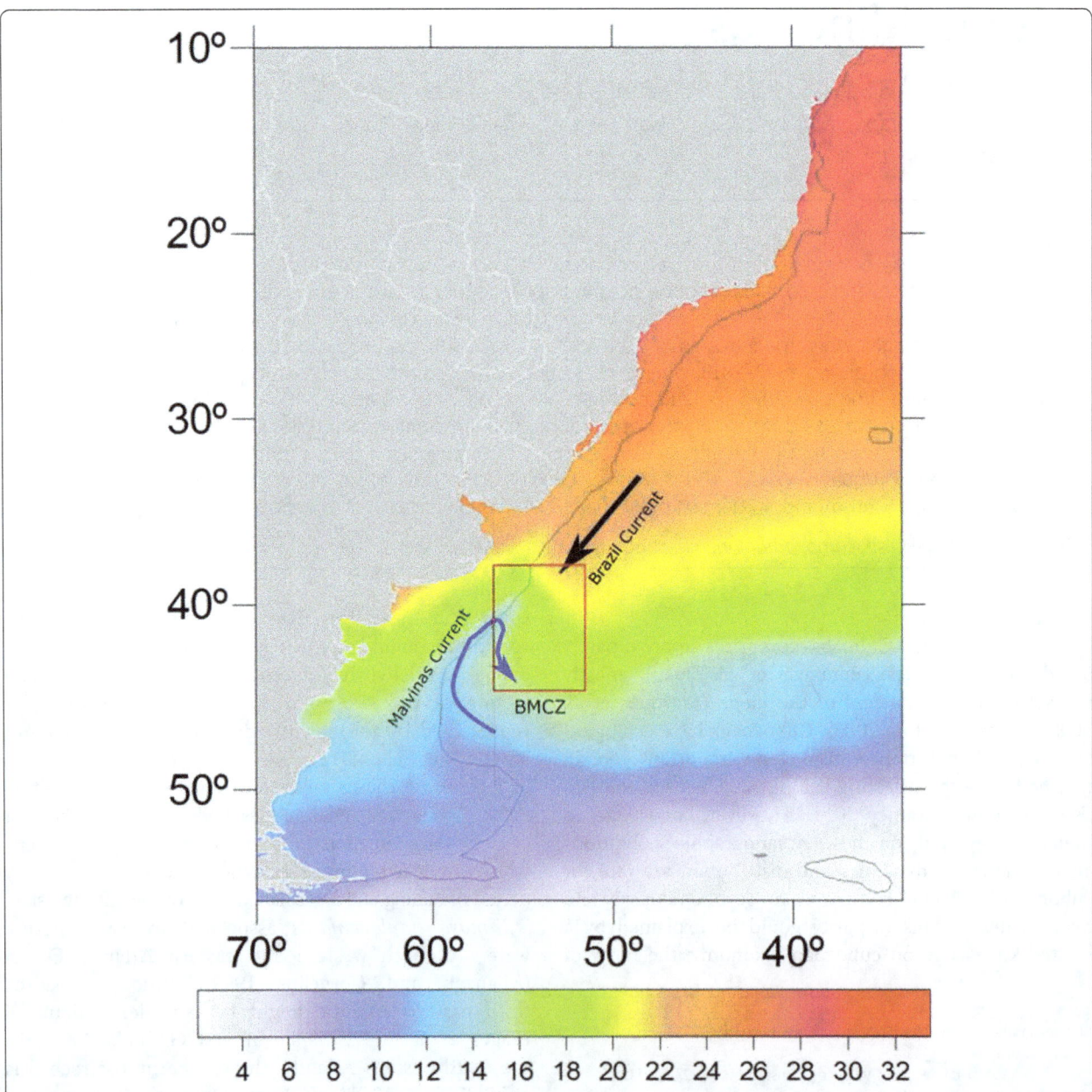

Fig. 1 Schematic representation of the most relevant surface currents of the western South Atlantic. The Brazil–Malvinas/Falklands Confluence Zone (BMCZ) is shown over Aqua MODIS Sea Surface Temperature (°C) austral summer climatology (2002–2013, http://oceancolor.gsfc.nasa.gov/cgi/l3); the 1000 m isobath is shown

austral spring-summer (Sepúlveda et al. 2004). Interannual changes in the precipitation regime mainly connected with El Niño Southern Oscillation (ENSO) (Cazes-Boezio et al. 2003) also influences discharges of Río de la Plata, increasing and decreasing during El Niño and La Niña events, respectively. Thus, the lower freshwater input during austral summer combined with a higher influence of Brazil Current waters in the region, favours the incursion of warm oceanic waters in the coastal area (Ortega and Martínez 2007) which could be enhanced during certain years in response to ENSO cold phase (La Niña).

Literature records, samples identification and measurements

Published records of cubozoans in the temperate zone of the Southwestern Atlantic Ocean are scarce and were analyzed in detail, even considering particularly cryptic/grey literature as regional conference abstracts. Samples consisted of: a) stranded individuals along the Uruguayan coast (February 1990 and 2007, January and February 2012, and January, February, March and September 2013), and b) specimens collected on board R.V. *"Aldebaran"* on November 2012. Such specimens were obtained as by-catch of a Engel-type bottom trawl net (100 mm stretched mesh in the wings, 50 mm stretched mesh in the cod ends, 28 m horizontal aperture and 4 m vertical opening), used for demersal fish assessment of the Dirección Nacional de Recursos Acuáticos (DINARA). Other records were made by sightings from lifeguards during summer (January 2012, February and March 2013), but specimens were not collected. Since 2011, lifeguards from the Uruguayan coast had been trained by specialists in courses and workshops about identification of gelatinous zooplankton species, receiving guides with photographs and key diagnostic features as well as standardized recording sheets (Failla Siquier and Dutra 2012; Failla Siquier and Dutra 2014).

In order to contextualize the historical presence of *T. haplonema* in the Uruguayan coast we considered: 1) regular efforts made since 1991 during the whole year in recording stranded fauna at Costa Azul - La Aguada in La Paloma (G. Fabiano, pers. comm.; M. Abreu, pers. obs.) and 2) during the last 11 years at La Esmeralda during the summer (A.G. Alonzo Campi, pers. obs.). These included at least two opportunities monthly recording unusual stranding and a weekly periodicity during the summer. These were carried on by professionals that lately found *T. haplonema* and assured the novelty of that finding since 2012. We also considered 12 trawl coastal surveys (each of ca. 40 tows of 30′each, see other details above) made all over the entire Uruguayan inner shelf since 1999 onboard the R.V. *"Aldebaran"*, personally inspected by two of us (S. González and F. Scarabino).

The specimens were preserved in 5 % formaldehyde solution in sea or freshwater. Identification was made following Morandini et al. (2005) after preservation and measurements were taken in well-preserved specimens. In all collected specimens the umbrella height and width (interpedalia distance, IPD) were measured (maximum, minimum and the average of the four IPD are reported). Pedalia were not preserved properly in some individuals and therefore measurements cannot be provided. The cnidome (set of types and sizes of cnidae from different body parts) was performed from a preserved specimen stranded in La Paloma (34°39′S 54°8′W) on February 2013 based on methods and nomenclature presented by Collins et al. (2011). Specimens were deposited in the collection of Invertebrate Zoology of the Museo Nacional de Historia Natural (Montevideo, Uruguay) and in the collection of Invertebrate Laboratory of Facultad de Ciencias, Universidad de la República (Montevideo, Uruguay).

Oceanographic and climatic data

Monthly Sea Surface Temperature Anomalies (SSTA) data were computed from the Extended Reconstructed Sea Surface Temperature, version 3 (ERSST_V3) dataset based on IRI/LDEO Climate Data Library (Xue et al. 2003; Smith et al. 2008). Monthly regional averages of SSTA from 1960 to 2013 were calculated for the area enclosed between latitudes 30° and 36°S and longitudes between 50° and 60°W. The analyses were performed using the Ingrid language provided by the IRI Data Library (http://iridl.ldeo.columbia.edu/). Weighted Anomaly Standardized Precipitation information from the IRI Climate Map Room (http://iridl.ldeo.columbia.edu/maproom/Global/Precipitation/WASP_Indices.html; Lyon 2004; Lyon and Barnston 2005) was used to analyze the relative deficit of precipitation during the 2011–2013 summer period.

During the survey performed in November 2012 on board R.V. *"Aldebaran"*, temperature and salinity data was obtained in situ with CTD (SBE-19 plus, Sea-Bird Electronics).

Results

Historical records of *T. haplonema* in Uruguayan and adjacent waters

The first record of a cubomedusa from Uruguayan coast was made by Barattini and Ureta (1961, sometimes dated as 1960), specifically for the coast of Maldonado, as being uncommon. These authors used (for the first time in cnidarian nomenclature) the name *Carybdea atlantica*, without providing any nomenclatural information like authority or indication that this was a new name. There was no illustration other than one drawing quoted as *"Charibdea marsupialis"* apparently given for illustrating the general aspect of a cubomedusa. Moreover, the characters provided in the two paragraphs under *C. atlantica* are non-informative and shared with several species. Mianzan et al. (1988), Mianzan and Cornelius (1999) and Pastorino (2001) commented the mention of Barattini and Ureta (1961) either as dubious and/or possibly conspecific with *Tamoya haplonema*, considering that this is the only cubomedusa recorded from Uruguayan waters or southward (Goy 1979; Mianzan and Cornelius 1999; Pastorino 2001; Failla Siquier 2006). Pastorino (2001) more specifically referred it as a *nomen nudum*. It is unclear if Barattini and Ureta (1961) had the intention to describe a new species or if even *C. atlantica* is a *lapsus*

calami for *Carybdea alata* Reynaud, 1830 (now *Alatina alata*, see Gershwin 2005; Lewis et al. 2013). However, we fully support the proposition of Pastorino (2001) to treat *Carybdea atlantica* Barattini and Ureta, 1961 as a *nomen nudum* and to refer definitively this record as being based on specimens of *Tamoya haplonema*. This conclusion is only based on geographical distribution (*T. haplonema* is the only species recorded from the region) considering that all characters and measurements

Table 1 New and historical records of *Tamoya haplonema* from Uruguayan and adjacent waters (Argentina and southernmost Brazil). Comparative measurements in cm (specimens lacking this data were damaged): umbrella and pedalia height, average, maximum and minimum interpedalia distance (IPD)

Date	Reference	Location	No. specimens	State	Umbrella height	IPD	Pedalia height
28th September 2013	Present work	La Aguada, Rocha (34°38'S 54°9'W)	1	Stranded	-	-	-
6th March 2013	Present work	Cabo Polonio Sur, Rocha	1	Stranded	-	-	-
4th March 2013	Present work	Arachania, Rocha	1	Stranded/Collected	6.3	4 (3.5–4.5)	3.6
23th February 2013	Present work	La Aguada, Rocha (34°38'S 54°9'W)	1	Stranded/Collected	9.4	4.0 (3.0–5.4)	5.0
22nd February 2013	**Present work**	**Playa Puerto de la Paloma, Rocha (34°39'S 54°8'W)**	**1**	**Stranded/Collected**	**10.0**	**4.0 (3.1–4.4)**	**5.6**
12nd February 2013	Present work	Aguas Dulces, Rocha	1	Stranded	-	-	-
10th February 2013	Present work	Cerro Verde, Rocha	1	Stranded/Photographed	-	-	-
16th January 2013	Present work	La Esmeralda, Rocha (34°10'S 53°40'W)	1	Stranded	-	-	-
28th November 2012	Present work	34°13'S 53°30'W	1	Collected onboard R.V. "*Aldebaran*"	15.8	6.3 (4.4–8.6)	5.0
28th November 2012	Present work	34°13'S 53°30'W	1	Collected onboard R.V. "*Aldebaran*"	14.9	6.5 (5.9–7.1)	5.2
16th February 2012	Present work	La Barra, Maldonado	1	Stranded/Collected	14	4.4 (3.5–5.5)	5.7
6th February 2012	Present work	Port of Punta del Este, Maldonado	1	Alive/Photographed	10	6.0	5.0
January 2012	Present work	La Esmeralda, Rocha (34°10'S 53°40'W)	2	Stranded/1 collected	8.4	4.5 (4.1–5.0)	4.0
30th January 2012	Present work	La Paloma, Rocha	1	Stranded	-	-	-
23rd February 2007	Present work	La Paloma, Rocha	1	Stranded/Collected	8.75	4.7 (4.2–5.0)	4.8
December 2004	Failla-Siquier 2006	Valizas, Rocha	1	Stranded/Collected	6.5	3.7 (3.0–4.5)	4.0
26th May 2000	Pastorino 2001	Off Puerto Quequén, Argentina (38°39'S 58°40'W)	1	Collected onboard	9.0	7.2	3.0
26th May 2000	Pastorino 2001	Off Puerto Quequén, Argentina (38°39'S 58°40'W)	1	Collected onboard	8.3	6.6	2.7
January 1993	Failla-Siquier 2006	Rocha	1	Stranded	-	-	-
20th February 1990	Present work	La Paloma (Bahía Chica), Rocha	1	Stranded/Collected	6	4.5	-
8th January 1962	Goy 1979	Southernmost Brazil (33°48'S 53°08'W)	1	Collected onboard R.V. "*Calypso*"	11.0	8.5	-
21st December 1961	Goy 1979	Off Punta del Diablo, Rocha (34°07'S 53°12'W)	2	Collected onboard R.V. "*Calypso*"	11.0	8.5	-
-	Barattini and Ureta 1961	Maldonado	-	-	-	-	-

Bold: Specimen with cnidome studied

given by Barattini and Ureta (1961) are vague. It must be pointed out that the closest geographical record of *Alatina alata* (as *Carybdea alata*) is from deep trawls (1067 m depth) off Bahia state (Brazil) at 14°S 38°W approximately 3,000 km away from Uruguay (Morandini 2003).

Reliable records of *T. haplonema* for the Uruguayan coast are indeed scarce and consists of only two specimens recorded by Goy (1979) and collected off Punta del Diablo (34°07′S, 53°12′W) on December 21st 1961 during the cruise of the R.V. *Calypso* at a depth of 20–22 m and a surface temperature of 20.8 °C (Forest 1966). Another record of this species from southernmost Brazilian waters, very close to Uruguay, also comes from the R.V. *Calypso* collected on January 8th 1962 (33°48′S, 53°08′W, 19 m) (Goy 1979). Failla Siquier (2006) recorded cubomedusae specimens found stranded in the coast of Rocha, Uruguay on January 1993 and December 2004 identifying them with doubts (due to the poor condition of the material) as *T. haplonema*. Following the same rationale applied to the record of Barattini and Ureta (1961), we consider these records as *T. haplonema*.

New records of *Tamoya haplonema* from Uruguayan coast (Southwestern Atlantic Ocean)

The following diagnostic characters were observed in well preserved specimens: 1) bell elongate cuboid, longer than wide, and translucid, 2) four lateral perradial rhopaliar with horizontal rhopaliar niche opening, 3) four simple inter-radial pedalia and one tentacle by pedalium and 4) four vertically arranged phacellae. The extreme values of two measures for all specimens collected are 6 – 15.8 cm (for umbrella height) and 3.6 – 5.7 cm (for pedalium height) (Table 1). Characterization of cnidome of different body parts of a specimen stranded in La Paloma was made, measuring capsule lengths and widths (Table 2). Nowadays, the absence of small oval amastigophore nematocysts in the tentacle base of *T. haplonema* is a feature that distinguishes the species.

The new records of *T. haplonema* consist of 17 specimens including: a) 14 stranded individuals in Rocha and Maldonado, b) two specimens collected on board R.V. "*Aldebaran*" on November 2012 (34°13′S-53°30′W, 22 m, i.e., off Punta del Diablo, Rocha), and c) one sighting on board on February 2012 in Maldonado (Figs. 2 and 3, Table 1). Three records of the stranded specimens were made by lifeguards in Rocha during summer months (January to March) but those were not collected neither photographed.

Oceanographic and climatic context of regional water records

The long term analysis of regional mean SSTA, showed an increasing trend from 1960 to 2013 in the study area (Fig. 4), with a predominance of positive anomalies

Table 2 Cnidome of different body parts of a *Tamoya haplonema* specimen from Uruguay

Tamoya haplonema (La Paloma, Uruguay)				
Tentacle tip	Mean (range) L	Mean (range) W	N	Abundance
Macrobasic p-eurytele	41.6 (35.5 – 49.6)	16.1 (14.1 – 19.2)	30	Abundant
Microbasic birhopaloid	23.1 (21.2 – 26.7)	14.1 (12.1 – 16.2)	30	Rare
Holotrichous isorhiza	28.2 (24.6 – 31.2)	22.4 (18.1 – 24.8)	30	Fair
Tentacle Base				
Macrobasic p-eurytele	51.2 (46.2 – 53.4)	16.5 (14.4 – 18.2)	30	Abundant
Oval isorhiza	14.6 (9.2 – 16.6)	9.5 (8.1 – 12.2)	20	Rare
Holotrichous isorhiza	29.6 (25.2 – 33)	23.6 (20 – 25.8)	30	Fair
Microbasic p-birhopaloid	25.5 (22.2 – 27.3)	18.2 (14.5 – 20.5)	30	Fair
Holotrichous isorhiza	27.5 (25.2 – 30)	22 (18.5 – 23.1)	30	Fair
Phacellae				
Microbasic p-birhopaloid	21.2 (19.5 – 22.9)	15.8 (12.8 – 18.9)	30	Fair
Holotrichous isorhiza	32.5 (28.5 – 34.8)	26.5 (23.1 – 29.4)	10	Rare
Pedalial Warts				
Holotrichous isorhiza	34.5 (27.5 – 36.6)	29.8 (25.5 – 31.6)	30	Fair
Oval isorhiza	11.2 (8.6 – 12.9)	7.8 (6.5 – 8.9)	10	Rare
Bell Warts				
Holotrichous isorhiza	33.6 (29.5 – 36.4)	28.2 (26.7 – 30.1)	30	Fair
Oval isorhiza	10.5 (9.6 – 13.2)	8.7 (7 – 10.2)	10	Rare
Apex Warts				
Holotrichous isorhiza	31.2 (28.4 – 36.5)	28.6 (26.9 – 32.2)	20	Rare
Oval isorhiza	11.1 (8.9 – 12.8)	8.5 (6.8 – 10.4)	10	Rare

L and W denote capsule lengths and widths, respectively, in μm. *N* means number of undischarged capsules measured

mainly after 2000. Most of the records match with the occurrence of positive SSTA (ca. 80 %) in the region during the month of the finding (Fig. 4) or during spring in a period characterized by a strong La Niña event. Moreover, nearly 80 % of the records of *T. haplonema* correspond to austral summer when the influence of warm oceanic waters from the Brazil Current is higher in the area.

The specimens collected on board R.V. "*Aldebaran*" in November 2012 in the coastal area of Rocha were associated with relatively warm and salty waters (mean water column temperature = 18.16 (SD ± 1.32) and salinity 30.64 (SD ± 0.28)) whereas surface temperature and salinity were 19.10 °C and 30.62 respectively, suggesting the

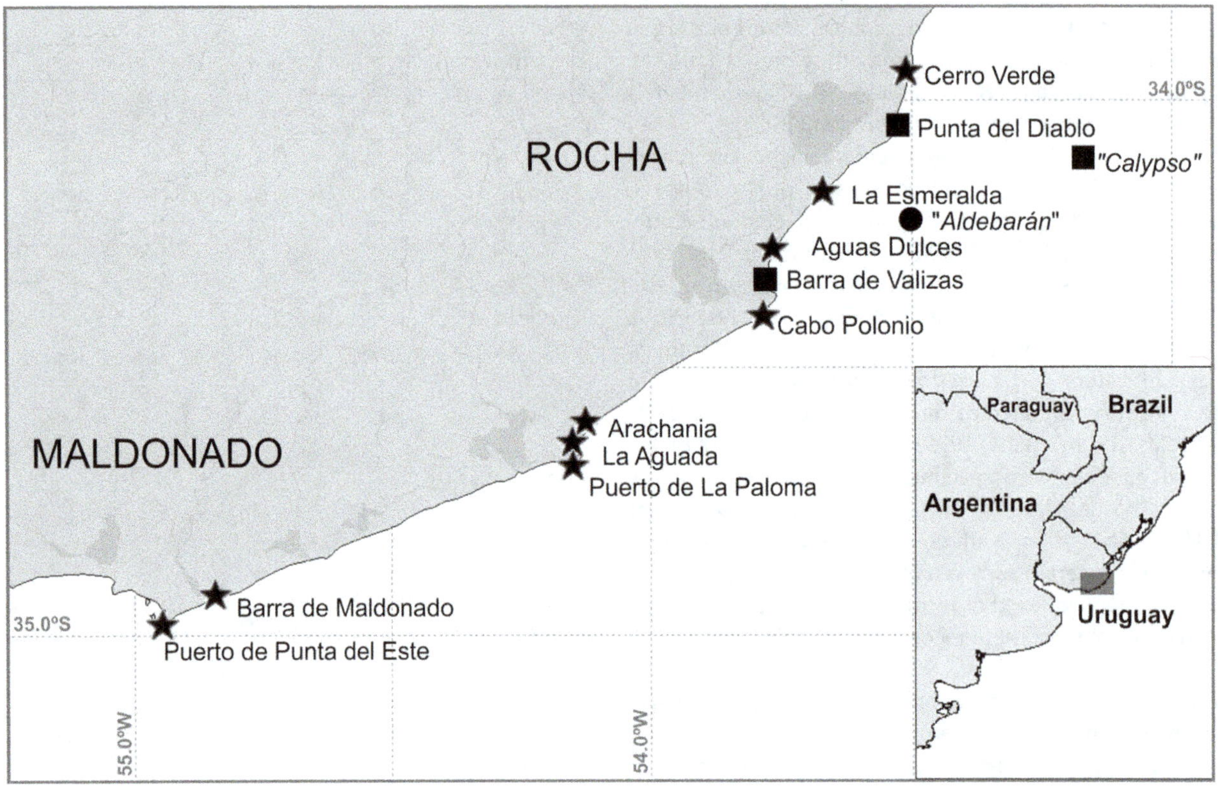

Fig. 2 Locations of new (stars and point) and historical records (square) of Tamoya haplonema from Uruguayan waters

early influence of warm oceanic waters in the coastal area (November surface temperature climatology from 1960–2013 period = 18.75 °C). This finding belongs to a period of several records along 2012–2013, which includes the majority of the findings so far recorded from the area.

Discussion

All new records belong to the coast/inner shelf, i.e., matching the habitat already reported for *Tamoya haplonema* (Mianzan and Cornelius 1999). Morphological measurements also fall within those values previously stated for this species (see Mianzan and Cornelius 1999; Morandini et al. 2005). The cnidome data here presented (Table 2) for a single specimen of *T. haplonema* in comparison with the one shown by Collins et al. (2011, Table 1) for *Tamoya ohboya* demonstrate that in some cases there is superposition of values.

Despite an increasing effort of jellyfish monitoring on the Uruguayan coast in the last years, and also because cubozoans are easy to detect between other invertebrates, *T. haplonema* can be currently considered occasional in this region with a good basis. This is reinforced

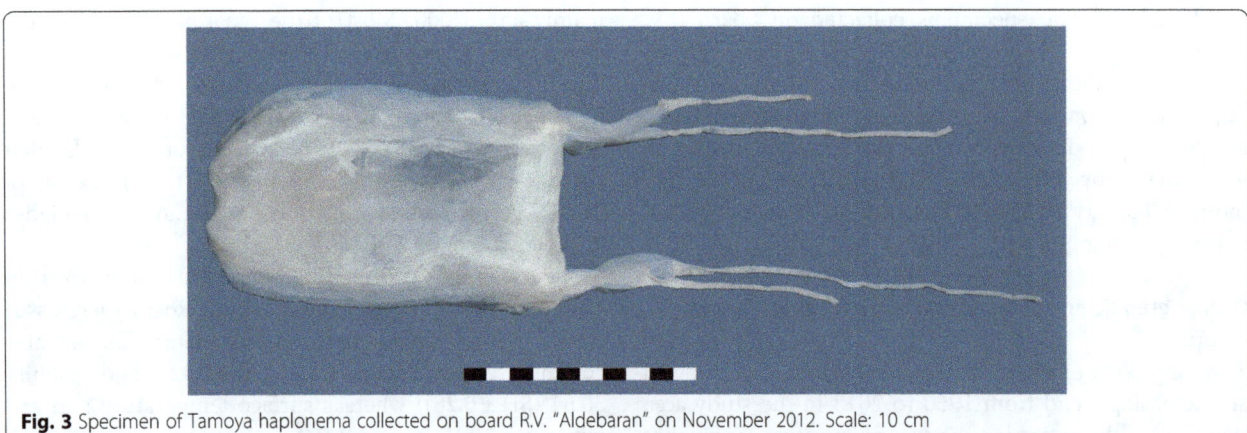

Fig. 3 Specimen of Tamoya haplonema collected on board R.V. "Aldebaran" on November 2012. Scale: 10 cm

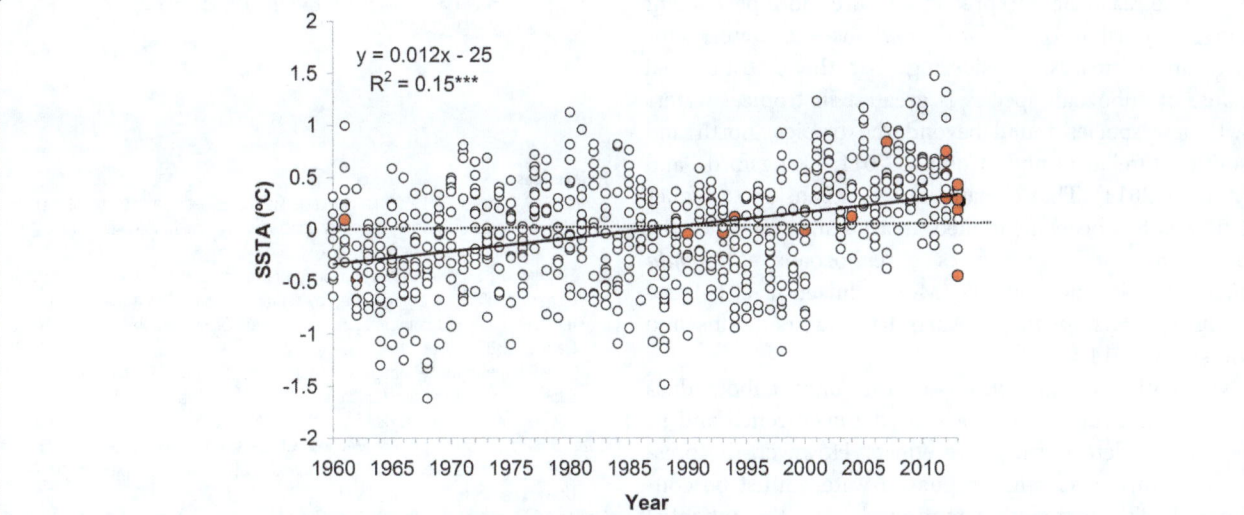

Fig. 4 Long term variation in monthly Sea Surface Temperature Anomaly (°C, SSTA) for the region comprised between latitudes 30° and 36°S and longitudes between 50° and 60°W. Filled dots correspond to SSTA values for the months with records of Tamoya haplonema in Uruguayan and adjacent waters. *** p<0.001

by the absence of records of this species during 2014 and 2015 on the Uruguayan coast despite the increasing record of stranded jellyfishes since 2011. Thus, the presence of *T. haplonema* could be associated with the occurrence of positive SSTA in the region (ca. 80 % of records) especially during summer months, when intrusion of warm oceanic waters to the Uruguayan coastline is stronger. The occurrence was particularly and unusually strong during 2012–2013. Spring of 2011 and summer of 2012 were under the influence of a cold ENSO episode (La Niña) that is characterized by negative sea surface temperature anomalies (SSTA) in the Niño 3.4 region (5°N-5°S, 120°–170°W). These conditions continued over summer 2013 but not surpassing the threshold needed to declare a cold ENSO episode. The aforementioned period was categorized as a dry one for the La Plata basin (http://iridl.ldeo.columbia.edu/maproom/Global/Precipitation). This could have determined an enhanced marine influence during 2012–2013 and the concomitant advection of tropical fauna.

The historical sporadic presence of *T. haplonema* in the Uruguayan coast and its common presence during 2012–2013 are supported by the observations made since 1991 during the whole year in recording stranded fauna at La Paloma and during the last 11 years on summer at La Esmeralda. This was also confirmed by our colleague M. Demicheli (pers. comm.), who made similar observations at La Paloma between 1965 and 2002 and found *T. haplonema* only twice. Moreover, the specimens obtained in November 2012 constitutes the firsts records of *T. haplonema* in at least 12 coastal trawl surveys of similar characteristics made all over the entire Uruguayan inner shelf since 1999 (S. González and F. Scarabino, pers. obs.).

The occurrence of *T. haplonema* in May 2000 in the coast off Buenos Aires (i.e., southernmost record of the

species) was within a period characterized by a strong La Niña event (http://www.cpc.ncep.noaa.gov/products/analysis_monitoring/ensostuff/ensoyears.shtml). This also coincided with the occurrence of other unusual subtropical species as the brachyuran crab *Arenaeus cribrarius* and the cephalopod *Argonauta nodosa* (Scelzo 2001; Pastorino and Tamini 2002), reported for the same period and region. The presence of this subtropical fauna at higher latitudes could be ascribed to the particular atmospheric and oceanographic conditions of that year. The combined effects of onshore winds and a strong La Niña event, implying a decreased freshwater runoff (Cazes Boezio et al., 2003), could enhance the influence of warm oceanic waters from the north (Ortega and Martínez 2007).

The observed increasing trend in SSTA is in agreement with the observations of Zavialov et al. (1999) who reported a positive secular trend of Sea Surface Temperature (SST) for this region (significantly above the global average of SST increase) with most of this warming occurring after 1940. Ortega et al. (2013) also reported a regime shift from a cold to a warm period after 1998 for the study area with a regional predominance of positive SSTA thereafter. This increasing trend in SST and predominance of positive SSTA could be associated with decadal cycles in climate that favour the intrusion of warm oceanic waters in the area which is supported by the occurrence in recent years of other species commonly associated to tropical or subtropical waters as the fishes *Stellifer rastrifer* (Segura et al. 2009) and *Aluterus scriptus* (Izzo et al. 2010). Moreover, Ortega et al. (2013) reported for this area an increase in the frequency and speed of onshore winds that could enhance the advection of warm oceanic waters and its associated biota to the coast during austral summer. In this context of increasing frequency of positive SSTA, it would be reasonable to predict a future more permanent range extension of *T. haplonema* as sea water temperature increases, considering that the distributional range of cubozoan species is greatest in tropical waters with few species found beyond the tropics, north and south (Orellana and Collins 2011; Kingsford and Mooney 2014). That climate-related expansion to higher latitudes has been mentioned as a possible cause of the occurrence of other species of cubozoans as *Copula sivickisi* in Japanese waters (Morandini et al. 2014) and *Tripedalia cystophora* in Australian waters (Ekins and Gershwin 2014).

Currently, *T. haplonema* is the only cubomedusa present in Uruguay, with a sporadic occurrence and so far has no detected negative effects. However, the presence of this species in Uruguayan waters must be considered with particular attention due to the potential impact on tourism and general public health. This is especially important considering that its occurrence in the coast is associated to a stronger influence of warm subtropical waters during austral summer, i.e. when the influx of tourist to the beach is higher. This also applies to other jellyfishes in Uruguay (considering also its economic and ecological interest) and indicates the need of supporting research and outreach programs focusing on gelatinous zooplankton, mainly to get baseline data on the abundance of different species in the country.

Acknowledgments
We dedicate this work to the memory of our colleague and friend Hermes W. Mianzan (1957–2014) who took the first step in the study of jellyfish in Uruguay. We are especially grateful to Graciela Fabiano (DINARA) for the samples, data and discussions shared as well as to the entire crew of the R.V. "Aldebaran" for their help during the fieldwork, particularly to E. Chiesa for his permanent support. We are also grateful to lifeguards for their contribution in the monitoring of this group on the beach, and to M. Demicheli for sharing his experience about the presence of *T. haplonema* in La Paloma. R. Capitoli (†) generously donated literature relevant to this note. We want to thank to the two anonymous reviewers for their valuable suggestions. ACM was supported by grants 2010/50174–7 and 2011/50242–5 São Paulo Research Foundation (FAPESP), and by CNPq (301039/2013–5). SNS was supported by grant CNPq (481549/2012–9). This is a contribution of NP-BioMar, USP.

Authors' contributions
All authors read and approved the final manuscript.

Competing interests
The authors declare that they have no competing interests.

Author details
[1]Dirección Nacional de Recursos Acuáticos, Constituyente 1497, C.P 11200 Montevideo, Uruguay. [2]Museo Nacional de Historia Natural, C.C. 399, C.P 11000 Montevideo, Uruguay. [3]InvBiota. Invertebrados del Uruguay, Montevideo, Uruguay. [4]Centro Universitario Regional Este, Sede Rocha, Universidad de la República, Ruta 9, Km 208, C.P 27000 Rocha, Uruguay. [5]Lab. Zoología de Invertebrados, Dpto. de Biología Animal, Facultad de Ciencias, Universidad de la República, CP 11400 Montevideo, Uruguay. [6]Prof. de Ciencias Biológicas egresada del I.P.A, Montevideo, Uruguay. [7]COENDU, Conservación de Especies Nativas del Uruguay, Montevideo, Uruguay. [8]Departamento de Ciências Biológicas, Faculdade de Ciências e Letras, Unesp - Univ Estadual Paulista, Assis, Av. Dom Antonio, 2100, Assis 19806-900, Brazil. [9]Departamento de Zoología.

Instituto de Biociências, Universidade de São Paulo, Rua do Matão, Trav. 14, n. 101, São Paulo, SP 05508-090, Brazil.

References
Barattini LP, Ureta EH. ("1960") La fauna de las costas del este (invertebrados). Montevideo: Publicaciones de Divulgación Científica, Museo «Dámaso Antonio Larrañaga»; 1961. p. 195.

Barré N, Provost C, Saraceno M. Spatial and temporal scales of the Brazil-Malvinas Current confluence documented by simultaneous MODIS Aqua 1.1-km resolution SST and color images. Nat Hazards Oceanographic Proc Satellite Data. 2006;37:770–86.

Bentlage B, Cartwright P, Yanagihara AA, Lewis C, Richards GS, Collins AG. Evolution of box jellyfish (Cnidaria: Cubozoa), a group of highly toxic invertebrates. Pro R Soc Biol Sci. 2010;1707:1–10.

Cazes-Boezio G, Robertson AW, Mechoso CR. Seasonal dependence of ENSO teleconnections over South America and relationships with precipitation in Uruguay. J Clim. 2003;16:1159–76.

Collins AG, Bentlage B, Gillan W, Lynn TH, Morandini AC, Marques AC. Naming the Bonaire banded box jelly, *Tamoya ohboya*, n.sp. (Cnidaria: Cubozoa: Carybdeida: Tamoyidae). Zootaxa. 2011;2753:53–68.

Ekins M, Gershwin L. First record of the Caribbean box jellyfish *Tripedalia cystophora* in Australian waters. Mar Biodivers Rec. 2014;7:e127.

Failla Siquier MG. Zooplancton gelatinoso de la costa uruguaya. In: Menafra R, Rodríguez-Gallego L, Scarabino F, Conde D, editors. Bases para la Conservación y el Manejo de la Costa Uruguaya. Montevideo: Vida Silvestre Uruguay; 2006. p. 97–103.

Failla Siquier MG, Dutra AA. Experiencia preliminar de relevamiento de organismos gelatinosos (Cnidaria y Ctenophora) con la colaboración de los Guardavidas de las playas de los Dptos. de Maldonado y Rocha. Montevideo: Uruguay. Segundo Congreso Uruguayo de Zoología. Facultad de Ciencias; 2012. p. 163.

Failla Siquier G, Dutra Alburquerque A. Avances del relevamiento de organismos gelatinosos costeros en Uruguay. Montevideo: Tercer Congreso Uruguayo de Zoología. Facultad de Ciencias; 2014. p. 193.

Forest J. Campagne de la Calypso au large des côtes de l'Amérique du Sud (1961–1962) (Première partie). 1. Compte rendu et liste des stations. Ann Inst Oceanogr. 1966;44:329–50.

Gershwin L. *Carybdea alata* auct. and *Manokia stiasnyi*, reclassification to a new family with description of a new genus and two new species. Mem Qld Mus. 2005;51:501–23.

Gershwin L, Gibbons MJ. *Carybdea branchi*, sp. nov., a new box jellyfish (Cnidaria: Cubozoa) from South Africa. Zootaxa. 2009;2088:41–50.

Gordon AL. South Atlantic thermocline ventilation. Deep-Sea Res. 1981;28:1239–64.

Gordon AL. Brazil-Malvinas Confluence. 1984. Deep-Sea Res. 1989;36:359–84.

Goy J. Campagne de la Calypso au large des côtes atlantiques de l'Amérique du sud (1961–1962). 35, vol. 55. Meduses: Résultats Scientifiques des Campagnes de la Calypso 11. Annales de l'Institut Oceanographique; 1979. p. 263–96.

Haddad V, Lupi O, Lonza JP, Tyring SK. Tropical dermatology: Marine and aquatic dermatology. J Am Acad Dermatol. 2009;61:733–59.

Izzo P, Milessi AC, Ortega L, Segura AM. First record of *Aluterus scriptus* (Monacanthidae) in Mar del Plata, Argentina. Mar Biodivers Rec. 2010;3:1–2.

Kingsford MJ, Mooney CJ. The ecology of box jellyfishes (Cubozoa). In: Pitt KA, Lucas CH, editors. Jellyfish Blooms. Dordrecht: Springer; 2014. p. 267–302.

Lewis C, Bentlage B, Yanagihara AA, Gillan W, van Blerk J, Keil DP, Belly AE, Collins AG. Redescription of *Alatina alata* (Reynaud, 1830) (Cnidaria: Cubozoa) from Bonaire, Dutch Caribbean. Zootaxa. 2013;3737:473–87.

Lyon B. The strength of El Niño and the spatial extent of tropical drought. Geophysical Res Lett. 2004;31:L21204.

Lyon B, Barnston AG. ENSO and the spatial extent of interannual precipitation extremes in tropical land areas. J Clim. 2005;18:5095–109.

Maamaatuaiahutapu K, Garçon VC, Provost C, Boulahdid M, Bianchi AA. Spring and winter water mass composition in the Brazil-Malvinas Confluence. J Mar Res. 1994;52:397–426.

Marques AC, Morandini AC, Migotto AE. Synopsis of knowledge on Cnidaria Medusozoa from Brazil. Biota Neotrop. 2003;3:1–18.

Mianzan HM, Cornelius PFS. Scyphomedusae and Cubomedusae of the south Atlantic. In: Boltovskoy D, editor. South Atlantic Zooplankton, vol. 1. Leiden: SPB Academic Publishing; 1999. p. 513–59.

Mianzan HM, Olague G, Montero R. Scyphomedusae de las aguas uruguayas. Spheniscus. 1988;6:1–9.

Morandini AC. Deep-sea medusae (Cnidaria: Cubozoa, Hydrozoa and Scyphozoa) from the coast of Bahia (western South Atlantic, Brazil). Mitteilungen aus dem Hamburgischen Zoologischen Museum und Institut. 2003;100:13–25.

Morandini AC, Marques AC. "Morbakka" syndrome: first report of envenomation by cubozoa (cnidaria) in Brazil. In: VII Congreso Latino-Americano sobre Ciencias do Mar, Resumos Expandidos, vol. 2. Santos: Inst. Oceanogr. USP; 1997. p. 188–9.

Morandini AC, Ascher D, Stampar SN, Ferreira JFV. Cubozoa e Scyphozoa (Cnidaria: Medusozoa) de águas costeiras do Brasil. Iheringia, Série Zoologia. 2005;95:281–94.

Morandini AC, Stampar SN, Kubota S. Mass occurrence of the cubomedusa Copula sivickisi (Cnidaria: Cubozoa) at Seto Harbor, Shirahama, Wakayama, Japan in summer 2013; a possible recent example of global warming. Publ Seto Mar Biol Lab. 2014;42:108–11.

Orellana ER, Collins AG. First report of the box jellyfish Tripedalia cystophora (Cubozoa: Tripedaliidae) in the continental USA, from Lake Wyman, Boca Raton, Florida. Mar Biodivers Rec. 2011;4:1–3.

Ortega L, Martínez A. Multiannual and seasonal variability of water masses and fronts over the Uruguayan shelf. J Coastal Res. 2007;23:629–81.

Ortega L, Celentano E, Finkl C, Defeo O. Effects of climate variability on the morphodynamics of Uruguayan sandy beaches. J Coastal Res. 2013;29:747–55.

Pagès F, Gili JM, Bouillon J. Medusae (Hydrozoa, Scyphozoa, Cubozoa) of the Benguela Current (southeastern Atlantic). Sci Mar. 1992;56:1–64.

Pastorino G. New record of the cubomedusae Tamoya haplonema Müller, 1859 (Cnidaria: Scyphozoa) in the South Atlantic. Bull Mar Sci. 2001;68:357–60.

Pastorino G, Tamini L. Argonauta nodosa Solander, 1786 (Cephalopoda: Argonautidae) in Argentine waters. J Conchology. 2002;37(5):477–82.

Scelzo MA. First record of the portunid crab Arenaeus cribrarius (Lamarck, 1818) (Crustacea: Brachyura: Portunidae) in marine waters of Argentina. Proc Biol Soc Wash. 2001;114:605–10.

Segura AM, Carranza A, Rubio LE, Ortega L, García M. Stellifer rastrifer (Pisces: Sciaenidae): first Uruguayan records and a 1200 km range extension. Mar Biodivers Rec. 2009;2:e67. doi:10.1017/S1755267209000852.

Sepúlveda H, Valle-Levinson A, Framinan M. Observations of subtidal and tidal flow in the Rio de la Plata Estuary. Cont Shelf Res. 2004;24:509–25.

Smith TM, Reynolds RW, Peterson TC, Lawrimore J. Improvements to NOAA's Historical Merged Land-Ocean Surface Temperature Analysis (1880–2006). J Clim. 2008;21:2283–96.

Xue Y, Smith TM, Reynolds RW. Interdecadal changes of 30-yr SST normals during 1871–2000. J Clim. 2003;16:1601–12.

Zavialov P, Wainer I, Absy JM. Sea surface temperature variability off southern Brazil and Uruguay as revealed from historical data since 1854. J Geophys Res. 1999;104:21021–32.

3

New records of four reef fish species for Hong Kong

Allen W. L. To[1*] and Stanley K. H. Shea[2]

Abstract

Background: A total of four reef fish species: *Amblyeleotris japonica*, *Halichoeres hartzfeldii*, *Canthigaster papua* and a *Parapriacanthus* species, are reported for the first time in Hong Kong.

Methods: The specimens were discovered during underwater reef fish surveys. About 20 individuals of one *Parapricanthus* species were found at Tai Chau. A single *Amblyeleotris japonica* and a single *Halichoeres hartzfeldii* were both observed at the west coast of north Ninepin Island. One individual of *Canthigaster papua* was found at Clearwater Bay.

Results: The likelihood of these four species having been artificially introduced into Hong Kong waters is discussed and eliminated. Comparison of these four species is made against six other species previously encountered in Hong Kong waters that were not already documented in local records and suspected to be artificially introduced.

Conclusions: Four reef fish species, including *Amblyeleotris japonica*, *Halichoeres hartzfeldii*, *Canthigaster papua* and a *Parapriacanthus* species, are reported for the first time in Hong Kong. Details on the sighting records, locations, habitats and behavior of the observed individuals are described. The validity of the natural origin of these fish is discussed.

Keywords: Hong Kong, Reef fish, Native species, South China Sea, *Amblyeleotris japonica*, *Halichoeres hartzfeldii*, *Canthigaster papua*, *Parapriacanthus*

Background

Hong Kong Special Administrative Region of the People's Republic of China, located in the northern sector of the South China Sea, has about 1650 km^2 of territorial waters and more than 200 offshore islands (hk-fish.net 2016a). The area of Hong Kong's territorial water is only about 0.05 % of the South China Sea's, but marine fish diversity is about 30 % of the latter (hk-fish.net 2016a). The sea in Hong Kong is host to a range of important fish habitats, including rocky reefs, coral communities and adjacent rubble and sandy areas (Morton and Morton 1983; Sadovy and Cornish 2000). Situated at the mouth of the Pearl River, the western coast of Hong Kong regularly receives large volumes of freshwater, increasing the turbidity and decreasing the salinity of marine water in the area. The effect of the Pearl River is progressively reduced towards the eastern waters of Hong Kong, leaving the south-eastern,

eastern and north-eastern coast of Hong Kong almost unaffected by the freshwater of the Pearl River (Morton and Morton 1983; Sadovy and Cornish 2000). Being more oceanic in nature, these eastern waters are home to the majority of more than 80 scleractinian coral species locally recorded (Chan et al. 2005).

Field studies on the reef fish diversity in Hong Kong so far have produced records of around 340 reef fish species (Sadovy and Cornish 2000; To et al. 2013). Four new records of reef fish species for Hong Kong sighted during underwater reef fish surveys from Jun 2014 to Nov 2015 is reported here, including a *Parapriacanthus* species, *Amblyeleotris japonica* Takagi, 1957, *Halichoeres hartzfeldii* (Bleeker, 1852), and *Canthigaster papua* (Bleeker, 1848).

Three species under the genus *Parapriacanthus* are currently recognized (Mooi 2001), of which *Parapriacanthus dispar* (Herre, 1935) and *Parapriacanthus ransonneti* Steindachner, 1870 are similar-looking species. At least part of their known native range is overlapping (Allen and Erdmann 2012). *Parapriacanthus dispar* has been recorded

* Correspondence: allenwlto@yahoo.com
[1]WWF-Hong Kong, 15/F, Manhattan Centre, 8 Kwai Cheong Road, Kwai Chung, New Territories, Hong Kong
Full list of author information is available at the end of the article

in Australia, Indonesia, the Philippines, New Caledonia, Papua New Guinea and the Solomon Islands (Allen and Erdmann 2012; Froese and Pauly 2016a). It has not been recorded in mainland China or Taiwan (Liu 2008; Wu 2012). The natural distribution of *Parapriacanthus ransonneti* is much more widespread in the Indo-west Pacific: from Red Sea to South Africa, eastward to the Marshall Islands, northward to southern Japan, and southward to Australia (Froese and Pauly 2016b). It is also found in mainland China and Taiwan (Liu 2008; Wu 2012). Nonetheless, species of the genus of *Parapriacanthus* have not previously been recorded in Hong Kong (Sadovy and Cornish 2000; To et al. 2013; hk-fish.net 2016b). Both species occur in aggregations and are found in caves and under ledges (Allen and Erdmann 2012). The two species can be distinguished by a combination of the number of lateral-line scales, dorsal fin spines, dorsal fin rays, and scales between the midventral edge of coracoid and the midlateral edge of the pelvic bone (Mooi 2001).

Amblyeleotris japonica occurs in the Indo-west Pacific Ocean, and is reported to occur natively in Israel, Japan, New Caledonia, Papua New Guinea, Saudi Arabia, Solomon Islands and Taiwan (Froese and Pauly 2016c), as well as in mainland China (Liu 2008), and the South China Sea (Shao 2016a). The species can grow to about 8.5 cm in total length (Froese and Pauly 2016c). It utilizes more coastal sheltered waters in sandy bottoms with scattered shell fragments and coral rubbles, and was reported to associate with alpheid shrimps (Yanagisawa 1976; Shoichi 2014; Shao 2016a). Its appearance resembles *Amblyeleotris gymnocephala* (Bleeker, 1853), with a considerably overlapping native range (Allen et al. 2015). *Amblyeleotris japonica* can be distinguished from *Amblyeleotris gymnocephala* by the presence of a C-shaped dark blotch on the caudal fin (Froese and Pauly 2016c), and the first dorsal fin being relatively more triangular in shape (Hayashi and Shiratori 2013). Of these two species, only *Amblyeleotris gymnocephala* has previously been recorded in Hong Kong (Sadovy and Cornish 2000; hk-fish.net 2016b).

Halichoeres hartzfeldii is found from the East Indian region to Micronesia and Samoa, southward to Australia and northward to Japan (Allen and Erdmann 2012; Shoichi 2014; Allen 2014; Allen et al. 2015). It is also found in mainland China, the South China Sea and Taiwan (Liu 2008; Wu 2012; Shao 2016b). It has not been recorded in Hong Kong (Sadovy and Cornish 2000; To et al. 2013; hk-fish.net 2016b). *Halichoeres hartzfeldii* can grow to 18 cm in total length (Froese and Pauly 2016d). This species can form small groups and occurs in sand and rubble bottoms near reefs (Allen and Erdmann 2012; Allen et al. 2015; Froese and Pauly 2016d). *Halichoeres zeylonicus* (Bennett, 1833) is very similar in appearance to *Halichoeres hartzfeldii* (Allen and Erdmann 2012;

Allen et al. 2015). There are records of *Halichoeres zeylonicus* in mainland China and Taiwan (Liu 2008; Wu 2012; Shao 2016c). However *Halichoeres zeylonicus* was also reported to be closely related to *Halichoeres hartzfeldii*, with its distribution restricted to the Indian Ocean and western Indonesia (Randall et al. 1997; Allen and Erdmann 2012; Allen et al. 2015). The two species can be distinguished by the less vivid banding on the head and the presence of a small oblique dark spot, or orange banding on the back, behind the level of the pectoral fin tip in male *Halichoeres zeylonicus*, and the noticeably thinner midlateral orange or yellow stripe in female *Halichoeres zeylonicus* (Allen and Erdmann 2012).

As detailed in the review by Matsuura (2015) on the taxonomy and systematics of tetraodontiform fishes, *Canthigaster papua* was previously regarded by Allen & Randall (1977) as a synonym of *Canthigaster solandri* (Richardson, 1845). However they have recently been separated and recognized as two distinctive species (Allen and Erdmann 2012; Allen 2014; Allen et al. 2015). *Canthigaster papua* is described to have prominent orange colour around the mouth, and *Canthigaster solandri* has yellow-orange colouration on the tail (Allen and Erdmann 2012; Allen et al. 2015). The native range of *Canthigaster papua* is reportedly Indonesia, Philippines, Micronesia, Papua New Guinea, and from the Solomon Islands to the Great Barrier Reef and New Caledonia (Allen and Erdmann 2012; Allen et al. 2015). *Canthigaster solandri* is found in southwest Japan, Micronesia, and from east Australia to French Polynesia (Allen et al. 2015). Of the two species, only *Canthigaster solandri* are recorded to occur in mainland China and Taiwan (Liu 2008; Wu 2012), however it is unknown if these earlier records included *Canthigaster papua*, which was once considered as its synonym. Neither species have been recorded in Hong Kong (Sadovy and Cornish 2000; To et al. 2013; hk-fish.net 2016b). *Canthigaster papua* can grow to 9 cm in total length, and inhabits coral areas either solitarily or in pairs (Allen and Erdmann 2012; Allen et al. 2015).

Methods

A group of about 20 individuals of *Parapricanthus* species was noticed on 1 July 2015 by the first author (Fig. 1), at about 4 m depth off the coast of Tai Chau (22°24'13.2"N; 114°23'05.93"E), sometimes also known as Tsim Chau. Individuals were about 7 cm in total length. The group was observed for 10 min, and was seen to swim quite rapidly back and forth between two large boulders 3–4 m apart, staying close to the boulders most of the time. Photographs were taken.

On 18 July 2015, one individual of *Amblyeleotris japonica* was observed resting within a burrow built on rubble habitat (Fig. 2), at the western coast of north Ninepin Island (22°15'49.77"N; 114°20'43.45"E), at about

Fig. 1 A group of *Parapriacanthus* species observed at Tai Chau

Fig. 3 *Halichoeres hartzfeldii* (Bleeker, 1852) observed at north Ninepin Island

5 m in depth. The individual, which was estimated to be around 8 cm in total length, was observed for about 10 min during which half of the body remained outside the burrow. The fish looked very similar at first glance to *Amblyeleotris gymnocephala* which has been known to occur in Hong Kong waters. However, moving further out of the burrow, the fish displayed a prominent and distinctly triangularly shaped dorsal fin. A commensal shrimp was also noted to share the burrow with this individual.

One individual of *Halichoeres hartzfeldii* was observed on the same day (Fig. 3), 18 July 2015, at the same site: the western coast of north Ninepin Island (22°15'49.77"N; 114°20'43.45"E). The individual was observed at about 10 m in depth in rubble habitat, and swam around the authors for about 5 min during which observation was made and photographs were taken. The individual kept a distance of about 2–3 m away from authors and swam with a speed and form of the *Halichoeres nigrescens* (Bloch & Schneider, 1801), which is known to occur in Hong Kong. The individual was about 10 cm in total length. Its

body was pinkish near the dorsal with a prominent mid-lateral yellow stripe running behind the eye and throughout the body. The mid-lateral stripe on this individual was much thicker than that of the similar-looking *Halichoeres zeylonicus*, which is restricted to the Indian Ocean and western Indonesia (Randall et al. 1997; Allen and Erdmann 2012; Allen et al. 2015).

On 8 Nov 2015, one small individual of *Canthigaster papua* was observed at about 4 m deep over rocky reefs at Clearwater Bay (22°16'18.06"N; 114°18'10.22"E) (Fig. 4). The bay is an artificially constructed sea wall, protecting a pier and a mooring area for privately owned yachts. The observed individual was about 4 cm in total length and was seen swimming over boulders covered with *Diadema* urchins. The individual was wary of the first author and swam among the spines of the urchin during the 10-min of the observation. Photographs were taken. The small individual had prominent orange colour around the mouth, which is a distinctive feature of *Canthigaster papua*, and lacked the yellow-orange colouration on the

Fig. 2 *Amblyeleotris japonica* Takagi, 1957 observed at north Ninepin Island, staying at the entrance to its burrow

Fig. 4 *Canthigaster papua* (Bleeker, 1848) observed swimming among *Diadema* urchins at Clearwater Bay

tail which is the feature for *Canthigaster solandri* (Allen and Erdmann 2012; Allen et al. 2015).

Results and discussion

New records of reef fish for Hong Kong have been published in many previous studies about Hong Kong reef fish faunal diversity (Sadovy and Cornish 2000; To et al. 2013), despite the more than 160 years of ichthyological research in Hong Kong (hk-fish.net 2016a). Notably, some sightings of reef fish, which would otherwise be new records to Hong Kong, were very likely artificially introduced, as noted by To et al. (2013). The occurrence of these reef fish might have originated from release by aquarium hobbyists who no longer wish to keep the fish for various reasons (To and Situ 2005). Fish records falling into this category usually have a few observable characteristics.

Firstly, if Hong Kong does not fall into the species' documented natural range, then this may be an indication that the species was artificially introduced. An example is the sighting of *Zebrasoma xanthurus* (Blyth, 1852) in the same survey, which is a species from the Indian Ocean (To and Situ 2005).

Secondly, if juveniles of a species have not been recorded locally, but a fully-grown adult is sighted in a publicly accessible location or in a fish sanctuary such as marine parks, then there is a possibility that the individual was artificially introduced. Examples include a previous sighting of an adult *Pomacanthus imperator* (Bloch, 1787) in an easily-accessible marine park in Hong Kong (To and Situ 2005). The individual's previous human owners might have chosen the marine park for releasing the fish after having grown tired of keeping it, believing that the marine park would offer it protection (To and Situ 2005). Past reports of a large *Naso lituratus* (Forster, 1801) and a fully-grown *Acanthurus japonicus* (Schmidt, 1931) for instance, were both locally unprecedented sightings that were rejected as new species to Hong Kong, as they were found in one of Hong Kong's most easily accessible, popular dive sites with open-access to anyone, which pointed to the likelihood of introduction by aquarium fish release for those individuals (To et al. 2013).

Thirdly, if the species is popular and available in the aquarium trade in Hong Kong, then there is an increased probability for individual occurrences in the wild to be results of artificial introduction. Although there are no recent studies on the aquarium fish trade in Hong Kong, an earlier study revealed a large diversity of reef fish involved in the trade (Chan and Sadovy 2000). A species' availability in the local trade can serve as an additional circumstantial evidence to the possible origin of individuals sighted in the wild, including the above-mentioned sighting records of surgeonfish and angelfish.

An earlier sighting of a large *Aulostomus chinensis* (Linnaeus, 1766) individual of brown colour-type was regarded as a valid new fish record for Hong Kong, as it was observed in a relatively remote dive site, and was not found in the local aquarium trade (To et al. 2013). These key characteristics, together with other observations such as the presence of injuries which can link to recent captivity, lack of wariness of approaching divers and the suitability of the habitat to the fish observed, can further shed light on the likely origin of the observed reef fish species and their validity as new records for Hong Kong.

Notably, in addition to release by aquarium hobbyists, reef fish are sometimes released in religious activities (To and Situ 2005; To et al. 2013). Based on the authors' observations, reef fishes involved in this type of fish release mainly include species of Lutjanidae and Serranidae, but any other species available in local fish markets may also be purchased for release. The release of the fish can take place in near-shore areas, such as from public piers or jetties, or further from the shore areas through release from onboard fish-carrying vessels. Individuals introduced into the sea by such fish release activities may also be differentiated from naturally occurring fish by the same characteristics as described in the earlier section. For instance, the sighting of an individual of *Epinephelus fuscoguttatus* (Forsskål, 1775), which was observed to have injuries around the jaw, and encountered in shallow coral areas, was almost certainly a result of such fish release. The plate-size of that individual and the popularity of this species in the live reef food fish trade, which is centered in Hong Kong (Lau and Li 2000; Craig et al. 2011), further substantiates the evidence pointing towards fish release as the origin of that observed individual (To et al. 2013).

With reference to the above accounts of released fish characteristics, it arises that the present study represents the first valid documentation of the four reef fish species featured in this paper, in Hong Kong. Although these species may be available in the local aquarium trade, the individuals described were encountered in outlying islands (e.g. Tsim Chau and Ninepin Islands), or in sites that are relatively inaccessible to the public (e.g. Clearwater Bay), reducing the probability of release by aquarium fish hobbyists. In addition, fish released by aquarium fish hobbyists are typically bigger in size, abandoned by their keepers when they grow out of their tanks. The specimens observed by the authors, however, were relatively small. Furthermore, all four species presented here have previously been recorded in mainland China, meaning these species naturally occur in adjacent areas of the northern South China Sea regions. Hong Kong therefore falls within the species' natural ranges. Last but not least, the behaviour of the fishes and their interaction with the respective habitats appeared natural, further supporting the validity of these species as new records for Hong Kong.

Regarding the origin of these new fish records, a previous study on Hong Kong reef fish assemblage compared the reef fish species in Hong Kong with those of adjacent areas, and suggested that some of the rarer or sporadically occurring tropical species occurring in Hong Kong may have been transported as larvae from south of Hong Kong by the Hainan current (Cornish 1999). The presence of rare species at their juvenile stages in Hong Kong was suggested to further support the legitimacy of this possible pathway (Cornish 1999). Such transport of juveniles of tropical species into subtropical regions has also been noted in other places, such as the occurrence of tropical and subtropical fishes in temperate southeastern Australia (Booth et al. 2007), and in New Zealand (Francis et al. 1999). Likewise, the Taiwan current from the East China Sea invades Hong Kong in winter (Morton and Morton 1983), which potentially brings subtropical or temperate fish larvae species into Hong Kong (Cornish 1999).

Many reef fishes are known to produce pelagic eggs, including species in the families Pempheridae and Labridae, to which *Parapriacanthus* species and *Halichoeres hartzfeldii* belong (Sadovy and Cornish 2000; Allen et al. 2015). Species in the families Gobiidae and Tetraodontidae, although are known to produce demersal eggs (Sadovy and Cornish 2000), once hatched the pelagic fish larvae may also become dispersed (Sadovy and Cornish 2000; Lecchini 2005). Depending on the species, the larval stage of reef fishes can last for weeks or months (Leis and McCormick 2002). Notably, all four species reported in this study have been recorded in places nearby, such as mainland China and Taiwan. The close proximity of Hong Kong to places where these species naturally occur could have facilitated the transport and settling of fish larvae of these species in Hong Kong. Therefore it is reasonable to suggest that the occurrence of these four species may have been due to the transport of pelagic fish larvae. However, whether these rare or sporadically occurring reef fish species can establish and become self-sustaining populations in local waters is yet to be investigated.

Existing studies indicate that new fish records can be associated with the addition of new artificial structures. For example, a study conducted in the Mediterranean coast of Israel laid support to the possibility that impoverished areas offered "unsaturated" environments for new fish species, and in that particular case it was filled by Red Sea immigrants, such as sweepers (Diamant et al. 1986). New structures may have been placed in a way that intercepts incoming larvae originally destined for natural reefs down current (Cenci et al. 2011), therefore new fish species could be found in these artificial structures rather than in natural reefs. Notably, although fish assemblage in certain areas can largely be an extract of a common pool of larval recruits, the fish species diversity at a certain site can be affected by microhabitat selection by different species (Kaufman and Ebersole 1984). New artificial structures may offer to these recruits a suitable habitat to settle that is unique in the area concerned (Carr and Hixon 1997). Artificial structures therefore have been suggested to potentially facilitate the settling of new fish species.

Among the four new species found in Hong Kong as reported in this study, the single individual of *Canthigaster papua* was found in artificial structure–a man-made sea-wall for sheltering yachts. Although this study did not investigate the uniqueness of the micro-habitat offered by this particular sea-wall, it is worth noting that, from a macro point of view, such artificial structures is not unique within waters of Hong Kong where underwater reef fish surveys are conducted. In addition, this artificial structure has been put in place for at least 15 years. Therefore it is doubtful if this structure should be considered as a new artificial structure. Whether this artificial sea wall offers an "unsaturated" environment for new species to inhabit is difficult to judge. With limited information and research into this area, the current study is unable to quantify the significance of this artificial structure to the new fish species record.

Hong Kong's reef fish diversity, including other marine species, remains, to date, relatively unexplored. The discovery of the four species presented in this paper demonstrates the need for Hong Kong's reef fish, as well as other understudied marine fauna in Hong Kong, to be surveyed and documented, to produce an updated inventory of local marine species.

Currently, it is the priority of governments worldwide to manage the spread of non-native species (Bax et al. 2003; Thresher and Kuris 2004), and from these attempts the challenge of defining what a "native" species is, is realized (Gilroy et al. 2016). Without a comprehensive and up to date inventory of local species, Hong Kong will soon come upon the same roadblock in its marine conservation efforts.

Any future work in Hong Kong's marine biodiversity conservation, whether in the elimination of invasive species or the protection of native species, will greatly benefit from a clear understanding of the marine fauna present in local waters. Research into the presence and distribution of species within Hong Kong territorial waters, and the identification of biodiversity hotspots to be prioritized for protection, is hence recommended.

Conclusions

This study provides the first sighting records of *Amblyeleotris japonica*, *Halichoeres hartzfeldii*, *Canthigaster papua* and a *Parapriacanthus* species in Hong Kong.

Acknowledgements
The authors would like to thank Professor K.T. Shao (Biodiversity Research Center, Academia Sinica), Dr. Andy Cornish (WWF) and Dr. Mark Erdmann (Conservation International) for advice on species identification in this study. The authors would like to thank the two anonymous reviewers for valuable comments to improve the quality of the manuscript. The authors would like to express deep gratitude to Fion Cheung and Kathleen Ho as dive buddies in this study and for their helpful comments during the preparation of this manuscript. The financial and technical support from Ocean Park Conservation Foundation Hong Kong (OPCFHK) in piloting this underwater reef fish survey is greatly appreciated. The Swire Group Charitable Trust kindly supported the expense of this publication.

Funding
Ocean Park Conservation Foundation Hong Kong (OPCFHK) provided financial support in this fish survey. The Swire Group Charitable Trust supported the expense of this publication.

Authors' contributions
AWLT and SKHS are both involved in the field survey for reef fish in this study, and jointly identified the species included in this study. AWLT prepared the draft of the manuscript and processed photos. SKHS provided comments to the manuscript and contributed to drafting the manuscript. All authors read and approved the final manuscript.

Competing interests
The authors declare that they have no competing interests.

Author details
[1]WWF-Hong Kong, 15/F, Manhattan Centre, 8 Kwai Cheong Road, Kwai Chung, New Territories, Hong Kong. [2]BLOOM Association, c/o, ADMCF, 9 Queen's Road Central, Hong Kong.

References
Allen GR. Field guide to marine fishes of tropical Australia and south-east Asia. 4th ed. Western Australia: Western Australia Museum; 2014.

Allen GR, Erdmann MV. Reef fishes of the East Indies volumes I-III. Perth: Tropical Reef Research; 2012.

Allen GR, Randall JE. Review of the sharpnose pufferfishes (subfamily Canthigasterinae) of the Indo-Pacific. Rec Aust Mus. 1977;30:475–517.

Allen G, Steene R, Humann P, Deloach N. Reef fish identification tropical Pacific. 2nd ed. Florida: New World Publications; 2015.

Bax N, Williamson A, Aguero M, Gonzalez E, Geeves W. Marine invasive alien species: a threat to global biodiversity. Mar Policy. 2003;27:313–23.

Booth DJ, Figueira WF, Gregson MA, Brown L, Beretta G. Occurrence of tropical fishes in temperate southeastern Australia: role of the east Australian current. Estuar Coast Shelf Sci. 2007;72:102–14.

Carr MH, Hixon MA. Artificial reefs: the importance of comparisons with natural reefs. Fisheries. 1997;22:28–33.

Cenci E, Pizzolon M, Chimento N, Mazzoldi C. The influence of a new artificial structure on fish assemblages of adjacent hard substrata. Estuar Coast Shelf Sci. 2011;91:133–49.

Chan TC, Sadovy Y. Profile of the marine aquarium fish trade in Hong Kong. Aquarium Sci Conserv. 2000;2(4):197–213.

Chan ALK, Choi CLS, McCorry D, Chan KK, Lee MW, Put Jr A. Field guide to hard corals of Hong Kong. Hong Kong: Friends of the Country Parks; Cosmos Books Ltd; 2005.

Cornish A. Fish assemblages associated with shallow, fringing coral communities in sub-tropical Hong Kong: species composition, spatial and temporal patterns. PhD Thesis. Hong Kong: The University of Hong Kong; 1999.

Craig MT, de Mitcheson YJ S, Heemstra PC. Groupers of the world a field and market guide. South Africa: NISA (Pty) Ltd; 2011.

Diamant A, Ben Tuvia A, Baranes A, Golani D. An analysis of rocky coastal Eastern Mediterranean fish assemblages and a comparison with an adjacent small artificial reef. J Exp Mar Bio Ecol. 1986;97:269–85.

Francis MP, Worthington CJ, Saul P, Clements KD. New and rare tropical and subtropical fishes from northern New Zealand. New Zeal J Mar Fresh. 1999;33:571–86.

Froese R, Pauly D. Fishbase. 2016a. http://www.fishbase.org/summary/SpeciesSummary.php?ID=61107&genusname=Parapriacanthus&speciesname=dispar. Accessed date 12 Jan 2016.

Froese R, Pauly D. Fishbase. 2016b. http://www.fishbase.org/summary/Parapriacanthus-ransonneti.html. Accessed date 12 Jan 2016.

Froese R, Pauly D. Fishbase. 2016c. http://www.fishbase.org/summary/Amblyeleotris-japonica.html. Accessed 8 Jan 2016.

Froese R, Pauly D. Fishbase. 2016c. http://www.fishbase.org/summary/Halichoeres-hartzfeldii.html. Accessed 17 Jan 2016.

Gilroy J, Avery J, Lockwood, J. Seeking international agreement on what it means to be 'native'. Conserv Lett. 2016. doi:10.1111/conl.12246.

Hayashi M, Shiratori T. Gobies of Japanese waters. Tokyo: Hankyu Communications Co., Ltd.; 2013 (in Japanese).

hk-fish.net. Introduction & map of Hong Kong waters. 2016a. http://www.hk-fish.net/eng/database/intro/intro.htm. Accessed 12 Jan 2016.

hk-fish.net. Hong Kong marine fish database. 2016b. http://www.hk-fish.net/eng/database/index.htm. Accessed 12 Jan 2016.

Kaufman LS, Ebersole J. Microtopography and the organization of two assemblages of coral reef fishes in the West Indies. J Exp Mar Biol Ecol. 1984;78:253–68.

Lau PPF, Li LWH. Identification guide to fishes in the live seafood trade of the Asia-Pacific region. Hong Kong: WWF-Hong Kong and Agriculture, Fisheries and Conservation Department; 2000.

Lecchini D. Spatial and behavioural patterns of reef habitat settlement by fish larvae. Mar Ecol Prog Ser. 2005;301:247–52.

Leis JM, McCormick MI. The biology, behavior, and ecology of the pelagic, larval of coral reef fishes. In: Sale P, editor. Coral reef fishes: dynamics and diversity in a complex ecosystem. California: Academic; 2002. p. 171–200.

Liu J. Subphylum vertebrata Cuvier, 1812. In: Liu JY, editor. Checklist of marine biota of China seas. Beijing: China Science Publishing & Media Ltd; 2008. p. 886–1066.

Matsuura K. Taxonomy and systematics of tetraodontiform fishes: a review focusing primarily on progress in the period from 1980 to 2014. Ichthyol Res. 2015;62:72–113.

Mooi RD. Pempheridae. Sweepers (bullseyes). In: Carpenter KE, Niem VH, editors. FAO species identification guide for fishery purposes. The living marine resources of the Western Central Pacific. Volume 5. Bony fishes part 3 (Menidae to Pomacentridae). Rome: FAO; 2001. p. 2791–3380.

Morton M, Morton J. The sea shore ecology of Hong Kong. Hong Kong: Hong Kong University Press; 1983.

Randall JE, Allen GR, Steene C. Fishes of the Great Barrier Reef and Coral Sea. Bathurst: Crawford House Publishing Pty Ltd.; 1997.

Sadovy Y, Cornish AS. Reef fishes of Hong Kong. Hong Kong: Hong Kong University Press; 2000.

Shao KT. Taiwan fish database. 2016a. http://fishdb.sinica.edu.tw/chi/species.php?id=381703. Accessed 8 Jan 2016.

Shao KT. Taiwan fish database. 2016b. http://fishdb.sinica.edu.tw/chi/species.php?id=381968. Accessed 17 Jan 2016.

Shao KT. Taiwan fish database. 2016c. http://fishdb.sinica.edu.tw/chi/species.php?id=383396. Accessed 17 Jan 2016.

Shoichi K. Marine fish illustrated. Tokyo: Seibundo Shinkosha; 2014 (in Japanese).

Thresher RE, Kuris AM. Options for managing invasive marine species. Biol Invas. 2004;6:295–300.

To A, Situ A. New comers to the local fish list, or unwelcome exotics! Porcupine! 2005;22:6–7.

To A, Ching K, Shea S. Hong Kong reef fish photo guide. Hong Kong: Eco-Education and Resources Centre; 2013.

Wu H. Vertebrata. In: Huang Z, Lin M, editors. The living species and their illustrations in China's seas (part I) the living species in China's seas. Beijing: China Ocean Press; 2012. p. 919–1160.

Yanagisawa Y. Genus Amblyeleotris (Gobiidae) of Japan and geographical variations of A. japonica Takagi. Publ Seto Mar Biol Lab. 1976;23:145–68.

4

First record of the red alga *Griffithsia capitata* (Ceramiales, Rhodophyta) in the southwestern Caribbean Sea, Western Atlantic

M. Natalia Rincón-Díaz[1], Brigitte Gavio[1,2]*, Michael J. Wynne[3] and Adriana Santos-Martínez[1]

Abstract

The red algal species *Griffithsia capitata* (Ceramiales, Rhodophyta), native to the Canary Islands and Madeira, is reported for the first time from San Andres Island, International Biosphere Reserve *Seaflower*, Caribbean Sea. All specimens, which are small and easily overlooked, were found in coral reef habitats, at depths in the range of 9-17 m, growing mostly on coral rubble or as epiphytes on larger algae. The species was found with all reproductive stages. Its morphological and reproductive features are described and discussed. Whether this new finding is a recent introduction or a species with a natural amphiatlantic distribution still needs to be addressed.

Keywords: Atlantic Ocean, Colombia, *Griffithsia capitata*, Marine algae

Introduction

The Caribbean marine flora has received considerable attention (Miloslavich et al. 2010), and it is rather well known (Costello et al. 2010), if compared to other biogeographic regions. It is species-rich, and the flora is rather homogeneoeus across the whole basin. As in many other regions, in the Caribbean the marine flora is dominated by red algae, and most species are small and epiphytic. Due to the progressive degradation of coral reefs in the region, macroalgae have proliferated, especially turf algae, and now cover what several decades ago was coral reef.

San Andres Island is part of the International Biosphere Reserve *Seaflower*, one of the largest marine reserves in the world (CORALINA-INVEMAR 2012). The marine flora of the island is not thoroughly known because historically it has received little attention (Albis-Salas and Gavio 2011). In the past few years, macroalgal sampling has been undertaken in the whole

archipelago, revealing a flora much more diverse than previously appreciated (Albis-Salas and Gavio 2011; Ortiz and Gavio 2012; Reyes-Gómez et al. 2013; Gavio et al. 2013; 2015). Most of the algal diversity is composed of small and easily overlooked species, although most of these taxa are often relatively abundant and widespread in the Caribbean basin. One of the species recently found is *Griffithsia capitata* (Ceramiales, Rhodophyta), native to the Canary Islands and Madeira. Here we describe the vegetative and reproductive morphology of the plants observed, and discuss the findings as a possible recent introduction or a case of amphiatlantic distribution as a remnant of a Tethyan distribution. This is the first report of the species for the Western Atlantic.

Materials and methods

San Andres (12°28′55″N 81°40′49″W) is an oceanic island of coralline origin situated in the southwestern Caribbean, Colombia (Fig. 1). The island is surrounded by a calcareous platform with a windward barrier reef, a lagoon with patch reefs, including leeward and windward fore reef terraces with coral carpets (Díaz et al. 1995). The leeward, western margin is composed of two submerged terraces, parallel to the coast and slightly inclined, separated by a sandy trough: one terrace is

* Correspondence: bgavio@unal.edu.co
[1]Universidad Nacional de Colombia, Sede Caribe. San Luis Free Town # 52-44, San Andrés Isla, Colombia
[2]Departamento de Biología, Universidad Nacional de Colombia, sede Bogotá, Ciudad Universitaria, Bogotá, Colombia
Full list of author information is available at the end of the article

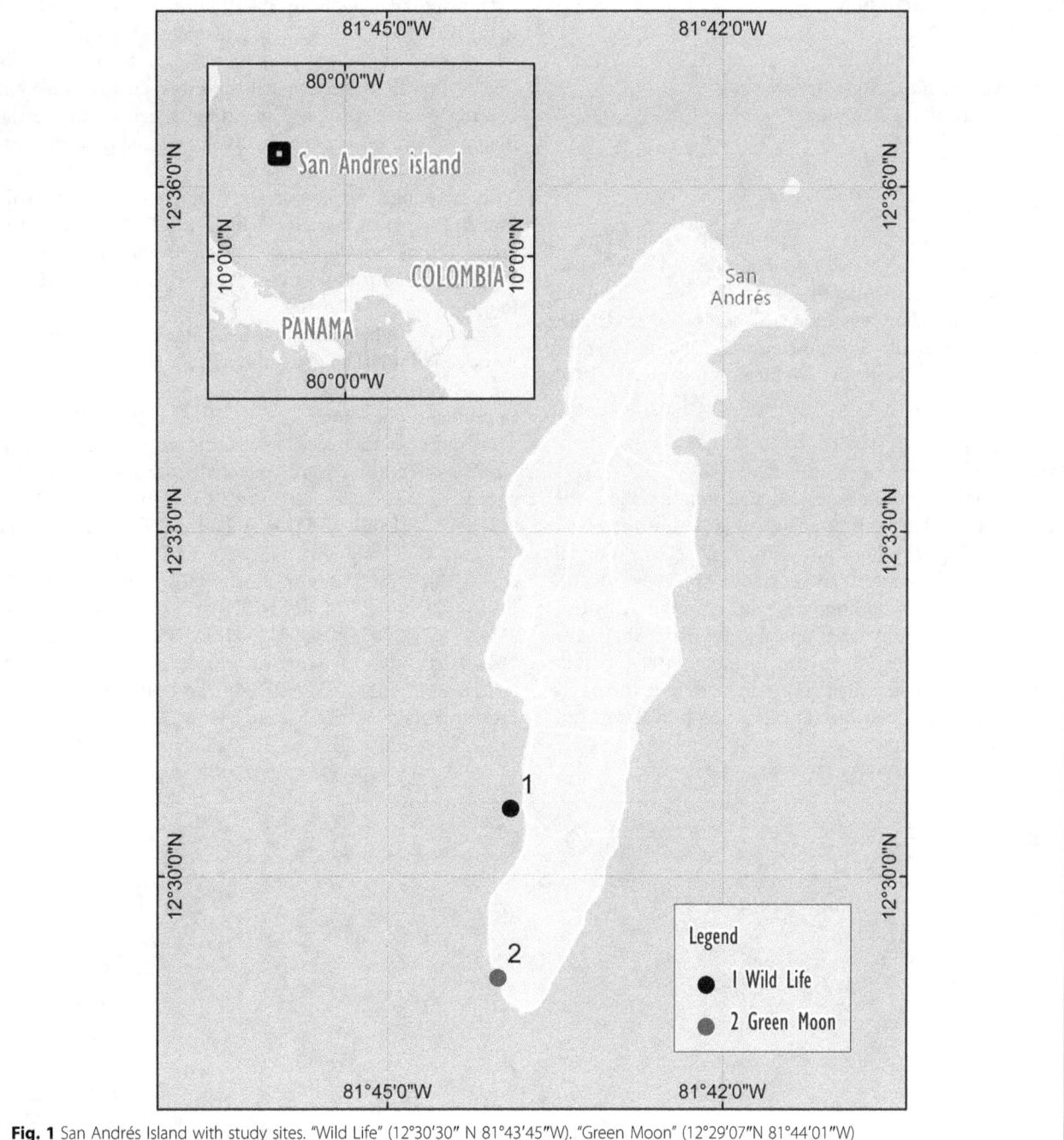

Fig. 1 San Andrés Island with study sites. "Wild Life" (12°30'30" N 81°43'45"W). "Green Moon" (12°29'07"N 81°44'01"W)

shallow (4-10 m depth) along the shore; the second is deeper (10–20 m depth), its outer edge dropping to the top of the insular slope below (Chaves-Fonnegra et al. 2007). All the specimens were collected on the western side of the island, at the localities "Wild Life" (12°30'30" N 81°43'45"W) and "Green Moon" (12°29'07"N 81°44'01"W). This sector is characterized by a reef terrace with coral mat formations starting at 10 m depth, characterized by high species diversity of corals, sponges, octocorals and algae (Díaz et al. 2000).

The algae were collected during a macroalgal survey of the sites by scuba diving, in the wet season (from November 2012 to January 2013 and June 2013 to August 2013), and dry season (from February 2013 to May 2013) at 9-17 m depth. Specimens were preserved in 4 % formalin/seawater solution. In the laboratory, algae were identified using an OLYMPUS BX51 microscope. Slide material was mounted in 50 % glycerin, after staining in aniline-blue solution. Specimens were deposited in COL, the Herbarium of the Universidad Nacional de Colombia.

Results and discussion

SYSTEMATICS

Family WRANGELIACEAE

Genus *Griffithsia* C. Agardh *nom. cons.*

Griffithsia capitata Børgesen

(Figure 2)

Distribution

This species, which has as type locality Gran Canaria, Canary Islands (Børgesen 1930), has a limited distribution in the north-eastern Atlantic Ocean. To date, it has been reported only from the Canary Islands (Gil-Rodriguez and Afonso-Carrillo 1980; Afonso-Carrillo and Sansón 1999) and the Madeira Archipelago (Levring 1974; Neto et al. 2001).

Description

Thallus delicate, pale rose, small, to 7 mm in height, epiphytic, erect (Fig. 2a), attached to the substrate by a basal disc (Fig. 2b); branching irregular. Cells cylindrical, 75–130 μm in diameter on vegetative branches (Fig. 2c) and up to 320 μm on branches with reproductive structures, 600–1000 μm long; apical cells dome-shaped in vegetative branches 75–90 μm diameter, 100 μm long (Fig. 2c), rounded when reproductive structures are present, 160 μm diameter, (Fig. 2d). Sterile filaments are not present.

Tetrasporangia globose, 55–70 μm diameter, 55–60 μm long, with basal cell 12.2 μm diameter, 15 μm long, in whorls encircling distal ends of subapical to median cells (Fig. 2e), with the upper cell of the vegetative filament bearing tetrasporangia. The ultimate cell of the fertile filament is often caducous. Involucral cells and trichoblasts absent.

Spermatangia develop on the ultimate cell of the fertile branch, as naked fascicles, which cover 25–30 % of the apical cell as cap-like masses (Fig. 2f). The fascicles are 500 μm × 350 μm, whereas the spermatangia are 5 μm long and 2.5 μm diameter.

Carposporophyte terminal to subterminal, surrounded by large kidney-shaped involucral cells (Fig. 2g).

Habitat and phenology

Tetrasporangiate, male, female, and vegetative plants of *Griffithsia capitata* were collected from the "Wild Life" site on 2 November 2012 at 12.5 m depth (NRD037, NRD042, NRD051, NRD104 and NRD123). On 22 January 2013 at 16.8 m depth (NRD410 and NRD499). On 4 April 2013 at 13 m depth (NRD992, NRD0030, NRD0070, NRD0074 and NRD0105). On 7 June 2013 at 14 m (NRD0209, NRD0271, NRD0273, NRD0319, and NRD0341). On 9 August 2013 at 14 m depth (NRD0406, NRD0465, NRD0487 and NRD0519). Vegetative plants from the "Green Moon" site were collected on 7 December 2012 at 10.2 m depth (NRD148 and

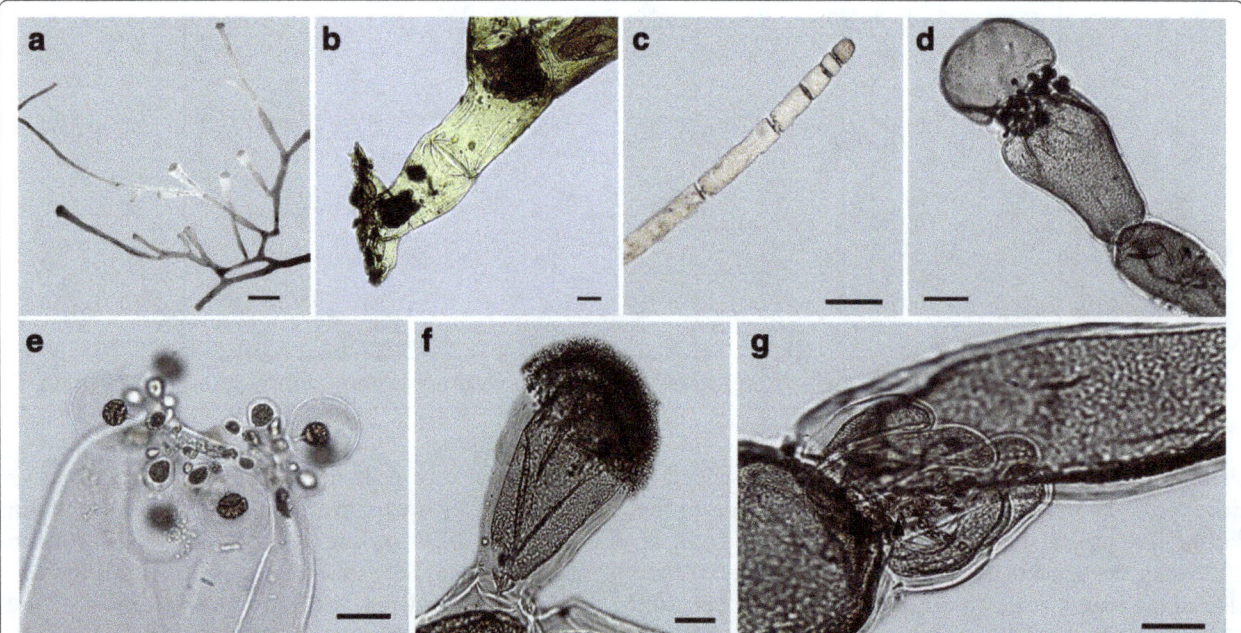

Fig. 2 *Griffithsia capitata.* **a** Habit of male gametophyte. **b** Basal disc. **c** Dome-shaped apical cell terminating vegetative branch. **d** Globose tetrasporangia in whorls encircling distal ends of subapical to median cells. **e** Tetrasporangia with loss of upper part. **f** Spermatangial cap developing on ultimate cell of a fertile branch. **g** Persistent kidney-shaped involucral cells which surrounded the carposporophyte (which has shed). Scale bars: **a**, 1 mm; **b**, 200 μm; **c**, 250 μm; **d**, 100 μm; **e**, 50 μm; **f**, 250 μm; **g**, 50 μm

NRD326). The plants were growing on coral rubble and on larger algae, especially species of *Halimeda*.

Remarks

To date, eight species of *Griffithsia* have been reported from the western North Atlantic: *G. aestivana* C.W. Schneider & C.E. Lane, *G. caribaea* Feldmann-Mazoyer, *G. globulifera* Harvey *ex* Kützing, *G. heteromorpha* Kützing, *G. opuntioides* J. Agardh, *G. radicans* Kützing, *G. schousboei* Mont. in Webb and *G. secundiramea* Vickers (Wynne 2011). Of these, only *G. aestivana* and *G. radicans* lack involucral cells around the tetrasporangia (Schneider and Lane 2007). However, *G. aestivana* is much larger in size, and the tetrasporangia are associated with trichoblasts, a feature not observed in *G. capitata* (Børgesen 1930).

Our male plants bearing terminal caps of spermatangia clearly align our species with the *Griffithsia capitata* group as proposed by Børgesen (1942). Only four species have been assigned to this group (Baldock 1976): *G. capitata* Børgesen (1930), *G. weber-van-bosseae* Børgesen (1942), *G. globulifera*, and *G. rhizophora* Weber-van Bosse (Børgesen 1942). *Griffithsia globulifera* can be separated from our specimens on the basis of its large thallus size, the large size of individual cells (to 1.5 mm diam), the presence of sterile filaments and also of involucral cells protecting the tetrasporangia (Taylor 1962; Littler and Littler 2000). We did not observe any trichoblasts or sterile filaments in our specimens (Fig. 2a).

Griffithsia rhizophora can be distinguished from our species because of its dichotomous branching habit, its extensive prostrate habit, and its frequent production of attachment rhizoids (Thivy and Iyengar 1963). Our specimens are totally erect and are attached to the substrate by means of a basal disc (Fig. 2b). Moreover, in *G. rhizophora* the cell length: width ratio is equal to 3 (Taylor 1960), while in our specimens the length: width ratio is >3. *Griffithsia weber-van-bosseae* differs from our specimen for its decumbent axis, with frequent rhizoids, and for the presence of numerous dwarf shoots forming at the upper ends of cells (Børgesen 1942). In general, our specimens fit well the description of *Griffithsia capitata* Børgesen. However, our thalli are smaller; our specimens were less than 1 cm in length, whereas Børgesen (1930) described the thalli as 2-3 cm high. Levring (1974) observed his specimens of this species from Madeira to be about 1 cm high and to carry tetrasporangia. Our specimens were fertile, and we collected the three stages of the reproductive cycle: male and female gametophytes, and tetrasporophytes. Therefore, we are confident that the observed thalli are adults and not just undeveloped juveniles.

This is the first record of *Griffithsia capitata* in the western Atlantic and outside its known range of distribution, which includes only the Canary Islands and Madeira in the eastern North Atlantic (Guiry and Guiry

2016). The present record expands the geographical range of the species to the Caribbean Sea, and adds to the record of species with amphiatlantic distribution. The alga is small and may have gone undetected, despite the attention that the Caribbean flora has received in the past decades. The Canary Islands, with Madeira, the Azores and the Cabo Verde islands, form the Macaronesian biogeographic region, and its marine flora has several representatives from the Caribbean basin (Prud'homme van Reine and van den Hoek 1990; Haroun et al. 1993). According to Prud'homme van Reine (1998), tropical-to-warm temperate species have a predominantly amphi-atlantic distribution. Therefore, it is possible that this species has been a natural resident of the Caribbean basin but has gone undetected due to its small size and turf habit. Long-distance dispersal might be possible due to the Canary current and the North Equatorial currents, which drift in an east to west direction (Mason et al. 2011). Pakker et al. (1996) demonstrated a lack of ecotypic differentiation between isolates from different localities on eastern and western Atlantic coasts, suggesting long-range dispersal events. Due to the presence of reproductive structures, we consider the species as well-established in the Caribbean basin; the alga is at present part of the macroalgal turf of the study sites, mainly epiphytic on *Halimeda* spp.

An alternative scenario includes a recent introduction of the species to the western Atlantic. Reports of small epiphytic red algae outside their range of introduction are common, and some species, within a few years after their first detection, become dominants at the demise of native turf algae (e.g. Kapraun and Searles 1990; Rindi et al. 1999; Nikolić et al. 2010).

A probable vector of introduction to San Andres island, due to the absence of aquaculture, is either ballast water or hull fouling. Various studies indicate that ballast water and sediments are one of the most important vectors of transoceanic and interoceanic movement of marine and coastal shallow water organisms (Shine et al. 2000). If this scenario results to be true, evidence of nutrient enrichment of the coastal waters of the island (Gavio et al. 2010) may favor the proliferation of turf algae (Vermeij et al. 2010), and indirectly promote the spread of *Griffithsia capitata*.

The origin of the species in the Caribbean basin, whether native of introduced, is not yet clear and should be addressed in the future to have a better understanding of the dynamics driving species diversity in the region.

Acknowledgements
The authors are grateful to Manuel Angarita (Landivers) and Francisco Javier Ramos for help in the field. We thank Julieta and Alejandro Rincón for assistance in managing Adobe Photoshop. Sven Zea and Paola Rodriguez provided insightful comments for field work. David Forero Parra elaborated the map of study sites, for which we acknowledge.

This research was funded by the Universidad Nacional de Colombia, sede Caribe, through the project Hermes # 12388. This work is contribution No. 422 of CECIMAR, Universidad Nacional de Colombia and Programa de Postgrado en Biología – Línea Biología Marina.

Authors' contribution

MNRD carried out field work, species identification and picture editing. BG carried out species identification and drafted the manuscript. MJW confirmed species identification, provided insights in the research and meliorate the manuscript. ASM provided financial and technical support for field work and laboratory analysis, and revised the manuscript. All authors read and approved the final manuscript.

Competing interests

The authors declare they do not have competing interests.

Author details

[1]Universidad Nacional de Colombia, Sede Caribe. San Luis Free Town # 52-44, San Andrés Isla, Colombia. [2]Departamento de Biología, Universidad Nacional de Colombia, sede Bogotá, Ciudad Universitaria, Bogotá, Colombia. [3]University of Michigan Herbarium, Ann Arbor, MI 48108, USA.

References

Afonso-Carrillo J, Sansón M. Algas, hongos y fanerógamas marinas de las Islas Canarias. Clave analítica. 1999. Materiales Didácticos Universitarios. Universidad de La Laguna.

Albis-Salas M, Gavio B. Notes on marine algae in the International Biosphere Reserve Seaflower, Caribbean Colombian I: new records of macroalgal epiphytes on the seagrass Thalassia testudinum. Botanica Marina. 2011;54: 537–43.

Baldock RN. The Griffithsieae group of the Ceramiaceae (Rhodophyta) and its southern Australian representatives. Australian Journal of Botany. 1976;24: 509–93.

Børgesen F. The marine algae from the Canary Islands especially from Tenerife and Gran Canaria. III. Rhodophyceae. Part III. Ceramiales. Kongelige Danske Videnskabernes Selskab, Biologiske Meddelelser. 1930;9(1):1–159.

Børgesen F. Griffithsia weber-van bosseae nov. spece. Dr. A. Weber-van Bosse Jubilee Volume. Blumea. 1942;Suppl. 2:15–20.

Chaves-Fonnegra A, Zea S, Gómez ML. Abundance of the excavating sponge Cliona delitrix in relation to sewage discharge at San Andrés Island, SW Caribbean, Colombia. Boletin de Investigaciones Marinas y Costeras. 2007;36: 63–78.

CORALINA-INVEMAR. In: Gómez- López DI, Segura-Quintero C, Sierra-Correa PC, Garay-Tinoco J, editors. Atlas de la Reserva de Biósfera Seaflower. Archipiélago de San Andrés, Providencia y Santa Catalina. Instituto de Investigaciones Marinas y Costeras "José Benito Vives De Andréis" -INVEMAR- y Corporación para el Desarrollo Sostenible del Archipiélago de San Andrés, Providencia y Santa Catalina -CORALINA-. Santa Marta: Serie de Publicaciones Especiales de INVEMAR # 28; 2012. p. 180.

Costello MJ, Coll M, Danovaro R, Halpin P, Ojaveer H, Miloslavich P. A Census of Marine Biodiversity Knowledge, Resources, and Future Challenges. PLoS One. 2010;5(8):e12110.

Díaz JM, Garzón-Ferreira J, Zea S. Los arrecifes coralinos de la isla de San Andrés, Colombia: Estado actual y perspectivas para su conservación. Academia Colombiana de Ciencias Exactas, Físicas y Naturales Colección Jorge Álvarez Lleras N° 7. Editora Guadalupe LTDA. Colombia. 1995. p. 150.

Díaz JM, Barrios LM, Cendales MH, Garzón J, Vargas-Angel B, Ferreira J, Geister J, Lopez-Victoria M, Ospina GH, Parra-Velandia F, Pinzo J, Zapata FA, Zea S. Áreas coralinas de Colombia. Serie Publicaciones Especiales No. 5. Santa Marta: INVEMAR; 2000.

Gavio B, Palmer-Cantillo S, Mancera JE. Historical analysis (2000–2005) of the coastal water quality in San Andrés Island, SeaFlower Biosphere Reserve, Caribbean Colombia. Mar Pollut Bull. 2010;60:1018–30.

Gavio B, Reyes-Gómez V, Wynne MJ. Crouania pumila sp. nov. (Callithamniaceae: Rhodophyta), a new species of marine red algae from the Seaflower International Biosphere Reserve, Caribbean Colombia. Revista Biologia Tropical. 2013;61:1015–23.

Gavio B, Cifuentes-Ossa MA, Wynne MJ. Notes on the marine algae of the International Biosphere Reserve Seaflower, Caribbean Colombia V: first preliminary study on the phycological flora of Quitasueño bank. Boletín de Investigaciones Marinas y Costeras. 2015;44:117–26.

Gil-Rodriguez MC, Afonso-Carrillo J. atalogo de las algas marinas bentonicas (Cyanophyta, Chlorophyta, Phaeophyta y Rhodophyta) para el Archipielago Canario. Santa Cruz de Tenerife: Act, Aula de Cultura de Tenerife; 1980.

Guiry MD, Guiry GM. AlgaeBase. Galway: World-wide electronic publication, National University of Ireland; 2016. http://www.algaebase.org; searched on 12 April 2016.

Haroun RJ, Prud'homme van Reine WF, Müller DG, Serrao E, Herrera R. Deep-water macroalgae from the Canary Islands: new records and biogeographical relationships. Hegoländer Meeresuntersuchungen. 1993;47:125–43.

Kapraun DF, Searles RB. Planktonic bloom of an introduced species of Polysiphonia (Ceramiales, Rhodophyta) along the coast of North Carolina, USA. Hydrobiologia. 1990;204(205):269–74.

Levring T. The marine algae of the Archipelago of Madeira. Boletim do Museu Municipal do Funchal. 1974;28:5–111.

Littler DS, Littler MM. Caribbean reef plants. Washington: OffShore Graphics, Inc.; 2000.

Mason E, Colas F, Molemaker J, Shchepetkin AF, Troupin C, McWilliams JC, Sangrà P. Seasonal variability of the Canary Current: A numerical study. J Geophys Res. 2011;116:C06001. dx.doi.org/10.1029/2010JC006665.

Miloslavich P, Díaz JM, Klein E, Alvarado JJ, Díaz C, Gobin J, Escobar-Briones E, Cruz-Motta JJ, Weil E, Cortés J, Bastidas AC, Robertson R, Zapata F, Martín A, Castillo J, Kazandjian A, Ortiz M. Marine Biodiversity in the Caribbean: Regional Estimates and Distribution Patterns. PLoS One. 2010;5(8):e11916.

Neto AI, Cravo DC, Haroun RT. Checklist of the benthic marine plants of the Madeira Archipelago. Botanica Marina. 2001;44:391–414.

Nikolić V, Žuljević A, Antolić B, Despalatović M, Cvitković I. Distribution of invasive red alga Womersleyella setacea (Hollenberg) R.E. Norris (Rhodophyta, Ceramiales) in the Adriatic Sea. 2010. Acta Adriatica. 2010;51(2):195–202.

Ortiz JF, Gavio B. Notes on the marine algae of the International Biosphere Reserve Seaflower, Caribbean Colombia II: diversity of drift algae in San Andres island, Caribbean Colombia. Caribbean Journal of Science. 2012;46: 313–21.

Pakker H, Breeman AM, Prud'homme Van Reine WF, Van Oppen MJH, Van Den Hoek C. Temperature responses of tropical to warm-temperate Atlantic seaweeds. I. Absence of ecotypic differentiation in amphi-Atlantic tropical-Canary Islands species. European Journal of Phycology. 1996;31:123–32.

Prud'homme Van Reine WF. Seaweeds and biogeography in the Macaronesian region. Bol Mus Mun Funchal. 1998;Suppl. no. 5:307–31.

Prud'homme van Reine WF, van den Hoek C. Biogeography of Macaronesian seaweeds. Courier Forsch-Inst Senckenberg. 1990;129:55–73.

Reyes-Gómez V, Gavio B, Velasquez H. Notes on the marine algae of the International Biosphere Reserve Seaflower, III. New records of Cyanophyta for the Caribbean coast of Colombia. Nova Hedwigia. 2013;97:349–60.

Rindi F, Guiry MD, Cinelli F. Morphology and reproduction of the adventive Mediterranean rhodophyte Polysiphonia setacea. Proceedings of the International Seaweed Symposium. 1999;16:91–100.

Schneider CW, Lane CE. Notes on the marine algae of the Bermudas. 8. Further additions to the flora, including Griffithsia aestivana sp. nov. (Ceramiaceae, Rhodophyta) and an update on the alien Cystoseira compressa (Sargassaceae, Heterokontophyta). Botanica Marina. 2007;50:128–40.

Shine C, Williams N, Gündling L. A Guide to Designing Legal and Institutional Frameworks on Alien Invasive Species, Environmental Policy and Law Paper No. 40 IUCN - Environmental Law Centre A Contribution to the Global Invasive Species Programme IUCN - The World Conservation Union. 2000.

Taylor WR. Marine algae of the eastern tropical and subtropical coasts of the Americas. Ann Arbor: The University of Michigan Press; 1960.

Taylor WR. Marine algae of the northeastern coast of North America. Revised edition Second printingth ed. Ann Arbor: The Unviersity of Michigan Press; 1962.

Thivy F, Iyengar ERR. A new record of Griffithsia rhizophora (Grunow) ex Weber-van Bosse, for India. Botanica Marina. 1963;5:33–7.

Vermeij MJA, van Moorselaar I, Engelhard S, Hörnlein C, Vonk SM, Visser PM. The effects of nutrient enrichment and herbivore abundance on the ability of turf algae to overgrow coral in the Caribbean. PLoS One. 2010;5(12):e14312.

Wynne MJ. A checklist of benthic marine algae of the tropical and subtropical western Atlantic: third revision. Nova Hedwigia Beiheft. 2011;140:1–166.

5

First records of two *Padina* species (Dictyotales, Phaeophyceae) from the Syrian coast (eastern Mediterranean)

H. Arraj[1*], H. Mayhoob[2] and A. Abbas[2]

Abstract

Padina ditristromatica and *Padina boryana* (Dictyotales, Phaeophyceae) are recorded for the first time from the Syrian coast (eastern Mediterranean). *Padina ditristromatica* has been previously reported in the west and the north east of the Mediterranean (Ni-Ni-Win et al. 2011) and *Padina boryana* from the southern Mediterranean (Geraldino et al. 2005). The morphological and anatomical characteristics have been used to confirm the new *Padina* spp.

Keywords: Dictyotales, Mediterranean Sea, *Padina*, *Padina boryana*, *Padina ditristromatica*, Phaeophyceae

Introduction

Marine macrophytes have been thoroughly studied in the western Mediterranean, but there are fewer studies for the eastern Mediterranean (Giaccone 1968; Mayhoob 1976; UNEP/IUCN/GIS 1990) especially of the Phaeophycean algae (Mayhoob 1989, 2004; Mayhoob and Billard 1991; Mayhoob and Hatoum 2005).

Specimens of the genus *Padina* Adanson were collected from Syria during a survey of Phaeophyceae.

This genus includes about 27 species (Papenfuss 1977) in tropical and sub-tropical waters. In the Mediterranean Sea only 5 species have been reported: *Padina pavonica* (L) Thivy, *P. boryana* Thivy (Ribera et al. 1992), *P. antillarum* (Kützing) Piccone (*P. tetrastromatica* Hauck (Mayhoob 2004)), *P. ditristromatica* Ni-Ni-Win & H. and *P. pavonicoides* Ni-Ni-Win & H. Kawai (Ni-Ni-Win et al. 2011; Cormaci et al. 2012). Two of these are found on the Syrian coast: *P. pavonica* and *P. tetrastromatica* (Mayhoob 2004).

This paper describes the recording of 2 species of the brown alga genus *Padina* (Dictyotales, Phaeophyceae) for the first time on the Syrian coast.

Materials and methods

Samples were collected from 2 sites: Al-madinah Al-riadiah (35° 17′ 38.07″ N, 35° 55′ 23.71″ E) and the coast of Tal Socass (35° 17′ 54.86″ N, 35° 55′ 17.67″ E) in spring-summer 2014 from the lower intertidal zone at a depth of 2 m. The samples were washed thoroughly and preserved in a 4 % formalin-seawater solution for further investigation. Some of these samples were preserved in the form of herbarium sheets and given series numbers with the date of collection. They were kept in the herbarium of the High Institute of Marine Researches (Latakia, Syria).

Transverse sections were made by freehand cutting with the help of shaving blades.

Results

Specimens examined

Twenty preserved specimens of *Padina boryana* and 50 specimens of *Padina ditristromatica* were studied morphologically and anatomically.

Padina boryana Thivy

Morphology The thalli are yellowish brown in colour, and are moderately calcified on the lower surface (opposite the inrolled margin), especially at the stipe and lightly calcified on the upper surface (facing the inrolled margin). They are composed of

* Correspondence: hadeel.arraj@gmail.com
[1]Department of Marine Biology, High Institute of Marine Researches (HIMR), Tishreen University, Latakia, Syria
Full list of author information is available at the end of the article

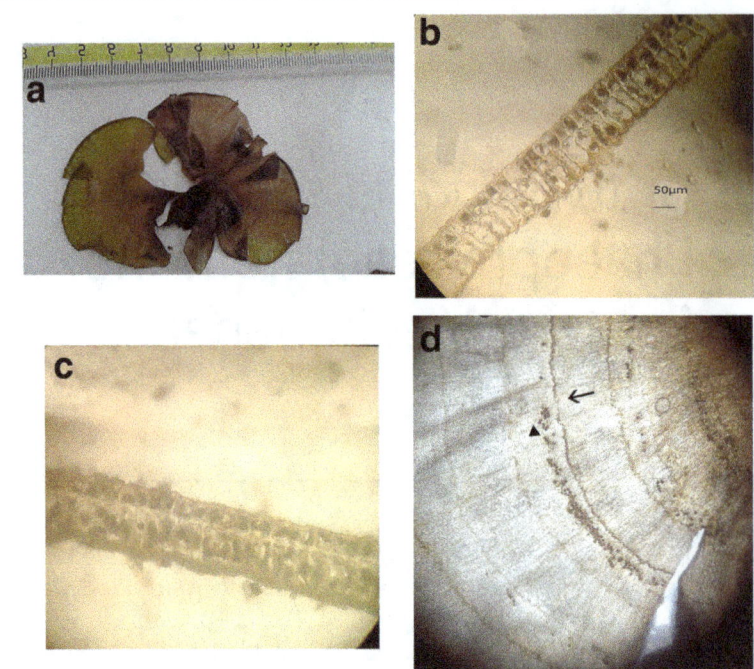

Fig. 1 *Padina boryana* Thivy: (**a**) Morphology of *Padina boryana* Thivy Sporophyte, (**b**) Transverse section of middle portion, showing 2 layers of cells, (**c**) Transverse section of basal portion, showing 2 layers of cells, (**d**) Surface view of the upper surface of the thallus, showing relationship of the hair lines (arrow) and the sporangial sori on the upper surface (triangle)

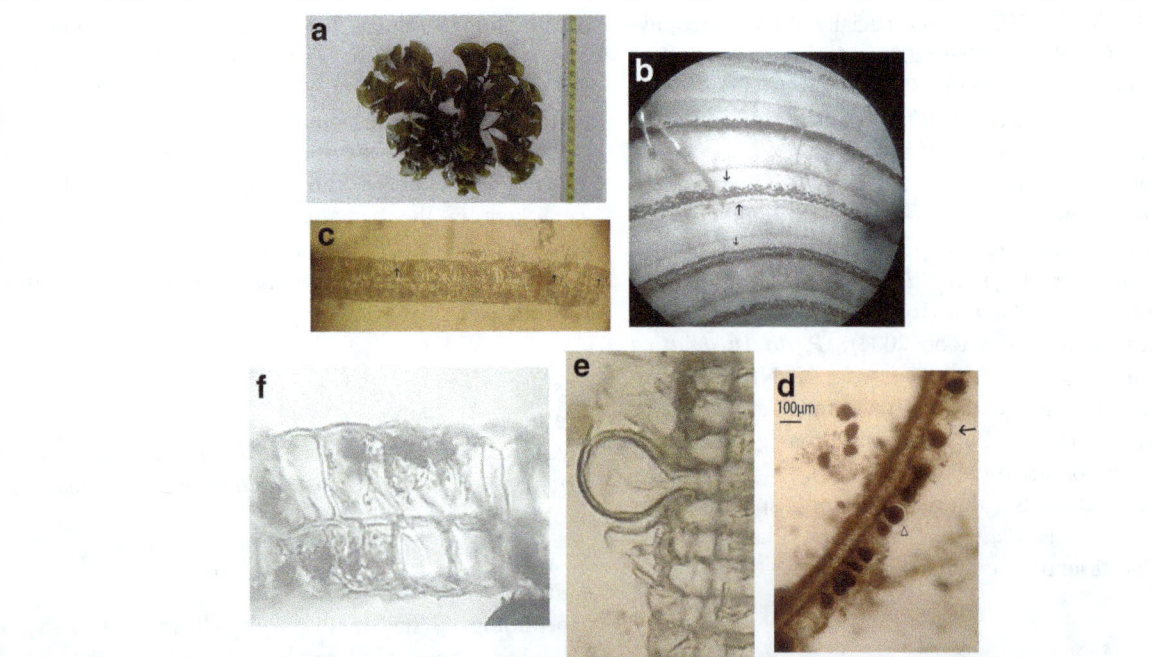

Fig. 2 *Padina ditristromatica* Ni-Ni-Win & H. Kawai. (**a**) Morphology of *Padina ditristromatica* Ni-Ni-Win & H. Kawai female gametophyte, (**b**) Surface view of the lower surface of the thallus, showing relationship of the alternating hair lines and the reproductive sori on lower surface (arrow), (**c**) Transverse section of middle portion, showing mixture of 2 to 3 layers (arrows), (**d**) Mature obovate oogonium (triangle) with empty oogonia (arrow), (**e**) Mature oogonium, showing spherical shape, (**f**) Transverse section of marginal portion showing two cell layers

Table 1 Comparing of the characteristics of five species of the Mediterranean *Padina*

	Height (cm)	Color	Surface	Hair lines		Sporangial sori		Holdfast		Cell layers	Indusium	References
				Position	Arrangemant	Position	Arrangemant	Length (mm)	Breadth (mm)			
P. tetrastromatica	8-16	Greenish brown	Lightly calcified	On both surfaces	Alternate	On both surfaces	Successive	7	4	4	Absent	(Gaillard 1967)
P. pavonica	6	Yellowish brown	Heavily calcified	On both surfaces	Alternate	On both surfaces	Alternate	7-20	5-15	3-4	Present	(Taylor, 1960)
P. pavonicoides	-	-	-	On both surfaces	Alternate	Lower surface	Successive	-	-	2-3	Present	(Ni-Ni-Win et al. 2011
P. boryana	7	Yellowish brown	Moderately calcified on the lower surface and lightly calcified on the upper surface	upper surface	successive	Upper surface	Successive	7	4	2	Absent	(this study)
P. ditristromatica	5-10	Yellowish or greenish	Heavily calcified	On both surfaces	Alternate	Lower surface	Successive	7-10	5-7	2-3	Present	(this study)

fan-shaped blades with rhizoids forming the holdfast (Fig. 1a). The erect thalli are up to 7 cm in height. The sporangial rows alternate with hair rows at different intervals only on the upper surface of the thallus.

Anatomy A transverse section of the thallus shows 2 layers of cells (Fig. 1b). The outer cells are small and nearly square, measuring 36 µm in length. The inner cells are large and rectangular in shape, measuring 58 µm in length. The sporangial sori and hair lines are close to each other (Fig. 1d). The sporangial sori are not covered with an indusium.

Padina ditristromatica Ni-Ni-Win & H. Kawai

Morphology The thalli are composed of fan-shaped lobes with inrolled margins. These lobes are attached to each other at the base by a short stem (Fig. 2a).
The thallus is yellowish or greenish-brown between 5 and 10 cm high and moderately calcified on both surfaces.

On the lower surface of the thallus a number of semi-circular lines of 2 different types can be seen: reproductive sori (the larger and darker lines) and hair lines. The reproductive sori were found only on the lower surface and formed a single line of separated dark spots (Fig. 2b). And they are successive at equal distance.

The reproductive sori and the hair lines are close to each other or may be merged into single lines (Fig. 2b).

Hair lines were found on the lower and upper surface of the thallus alternating between the 2 surfaces (Fig. 2b). The long fibrous hairs are only on the lower surface of stem of the thallus. The 'Vaughaniella' stage was not found in this species.

Anatomy Figure 2f shows the 2 cell layers at 68–72 mm from the margin of the thallus. In other parts of the thallus, we found layers of 2 and 3 cells at 80–120 mm in the transverse section (Fig. 2c).

The oogonial sori are located near to the hair lines (Fig. 2b) and formed spots in narrow lines only on the lower surface (Fig. 2a). Each 1 of them is surrounded by an indusium (Fig. 2d). The mature oogonium has a spherical shape and is 100 µm in diameter (Fig. 2e).

Discussion

According to the morphological and anatomical observations of these specimens and previous research (Gaillard 1967; Geraldino et al. 2005; Coppejans et al. 2009; Ni-Ni-Win et al. 2011; Abbas and Shameel 2013) the species are identified as *P. boryana* Thivy and *P. ditristromatica* Ni-Ni-Win & H. Kawai.

They are clearly distinguished from each other in terms of morphological and anatomical characteristics, mainly in the numbers of cell layers in the thallus, the degree of calcification on the thallus surface, and the structure of the reproductive sori.

P. pavonica is mostly 3 layers thick and occasionally 4 layers at the base of the thallus (Taylor 1960; Ni-Ni-Win et al. 2011), but there is a mixture of 2 and three layers in *P. ditristromatica* (Ni-Ni-Win et al. 2011; this study).

Padina ditristromatica is similar to *P. tetrastromatica* Hauck in respect of the dioecious gametophyte and the presence of an indusium (Gaillard 1967; Ni-Ni-Win et al. 2011; this study) (Table 1), but they differ in the numbers of cell layers in the thallus (Table 1) where *P. tetrastromatica* is mostly 4 layers thick (Gaillard 1967).

In addition, *P. ditristromatica* is different from *P. pavonica* and *P. tetrastromatica* in terms of the structure and arrangement of sporangial sori which in this species are located distally and adjacent to the hair lines only on the lower surface (Ni-Ni-Win et al. 2011). Mean while in *P. pavonica* and *P. tetrastromatica* they are located in concentric rows girdling the hair lines on both surfaces (Taylor 1960; Gaillard 1967; Ni-Ni-Win et al. 2011).

Padina pavonicoides is different from *P. ditristromatica* in that its thallus is composed of 3 cell layers from the base to the marginal portion and 2 layers at the inrolled margin and by the alternating hair lines that are spaced at equal distances between the upper and lower surfaces (Ni-Ni-Win et al. 2011).

Padina boryana differs from *P. ditristromatica* in terms of the number of cell layers in that it is mostly 2-layers thick throughout the thallus and it also lacks an indusium (Farrant and King 1989; Geraldino et al. 2005; this study) (Table 1).

Marine vegetation in the eastern Mediterranean, including the coast of Syria, belongs to the Atlantic-Mediterranean province (Giaccone 1968; Mayhoob, 1976). However, some circumtropical species are establishing permanent populations so that some tropical characteristics can be attributed to this region.

According to the world-wide distribution of these 2 species and the lack of information concerning the biodiversity of macroalgae in the eastern Mediterranean they might be endemic, relics of the Sea of Tethys, or alien species that have recently been introduced to the Mediterranean sea (Occhipinti-Ambrogi 2000; Boudouresque and Verlaque 2002; Streftaris et al. 2005, 2007; Galil and Zenetos 2008).

Acknowledgements
We thank Tishreen University for funding this project and the High Institute of Marine Research, Latakia for logistical support.

Author details
[1]Department of Marine Biology, High Institute of Marine Researches (HIMR), Tishreen University, Latakia, Syria. [2]Department of Botany, Faculty of Science, Tishreen University, Latakia, Syria.

References

Abbas A, Shameel M. Morpho-anatomical studies on the genus Padina (Dictyotales, Phaeophycota) from the Coast of Karachi, Pakistan. Proc Pak Acad Sci. 2013;50:21–36.

Boudouresque CF, Verlaque M. Assessing scale and impact of ship-transported alien macrophytes in the Mediterranean Sea. In: Briand F, editor. Alien marine organisms introduced by ships in the Mediterranean and Black seas. Monaco: CIESM Workshop Monographs 20; 2002. p. 53–62.

Coppejans E, Leliaert F, Dargent O, Gunasekara R, De Clerck O. Abc Taxa. Sri Lankan Seaweeds Methodologies and field guide to the dominant species. *Divisions of Algae from Sri Lanka and general remarks*. Belgium. 2009;6:73–221.

Cormaci M, Furnari G, Catra M, Alongi G, Giaccone G. Marine benthic flora of the mediterranean: Phaeophyceae. Bollettine Accad Gioenia. 2012;45:1–508.

Farrant PA, King RJ. The Dictyotales of New South Wales. Proc Linnean Soc New South Wall. 1989;110:369–405.

Gaillard J. Monographic study of *Padina tetrastromatica* Hauck. Bulletin Fondamental Inst Black Afr Seri Sci Nat. 1967;29:447–63.

Galil BS, Zenetos A. Alien species in the Mediterranean Sea: which, when, where, why? Hydrobiologia. 2008;606:105–16.

Geraldino PJL, Lawrence ML, Boo SM. Morphological study of the marine algal genus Padina (Dictyotales, Phaeophyceae) from the southern Philippines: 3 species new to the Philippines. Algae. 2005;20:99–112.

Giaccone G. New and interesting species of Rhodophyceae from the eastern Mediterranean Sea. G Bot Ital. 1968;102:397–414.

Mayhoob H. Research of the marine vegetation at the Syrian coast. Experimental study on the morphogenesis and development of some species which less Known, PhD Thesis. France: University Caen; 1976. p. 286.

Mayhoob H. The invasion of the Syrian coast by the brown alga of the red sea. J Damascus Univ. 1989;5:65–79.

Mayhoob H, Billard C. Contribution of knowledge of *Stypopodium* Sp. installed at the Syrian coast. Cryptogamie Algology. 1991;12:125–36.

Mayhoob H. The presence of tropical alga *Padina tetrastromatica* Hauck near Latakia (Syria). J Damascus Univ. 2004;2:77–89.

Mayhoob H, Hatoum O. The presence of *Cystoseira Balearica* Sauv. Et *C. barbatula* Kg. Emend Cormaci et Al., Syria. J Tishreen Univ. 2005;27:207–18.

Ni-Ni-Win, Hanyuda T, Stefano GAD, Furnari G, Meinesz A, Kawai H. *Padina ditristromatica* sp. nov. and *Padina pavonicoides* sp. nov. (Dictyotales, Phaeophyceae), two new species from the Mediterranean Sea based on morphological and molecular markers. Eur J Phycology. 2011;46:327–41.

Occhipinti-Ambrogi A. Biotic invasions in a Mediterranean lagoon. J Biol Invasions. 2000;2:165–76.

Papenfuss GF. Review of the genera of Dictyotales (Phaeophycophyta). J Bulletin Japanese Soc Phycology. 1977;25:271–87.

Ribera MA, Go'mez-Garreta A, Gallardo T, Cormaci M, Furnari G, Giaccone G. Check-list of Mediterranean seaweeds. I. Fucophyceae (Warming 1884). J Bot Marina. 1992;35:109–30.

Streftaris NN, Zenetos A, Papathnassiou E. Globalistaion in marine ecosystems: the story of non-indigenous marine species across European seas. Greece. Oceanogr Mar Biol. 2005;43:419–53.

Taylor WR. Marine algae of the northeastern coast of North America. Ann Arbor: University of Michigan, Press; 1960.

UNEP/IUCN/GIS. Posidonia: Red Book "Gérard Vuignier", marine plants, populations and landscapes threatened in the Mediterranean, MAP Technical Reports Series. No. 43. Athens: UNEP; 1990. p. 250. French only.

6

Species composition of extant coccolithophores including twenty six new records from the southwest Pacific near New Zealand

F. Hoe Chang[1*] and Lisa Northcote[2]

Abstract

Background: Coccolithophores are one of the major components of marine phytoplankton and also one of the most prominent members of haptophyte algae. Studies of the extant coccolithophores started more than half a century ago in New Zealand waters, and with two exceptions, were limited to only a few relatively small areas close to shore. In this study the diversity of these 'calcium carbonate scale-producers' were updated from specimens collected in oceanic waters around the wider region of New Zealand.

Methods: Water samples collected from 156 stations on 10 voyages between 2009 and 2011 were filtered through Nuclepore polycarbonate membrane filters. Coccolithophores retained on these filters were identified using scanning electron microscopy.

Results: A total of 46 extant coccolithophore taxa were identified from 160 samples collected around New Zealand. The total number of coccolithophore taxa identified was greatest to the east (46), intermediate to the west (15) and least to the northeast and south (4 each) of New Zealand. These coccolithophores were classified into seven families in the four orders, with three families in *incertae sedis*, and one in a nannolith family. Forty two taxa were heterococcolithophores and four were holococcolithophores.

Conclusion: Approximately 57 % of the extant coccolithophores recorded were first-time records for the region. Even though *Syracosphaera* taxa generally occurred at low frequencies, they were the largest group and made up *c.* 31 % of all extant coccolithophores recorded in this study. Our findings provide updated information on the species composition and biogeography of coccolithophores in the southwest Pacific near New Zealand.

Keywords: *Emiliania huxleyi*, Extant coccolithophores, New records, *Syracosphaera pemmadiscus*, Species composition

Background

Coccolithophores (Coccolithophyceae) are marine, unicellular, golden-brown haptophytes (Probert et al. 2007; Ruggiero et al. 2015). At some stage in their life cycle they produce very small calcium carbonate-scales called coccoliths. This group has been a major producer of calcite in the open ocean since the late Jurassic (Hay 2004). These algae have gained increased attention as they play an important role in the global carbon cycle (e.g., Hiramatsu and Deckker 1997; Baumann et al. 2000) and possibly are also susceptible to ocean acidification as are other calcifiers (e.g., Doney et al. 2009; Beaufort et al. 2011).

The most common form of coccolith is the heterococcolith (e.g., Cros and Fortuño 2002; Young et al. 2003). They are formed by crystal units of variable shape and size, and their biomineralisation occurs intracellularly (Manton and Leadale 1969). Another less common form is the holococcolith which is constructed from numerous minute crystallites that appear to get calcified extracellularly (e.g., Rowson et al. 1986; Kleijne 1991). Quite often both forms

* Correspondence: h.chang@niwa.co.nz
[1]Biodiversity and Biosecurity Group, National Institute of Water & Atmospheric Research Ltd., P. O. Box 14-901, Kilbirnie, Wellington 6241, New Zealand
Full list of author information is available at the end of the article

of coccolith of the same species are found in the same samples (e.g., Cros and Fortuño 2002; Young et al. 2003). The taxonomy of coccolithophores is primarily based on the morphology of the exquisite calcium carbonate-coccoliths of either heterococcolith or holococcolith form.

Living coccolithophores are distributed widely around the globe, from tropical to polar region (e.g., Thomsen 1981; Winter and Siesser 1994; Baumann et al. 2000; Findlay et al. 2005). Their biogeography has been studied most extensively in Mediterranean Sea, Indian, Pacific and Atlantic Oceans (e.g., Okada and McIntyre 1977; Kleijne 1993; Cros and Fortuño 2002; Young et al. 2003; Tyrrel and Merico 2004; Wang et al. 2012), and in relatively recent times in high latitudes of the Southern Ocean (e.g., Nishida 1986; Eynaud et al. 1999; Findlay and Giraudeau 2000; Saavedra-Pellitero et al. 2014).

In New Zealand, initial studies were limited to coccoliths collected from deep sediments over the Challenger Plateau (Murray and Renard 1891) and also in other parts of the New Zealand region (Edwards 1968, 1982; McIntyre et al. 1970; Burns 1973, 1975). Studies of living coccolithophores started more than half a century ago, however, were limited to several local areas near New Zealand (Cassie 1961; Burns 1977; Rhodes et al. 1993, 1994). The most extensive survey was conducted by Norris (1961) along a transect from northeast New Zealand to Tonga. Samples collected from this voyage were examined using light microscopy and records were not illustrated. Very recently a survey carried out in the Pacific sector of the Southern Ocean by Saavedra-Pellitero et al. (2014) examined some coccolithophore samples collected from the Subtropical Front (STF) and Subantarctic (SA) waters, east of New Zealand. But this study was limited to three sites in STF and SA waters in a summer. Up until now, no study has been conducted on species composition of extant coccolithophores in oceanic waters around the wider region of New Zealand.

The present study is a component of New Zealand research into the impacts of ocean acidification on pelagic plankton. The aims of this study were to determine the species composition of extant coccolithophores and to update the diversity (checklist) and distribution of this group including the wider region of the southwest Pacific near New Zealand. As a substantial number of taxa were new records, SEM images of all taxa are presented here. Cell abundance in relation to environmental variables is not included in this study but will be reported separately elsewhere (Law et al., manuscript in preparation).

Methods
Coccolithophore samples
A total of ten surveys were conducted aboard the RV's *Tangaroa* and *Kaharoa* from January 2009 to February 2012 in the southwest Pacific Ocean near New Zealand (Fig. 1, Table 1). Seven of these surveys were conducted in spring, summer and autumn, in the subtropical (ST), STF and SA waters over the Chatham Rise and off Kaikoura coast, east of New Zealand (Fig. 1). The remaining three surveys were conducted on the northeast, west (along two transects in the Tasman Sea), and south of New Zealand. A combined total of 160 coccolithophore samples were collected from 156 stations around New Zealand (Fig. 1).

On each survey discrete water samples were collected from the upper 10 m using a 10-litre Niskin bottles mounted on a CTD rosette system. Immediately, 1.5 to 3 l water was filtered through a 47 mm diameter (0.8 μm pore size) Nuclepore Polycarbonate Track-Etch membrane filter (Whatman 111109). To minimise mechanical disruption of coccospheres, a vacuum of less than 100 mm Hg (low pressure) was applied below the filter with an electrical vacuum pump. These filters were individually placed in a labelled plastic Millipore petri dish (PF10266, 47 mm diameter) and air dried. The filters were then stored in a sealed plastic storage container with desiccant until analysis.

Scanning electron microscope
Using a cork borer (10 mm in diameter) a small circular piece was cut out of the 47 mm Nuclepore polycarbonate membrane filter and mounted on JEOL 12x10 mm aluminium slug using double-sided adhesive tape. These specimens were then coated with either platinum or carbon (15 nm thick) and examined with either a JEOL JSM-5300LV (Tokyo, Japan) or Quanta 450 (Oregon, U.S.A.) scanning electron microscopes (SEM) (20 kV) as described in Chang (2013). For more than half of samples, high resolution images were further taken using a JEOL JSM6500F FEG-SEM (10 kV) (Tokyo, Japan). The entire coated, cut-out membrane filter was carefully examined using either JEOL or Quanta SEMs. The diameter/ length of coccospheres and coccoliths were measured using individual scales on the SEM images.

Species identification
Taxonomic identification employed the work of Okada and McIntyre (1977), Hallegraeff (1984, 2010), Kleijne (1992, 1993), Winter and Siesser (1994), Cros and Fortuño (2002), Young et al. (2003), Kleijne and Cros (2009) and Frada et al. (2010). The species list used here was based on that of Young et al. (2003) and Jordan et al. (2004).

Results
Species composition
A total of 46 extant coccolithophore taxa, two of which are represented by both hetero- and holococcolith forms, were identified from 160 samples collected from

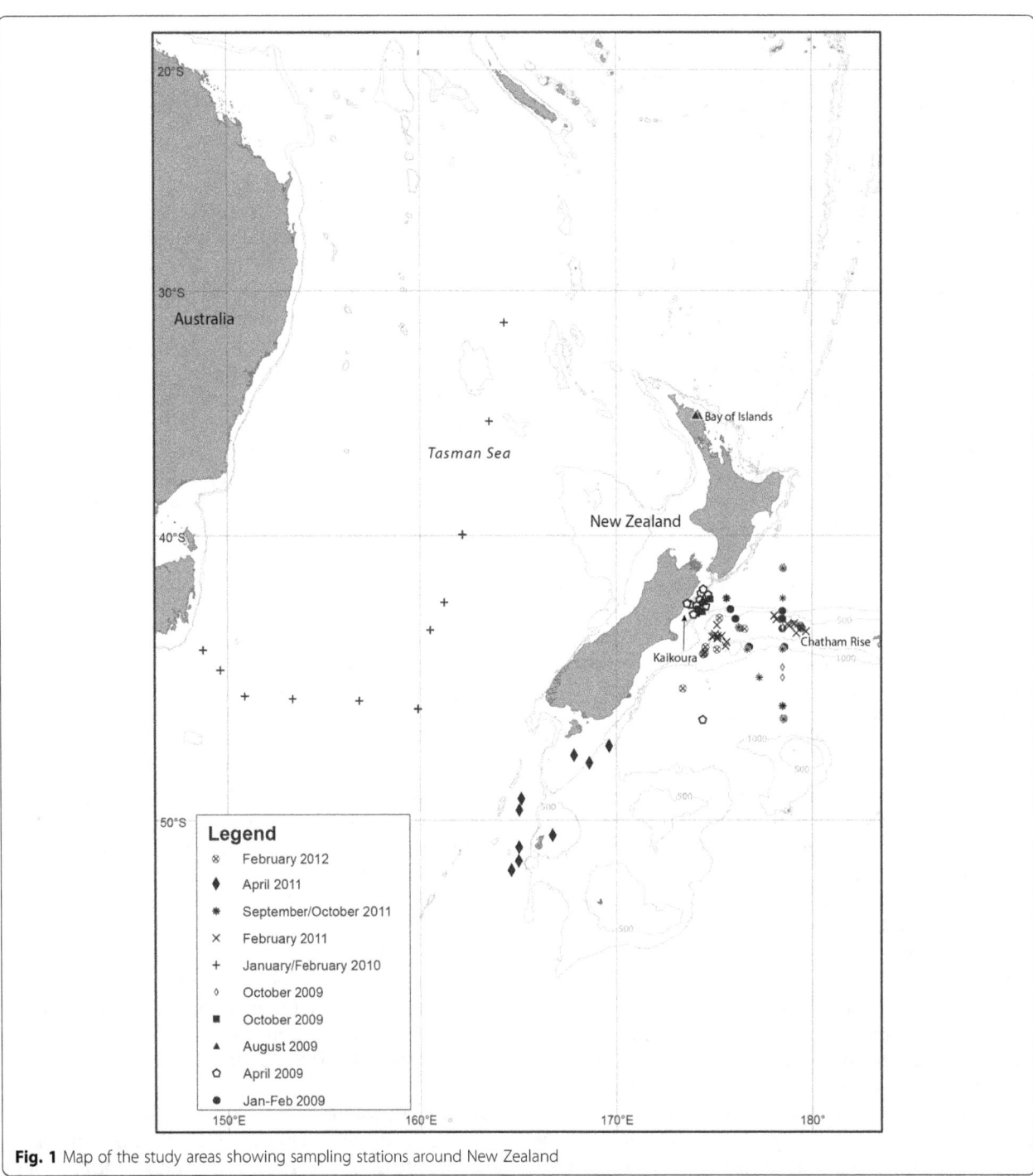

Fig. 1 Map of the study areas showing sampling stations around New Zealand

156 stations of ten surveys between 2009 and 2012 (Table 2; Figs. 2, 3 and 4). Forty two taxa were robust heterococcolithophores and 35 of them were placed in 7 families of the four orders, Coccosphaerales, Isochrysidales, Syracosphaerales and Zygodiscales, with the rest placed in *incertae sedis* (4 families) and in the holococcolith group according to the scheme of Young et al. (2003) and Jordan et al. (2004).

Syracosphaerales was by far the largest group, with 23 species/ taxa recorded in three families, Calciosoleniaceae (2), Rhabdosphaeraceae (7) and Syracosphaeraceae (14) (Table 2). Isochrysidales was intermediate, with 7 taxa in the family Noëlaerhabdaceae. Both Coccosphaerales (two species in the Calcidiscaceae) and Zygodiscales (three species plus one holococcolith form in Helicosphaeraceae and one species in Pontosphaeraceae) were

Table 1 Number of stations where water samples were collected from each voyage during the period from 2009 to 2012 near New Zealand

	Voyage	Location	Date	Number of stations
1	TAN0902	Chatham Rise	28-01-09 to 3-02-09	15
2	TAN0904	Kaikoura	21-04-09 to 30-04-09	21
3	TAM0908	Kaikoura	1-10-09 to 21-10-09	4
4	TAN0909	Chatham Rise	16-10-09 to 30-10-09	19
5	KAH0907	Bay of Islands	20-8-09 to 22-8-09	3
6	PINTS	Tasman Sea	31-01-10 to 15-02-10	36
7	TAN1102	Chatham Rise	6-02-11 to 11-02-11	16
8	TAN1107	Chatham Rise	28-09-11 to 1-10-11	9
9	TAN1106	South of NZ	19-04-11 to 29-04-11	8
10	TAN1203	Chatham Rise	5-02-12 to 28-02-12	25

the least speciose. Six species (plus a holococcolith form of *Coronosphaera mediterranea*) were, however, placed in the three different groups of *incertae sedis*, with 4 species placed in the holococcolithophore group.

Morphological characteristic

As the majority of extant coccolithophores previously reported in New Zealand were examined using light microscopy and some with only low resolution SEM (e.g., Cassie 1961; Norris 1961; Burns 1973, 1975, 1977; Rhodes et al. 1993, 1994), the morphological characteristics of all 46 taxa recorded here were illustrated using mostly high resolution scanning electron microscopy. SEM images of 42 heterococcolithophores, including the two holococcolith forms [*Helicosphaera carteri* HOL = (*Syracolithus catilliferus* (Kamptner 1937) Deflandre 1952) and *Coronosphaera mediterranea* HOL *gracillima*-type (*Calyptrolithophora gracillima* (Kamptner 1941) Heimdal in Heimdal and Gaarder 1980)], were displayed in Figs. 2a-t, 3a-t and 4a-p, while the 4 holococcolithophores were illustrated in Fig. 4q-t.

Of the 46 taxa identified from samples collected between 2009 and 2012, 26 taxa/ forms were first-time records (*c.* 57 %). Four of these first-time records were members of the order Isochrysidales, twelve were members of the order Syracosphaerales, three were members of the order Zygodiscales, another three were in the four families of *insertae sedis*, and four were holococcolithophores. In terms of size, coccoliths produced by the 26 first-time records of the New Zealand specimens (Table 3) were found to be generally reminiscent of those (with some variations) reported by Cros and Fortuño (2002) and Young et al. (2003). In the following only simple descriptions of the first-time records and a couple of new morphotypes (e.g., supercalcified and highly calcified) of *Emiliania huxleyi* (Lohmann 1902)

Hay and Mohler in Hay et al. 1967 found in the New Zealand region are presented here.

Gephyrocapsa muellerae Bréhéret 1978 (Fig. 2i), coccosphere spherical to subspherical (5–6 µm). Elliptical placoliths (2.9–3.8 µm long) are similar in construction to those of *Gephyrocapsa oceanica* Kamptner 1943, with bridge at low angle to long axis; central area rather small.

Gephyrocapsa ornata Heimdal 1973 (Fig. 2k-l), coccosphere spherical to subspherical (4–5 µm) (Table 2), coccoliths similar to *G. ericsonii* McIntyre and Bé 1967 (Fig. 2h) but with a conspicuous ring of spines around the central area and with much higher bridge of two diametrically opposite plates of varying shape. The bridge of *G. ornata* is at low angle to the long axis.

Reticulofenestra parvula (Okada and McIntyre 1977) Biekart 1989 (Fig. 2m) and *R. parvula* var. *tecticentrum* (Okada and McIntyre 1977) Jordan and Young 1990 (Fig. 2n), coccospheres (4–6 µm) of both species/variety similar to *Emiliania huxleyi* (Fig. 2c-g; Table 2). Coccoliths of *R. parvula* differed from those of *E. huxleyi* in not having slits between distal shield elements while *R. parvula* var. *tecticentrum* differed from the latter species in having coccolith with over calcified central area.

Anacanthoica acanthos (Schiller 1925) Deflandre 1952 (=*Acanthoica*) (Fig. 2s), coccosphere ovoid (*c.* 6 µm), monomorphic with no spines (Table 2). Coccoliths elliptical (1.9–2.1 µm) with relatively wide rim; ring of radial laths with central, wide, low protrusion (Table 3).

Cyrtosphaera aculeata (Kamptner 1941) Kleijne 1992 (=*Acanthoica*) (Fig. 2t), coccosphere (8–10 µm) subspherical to elongate (Table 2). Coccoliths (2.5–2.8 µm) varimorphic with rim somewhat bent upwards; central area with ring of radial laths (Table 3). Conical inner central area with protrusion ending in papilla; protrusion of some apical coccoliths modified into spine.

Palusphaera vandelii Lecal 1965 emend. Norris 1984 (Fig. 3b), coccosphere subspherical (without spines, 4.4–4.7 µm) (Table 2). Coccoliths with relatively long, distal spines and almost circular proxima disc (1.5–2.0 µm) with smooth central area towards base of spine (Table 3).

Rhabdosphaera xiphos (Deflandre and Fert 1954) Norris 1984 (Fig. 3c-d), coccosphere subspherical (4.0–5.8 µm) and dimorphic (Table 2). There is a distinctive star pattern on distal face of circular body coccolith (1.3–1.9 µm) with delicate spine and short collar at its base (Fig. 3d) (Table 3).

Syracosphaera anthos (Lohmann 1912) Janin 1987 (=*Deutshlandia*) (Fig. 3e-f), coccosphere subspherical (12.3–15.3 µm) and dithecate. Body coccolith (4.5–5.8 µm) dimorphic, circum-flagellar coccolith with large spine (Tables 2 and 3).

Syracosphaera bannockii (Borsetti and Cati 1976) Cros et al. 2000 (Fig. 3g-h), coccosphere ovoid (6.2–8.2 µm) and dithecate; body coccoliths (1.7–2.8 µm) elliptical

Table 2 List of extant coccolithophores recorded from 160 samples collected from 156 stations during 2009 and 2012: a) over the Chatham Rise and Kaikoura coast, east of New Zealand (NZ); b) in Bay of Plenty, northeast NZ; c) along two transects in the Tasman Sea, west of NZ; d) south of NZ

Taxa	Figure No.	Cell size[§] (µm)	Chatham Rise					Kaikoura		Bay of Plenty	Tasman Sea	South of NZ
			SP 09	SU 09	SU 11	AU 11	SU 12	SP 09	AU 09	SP 09	SU 10	AU 11
Heterococcolith Group												
Order Coccosphaerales												
Family Calcidiscaceae												
Calcidiscus leptoporus	2A	10–15	+	-	-	-	+	-	+	-	+	+
Umbilicosphaera hulburtiana	2B	5–6	-	-	-	+	-	-	-	-	-	-
Order Isochrysidales												
Family Noëlaerhabdaceae												
Emiliania huxleyi	2C-G	3–7	+[*]	+	+[*]	+	+	+	+	+	+	+
Gephyrocapsa ericsonii	2H	4–5	+	+	+	+	+	+	+	+	+	-
Gephyrocapsa muellerae	2I	5–6	-	+	-	+	+	+	+	+	+	-
Gephyrocapsa oceanica	2J	8–10	+	+	+	+	+	+	+	-	+	-
Gephyrocapsa ornata	2K-L	4–5	-	+	-	-	+	-	+	-	-	-
Reticulofenestra parvula	2M	4–6	-	-	+	+	-	+	+	-	-	-
R. parvula var. *tecticentrum*	2N	4–6	-	-	-	+	-	+	+	-	-	-
Order Syracosphaerales												
Family Calciosoleniaceae												
Calciosolenia brasiliensis	3G-H	6–8	-	+	-	-	-	-	+	-	+	-
Calciosolenia murrayi	3I	11–13	-	-	-	+	-	-	-	-	-	-
Family Rhabdosphaeraceae	3J-K	6–7	-	+	-	-	-	-	-	-	-	-
Acanthoica quattrospina	3L-M	6–9	-	+	-	+	+	-	+	-	+	-
Algirosphaera robusta	3N	6–7	-	+	-	+	+	-	-	-	-	-
Anacanthoica acanthos	3O	5–6	-	+	-	-	-	-	+	-	+	-
Cyrtosphaera aculeata	3P-Q	c. 11	-	+	-	-	-	-	-	-	-	-
Discosphaera tubifera	3R	c. 7	-	-	-	+	-	-	+	-	-	-
Palusphaera vandelii	3S	6–7	-	-	-	+	+	-	-	-	-	-
Rhabdosphaera xiphos	3T	7–8	-	+	-	+	-	-	-	-	-	-
Family Syracosphaeraceae	4A	10–12	+	+	-	+	+	-	+	-	+	-
Syracosphaera anthos	4B-C	c. 9	-	-	-	-	-	-	-	-	-	-
Syracosphaera bannockii	4D	n.d.	-	-	-	+	-	-	-	-	-	-
Syracosphaera corolla												
Syracosphaera leptolepis	4E	12–15	-	+	-	+	-	-	+	+	+	+
Syracosphaera molischii	3G-H	6–8	-	+	-	-	-	-	+	-	+	-
Syracosphaera nana	3I	11–13	-	-	-	+	-	-	-	-	-	-
Syracosphaera nodosa	3J-K	6–7	-	+	-	-	-	-	-	-	-	-
S. nodosa aff. *S.* sp. type 2	3L-M	6–9	-	+	-	+	+	-	+	-	+	-
Syracosphaera cf. *orbiculus*	3N	6–7	-	+	-	+	+	-	-	-	-	-
Syracosphaera ossa	3O	5–6	-	+	-	-	-	-	+	-	+	-
Syracosphaera pemmadiscus	3P-Q	c. 11	-	+	-	-	-	-	-	-	-	-
Syracosphaera pulchra	3R	c. 7	-	-	-	+	-	-	+	-	-	-
Syracosphaera serrata	3S	6–7	-	-	-	+	+	-	-	-	-	-
Syracosphaera tumularis	3T	7–8	-	+	-	+	-	-	-	-	-	-

Table 2 List of extant coccolithophores recorded from 160 samples collected from 156 stations during 2009 and 2012: a) over the Chatham Rise and Kaikoura coast, east of New Zealand (NZ); b) in Bay of Plenty, northeast NZ; c) along two transects in the Tasman Sea, west of NZ; d) south of NZ *(Continued)*

Genus *incertae sedis*	4A	10–12	+	+	-	+	+	-	+	-	+	-
Coronosphaera mediterranea	4B-C	*c.* 9	-	-	-	-	-	-	-	-	-	-
Coronosphaera mediterranea												
HOL *gracillima*-type	4F	n.d.	-	-	-	-	-	-	-	-	+	-
Order Zygodiscales												
Family Helicosphaeraceae												
Helicosphaera carteri	4G	17–19	-	+	-	-	+	-	+	-	+	-
Helicosphaera carteri HOL	4H	*c.* 12	-	+	-	-	-	-	-	-	-	-
Helicosphaera hyalina	4I	*c.* 12	-	+	-	-	-	-	-	-	-	-
Helicosphaera wallichii	4J	*c.* 15	-	+	-	-	-	-	-	-	-	-
Family Pontosphaeraceae												
Scyphosphaera apsteinii	4K	*c.* 26	-	+	-	-	-	-	+	-	-	-
Coccolith families *incertae sedis*												
Family Alisphaeraceae												
Alisphaera pinnigera	4L	*c.* 7	-	-	-	+	-	-	+	-	-	-
Polycrater galapagensis	4M	*c.* 8	-	+	-	-	-	-	-	-	-	-
Family Papposphaeraceae												
Papposphaera lepida	4N	*c.* 5	-	+	-	+	-	-	-	-	-	-
Family Umbellosphaeraceae												
Umbellosphaera tenuis Type II	4O	8–14	+	+	+	+	-	-	-	-	+	+
Nannolith family *incertae sedis*												
Family Braarudosphaeraceae												
Braarudosphaera bigelowii	4P	*c.* 5–6	-	+	-	-	-	-	-	-	-	-
Holococcolith Group												
Corisphaera gracilis	4Q	*c.* 5	-	+	-	+	-	-	-	-	-	-
Holococcolithophora sphaeroidea	4R	8-10	-	+	-	-	-	-	-	-	-	-
Homozygosphaera arethusae	4S	8–12	-	+	-	-	-	-	-	-	-	-
Poricalyptra aurisinae	4T	*c.* 8	-	+	-	-	-	-	-	-	-	-

[§]Cell size is a measurement of either the diameter or length of cell in μm; [*] *Emiliania huxleyi* blooms confirmed by the MODIS sensor on NASA'S satellite (NASA Earth Observatory 2009, 2011); '+' = present, '-'= absent; SP 09 = Spring 2009; SU 09 = Summer 2009; AU 09 = Autumn 2009; SU 11 = Summer 2011; SU 10 = Summer 2010; AU 11 = Autumn 2011; SU 12 = Summer 2012

with tube-like structure; circum-flagellar coccoliths with pointed spine, sometimes slightly curved (Tables 2 and 3).

Syracosphaera leptolepis Kleijne and Cros 2009 (Fig. 3j-k), coccosphere subspherical (*c.* 7.0 μm); body coccoliths (1.8–2.0 μm) broadly elliptical dishes with slight distal edge and straight wall; central area slightly vaulted with slightly raised, inner central structure. Exothecal planoliths circular discs, central part consisting of dextrally oblique elements (Fig. 3k); loosely attached to the coccospheres (Tables 2 and 3).

Syracosphaera nodosa Kamptner 1941 (Fig. 3o), coccosphere ovoid (6.2–6.6 μm) and dithecate. Body coccolith elliptical (1.3–2.7 μm), wall relatively deep but with no distal flange and with characteristic vertical ribs on outer surface of wall (Tables 2 and 3). Circum-flagellar

coccolith has strong spine, with sheath-like structure which covers about 80 % of proximal part of the spine.

Syracosphaera nodosa aff. *S.* sp. type 2 of Kleijne 1993 (Fig. 3p-q), coccosphere almost spherical (c. 11 μm), and dithecate (Table 2). Thin and subcircular outer coccoliths form complete outer layer over body coccoliths. Central area of elliptical body coccolith (2.8–2.9 μm) with radial laths and elongated mound as central connecting structure; thin, subcircular exothecal planoliths loosely attached to coccosphere (Fig. 3q; Table 3).

Syracosphaera cf. *orbiculus* Okada and McIntyre 1977 (Fig. 3r), coccosphere subspherical (c. 7.0 μm). Body coccoliths (2.2–2.6 μm) with relatively thin smooth wall (Tables 2 and 3); central area with well-developed connecting external ring and flat, elongated internal

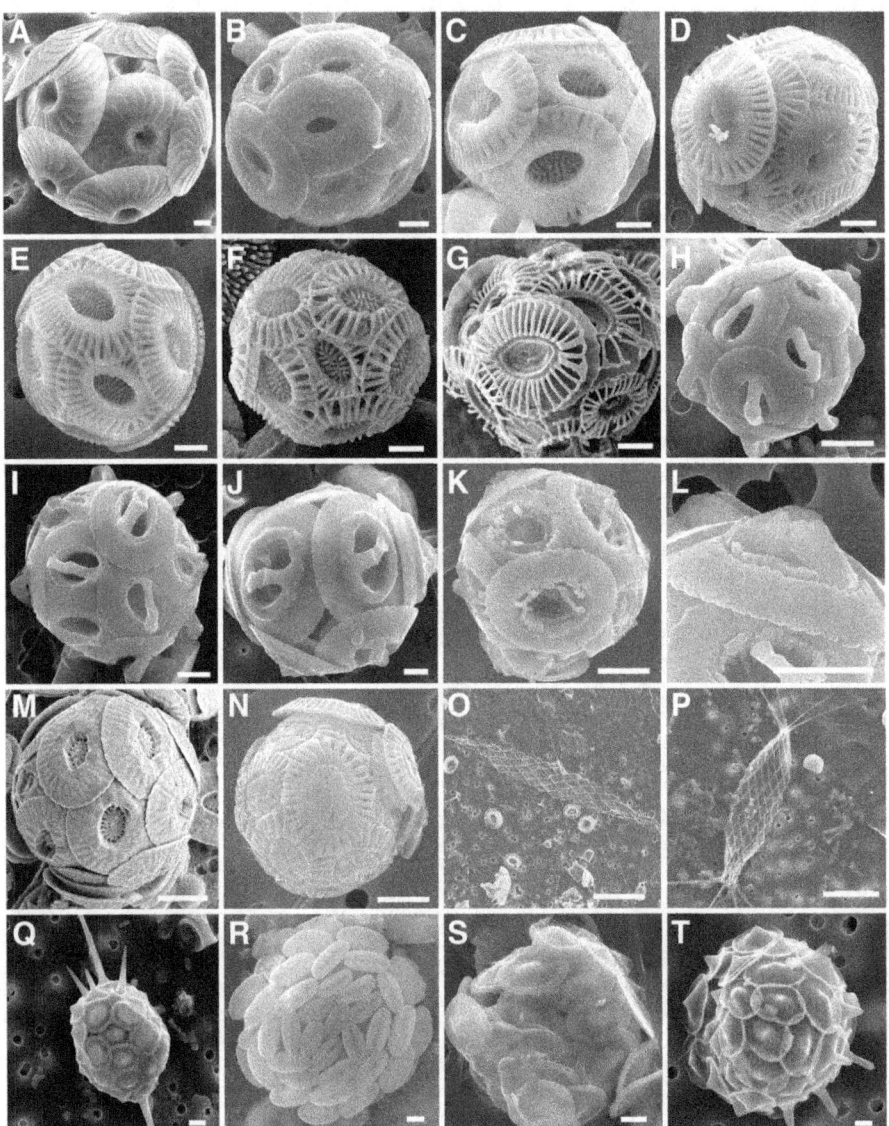

Fig. 2 Heterococcolith and holococcolith forms: (**a**) *Calcidiscus leptoporus*; (**b**) *Umbilicosphaera hulburtiana*; (**c-g**) *Emiliania huxleyi*, showing five different morphotypes (**c**, supercalcified; **d**, over-calcified; **e**, type A; **f**, type B; **g**, type C); (**h**) *Gephyrocapsa ericsonii*; (**i**) *Gephyrocapsa muelerae*; (**j**) *Gephyrocapsa oceanica*; (**k**) *Gephyrocapsa ornata*; (**l**) Coccolith of *G. ornata*; (**m**) *Reticulofenestra parvula*; (**n**) *Reticulofenestra parvula* var. *tecticentrum*; (**o**) *Calciosolenia brasiliensis*; (**p**) *Calciosolenia murrayi*; (**q**) *Acanthoica quattrrospina*; (**r**) *Algirosphaera robusta*; (**s**) *Anacanthoica acanthos*; (**t**) *Cyrtosphaera aculeata*. Scale bars: **a-n** and **q-t**, 1 μm; **o-p**, 10 μm

connecting structure. Circum-flagellar coccoliths with long and somewhat bent spine (Cros and Fortuño 2002).

Syracosphaera serrata Kleijne and Cros 2009 (Fig. 4b-c), coccosphere (*c.* 9 μm) dithecate (Table 2). Body coccoliths broadly elliptical (2.1–2.2 μm) with irregular outline and low, thin flaring wall; central area made up of radial laths and a flat inner centre (Table 3). Exothecal coccoliths are wheel-like planoliths with serrate margin (Fig. 4c).

Coronosphaera mediterranea HOL *gracillima*-type (=*Calyptrolithophora gracillima* (Kamptner 1941) Heimdal in Heimdal and Gaarder 1980) (Fig. 4f), holococcolith

form. Body coccoliths (2.8–3.2 μm) have rounded distal protrusion with flared tube (Table 3), no flange, hexagonal mesh fabric without large pores; discontinuous rim formed from two rows of crystallites.

Helicosphaera carteri HOL [=*Syracolithus catilliferus* (Kamptner 1937) Deflandre, 1952 (=*Syracosphaera*)] (Fig. 4h), holococcolith form produced only small elliptical coccoliths (2.4–2.7 μm) (Table 3) with central pyramidal spine.

Helicosphaera hylina Gaarder 1970 (Fig. 4i) and *H. wallichii* (Lohmann 1902) Okada and McIntyre 1977 (Fig. 4j), coccospheres of the heterococcolith form of

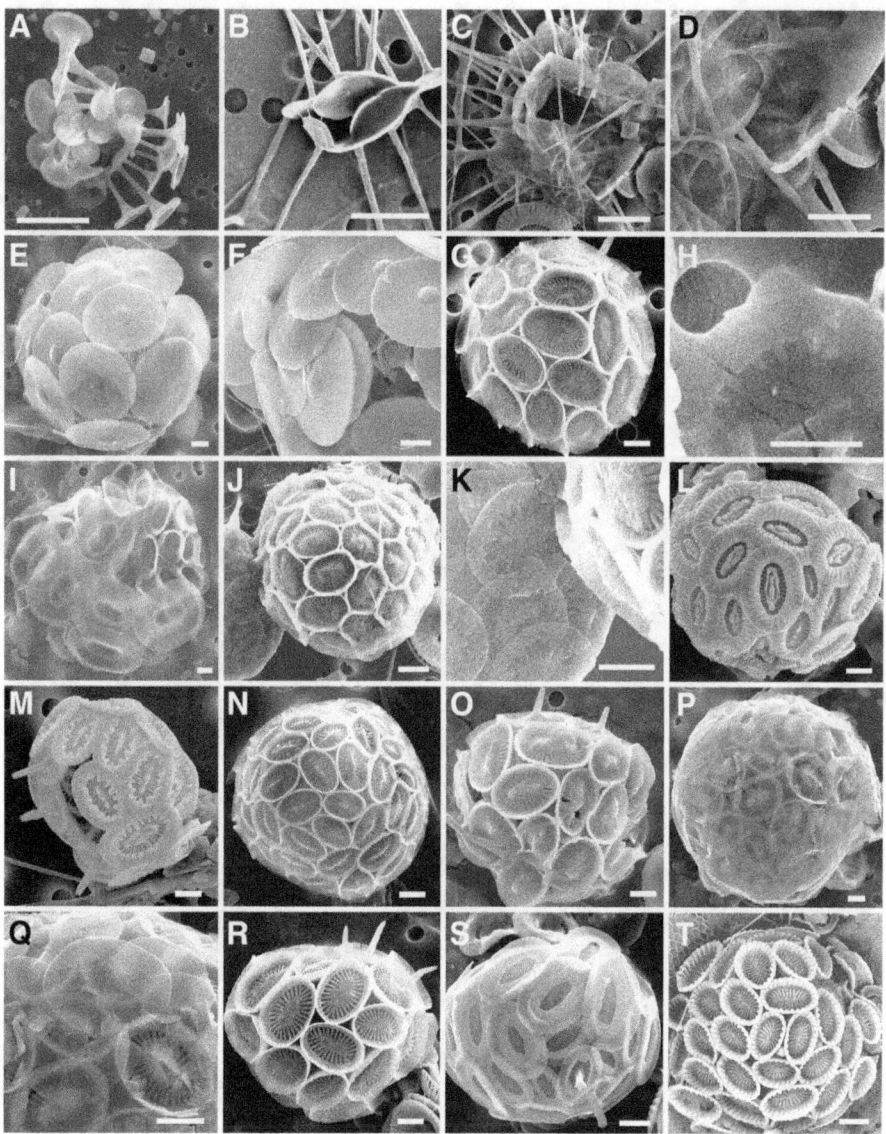

Fig. 3 Heterococcolith form: (**a**) *Discosphaera tubifera*; (**b**) *Palusphaera vandelii*; (**c**) *Rhabdosphaera xiphos*; (**d**) Coccoliths of *R. xiphos*, showing a distinctive star pattern on the distal surface; (**e**) *Syracosphaera anthos*; (**f**) Circum-flagellar coccoliths of *S. anthos*, showing a prominent spine; (**g**) *Syracosphaera bannockii*; (**h**) Exothecal coccolith of *S. bannockii*; (**i**) *Syracosphaera corolla*; (**j**) *Syracosphaera leptolepis*; (**k**) Exothecal coccoliths of *S. leptolepis*; (**l**) *Syracosphaera molischii* type 1; (**m**) *S. molischii* type 2; (**N**) *Syracosphaera nana*; (**o**) *Syracosphaera nodosa*; (**p**) *Syracosphaera* sp. aff. *S. nodosa* type 2 of Kleijne; (**q**) Thin circular exothecal coccoliths of *S.* sp. aff. *S. nodosa* type 2; (**r**) *Syracosphaera* cf. *orbiculus*; (**s**) *Syracosphaera ossa*; (**t**) *Syracosphaera pemmadiscus*. Scale bars: **a-t**, 1 μm

both species (<15 μm) produced relatively large coccoliths (5.3–6.0 μm and 8.4–10 μm respectively) (Tables 2 and 3) that wedged into one another by winged flanges, reminiscent of those of *H. carteri* (Wallich 1877) Kamptner 1954. There were, however, no slits in central area of *H. hylina*, while central area of *H. wallichii* has oblique twisted slits, rather than two inline slits as with *H. carteri*.

Alisphaera pinnigera Kleijne et al. 2002 (Fig. 4l), coccosphere (*c.* 7 μm) dimorphic. Central area of coccolith (1.2–1.5 μm) has horizontal fissure (Tables 2 and 3).

Some coccoliths have either tooth-like or flat, triangular-like protrusion along their inner margin.

Polycrater galapagensis Manton and Oates 1980 (Fig. 4m), coccosphere (15–16 μm) with numerous very small coccoliths (0.6–0.7 μm) (Tables 2 and 3). Quadrate in plan-view and upside-down triangle-shaped in lateral view. Coccoliths of aragonite (Manton and Oates 1980).

Papposphaera lepida Tangen 1972 (Fig. 4n), cell spherical (5–6 μm), diameter of coccosphere 14–15 μm. Base of coccolith (1.2–1.6 μm) subcircular to elliptical, central

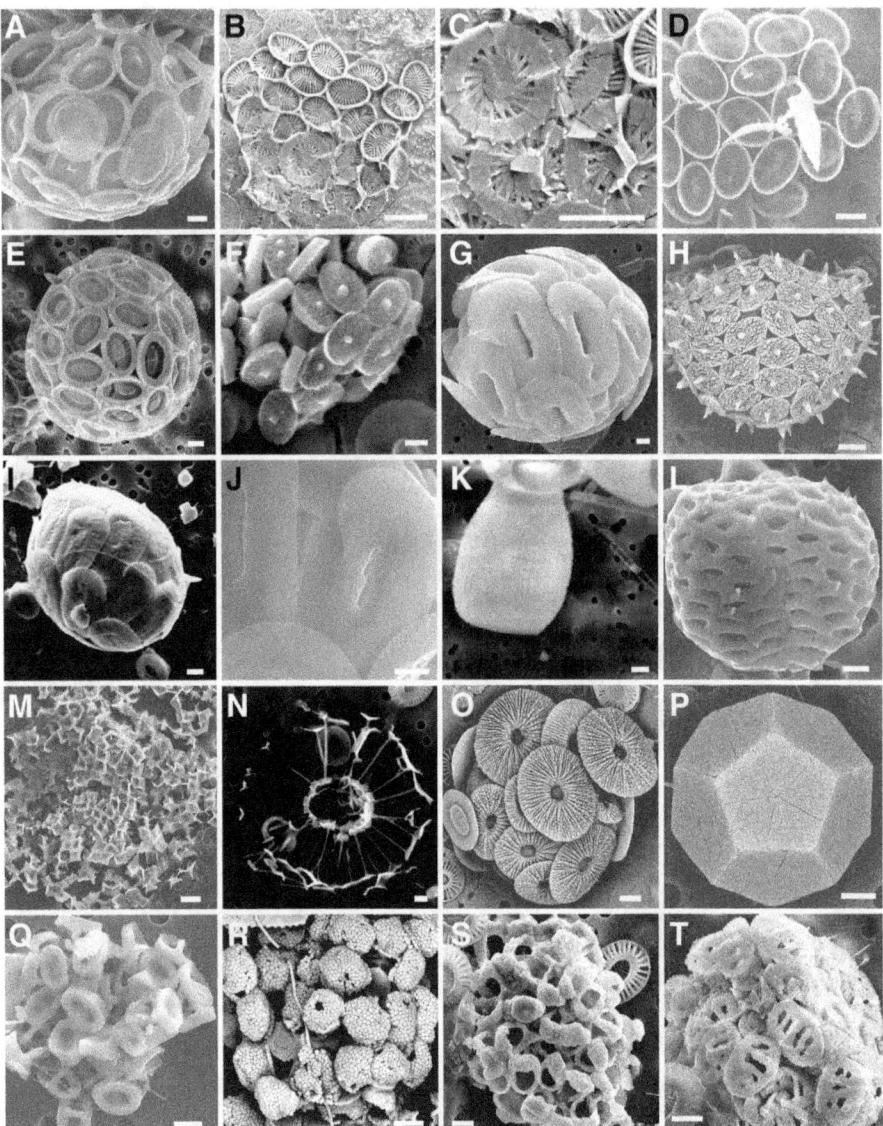

Fig. 4 Heterococcolith and holococcolith forms: (**a**) *Syracosphaera pulchra*; (**b**) *Syracosphaera serrata*; (**c**) Wheel-like exothecal planoliths of *S. serrata*; (**d**) *Syracosphaera tumularis*; (**e**) *Coronosphaera mediterranea*; (**f**) *Coronosphaera mediterranea* HOL *gracillima*-type; (**g**) *Helicosphaera carteri*; (**h**) *Helicosphaera carteri* HOL (= *Syracolithus catilliferus*) (**i**) *Helicosphaera hylina*; (**j**) *Helicosphaera wallichii*; (**k**) *Scyphosphaera apsteinii*; (**l**) *Alisphaera pinnigera*; (**m**) *Polycrater galapagensis*; (**n**) *Papposphaera lepida*; (**o**) *Umbellosphaera tenuis* type II; (**p**) *Braarudosphaera bigelowii*; (**q**) *Corisphaera gracilis*; (**r**) *Holococcolithophora sphaeroidea*; (**s**) *Homozygosphaera arethusae*; (**t**) *Poricalyptra aurisinae*. Scale bars: A-T, 1 µm

area with long central stem which supports flat cone of four large elements giving rise to almost continuous outer layer of cell.

Corisphaera gracilis Kamptner 1937 (Fig. 4q), coccosphere subspherical (*c.* 5 µm); coccoliths (1.2–1.5 µm) with low wall and transverse arched bridge across open distal end (Tables 2 and 3). Proximal end appears to be sealed by thin-layer of crystallites.

Holococcolithophora sphaeroidea (Schiller 1913) Jordan et al. 2004 (=*Calyptrosphaera*) (Fig. 4r), coccosphere ovoid (*c.* 10–12 µm long) (Table 2). Coccoliths conical with basal flange; distal end tapers abruptly into

small projection. Microcrystallites irregularly arranged, separated by small perforations.

Homozygosphaera arethusae (Kamptner 1941) Kleijne 1991 (=*Corisphaera*) (Fig. 4s), coccosphere ovoid (8–12 µm) (Table 2); coccoliths (1.7–2.1 µm) have proximal tube and robust, arched distal bridge (Table 3).

Poricalyptra aurisinae (Kamptner 1941) Kleijne 1991 (=*Helladosphaera*) (Fig. 4t), coccosphere ovoid (*c.* 9.0 µm) (Table 2). Coccoliths elliptical (2.3–2.5 µm) (Table 3), with four oblong openings and transverse, virtually one-layered ridge on distal surface.

Table 3 Comparison of the diameter/ length of body coccoliths of the twenty six first-time records in this study with those corresponding taxa reported by Cros & Fortuño (2002) and Young et al. (2003)

Taxa	Diameter/length of body coccolith (µm)		
	This Study	Cros & Fortuño (2002)	Young et al. (2003)
Alisphaera pinnigera	1.2–1.5	1.5–1.6	1.3–2.0
Anacanthoica acanthos	1.9–2.1	2.1–2.6	n.d.
Corisphaera gracilis	1.2–1.5	1.4–1.6	c. 1.5
Coronosphaera mediterranea HOL gracillima-type	2.8–3.2	2.1–2.3	n.d.
Cyrtosphaera aculeata	2.4–2.5	2.5–2.8	1.8–2.5
Gephyrocapsa muellerae	2.9–3.8	3.1–3.9	3–4
Gephyrocapsa ornata	2.8–3.6	2.2–2.5	n.d.
Helicosphaera carteri HOL	2.4–2.7	2.7–3.0	n.d.
Helicosphaera hylina	5.3–6.0	6.2–6.8	n.d.
Helicosphaera wallichii	8.4–10	c. 9.0	n.d.
Holococcolithophora sphaeroidea	1.5–2.0	1.6–1.8	1.8
Homozygosphaera arethusae	1.7–2.1	1.6–1.8	1.5
Palusphaera vandelii	1.2–1.8	1.5–1.9	n.d.
Papposphaera lepida	1.2–1.6	0.7–1.5	1–1.5
Polycrater galapagensis	0.6–1.0	0.6–0.7	n.d.
Poricalyptra aurisinae	2.3–2.5	2.3–2.4	n.d.
Rhabdosphaera xiphos	1.3–1.9	1.1–1.3	n.d.
Reticulofenestra parvula	1.9–2.8	1.4–1.9	1.2–2.0
Reticulofenestra parvula var. tecticentrum	2.5–3.0	1.4–1.9	1.2–2.0
Syracosphaera anthos	4.5–5.8	2.2–2.5	3.0–5.5
Syracosphaera bannockii	1.7–2.8	1.5–1.7	n.d.
Syracosphaera leptolepis	1.8–2.0	1.5–2.0	n.d.
Syracosphaera nodosa	1.3–2.7	2.3–2.5	1.5–2.5
Syracosphaera nodosa aff. S. nodosa type 2	2.8–2.9	2.7–2.9	n.d.
Syracosphaera cf. orbiculus	2.2–2.6	2.0–2.2	1.5–2.5
Syracosphaera serrata	2.1–2.2	n.d.	1.5–2.5[a]

[a]*Syracosphaera serrata* [=S. *nodosa* type B, Young et al., 2003, p. 36, Plate 15]; n.d. = no da

Emiliania huxleyi (Lohmann 1902) Hay and Mohler in Hay et al. 1967 (Fig. 2c-g), small coccosphere (3.5–5.0 µm), spherical to subspherical, with relatively large elliptical coccoliths (2.7–3.9 µm) (Table 2). Cells of the supercalcified and highly calcified forms more or less the same as other morphotypes; both forms recorded for the first time among Type A, B, and C, in particular during the spring 2009 blooms east of New Zealand (Fig. 2c-e).

Species diversity and distribution
The total number of coccolithophores identified from samples collected between 2009 and 2012 was greatest to the east (46), intermediate to the west (along two transects in the Tasman Sea) (15), and least to the northeast (Bay of Plenty) and south (4 each) of New Zealand (Table 2). During the 3 year study two massive, almost monospecific, *Emiliania huxleyi* blooms were recorded in the vicinity of

the STF (in the spring 2009 and summer 2011) and both were confirmed by the MODIS sensor on NASA's Aqua satellite (NASA Earth Observatory 2009, 2011). The total number of coccolithophores found on these two occasions was 4 and 6 taxa respectively.

In non-bloom conditions, based on the frequency of occurrence, *Emiliania huxleyi* was found to be most abundant at virtually all sampling stations. The only exception was in the autumn 2009, when *Reticulofenestra parvula* dominated the coccolithophore assemblage along the Kaikoura coast, east of New Zealand. In non-bloom conditions the average total number of taxa (22) recorded on five occasions, in the spring, summer and autumn of 2009, autumn 2011 and summer 2012, to the east, was greater than the average (7.6) recorded on three other occasions, in the summer 2010, spring 2009 and autumn 2011, to the rest of New Zealand (Table 2).

The few members of the order Isochrysidales, e.g., *Emiliania huxleyi*, *Gephyrocapsa ericsonii*, *G. muellerae* and *G. oceanica*, were most widely distributed, being recorded at most sampling sites (Table 2). Even though members of the genus *Syracosphaera* generally occurred in low frequencies (e.g., *S. pemmadiscus* Chang 2013, *S. bannockii*, *S. leptolepis*, *S. nana* [Kamptner 1941] Okada and Mcintyre 1977 [=*S.* sp. type A of Kleijne 1991, 1993], *S. nodosa*, *S. nodosa* aff. *S.* sp. type 2, *S.* cf. *orbiculus*, *S. ossa* [Lecal 1966] Loeblich and Tappan 1968 [*Syracolithus*], *S. serrata* and *S. tumularis* Sánchez-Suárez 1990 [=*S.* sp. type C of Kleijne 1993]), they were found to be more widespread to the east (14) than to the west (4), northeast and south (none) of New Zealand (Table 2). All the remaining taxa appeared to be more sparsely distributed at fewer stations than the other two groups but generally they were also more common to the east than to the rest of New Zealand.

Discussion

The total of 46 living coccolithophores recorded here in the 3-year study is greater than the number previously documented (40) in New Zealand (Norris 1961; Cassie 1961; Burns 1977, Rhodes et al. 1993, 1994) and also is greater than the number (27) recently recorded on the three offshore stations, southeast of New Zealand (Saavedra-Pellitero et al. 2014). Nevertheless, more than half (*c.* 57 %) of these 46 taxa are first-time records. Adding 26 new records plus a recently described species, *Syracosphaera pemmadiscus* (Chang 2013) of this study, to the 40 previously documented taxa in New Zealand, and about 10 newly recorded taxa from southeast New Zealand, the total number of coccolithophores found in the New Zealand region is now estimated to be 77. This is greater than the total of about 42 taxa recognised in the tropical, subtropical and temperate waters of Australia (about 16 are different from those recorded in the New Zealand region) (Conley 1979; Hallegraeff 1984, 2010; Callaghan 1992). By combining all the taxa identified in both New Zealand and Australia waters, the grand total of the current living coccolithophores in the southwest Pacific stands at about 93.

Most of the surveys (seven out of ten) conducted from 2009 to 2012 in this study were centred in the vicinity of the STF near 43 S, east of New Zealand. This is a region of two contrasting water masses – warm, macronutrient-poor, iron-rich ST water in the north meeting cool, fresher, macronutrient-rich, iron-poor SA water in the south (e.g., Heath 1985; Chang and Gall 1998; Boyd et al. 1999) and is part of the so-called "The Great Calcite Belt" (Balch et al. 2011). The greater number of taxa plus two massive coccolithophore blooms (NASA Earth Observatory 2009, 2011) observed in the vicinity of STF compared with other areas of New Zealand, could reflect the wider seasonal range of sampling and greater number of surveys conducted near the STF. Nevertheless, it is likely that the physico-chemical conditions in this region favoured coccolithophores (e.g., Balch et al. 2011; Sadeghi et al. 2012; Saavedra-Pellitero et al. 2014).

In this study *Emiliania huxleyi* was virtually the only species observed to dominate in both bloom and non-bloom conditions. This observation is consistent with those made worldwide (e.g., Birkenes and Braarud 1952; Okada and Honjo 1975; Holligan et al. 1993; Cokacar et al. 2001; Wang et al. 2012; Saavedra-Pellitero et al. 2014). The build-up of *Reticulofenestra parva* as a dominant species in the spring 2009 off Kaikoura coast, however, was a unique event not just to New Zealand, but also in the southwest Pacific Ocean. Previously *Reticulofenestra parva* was only found to be dominant in surface sediments of the eastern Mediterranean Sea (Ziveri et al. 2000).

In terms of taxa, the genus *Syracosphaera* was the most diverse group. Fourteen taxa, about 31 % of all extant coccolithophores, were recorded east of New Zealand alone. This number exceeded records of extant coccolithophores previously documented in New Zealand (12) (e.g., Cassie 1961; Norris 1961; Burns 1977; Saavedra-Pellitero et al. 2014), Australia (8) (Hallegraeff 1984, 2010) and also the Australian Sector of the Southern Ocean (9) (Findlay and Giraudeau 2000). With only a few exceptions, most of these *Syracosphaera* spp. occurred at low frequencies similar to reports elsewhere in the world (e.g., Kleijne 1993; Jordan and Kleijne 1994; Cros and Fortuño 2002; Young et al. 2003; Jordan et al. 2004).

Conclusions

Almost all 46 coccolithophores identified in this study were recorded from samples collected in the vicinity of Chatham Rise and Kaikoura coast, east of New Zealand alone. In contrast, a relatively small number of coccolithophores were recorded from samples collected to the west (in the Tasman Sea), northeast (Bay of Plenty) and south of the country. About 57 % of these taxa/ forms are first-time records. *Emiliania huxleyi* was the dominant species observed not only in the two almost monospecific blooms east of New Zealand, but also in most cases, in non-bloom conditions. *Reticulofenestra parva* was, however, found on one occasion, to be dominant off Kaikoura coast. Even though *Syracosphaera* taxa generally occurred at low frequencies, they were the largest group and made up *c.* 31 % of all extant coccolithophores recorded in this study. The total number of coccolithophores currently estimated in the Southwest Pacific near New Zealand and Australia is 93.

Abbreviations
aff., species affinis or akin to; *cf.*, see also; HOL, holococcolithophore; MODIS, Moderate Resolution Imaging Spectroradiometer; var., variety

Acknowledgments

We thank Dr Janet Grieve of NIWA, Wellington, for her constructive criticisms of this manuscript, Dr Cliff Law for initiating and coordinating the coccolithophore surveys, and last but not least, Dr Annelies Kleijne of Geomarine Centre, Vrije Universiteit, The Netherland, and Dr Luïsa Cros of Institut de Ciéncies del Mar (CSIC), for the clarification of the taxonomy of several *Syracosphaera* taxa. The assistance of the crew and officers of the NIWA *Research Vessels* Tangaroa and Kaharoa during all surveys is gratefully acknowledged. Funding from the New Zealand Ministry for Primary Industries and from NIWA under Coasts and Oceans Research Programme 2 (2013/14 SCI) provided support for this work.

Authors' contributions

FHC: conceived of the study, conducted the SEM examination, identified all the taxa of extant coccolithophores and drafted the manuscript. LN: collected the samples from each voyage. Both authors read and approved the final manuscript.

Competing interests

The authors declare that they have no competing interests.

Author details

[1]Biodiversity and Biosecurity Group, National Institute of Water & Atmospheric Research Ltd., P. O. Box 14-901, Kilbirnie, Wellington 6241, New Zealand. [2]Ocean Sediments Group, National Institute of Water & Atmospheric Research Ltd., P. O. Box 14-901, Kilbirnie, Wellington 6241, New Zealand.

References

Balch WM, Drapeau DT, Bowler BC, Lyczskowski E, Booth S, Alley D. The contribution of coccolithophores to the optical and inorganic carbon budgets during the Southern Ocean Gas Exchange Experiments: New evidence in support of the Great Calcite Belt hypothesis. J Geophys Res. 2011;116:C00F06. doi:10.1029/2011JC006941.

Baumann K-H, Andruleit HA, Samtleben C. Coccolithophores in the Nordic Seas: comparison of living communities with surface sediment assemblages. Deep-Sea Res II. 2000;47:1743–72.

Beaufort L, Probert I, De Garode-Thoron T, Bendif EM, Ruiz-Pino D, Metzl N, Goyet C, Buchet N, Coupel P, Grelaud M, Rost B, Rickaby REM, De Vargas C. Sensitivity of coccolithophores to carbonate chemistry and ocean acidification. Nature. 2011; 476:80–3.

Biekart JW. The distribution of calcareous nannoplankton in late Quaternary sediments collected by the Snellius II Expedition in some southeast Indonesian basins. Proceedings Koninklijke Nederlandse Akadem van Wetenschappen, B, 1989;92:77–141.

Birkenes E, Braarud T. Phytoplankton in the Oslo Fjord during a "Coccolithus huxleyi – summer". Avhabdkubger ytgutt av det Norske Videnskap- Akademi Oslo (Matem Naturvid Klasse); 1952. p. 1–23.

Borsetti AM, Cati F. Il nannoplancton calcareo vivente nel Tirreno centro-meridionale. Parte II. Giornale Geol. 1976;40:209–40.

Boyd PD, LaRoche J, Gall M, Frew R, McKay RML. The role of iron, light and silicate in controlling algal biomass in sub-Antarctic water SE of New Zealand. J Geophys Res. 1999;104(C6):13395–408.

Bréhéret JG. Formes nouvelles quaternaries et actuelles de la famille des Gephyrocapsaceae (Coccolithophorides). C R Acad Sci, Paris Série D. 1978;287:447–9.

Burns DA. The latitudinal distribution and significance of calcareous nannofossils in the bottom sediments of the South-west Pacific Ocean (Lat. 15-55S) around New Zealand. In: Fraser R, editor. Oceanography of the South Pacific for UNESCO, Wellington; 1973. p. 221-8.

Burns DA. The abundance and species composition of nannofossil assemblages in sediments from continental shelf to offshore basin, western Tasman Sea. Deep-Sea Res. 1975;22:425–31.

Burns DA. Phenotypes and dissolution morphotypes of the genus *Gephyrocapsa* Kamptner and *Emiliania huxleyi* Lohmann. New Zealand J Geol Geophys. 1977;20:143–55.

Callaghan KM. A taxonomic survey of the coccolithophorids of Bass Strait, with life history observations. B.Sc. (Hons.) thesis, University of Melbourne; 1992.

Cassie V. Marine phytoplankton in New Zealand waters. Bot Mar. 1961;2(Suppl):1–54.

Chang FH. *Syracosphaera pemmadiscus* sp. nov. (Prymnesiophyceae), an extant coccolithophore from the southwest Pacific Ocean near New Zealand. Phycologia. 2013;52(6):618–24.

Chang FH, Gall M. Phytoplankton assemblages and photosynthetic pigments during winter and spring in the Subtropical Convergence region near New Zealand. New Zealand J Mar Freshw Res. 1998;32:515–30.

Cokacar T, Kubilay N, Oguz T. Structure of *Emiliania huxleyi* blooms in the Black Sea surface waters as detected by SeaWiFS imagery. Geophys Res Lett. 2001; 28:4607–10.

Conley SM. Recent coccolithophores from the Great Barrier Reef-Coral Sea region. Micropaleontology. 1979;25:20–40.

Cros L, Kleijne A, Zeltner A, Billar C, Young JR. New examples of holococcolith-heterococcolith combination coccospheres and their implications for coccolithophorid biology. Mar Micropaleontol. 2000;39:1–34.

Cros L, Fortuño J-M. Atlas of Northwestern Mediterranean Coccolithophores. Sci Mar. 2002;66 Suppl 1:7–182.

Deflandre G. Glasse des coccolithophoridés. (Coccolithophoridae Lohmann, 1902). In: Grassé P-P, editor. Traité de Zoologie 1. Paris: Masson; 1952. p. 439–70.

Deflandre G, Fert C. Observations sur les coccolithophoridés actuels et fossils en microscopie ordinaire et électronique. Ann de Paléontol. 1954;40:115–76.

Doney SC, Fabry VJ, Feely RA, Kleypas JA. Ocean acidification: the other CO_2 problem. Ann Rev Mar Sci. 2009;1:169–92.

Edwards AR. The calcareous nannoplankton – evidence for New Zealand Tertiary marine climate. Tuatara. 1968;16:26–31.

Edwards AR. Calcareous nannofossils. In: Hoskins RH, editor. Stages of the New Zealand Cenozoic: a synopsis, New Zealand Geol Surv Rep, vol. 107. 1982. p. 23–7.

Eynaud F, Giraudeau J, Pichon JJ, Pudsey CJ. Sea surface distribution of coccolithophores, diatoms, silicoflagellates and dinoflagellates in the South Atlantic Ocean during the late austral summer 1995. Deep-Sea Res Part I. 1999;46:451–82.

Findlay CS, Giraudeau J. Extant calcareous nannoplankton in the Australian Sector of the Southern Ocean (Austral summers 1994 and 1995). Mar Micropaleontol. 2000;40:417–39.

Findlay CS, Young JR, Scott FJ. 6. Haptophytes: order Coccolithophorales. In: Scott FJ, Marchant HJ, editors. Antarctic marine protists. Canberra: Australian Biological Resource Study; 2005. p. 276–94.

Frada M, Young J, Cacháo M, Lino S, Marcosp A, Probert I, Vargas CD. A guide to extant coccolithophores (Calcihaptophycidae, Haptophyta) using light microscopy. J Nannoplankton Res. 2010;31:58–112.

Gaarder KR. Three New Taxa of Coccolithineae. Nytt Mag Bot. 1970;17:113–26.

Hallegraeff GM. Coccolithophorids (Calcareous nanoplankton) from Australian Waters. Bot Mar. 1984;27:229–47.

Hallegraeff GM. Coccolithophorids (Haptophyta). In: Hallegraeff GM, Bolch CJS, Hill DRA, Jameson I, LeRoi J-M, Murray S, De Salas MF, Saunders K, editors. Algae of Australia phytoplankton of temperate coastal waters. Department of the Environment, Water, Heritage and the Arts, CSIRO Publishing; 2010. p. 342–60.

Hay WW. Carbonate fluxes and calcareous nannoplankton. In: Thierstein HR, Young JR, editors. Coccolithophores: from molecular processes to global impact. Singinger-Verlag Berlin Heidelberg; 2004. p. 509–29.

Hay WW, Mohler HP, Roth PH, Schmidt RR, Boudreaux JE. Calcareous nanoplankton zonation of the Cenozoic of the Gulf Coast and Caribbean-Antillean area, and transoceanic correlation. Trans Gulf Coast Assoc Geol Soc. 1967;17:428–80.

Heath RA. Large-scale influence of the New Zealand seafloor topography on western boundary currents of the South Pacific Ocean. Aust J Mar Freshw Res. 1985;36:1–14.

Heimdal BR. Two new taxa of recent coccolithophorids. "Meteor" Forsch–Ergebn, Reihe D, Biol. 1973;13:70–5.

Heimdal BR, Gaarder KR. Coccolithophorids from the northern part of the eastern central Atlantic I. Holococcolithophorids. "Meteor" Forsch-Ergebn, Reihe D, Biol. 1980;32:1–14.

Hiramatsu C, De Deckker P. Distribution of calcareous nannoplankton near the Subtropical Convergence, south of Tasmania, Australia. Aust J Mar Fresh Res. 1997;47:707–13.

Holligan PM, Fernández E, Aiken J, Balch WM, Boyd P, Burkill PH, Finch M, Groom SB, Malin G, Muller K, Purdie DA, Robinson C, Trees CC, Turner SM, van der Wal P. A biogeochemical study of the coccolithophore, Emiliania huxleyi, in the North Atlantic. Global Biogeochem Cycles. 1993;7(4):879–900.

Janin M-C. Micropaléontologie de concretions polymétaliques du Pacifique central: zone Clarion-Clipperton, chaine Centre-Pacifique, Iles de la Ligne et archipel des Tuoamotou (Eocéne-Actuel). Mém Soc Géol Fr. 1987;152:1–317.

Jordan RW, Young JR. Proposed changes to the classification system of living coccolithophorids. Int Nannoplankton Assoc Newslett. 1990;12:15–8.

Jordan RW, Kleijne A. A classification system for living coccolithophores. In: Winter A, Siesser WG, editors. Coccolithophores. Cambridge University Press; 1994. p. 83–105.

Jordan RW, Cros L, Young JR. A revised classification scheme for living haptophytes. In: Triantaphyllou MV, editor. Advances in biology, ecology and taxonomy of extant calcareous nannoplankton. New York: Micropaleontology Press, Micropaleontology 2004;50 Suppl 1:55–79.

Kamptner E. Neue und bemerkenswerte Coccolithineen aus dem Mittelmeer. Arch Protistenk. 1937;89:279–316.

Kamptner E. Die Coccolithineen der Südwestküste von Istrien. Ann Naturh Mus Wien. 1941;51:54–149.

Kamptner E. Zur Revision der Coccolithineen-Spezies Pontosphaera huxleyi Lohm. Akad Wiss Wien, Math-Naturw Kl. 1943;80:43–9.

Kamptner E. Untersuchungen über den Feinbau der Coccolithen. Arch Protistenk. 1954;100:1–90.

Kleijne A. Holococcolithophorids from the Indian Ocean, Red Sea, Mediterranean Sea and North Atlantic Ocean. Mar Micropaleontol. 1991;17:1–76.

Kleijne A. Extant Rhabdosphaeraceae (coccolithophorids, Class Prymnesiophyceae) from the Indian Ocean, Red Sea, Mediterranean Sea and North Atlantic Ocean. Scripta Geol. 1992;100:1–61.

Kleijne A. Morphology, taxonomy and distribution of extant coccolithophorids (calcareous nannoplankton). Drukkerij FEBO B.V.; 1993. 321 pp.

Kleijne A, Jordan RW, Heimdal BR, Samtleben C, Chamberlain AHL, Cros L. Five new species of the coccolithophorid genus Alisphaera (Haptophyta), with notes on their distribution, coccolith structure and taxonomy. Phycologia. 2002;40:583–601.

Kleijne A, Cros L. Ten new extant species of the coccolithophore Syracosphaera and a revised classification scheme for the genus. Micropaleontology. 2009; 55:425–62.

Lecal J. Un nouvel Hymenomonas: H. prenantii n. sp. (Cocccolithophoridés). Ann Limnol. 1965;1:155–62.

Lecal J. Coccolithophorides littoraux de Banyuls. Vie et Millieu. 1966;16:251–70.

Loeblich AR, Tappan H. Annotated index and bibliography of the calcareous nannoplankton II. J Paleontol. 1968;42:584–98.

Lohmann H. Die coccolithophoridae, eine monographic der coccolithen bildenden flagellaten. Zugleich ein Beitrag zur Kenntnis des Mitterlmeerauftriebs. Arch Protistenk. 1902;1:89–165.

Lohmann H. Untersuchungen über das Pflanzen- und Tierleben der Hochsee, zugleich ein Bericht über die biologischen Arbeiten auf der Fahrt der "Deutschland" von Bremerhaven nach Buenos-Aires in der Zeit vom 7, Mai bis 7. September 1911. Univ Berlin, Veröff Inst Meereschundund. Geograp-Nat wiss. 1912;1:1–92.

Manton I, Leadale GF. Observations on the microanatomy of Coccolithus pelagicus and Cricosphaera carterae, with special reference to the origin and nature of coccoliths and scales. J Mar biol Ass UK. 1969;49:1–16.

Manton I, Oates K. Polycrater galapagensis gen. et sp. nov., a putative coccolithophorid from the Galapagos Islands with an unusual aragonitic periplast. Br Phycol J. 1980;15:95–103.

McIntyre A, Bé AWH. Modern coccolithophoridae of the Atlantic Ocean – I. Placoliths and cyrtholiths Deep-Sea Res. 1967;14:561–9.

McIntyre A, Bé AWH, Roche B. Modern Pacific coccolithophorida: a paleontological thermometer. New York: Academic of Science. Trans Series II. 1970;32(6):720–31.

Murray J, Renard AF. Report on the Scientific Results of the Voyage of H.M.S. Challenger during the year 1873-76. Deep-Sea Deposits. HMSO, London xxxiv; 1891. + 525 p., 29 pls., 43 charts, 22 diagrams.

NASA. Earth Observatory. In: earthobservatory.nasa.gov/IOTD/view.php?id = 40924; 2009.

NASA Earth Observatory. In: earthobservatory.nasa.gov/IOTD/view.php?id = 49459; 2011.

Nishida S. Nannoplankton flora in the Southern Ocean, with special reference to siliceous varieties. Mem Nat Inst Polar Res, Spec Issue. 1986;40:56–68.

Norris RE. Observations on phytoplankton organisms collected on the N.Z.O.I. Pacific Cruise, September 1958. N Z J Sci. 1961;4:162–88.

Norris RE. Indian Ocean nannoplankton. I. Rhabdosphaeraceae (Prymnesiophyceae) with a review of extant taxa. J Phycol. 1984;20:27–41.

Okada H, Honjo S. Distribution of coccolithophores in Marginal Seas along the Western Pacific Ocean and in the Red Sea. Mar Biol. 1975;31:271–85.

Okada H, McIntyre A. Modern coccolithophores of the Pacific and north Atlantic Oceans. Micropaleontology. 1977;23:1–55.

Probert I, Fresnel J, Billard C, Geisen M, Young JR. Light and electron microscope observation of Algirosphaera robusta (Prymnesophyceae). J Phycol. 2007;43: 319–32.

Rhodes LL, Haywood A, Ballantine WJ, MacKenzie AL. Algal blooms and climate anomalies in north-east New Zealand, August- December 1992. N Z J Mar Freshw Res. 1993;27:419–30.

Rhodes LL, Peake B, MacKenzie AL, Marwick S. Coccolithophores Gephyrocapsa oceanica and Emiliania huxleyi (Prymnesiophyceae = Haptophyceae) in New Zealand's coastal waters: Characteristics of blooms and growth in laboratory culture. N Z J Mar Freshw Res. 1994;29:345–57.

Rowson JD, Leabeater BSC, Green JC. Calcium carbonate deposition in the motile (Crystallolithus) phase of Coccolithus pelagicus (Prymnesiophyceae). Br Phycol J. 1986;21:359–70.

Ruggiero MA, Gordon DP, Orrell TM, Bailly N, Bourgoin T, Brusca RC, Gavalier-Smith T, Guiry MD, Kirk PM. A higher level classification of all living organisms. Plos One 2015; p 60. DOI: 10.1371

Sánchez-Suárez IG. Three new Coccolithophorids (Haptophyta) from the South-Eastern Caribean Sea: Cyclolithellla ferrazae sp. nov. Syracosphaera florida sp. nov. Syracosphaera tumularis sp. nov. Biol Mar Acta Cientifica Venezolana. 1990;41:152–8.

Saavedra-Pellitero M, Baumann K-L, Flores J-A, Gersonde R. Biogeographic distribution of living coccolithophores in the Pacific sector of the Southern Ocean. Mar Micropaleontol. 2014;109:1–20.

Sadeghi A, Dinter T, Gountas M, Taylor B, Altenburg-Soppa M, Bracher A. Remote sensing of coccolithophore blooms in selected oceanic regions using the PhytoDOAS method applied to hyper-spectral satellite data. Biogeosciences. 2012;9:2127–43.

Schiller J. Vorläufige Ergebnisse der Phytoplankton-Untersuchungen auf den Fahrten S.M.S. Najade in der Adria 1911/12. I. Die Coccolithophoriden- K Akad Wiss, Wien, Sitzber, Math Naturw Klasse. 1913;122(1):597–617

Schiller J. Die planktonischen vegetationen des adriatischen Meeres. A. Die coccolithophoriden-Vegetation in den Jahren 1911-14. Arch Potistenk. 1925; 51:1–130.

Tangen K. Papposphaera lepida, gen. nov. sp., a new marine coccolithophorid from Norwegian coastal waters. Nor J Bot. 1972;19:171–8.

Thomsen HA. Identification by electron-microscopy of nannoplanktonic coccolithophorids (Prymnesiophyceae) from West Greenland, including the description of Papposphaera sarion sp. nov. Br Phycol J. 1981;16(1):77–94.

Tyrrell T, Merico A. Emiliania huxleyi: bloom observations and the conditions that induce them. In: Thierstein HR, Young JR, editors. Coccolithophores: from molecular processes to global Impact. Singinger-Verlag Berlin Heidelberg; 2004. p.75–97.

Wallich GC. Observations on the coccosphere. Ann Magaz Nat Hist. 1877;9:342–50.

Wang J, Luan Q, Zuo T, Chen R, Sun J. Taxonomic composition of marine-living cocolithophores in the Yellow Sea and East China Sea – new records and a species list. Mar Biodivers Rec. 2012;5:1–8.

Winter A, Siesser WG. Atlas of living coccolithophores. In Winter A, Siesser WG, editors. Coccolithophores. Cambridge Unversity Press; 1994. p. 107–59.

Young JR, Geisen M, Cros L, Kleijne A, Sprengel C, Probert I, Østergaard J. A guide to extant coccolithophore taxonomy. J Nannoplankton Res Spec Issue. 2003;1:125.

Ziveri P, Rutten A, de Lange GJ, Thomson J, Corselli C. Present-day coccolith fluxes recorded in central eastern Mediterranean sediment traps and surface sediments. Palaeogeog Palaeoclimatol Palaeoecol. 2000;158:175–95.

7

New record of *Carcharhinus leucas* (Valenciennes, 1839) in an equatorial river system

Leonardo Manir Feitosa[1*], Ana Paula Barbosa Martins[2] and Jorge Luiz Silva Nunes[3]

Abstract

Bull sharks are a cosmopolitan shark species frequently found in shallow shelf ocean waters and, occasionally, in several tropical river systems around the world. Due to bull shark's capability to enter riverine systems, the documentation of its occurrence is essential for future fisheries inspections and studies. In this way, this study aims to report the presence of a medium sized specimen of *C. leucas* in an equatorial river system. The specimen was caught by fishermen at Mearim River, located in Northern Brazil and well known for the occurrence of tidal bores during the highest spring tides of the dry season. The event coincided with the occurrence of one of the strongest spring tides of 2015. The captured female specimen measured approximately 1300 mm and weighted 35 kg. The occurrence of this species was not known in this river basin until now. We recommend and support future ichthyologic studies in the Mearim River basin in order to provide data for the delimitation of the territory used by *C. leucas* in Maranhão State, specially looking into its age, growth, diet, spatial, and temporal movement patterns in this area.

Keywords: Euryhaline shark, Elasmobranch, North Brazil

Introduction

Bull sharks are a coastal cosmopolitan species frequently found in shallow shelf waters, present in all of the world's oceans, occasionally entering warm river systems (Cervigón & Alcalá, 1999; Compagno et al., 2005). It has been known to occur in several other rivers in all continents (Ballantyne & Fraser, 2013). Its presence is known in the Amazon basin since the early 1900s (Thorson, 1972), where specimens were caught about 4200 km into the river (Carvalho and McEachran 2003). This species has also been registered 1200 km into the Mississippi River (Moss, 1984); in Matawan Creek, New Jersey (Klimley, 2013); 175 km into the San Juan River, as well as in other rivers and lakes of Mexico (Helfman et al., 2009); in the Iquitos and Ucayali rivers, both in Peru, among other freshwater systems (Carvalho and McEachran 2003).

Furthermore, it is also known to give birth in estuaries and rivers and to move towards coastal ecosystems when reaches a larger size (Compagno et al., 2005). This behavior is displayed by several other coastal shark species, and is believed to be related to a lower predation risk for the young in these areas (Grubbs, 2010).

This study aims to report the presence of a medium sized specimen of *C. leucas* in the Mearim River basin, Maranhão State, Brazil, extending the number of elasmobranchs listed for this area.

Materials and methods

Mearim River is located in central Maranhão State, geographically considered as Northern Brazil by Programa Revizee (2006) (Fig. 1). It extends for 930 km until its mouth located in the meridional edge of *Ilha dos Caranguejos* (Soares, 2005). Its final portion is known for slow currents, leading to turbid waters and concentration of nutrients and muddy sediments. Perhaps its most unique feature is the occurrence of tidal bores, locally known as *pororoca*, during the highest spring tides of the dry season (August to December). The effect of the high tide can be seen until 256 km into the river basin (Soares, 2005) and causes a mixture between the salty and freshwaters, as well as fine

* Correspondence: lmfeitos@gmail.com
[1]Departamento de Biologia, Universidade Federal do Maranhão, São Luís, Maranhão 65080-805, Brasil
Full list of author information is available at the end of the article

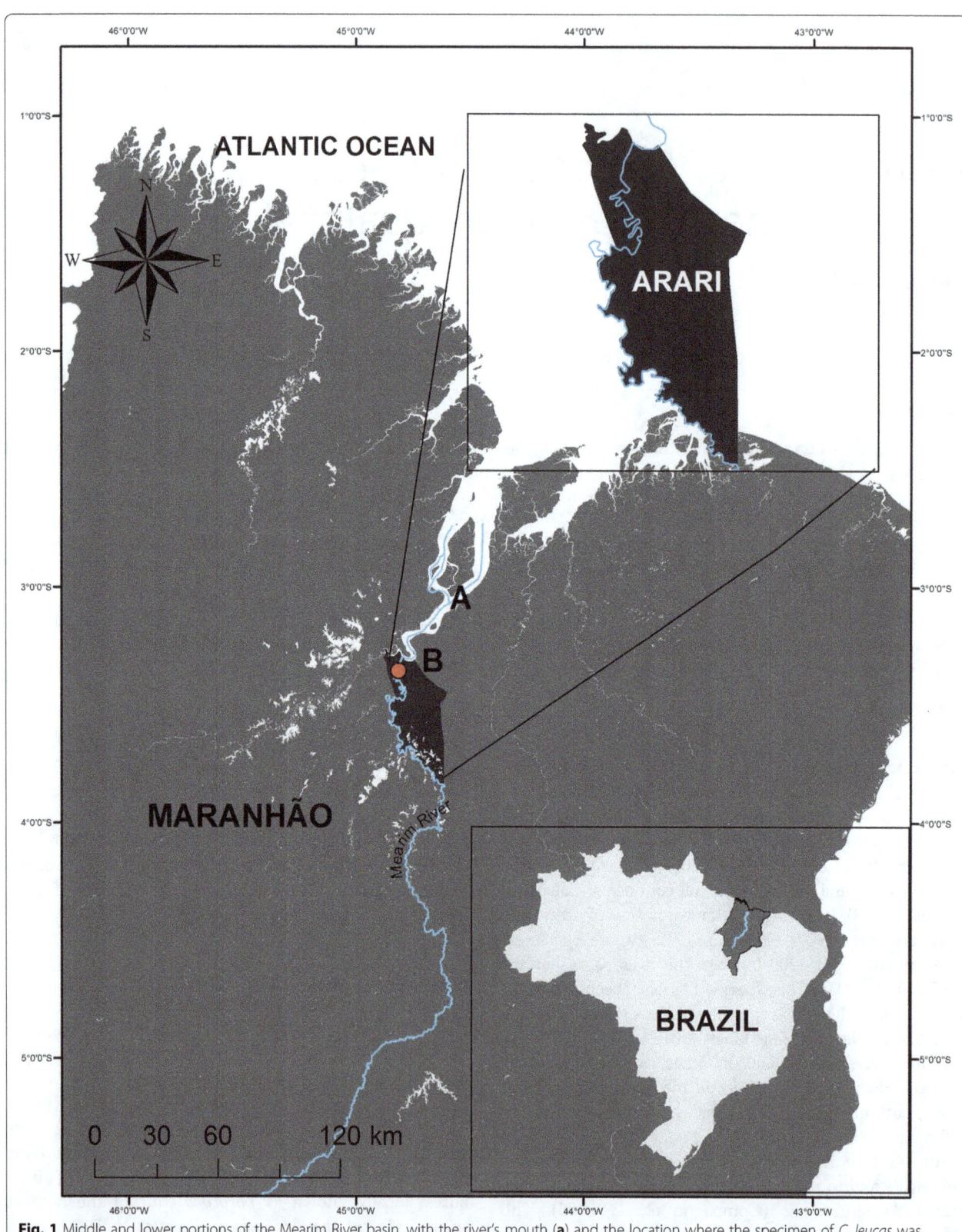

Fig. 1 Middle and lower portions of the Mearim River basin, with the river's mouth (**a**) and the location where the specimen of *C. leucas* was captured (**b**)

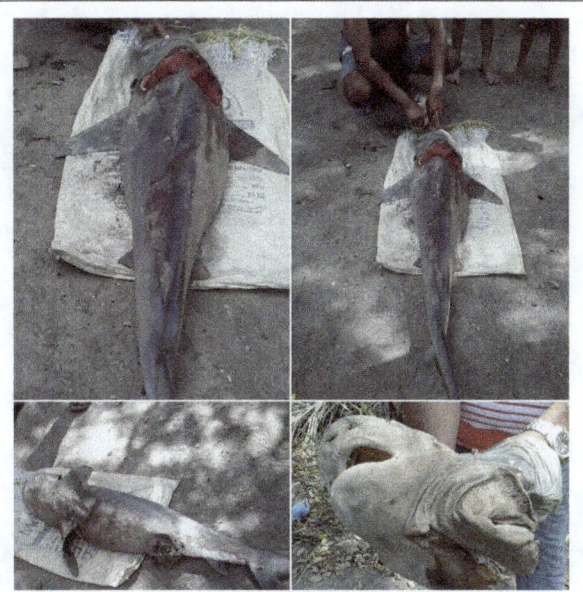

Fig. 2 Different views of the juvenile *C. leucas* specimen captured and processed by fishermen in Arari city

sediments (Chanson, 2005). According to Kjerfve & Ferreira (1993), the Mearim River tidal bore, whose effects gradually decrease upstream, causes an increase in salinity up to 18 % and a small decrease in water temperature, reaching its furthest extent in Arari city.

On 3 September of 2015 a juvenile female *Carcharhinus leucas* was caught by local fishermen in the municipality of Arari (3°23′14″S and 44°49′55″W), Maranhão State, on the margin of the Mearim River, located approximately 80 km far from the river's mouth. According to the fishermen, the shark tried to attack a dog on the margin and got stranded due to the low depth. The event coincided with the occurrence of one of the strongest spring tides of the year in the Maranhão State, which has one of the largest tidal variations in Brazil, reaching up to 7 m.

Results

One female specimen of *Carcharhinus leucas* measuring approximately 1300 mm in total length (TL) and weighing 35 kg was caught by local fishermen (Fig. 2). Its identification was carried out following Compagno et al. (2005). It was not possible to analyze the specimen before the fishermen processed it. However, based on studies of Compagno et al. (2005) and Cruz-Martinez et al. (2005) female individuals mature between 1800 mm and 2300 mm TL. Taking this into account, the approximate TL or this specimen suggests that it was not yet mature.

Discussion

Only 5 % of all elasmobranchs can tolerate some sort of salinity range during their lifetime (Helfman et al., 1997). The tolerance level varies according to the age class and

specific features of habitat use by each individuals and/or species (Cervigón & Alcalá, 1999; Compagno et al., 2005; Ballantyne & Fraser, 2013). In Maranhão State, several marine species of elasmobranchs have been captured in estuarine areas, such as *Isogomphodon oxyrhynchus*, *Carcharhinus porosus*, *Sphyrna tiburo*, *Rhizoprionodon porosus* (Lessa, 1997; Almeida & Vieira, 2000) and a juvenile specimen of *Pristis pristis* [see Faria et al. (2013) for updated taxonomic nomenclature] measuring weighing 20 kg was incidentally caught in the same area in 1999 (Soares, 2005).

The ability that *C. leucas* has to enter riverine systems further than any other shark species is related to its osmotic acclimation to salinity gradients (Klimley, 2013; Pillans & Franklin, 2004). It is believed to be related to the rectal gland activity plasticity (Pillans et al., 2005), urea and trimethylamine oxide (TMAO) reabsorption by the kidney (Pillans et al., 2008), and ion uptake by the gills (Ballantyne & Robinson, 2010), which is enhanced when in freshwater. With all this put together, according to these authors, *C. leucas* can maintain its body hyperosmotic in freshwater environments, but loses much more water and ions due to large amounts of urine produced.

This record is important to direct future more thorough ichthyologic studies in the Mearim River basin that look into the spatial and temporal scales in which these animals can be found, their size range in the river, diet, age, growth, spatial, and temporal movements to provide useful ecological and biological data for the assessment of distribution of bull shark individuals in the Mearim River.

Acknowledgements
We would like to thank the officials from Arari city's environmental agency *Secretaria de Meio Ambiente e Recursos Naturais* (SEMA) for contacting us about the shark specimen capture. Also, we would like to thank Taissa Caroline Silva Rodrigues, Clarisse Mendes Éleres de Figueiredo, and Osmann Cid Conde Oliveira for the help with the figures presented in this report.

Authors' contributions
JLSN and LMF gathered the data presented in this study. LMF, APBM and JLSN participated in the conception and elaboration of the manuscript. All of them have read and approved the final version of this paper and agree to be accountable for all aspects of the work.

Competing interests
The authors declare that they have no competing interests.

Author details
[1]Departamento de Biologia, Universidade Federal do Maranhão, São Luís, Maranhão 65080-805, Brasil. [2]Centre for Sustainable Tropical Fisheries and Aquaculture, College of Science and Engineering, James Cook University, Townsville, QLD 4811, Australia. [3]Departamento de Oceanografia e Limnologia, Universidade Federal do Maranhão, Cidade Universitária do Bacanga, São Luís, Maranhão 65080-805, Brasil.

References
Almeida ZS, Vieira HCP. Distribuição e abundância de elasmobrânquios no litoral maranhense, Brasil. Pesquisa em Foco. 2000;8(11):89–104.

Ballantyne JS, Robinson JW. Freshwater elasmobranchs: a review of their physiology and biochemistry. J Comp Physiol B. 2010. doi:10.1007/s00360-010-0447-0.

Ballantyne JS, Fraser DI. Euryhaline elasmobranchs. In: McCormick SD, Farrell AP, Brauner CJ, editors. Fish Physiology: Euryhaline Fishes: Fish Physiology. New York: Academic Press; 2013. p. 125–98.

Carvalho MR, McEachran J. Family Carcharhinidae (Requiem Sharks). In: Reis RE, Kullander SO, Ferraris CJ, editors. Checklist of the freshwater fishes of South and Central America. Porto Alegre: EDIPUCRS; 2003. p. 13–7.

Cervigón F, Alcalá A. Los peces marinos de Venezuela: Tiburones y rayas. Estado Nueva Esparta, Venezuela: Fundación Museo del Mar; 1999.

Chanson H. Tidal Bore, Aegir, Pororoca, Mascaret. What? Where? When? How? Why? La Houille Blanche. 2005. doi:10.1051/lhb:200503014.

Compagno LV, Dando M, Fowler S. A Field Guide to the Sharks of the World. London. Princeton Field Guide: Harper Collins Publishers Ltd.; 2005.

Cruz-Martinez A, Chiappa-Carrara X, Arenas-Fuentes V. Age and growth of the bull shark, *Carcharhinus leucas*, from southern Gulf of Mexico. Journal of Northwest Atlantic Fisheries Science. 2005. doi:10.2960/J.v35.m481.

Faria VV, McDavitt MT, Charvet P, Wiley TR, Simpfendorfer CA, Naylor GJP. Species delineation and global population structure of critically endangered sawfishes (Pristidae). Zool J Linnean Soc. 2013. doi:10.1111/j.1096-3642.2012.00872.x.

Grubbs RD. Ontogenetic shifts in movements and habitat use. In: Carrier JC, Musick JA, Heithaus MR, editors. Sharks and their relatives II: Biodiversity, adaptive physiology, and conservation. Boca Raton: CRC Press; 2010. p. 319–50.

Helfman GS, Collette BB, Facey DE. The Diversity of Fishes. 3rd ed. Malden: MA: Blackwell Science; 1997. p. 528.

Helfman GS, Collete BB, Facey DE, Bowen BW. The diversity of fishes: Biology, Evolution, and Ecology. West Sussex: Wiley Blackwell; 2009. p. 720.

Kjerfve B, Ferreira HO. Tidal bores: first ever measurements. Ciência e Cultura. 1993;45:135–7.

Klimley AP. The biology of sharks and rays. Chicago: The University of Chicago Press; 2013.

Lessa R. Sinopse dos estudos sobre elasmobrânquios da costa do Maranhão. Boletim Laboratório de Hidrobiologia. 1997;10:19–36.

Moss SA. Sharks: an introduction for the amateur naturalist. Englewood Cliffs, New Jersey: Prentice Hall; 1984.

Pillans RD, Franklin CE. Plasma osmolyte concentrations and rectal gland mass of bull sharks *Carcharhinus leucas*, captured along a salinity gradient. Comp Biochem Physiol A Mol Integr Physiol. 2004. doi:10.1016/j.cbpb.2004.05.006.

Pillans RD, Good JP, Anderson WG, Hazon N, Franklin CE. Freshwater to seawater acclimation of juvenile bull sharks (*Carcharhinus leucas*): plasma osmolytes and Na+/K + – ATPase activity in gill, rectal gland, kidney and intestine. J Comp Physiol B. 2005. doi:10.1007/s00360-004-0460-2.

Pillans RD, Good JP, Anderson WG, Hazon N, Franklin CE. Rectal gland morphology of freshwater and seawater acclimated bull sharks *Carcharhinus leucas*. J Fish Biol. 2008. doi:10.1111/j.1095-8649.2008.01765.x.

Programa Revizee. Relatório Executivo: Avaliação do Potencial Sustentável de Recursos Vivos na Zona Econômica Exclusiva. Brasília/DF: MMA/SQA/PGT/GERCOM; 2006.

Soares EC. Peixes do Mearim. São Luís: Instituto Geia; 2005. 131.

Thorson TB. The status of the bull shark, *Carcharhinus leucas*. Amazon River: Copeia; 1972. p. 601–5.

8

First documented presence of *Galeocerdo cuvier* (Péron & Lesueur, 1822) (ELASMOBRANCHII, CARCHARHINIDAE) in the Mediterranean basin (Libyan waters)

Ibrahim M. Tobuni[1], Ben-Abdallah R. Benabdallah[1], Fabrizio Serena[2,3] and Esmail A. Shakman[1*]

Abstract

One male and one female specimen of tiger shark, Galeocerdo cuvier (Péron & Lesueur, 1822), were accidentally caught by a drifting longline for swordfish in the south Mediterranean (Libyan waters). This finding confirms beyond any doubt that the tiger shark may be encountered in the waters of the Mediterranean Sea. Although records of this species has previously been reported, the information is partial or dubious, due to the lack of a description of the individuals found or the uncertain provenance of preserved material. Our finding confirms the record of this species in the southern part of the Mediterranean basin. Images, as well as morphometrics and information on stomach contents are given. Based on the size of the individuals, it is considered that the two specimens were born recently, presumably inside the Mediterranean Sea and likely close to the area where the individuals were found.

Keywords: Tiger shark, Galeocerdo cuvier, Carcharhinidae, Mediterranean Sea

Background

On 7th January 2015, two juvenile specimens, one male and one female of *Galeocerdo cuvier* (Péron & Lesueur, 1822) (tiger shark) were caught in the Libyan waters north east of Tripoli on the west part of the Gulf of Sidra (Fig. 1). Tiger shark is a species not usually considered resident in the Mediterranean basin. In line with this, the first two records of *G. cuvier*, one from Malaga, Spain (Pinto de la Rosa, 1994) and the second from Sicily, Italy (Celona, 2000) have always been considered doubtful, being based solely on the description of the recovered jaws and no reports of live individuals (Serena, 2005).

Our report is the first tiger shark record documented by photographic evidence, in the Mediterranean Sea. The new findings, in this particular case of two recently born individuals, may contribute to addressing the question of whether the presence of the tiger shark can be considered occasional (as vagrant or isolated), or proof of its stable occurrence in the region.

Tiger sharks belong to the Family of CARCHARHINI-DAE Jordan & Evermann, 1896. This species can be considered circumglobal, found at all latitudes excluding Polar Regions. They generally live in warm, temperate and tropical seas and are considered vagrant in the Eastern North Atlantic (Ebert & Stehmann, 2013) including the Mediterranean Sea (Golani et al. 2002). Tiger shark normally resides further south; including the Azores, Morocco and Canary Islands as far as the Ghana coasts, but it is probable that the species has a wider distribution range in the area. It is also present from the coast of the U.S.A. stretching to Uruguay, including the Gulf of Mexico and Caribbean islands and in the Pacific and Indian oceans and Red Sea (Compagno, 1984; Randall, 1992; Bonfil & Abdallah, 2004; Ebert & Stehmann, 2013).

The maximum reported size of the tiger shark is 740 cm in total length (TL) and 3110 kg (Fourmanoir, 1961). However, tiger sharks are generally not longer than 500 cm in TL. The tiger shark is the only viviparous carcharhinid shark to exhibit internal incubator that does not develop a yolk-sac placenta (Bram et al., 2005); it is able to produce one of the larger litter sizes with

* Correspondence: shugmanism@yahoo.com
[1]Zoology Department, Tripoli University, Tripoli, Libya
Full list of author information is available at the end of the article

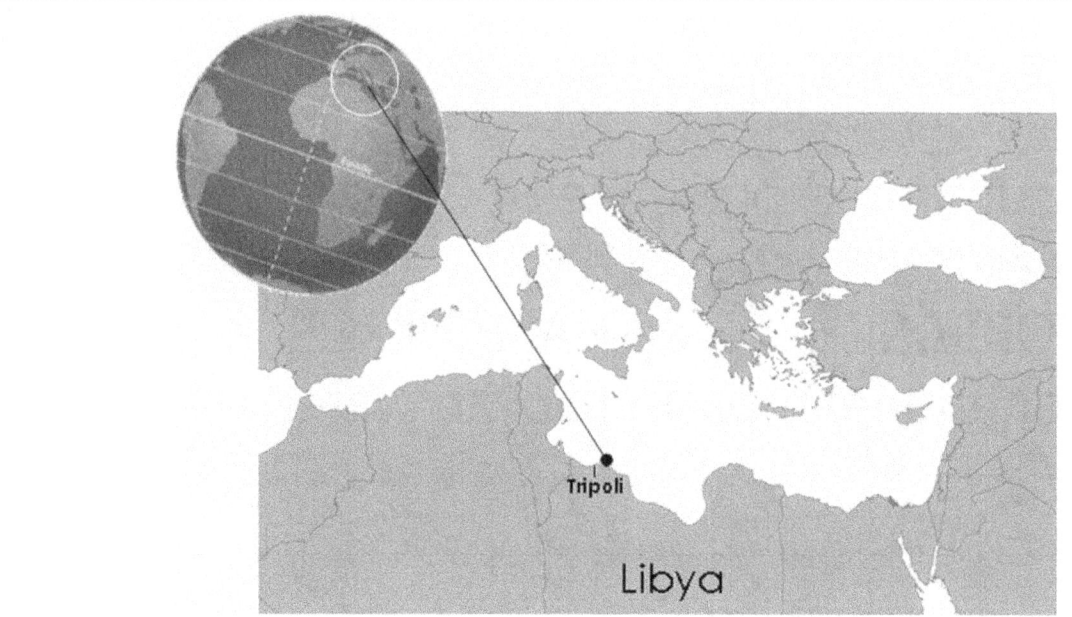

Fig. 1 Map of the Mediterranean Sea and location where the two specimens of *Galeocerdo cuvier* were caught. The temperature of the water was 13 °C

10–82 young per litter (Tester, 1969; Bass et al., 1975; Simpfendorfer, 1992) with a gestation period between 12 to 16 months in the Northern Hemisphere (Clark & von Schmidt, 1965). Mating normally occurs in the spring and pupping takes place throughout spring and early summer, usually from April to June. Size at birth ranges between 46 and 90 cm TL (Compagno, 1984; Randall, 1992; Simpfendorfer, 1992).

Materials and methods

The tiger shark specimens were caught by Tripoli fishers using artisanal drifting longlines baited with sardines in order to target swordfish. This kind of gear, locally named "bringali sayeb", fish between 10–500 m in the water column, and is especially utilized by "mator" boats. These vessels have an average power of 214,86 kW and range in length from 5 to over 18 m (12 m on average). In the early 2000s, mator boats constituted the 28 % of the 1800 artisanal fishing boats operating in Libya. Only few mator boats target swordfish (Lamboeuf, 2000). The longline gear is made of a main monofilament (Ø 2–3 mm) of about 14 km long. One double line snood, consisting of a monofilament (Ø 1.3 mm) of 1.5 m with hooks (size number 01), is connected to the main line every 50 m. Every 5 hooks a float is connected the main line by a mono-filament 3.5 m long.

The two tiger sharks caught are documented by pic-tures taken in the laboratory of the University of Tripoli. In this report, for chondrichthyan taxonomy and no-menclature, the guidelines in Serena (2005), Ebert *et al.*, (2013) and Eschmeyer *et al.*, (2016) were followed. The

morphometric measurements shown in Table 1 refer to MEDLEM protocol format (Serena *et al.*, 2006). The stomach content was analyzed in order to identify the food composition (Table 2). Examined material was pre-served in formalin 10 % and individuals stored in the col-lection of the Zoological Museum of Tripoli University with the following code: (ZST, FISH0010; FISH0011). The specimens are available for inspection on request.

Results

The two juvenile specimens of *G. cuvier,* one male and one female, were caught at 33°03,710 Latitude North-013°44,529 Longitude Est off Gasr Garabulli, at a depth of 180 m, using artisanal drifting longlines for swordfish *Xiphias gladius* as target and with *Thunnus alalunga*, *Carcharhinus plumbeus*, *Isurus oxyrinchus* juv. and *Pteroplatytrygron violacea* as by catch (Fig. 1). The total lengths of the individuals were 95.8 cm and 97.4 cm for the male and female respectively (Fig. 2). Both speci-mens had a large head with short and blunt snouts, large mouths, long upper labial furrows that go beyond the eye lines, large spiracles and small nostrils (Fig. 3). The eyes were fairly large without posterior notches. The first dorsal fin origin was above the pectoral fin inner margins in both specimens. The interdorsal ridge was present and prominent; the body part of the speci-mens analyzed behind the pectoral fins was, in both cases, rather thin with small keels on the caudal peduncles. The upper pre-caudal pits were transverse and semilunar. The caudal fins were slender with acutely pointed tips. Along the body, including the fins, characteristic bold dark and

Table 1 Morphometric measurements for tiger Shark *Galeocerdo cuvier* caught in the Libyan coasts

Measurements		Female	Male
Total length	TOT	97.4	95.8
Fork length	FOR	71.3	69.0
Trunk length	TL	28.0	27.8
Precaudal tail	PCT	18.0	17.7
Percaudal length	PRC	64.8	64.7
Presecond dorsal length	PD2	54.1	53.8
Prefirst dorsal length	PD1	24.7	24.6
Head length	HDL	19.0	18.8
Prebranchial length	PGL	14.0	13.5
Preorbital length	POB	5.2	5.0
Prepectoral length	PP1	17.1	16.9
Prepelvic length	PP2	42.4	42.3
Preanal length	PAL	52.0	51.5
Interdorsal space	IDS	22.8	22.5
Eye length	EYL	2.0	2.0
Eye height	EYH	1.6	1.6
Prenarial	PRN	1.8	1.7
Preoral length	POR	4.1	4.0
Intergill length	ING	5.0	5.1
Pectoral-fin length	P1L	12.1	11.7
Pectoral-base length	P1B	4.6	4.6
Pectoral-inner margin	P1I	4.2	4.1
1st dorsal anterior margin	D1A	10.8	10.7
1st dorsal base length	D1B	7.8	7.5
1st dorsal height	D1H	7.8	7.6
Dorsal caudal margin	CDM	32.5	32.3
Preventral caudal margin	CPV	10.1	10.0
Total weight	TOW	2840 gm	2750 gm

Fig. 2 Specimens of *Galeocerdo cuvier* caught in the Libyan coast. Male above, female below

Other morphometric measurements are reported in Table 1. The stomach contents of both the female and male specimens were not significant regarding to the possibility of recognizing any organisms present. Indeed, in both cases, the contents of the stomachs, cephalopods and fishes, were almost completely digested and difficult to determine at the species level (Table 2).

Discussion

The Mediterranean Sea has a long and complex history, which left traces in the actual status of its fauna and flora. After the opening of the Suez Canal and climate changes occurring at global and Mediterranean levels resulting in increased mean sea and surface temperatures, immigrant fauna and flora species, called lessepsian, moved into the Mediterranean from the Red Sea through the Suez Canal (Por, 1978). This phenomenon is still on-going. Continuous and important exchanges of marine organisms also occur at the Strait of Gibraltar between the Mediterranean Sea and the Atlantic Ocean. The Libyan coasts, positioned in the south of the Mediterranean, may be affected by immigration from both origins. This wide area is characterised by different habitats and topography.

vertical tiger-stripe markings were present, especially on the dorsal surface of body. Only the pectoral fins appeared uniformly dark. The bellies were yellowish white in colour. The teeth were similar in the upper and lower jaws, being typical cockscomb-shaped curved with heavy serrations and distal cusplets. The tooth count was 24 on the upper jaw and 21 on the lower jaw of both the male and female specimens.

Table 2 Stomach contents of tiger sharks, female and male, caught along the coasts of Libya

Female	Male
• Cephalopod beak	• Pair of cephalopod beaks
• Semi digested cephalopod	• Semi digested cephalopod
• Semi digested fish (9 cm) probably Clupeiformes	
• Semi digested fish (14 cm) probably Mackerel	

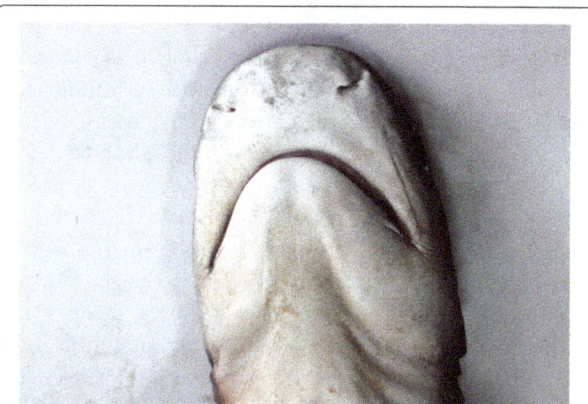

Fig. 3 Ventral view of the head of the male specimen of *Galeocerdo cuvier* caught in the Libyan coasts

In the Mediterranean, many elasmobranch populations are now rare due to intense fishing activity, in which they are target or by-catch. Such fishing pressure has produced local extinctions or stock reductions of some elasmobranchs making certain fishing activities no longer economically viable (Ferretti et al., 2008). Such phenomena are less intense in the southern part of the basin, especially along the Libyan coasts. In fact, in this area, some artisanal fisheries using fixed gillnets, bottom set nets and drifting longlines still target cartilaginous fish such as Carcharhinidae, Lamnidae, *Rhinobathos* spp. and *Squatina squatina* (Lamboeuf, 2000). For this reason, the catch of some top predators such as the tiger shark near the coast is likely. Tiger sharks, especially juveniles, are taken as by-catch in many fisheries around the world, including those using longlines targeting swordfish and tuna, particularly those operating close to the continental shelves (Anderson, 1985; Berkeley & Campos, 1988).

Conclusions

Our finding can be considered the first well-documented occurrence of *G. cuvier* in the southern Mediterranean. It confirms the sporadic occurrence of the species in the Mediterranean Sea. However this record cannot be taken as a proof of the stable occurrence of the species in the region or of a geographic extension of the species in to the Mediterranean Sea, even if, recently, 5 nm north of the Gorgona Island (North West Mediterranean), one of us has personally observed another specimen of about 2 m long, during a sportfishing on 29th of September 2015, unfortunately no photo was taken. As noted, the previous records led to doubts about the occurrence of tiger sharks in the Mediterranean basin (Serena, 2005) one from Malaga, Spain (Pinto de la Rosa, 1994) and the second from Sicily, Italy (Celona, 2000) were based only on the description of the jaws with no certified evidence of their provenance. The information in both publications is not accompanied by any photographic documentation of the whole individual and in one case (Pinto de la Rosa, 1994) the description was very limited.

Tiger sharks give birth generally along coastal areas at depths lower than 100 m. The growth of young tiger sharks is quite rapid and they nearly double their size over the first year of life (Branstetter et al., 1987; Randall, 1992; Smith et al., 1998). Therefore, the specimens described here (male 95.8 cm and female 97.4 cm) are likely to have been born recently in that area probably in the second part of 2014 and hence are assumed to be only a few months old. Juvenile tiger sharks are very slender with flexible bodies, swim with an inefficient motion and are very vulnerable to predators (Branstetter et al., 1987). It is unlikely that they were able to swim for very long distances as those needed to reach Libyan waters from either the Red Sea or the Atlantic. As a consequence,

there are two possible alternative explanations for the finding of those two juvenile specimens. The first one is to hypothesize that the two pups were delivered from a resident female. Alternatively, these pups could have been delivered by a pregnant vagrant female. It has been reported that tiger sharks undertake trips of thousands of kilometres (8,000 km in 99 days (Heithaus et al., 2007), or 1,800 km in 48 days (Holmes et al., 2014) and gestation periods of 12–16 months have been estimated (Clark & von Schmidt, 1965). Therefore it is possible that mating could have happened outside the Mediterranean. Indeed, more compelling arguments are needed to support an enlargement of the distribution range and stable presence of the tiger shark in the Mediterranean waters.

Increased tropical influx through the Gibraltar Strait and the Suez Canal has resulted in the so-called "tropicalization" of the Mediterranean Sea (Bianchi & Morri, 2003). Until the mid 20th century, alien species introduction, establishment, and expansion rates were low (Zenetos et al., 2008), mainly as a result of the water temperature and salinity barriers between the Red Sea and the Mediterranean Sea (Galil, 2006). The dispersal success and expansion of an invasive species depends upon many aspects after the suitability of the abiotic environment. The occasional specimens of tiger sharks recorded up to now are unable to support any real dispersal process linked to climate change.

Acknowledgements

The data used for the present short note were obtained thanks to the collaboration of artisanal fishery fishermen of the Tripoli harbour. We are very grateful to the fishermen of the Tripoli harbour for their collaboration. Thanks also to Alvaro Juan Abella and Cecilia Mancusi (ARPAT, Italy) for the invaluable suggestions and for the revision of the text. Special thanks to the National Authority for Scientific Research of Libya for funding the project of invasive marine species in the Libyan coasts. Thanks also to Anassuya Ramachandran (Francis Crick Institute Limited, London, UK) and Ronan Russell (UCSF School of Medicine, San Francisco, USA) for reviewing the English language used in the text. Finally we want to thank the anonymous referees for suggesting appropriate changes which made the text more comprehensive.

Authors' contributions

All authors read and approved the final manuscript.

Competing interests

The authors declare that they have no competing interests.

Author details

[1]Zoology Department, Tripoli University, Tripoli, Libya. [2]Environmental Protection Agency-Tuscany Region (ARPAT), Tripoli, Italy. [3]Institute for the Coastal Marine Environment (IAMC)-Italian National Research Council (CNR), Mazara, Italy.

References

Anderson ED. Analysis of various sources of pelagic shark catches in the northwest and western central Atlantic Ocean and Gulf of Mexico with comments on catches of other large pelagics. In: Shark catches from selected fisheries off the US East Coast. US East Coast: NOAA Technical Report NMFS; 1985. p. 1–14.

Bass AJ, D'Aubrey JD, Kistnasamy N. Sharks of the east coast of southern Africa. III. The families Carcharhinidae (excluding *Mustelus* and *Carcharhinus*) and Sphyrnidae. In: South African Association for Marine Biological Research. Oceanographic Research Institute. Investigational Reports. 1975.

Berkeley SA, Campos WL. Relative abundance and fishery potential of pelagic sharks along Florida's east coast. Mar Fish Rev. 1988;50:9–16.

Bianchi & Morri. Global sea warming and "tropicalization" of the Mediterranean Sea: Biogeographic and ecological aspects. Biogeographia. 2003; 24: 319–327.

Bonfil R, Abdallah M. Field identification guide to the sharks and rays of the Red Sea and Gulf of Aden. In: FAO Species Identification Guide for Fishery Purposes. Rome: FAO; 2004. p. 71. 12 colour plates.

Bram JB, Page HM, Dugan JE. Spatial and temporal variability in early successional patterns of an invertebrate assemblage at an offshore oil platform. J. Exp. Mar. Bio. Ecol. 2005;317:223–237.

Branstetter S, Musick JA, Colvocoresses JA. A comparison of the age and growth of the tiger shark, *Galeocerdo cuvieri*, from off Virginia and from the northwestern Gulf of Mexico. NMFS. Fish Bull. 1987;85:269–79.

Celona A. First record of a tiger shark *Galeocerdo cuvier* (Peron and LeSueur, 1822) in the Italian waters Annales for Istrian and Mediterranean studies, Series. Hist Nat. 2000;2(21):207–10.

Clark E, von Schmidt K. Sharks of the central gulf coast of Florida. Bull. Mar. Sci. 1965;15:13–83.

Compagno LJV. Sharks of the World: An Annotated and Illustrated Catalogue of Shark Species Known to Date. Volume 4, Part 1. Hexanchiformes to Lamniformes. Rome: FAO; 1984.

Ebert DA, Fowler S, Compagno LJV, Dando M. Sharks of the world: a Fully Illustrated Guide. Plymouth: Wild Nature Press; 2013. p. 528.

Ebert DA, Stehmann MFW. Sharks, batoids, and chimaeras of the North Atlantic FAO Species Catalogue for Fishery Purposes. No. 7. Rome: FAO; 2013. p. 523.

Eschmeyer WN, Fricke R, van der Laan R. (eds) (2016) Catalog of fishes: Genera, Species, References. (http://researcharchive.calacademy.org/research/ichthyology/catalog/fishcatmain.asp). Electronic version Accessed 16 Apr 2016.

Ferretti F, Myers RA, Serena F, Lotze HK. Loss of large predatory sharks from the Mediterranean Sea. Conservation Biology. 2008;22(4):952–964.

Fourmanoir P. Requins de la côte ouest de Madagascar. Ser. Oceanog. 1961;4:3–8.

Galil BS. Shipwrecked-shipping impacts on the biota of the Mediterranean Sea. In Davenport J.L. and Davenport J. (eds) The ecology of transportation: managing mobility for the environment. Dordrecht. The Netherlands: Springer-Verlag; 2006. p. 36–69.

Golani D, Orsi-Relini L, Massuti E, Quignad JP. CIESM Atlas of exotic species in the Mediterranean. Vol. 1. Fishes. In: Briand F, editor. CIESM. Publishers, Monaco. 2002. p. 256.

Jordan DS, Evermann BW. 1896. A check-list of the fishes and fish-like vertebrates of North and Middle America. Fish Comm. 5:207–584.

Heithaus MR, Frid A, Wirsing AJ, Dill LM, Fourqurean JW, Burkholder D, Thomson JA, Bejder L. State-dependent risk-taking by green sea turtles mediates top-down effects of tiger shark intimidation in a marine ecosystem Journal of Animal Ecology. 2007;76:837–844.

Holmes BJ, Pepperell JG, Griffiths SP, Jaine FRA, Tibbetts IR, et al. Tiger shark (Galeocerdo cuvier) movement patterns and habitat use determined b satellite tagging in eastern Australian waters. Mar Biol. 2014;161:2645–2658.

Lamboeuf M. Artisanal fisheries in Libya. Census of fishing vessels and inventory of artisanal fishery metiers. FAO-COPEMED-MBRC. Final Report. 2000. p. 42.

Pinto de la Rosa FJ. Tiburones del mar de Alboran. Diputacion de Malaga: Servicio publicaciones Centro de Ediciones; 1994. p. 115.

Por FD. Lessepsian migration. In: The influx of Red Sea biota into the Mediterranean by way of the Suez Canal. 23rd ed. Berlin: Ecological Studies Springer Verlag; 1978. p. 228.

Randall JE. Review of the biology of the tiger shark (*Galeocerdo cuvier*). Aust J Mar Freshwat Res. 1992;43:21–31.

Serena F. Field identification guide to the sharks and rays of the Mediterranean and Black Sea. Rome: Food and Agriculture Organization of the United Nations; 2005.

Serena F, Barone M, Mancusi C, Magnelli G, Vacchi M. The MEDLEM database application: a tool for storing and sarin the large sharks data collected in the Mediterranean countries. In: Basusta N, Keskin C, Serena F, Seret B, editors. The Proceedings of the Workshop on Mediterranean Cartilaginous Fish with Emphasis on Southern and Eastern Mediterranean. 23rd ed. Istanbul-Turkey: Turkish Marine Research Foundation; 2006. p. 118–27.

Simpfendorfer C. Biology of tiger sharks (*Galeocerdo cuvier*) caught by the Queensland shark meshing program off Townsville, Australia. Aust J Mar Freshwat Res. 1992;43:33–43.

Smith SE, Au DW, Show C. Intrinsic rebound potentials of 26 species of Pacific sharks. Mar Freshw Res. 1998;49:663–78.

Tester AL. Cooperative Shark Research and Control Program. Honolulu: University of Hawaii; 1969. Final Report.

Zenetos A, Meric E, Verlaque M, Galli P, Boudouresque C.-F, Giangrande A, Cinar E, Bilecenoglu M. Additions to the annotated list of marine alien biota in the Mediterranean with special emphasis on Foraminifera and Parasites. Mediterranean Marine Science. 2008;9:119-165.

9

First North Pacific records of the pointy nosed blue chimaera, *Hydrolagus* cf. *trolli* (Chondrichthyes: Chimaeriformes: Chimaeridae)

Amber N. Reichert[1*], Lonny Lundsten[2] and David A. Ebert[1,3]

Abstract

The occurrence of *Hydrolagus* cf. *trolli* is reported for the first time from the central and eastern North Pacific Ocean. This is a geographic range extension for this species, as it was previously only known to occur in the southern Pacific Ocean off of Australia, New Zealand, and New Caledonia.

Keywords: *Hydrolagus trolli*, Northern Hemisphere, Range extension

Introduction

Currently there are 38 recognized species of short-nosed chimaeras (family Chimaeridae), making it the most species rich family in the order Chimaeriformes (Kemper et al. 2015). The family has two recognized genera that are separated by the presence (*Chimaera*) or absence (*Hydrolagus*) of an anal fin. The genus *Hydrolagus* is the more diverse of the two genera with 22 species (Didier et al., 2012). Fifteen of these species are recognized as occurring in the Pacific Ocean, but only five species are known in the eastern Pacific (James et al. 2009). These five *Hydrolagus* species are geographically dispersed around the Galapagos Island, the southeastern Pacific along the coasts of Chile and Peru, and lastly in the northeastern Pacific (James et al. 2009). Until recently *Hydrolagus colliei* (Lay and Bennett, 1839) and *Hydrolagus melanophasma* (James, Ebert, Long, and Didier, 2009) were the only species confirmed to occur in the northeastern Pacific. *Hydrolagus colliei* is found from Alaska to Costa Rica, while *H. melanophasma* occurs off southern California, USA to northern Chile (Angulo et al., 2014; Aguirre-Villaseñor et al. 2013). However, a third species of *Hydrolagus*, first noted but not identified by Ebert (2003), had been observed by remotely operated vehicle (ROV) at

Davidson Seamount off central California at a depth greater than 2000 m. Lundsten et al. 2009 identify *Hydrolagus trolli* from the Davidson Seamount, however provide no descriptive information. Central North Pacific records of chimaeras are few, with only the purple chimaera, *Hydrolagus purpurescens* (Gilbert 1905), being reported from the Hawaiian Islands.

The pointy nosed blue chimaera, *Hydrolagus trolli* (Didier and Séret 2002), was described from 23 specimens. The holotype is a male specimen measuring 1030 mm (TL) that was captured by bottom trawl off New Caledonia (20°44.90'S, 167°43.10'E) at a bottom depth of 1246 m. *Hydrolagus trolli* is a little known chimaera species usually found at depths ranging 610–2000 m (Last and Stevens 2009). To date, this species has only been confirmed from the southwestern Pacific, off Australia, New Zealand, New Caledonia, the Lord Howe Rise and Norfolk Ridge (Last and Stevens 2009). This species, or a similar looking species, is known to occur over a much broader geographical range in the Southern Hemisphere, e.g. South Africa (Ebert and van Hees, 2015), but to date there are no descriptions of this species from the Northern Hemisphere. A specimen of *H. cf. trolli* was recorded off Chile (as *H. pallidus* in Andrade & Pequeño, 2006), (Bustamante et al. 2012). A potential specimen of *H. trolli* was collected off St. Paul Island in the Southern Indian Ocean but identification could not be confirmed due to poor condition (Didier and Séret 2002).

* Correspondence: areichert@mlml.calstate.edu
[1]Pacific Shark Research Center, Moss Landing Marine Laboratories, 8272 Moss Landing Road, Moss Landing, CA 95039, USA
Full list of author information is available at the end of the article

During a series of remotely operated vehicle (ROV) deep-sea surveys off the California coast and west of the Hawaiian Islands conducted by the Monterey Bay Aquarium Research Institute (MBARI), a large, bluish, short-nosed *Hydrolagus* species was observed on several occasions. Here we report on the occurrence of these *Hydrolagus* specimens that we have identified as *Hydrolagus* cf. *trolli*.

Materials and methods

Observations of *Hydrolagus* cf. *trolli* were videotaped in situ using either a Panasonic WVE550 3-chip standard definition or Ikegami HDL-40 high definition video camera during deep-water surveys using the ROV *Tiburon* off the coast of California and Hawaii. Using MBARI's Video Annotation and Reference System database (VARS, Schlining et al. 2006), videotaped observations were entered into a searchable database and merged with ancillary data so that latitude, longitude, depth, and CTD information is known for every observation. We queried VARS for observations identified as potentially being these species based on macroscopic characters. Due to inconsistencies in the calibrated laser system which is used for estimating organism size, we were unable to obtain lengths of our specimens.

During the surveys off central California (T0142-06, T1075-02, T1102-04, T0215-01), southern California (T0664-10), and west of the Hawaiian Islands, (T0296-12), unidentified *Hydrolagus* specimens were repeatedly observed (Fig. 1, Table 1). These records are of a large, bluish, short-nosed chimaera that had never before been observed previously in the central or eastern North Pacific. Surveys T0664-10, T0215-01, and, T0296-12 from the San Juan Seamount, Monterey Submarine Canyon, and from off the Hawaiian islands, respectively, were initially identified as *H.* cf. *trolli* by one of us (D.A. Ebert), and D.A. Didier (Millersville University) and L.A.K. Barnett (University California, Davis).

Results and discussion

Observations

Hydrolagus cf. *trolli* from dive T-0142 (Fig. 2a) was the first observation of these unknown bluish *Hydrolagus* specimens in the Northern Hemisphere, and the first of four observations from Monterey Bay, California. Specimen from trawl T-0142 was first reported by Lundsten et al. 2009, and recorded off the coast of central California at the Davidson Seamount on 9 May 2000 at approximately 2064 m. A similar specimen was observed near the same location on dive T-1102 (Fig. 2b) at the Davidson Seamount, on 20 June 2007, at approximately 1641 m. Both specimens from dives T-0142 and T-1102 have striking similarities including: a short, pointed snout, with preopercular and oral canals that share a common branch from the infraorbital

canal, bluish-grey body coloration, a concave first dorsal fin with keeled spine, large, triangular pectoral fins, and broad pelvic fins. Additionally, the dorsal, pectoral, and pelvic fins of these specimens all have a pale blue margin on the distal edge.

The third observation of this *Hydrolagus* species from Monterey Bay, CA was identified from dive T-0215 in the Monterey Canyon, on 5 October 2000, at approximately 1679 m (Fig. 2c). A fourth specimen was observed on dive T-1075 (Fig. 2d), also in the Monterey Canyon on 24 January 2007 at approximately 1658 m. Specimens from dives T-1075 and T-0215 had similar morphological features to the Davidson Seamount specimens. The specimen observed on dive T-1075 had preopercular and oral canals sharing a common branch from the infraorbital canal on the left hand side, while these canals branched separately on the right side of the head. This could not be observed in the specimen from dive T-0215 as the majority of the photos were from a dorsal view.

During a subsequent survey of seamounts off southern California, another large, bluish *Hydrolagus* specimen was observed. This specimen, also identified as *H.* cf. *trolli*, from dive T-0664 was recorded on 2 May 2004 on the San Juan Seamount at approximately 1629 m (Fig. 2e). The *Hydrolagus* specimen from this dive had a short pointed snout, large, triangular pectoral fins, and a bluish-grey coloration, consistent with those specimens observed on the Davidson Seamount.

A sixth observation of a *H.* cf. *trolli* specimen was during surveys off the western coast of the Hawaiian Islands. This was the first observation of *H.* cf. *trolli* from the central North Pacific. *Hydrolagus* cf. *trolli* from dive T-0296 was recorded on 16 April 2001 at approximately 1641 m (Fig. 2f). Although this *Hydrolagus* specimen was identified as *H.* cf. *trolli* it had several subtle differences from specimens T-0142 and T-0215 such as a short, blunt snout, and larger eyes. However, this specimen exhibited a few similar features to specimen T-0664. These similarities include: a keeled dorsal spine, longer than the height of the primary dorsal, and a short caudal filament.

Observations from all surveys show *H.* cf. *trolli* occurring over rocky substrates, sometimes over high relief outcrops. These observations suggest individuals typically occur over hard-bottom habitats or rocky, and mixed substrate patches with vertical relief. These observed substrate associations are in contrast to *Hydrolagus melanophasma*, which usually associates with soft bottom habitats (James et al. 2009). However, some eastern Pacific *Hydrolagus* species such as *Hydrolagus mccoskeri*, are known to associate with high relief habitats (James et al. 2009; Barnett et al. 2006).

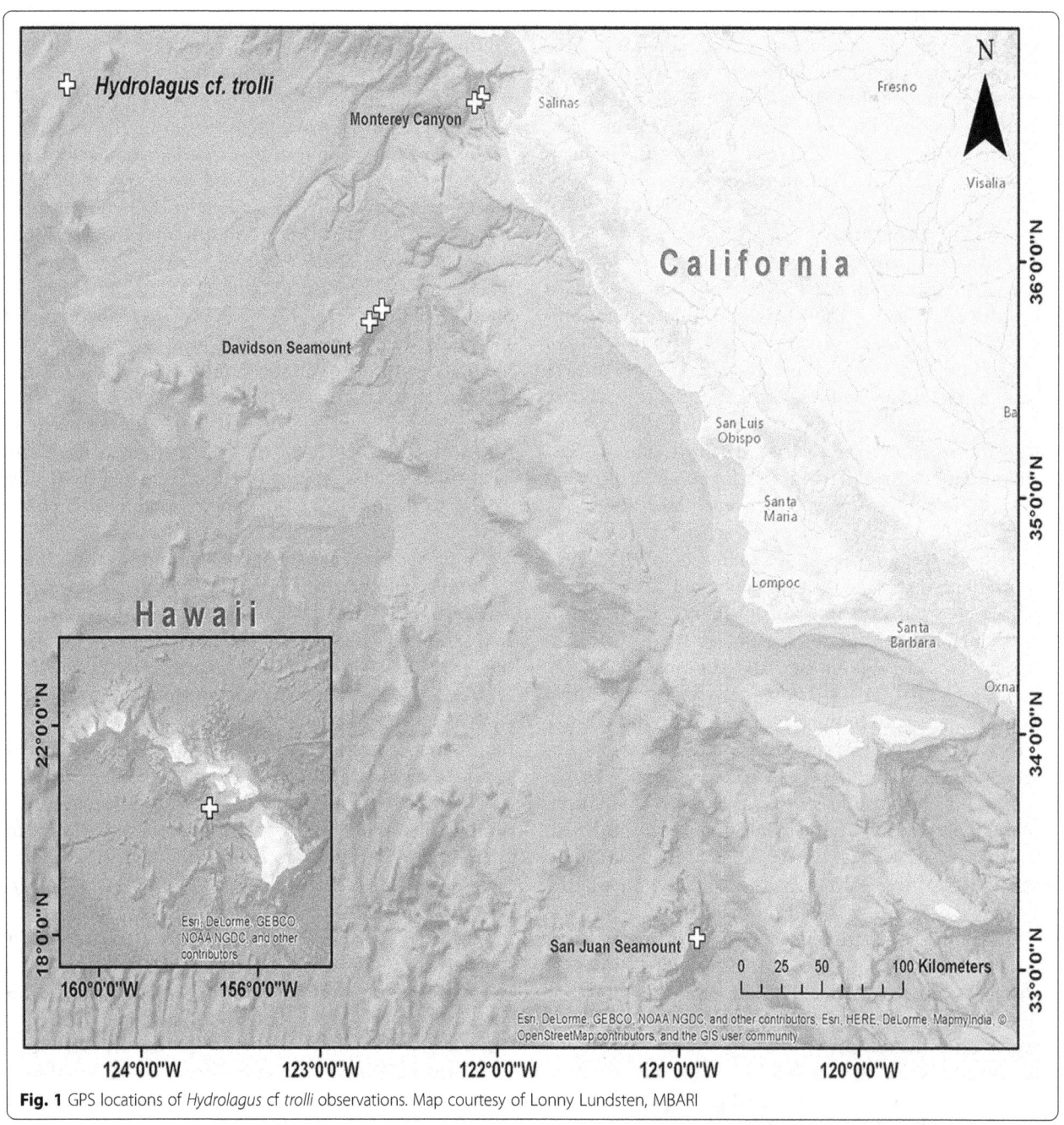

Fig. 1 GPS locations of *Hydrolagus* cf *trolli* observations. Map courtesy of Lonny Lundsten, MBARI

Description

Hydrolagus trolli is a highly distinctive chimaera species, often identified by a combination of the following characteristics: an even blue-gray to pale blue color, a pointed snout, a dark margin around the orbit with dark shadows along edges of the lateral line, and preopercular canal and oral canals usually sharing a common branch (Didier and Séret 2002; Compagno and Dagit 2006). *Hydrolagus trolli* is a large, although slender bodied species with a narrow head that evenly tapers to a whip-like tail. A caudal filament is present, although short and blunt. The pectoral fins of *H. trolli* are large and triangular, usually tapering to a point on the distal edge. The pelvic fins are broad, and square along distal edge. The first dorsal fin is triangular, with a concave distal edge (Didier and Séret 2002). The dorsal fin spine is curved anteriorly, with two small rows of serrations on the distal 1/3 of the posterior surface. The fin spine is usually just shorter than the height of dorsal fin in juveniles, and slightly longer in adults. The second dorsal fin is elongate, sloping, relatively even in height, and is connected to the dorsal caudal fin by a narrow piece of skin.

First North Pacific records of the pointy nosed blue chimaera, Hydrolagus cf. trolli...

55

Table 1 CTD, GPS coordinates, and additional comments for observed *Hydrolagus* cf. *trolli* specimens

Specimen	Conductivity (psu)	Temperature (C)	Depth (m)	Latitude/Longitude	Location	Comments
T0142-06	34.49	2.01	2063.2	35.797391/−122.650021	Davidson Seamount	POP and O lateral line appear to share a common branch. No claspers observed.
T1102-04	34.37	2.45	1641.7	35.742034/−122.71744	Davidson Seamount	POP and O lateral line canal share a common branch. No claspers observed.
T0215-01	34.46	2.47	1679.6	36.671808/−122.122993	Monterey Submarine Canyon	Coloration appears blue-purple. Short caudal filament. No claspers observed.
T1075-02	34.374	2.47	1658.3	36.695932/−122.08455	Monterey Submarine Canyon	POP and O lateral line canal branch separately on right side of head only. No claspers observed.
T0664-10	34.49	2.58	1629.4	33.133464/−120.90126	San Juan Seamount	Coloration appears blue-purple. Short caudal filament. No claspers observed.
T0296-12	34.57	2.77	1640.8	20.426773/−157.223329	Hawaii	Snout more compressed, similar to *H. purpurescens*. No claspers observed.

The caudal fin is rounded with dorsal and ventral lobes nearly equal in height, though ventral lobe is slightly longer (Didier and Séret 2002). Males have a deeply curved frontal tenaculum, which is distally upturned with spines along the dorsal edge; and pelvic claspers that have fleshy, pale, distal lobes, divided for 1/3 their length, with tips usually extending beyond distal edge of pelvic fin (Didier and Séret 2002).

Comparisons

Hydrolagus cf. *trolli* is the third species of *Hydrolagus* observed in the eastern North Pacific, and is the second species observed from the central North Pacific. *Hydrolagus* cf. *trolli* is morphologically distinct from the other two eastern North Pacific *Hydrolagus* species, *Hydrolagus colliei* and *Hydrolagus melanophasma*, in coloration. The overall brownish-red color and white spots of *H.*

Fig. 2 a *Hydrolagus* cf *trolli* specimen T-0142 full-length lateral view. **b** *Hydrolagus* cf *trolli* specimen T-1102 lateral view, close up. **c** *Hydrolagus* cf *trolli* specimen T-0215 dorsal view. **d** *Hydrolagus* cf *trolli* li specimen T-1075 lateral view. **e** *Hydrolagus* cf *trolli* specimen T-0664 full-length lateral view. **f** *Hydrolagus* cf *trolli* specimen T-0296 lateral view, over rocky substrate

colliei, and overall dark, black coloration of *H. melano-phasma* are easily distinguished from the even blue color of *H. cf. trolli* (Ebert 2003).

The only *Hydrolagus* species currently known from the central North Pacific is *Hydrolagus purpurescens* (Gilbert 1905) that was described from Hawaiian waters. It closely resembles *H. cf. trolli* specimens, but differs from it in body coloration, pectoral fin shape, and dorsal spine length and shape. *Hydrolagus purpurescens* is thought to be more widespread throughout the central and western North Pacific, though similar bottom depth ~1130 m. Despite possible overlapping geographically with *H. cf. trolli* these two species are morphologically different in several aspects. The head of *H. purpurescens* is robust and deeply compressed, with a snout that is high and compressed, while the observed specimens had pointy snouts (Gilbert 1905; Garman 1911). The pectoral fins of *H. purpurescens* are large, and broad (Gilbert 1905; Garman 1911), while our specimens had pectoral fins more pointed, and triangular. *Hydrolagus purpurescens* has a straight dorsal spine that is longer than the height of the first dorsal fin, with no serrations (Gilbert 1905; Garman 1911). Dorsal fin spines of our *H. cf. trolli* specimens were keeled, except for the specimens observed on dives T-0142, and T-0215 where the fin spines could not be observed in detail. Finally, the even black-purple to purplish-plum coloration of *H. purpurescens* is distinct from the grey and blue coloration of our *H. cf. trolli* specimens. All the observed *H. cf. trolli* specimens have characteristics that more closely resemble *H. trolli* than *H. purpurescens*.

Remarks

The presence of *Hydrolagus* cf. *trolli* increases the number of known *Hydrolagus* species to three off California, and to two species off the Hawaiian Islands. Our specimens cannot yet be confirmed as *Hydrolagus trolli* until morphometric data and or DNA samples from preserved specimen have been collected and analyzed. However, these observations by ROVs suggest that even in relatively well-known areas much remains to be elucidated on the Chondrichthyan fauna from these regions.

Acknowledgements

We are grateful for support from the pilots of the ROV *Tiburon* and the crew of the R/V *Western Flyer*. We thank D. Clague, S. Haddock, and P. Lonsdale for use of video observations. We thank D.A. Didier (Millerville University), L.A.K. Barnett (University California, Davis) for taxonomic assistance and L. Kuhnz (MBARI) for technical assistance. This work was partially supported by the David and Lucile Packard Foundation's funding of MBARI.

Authors' contributions

All authors read and approved the final manuscript.

Competing interests

The authors declare that they have no competing interests.

Author details

[1]Pacific Shark Research Center, Moss Landing Marine Laboratories, 8272 Moss Landing Road, Moss Landing, CA 95039, USA. [2]Monterey Bay Aquarium Research Institute, 7700 Sandholdt Road, Moss Landing, CA 95039, USA. [3]Department of Ichthyology Research Associate, California Academy of Sciences, 55 Music Concourse Drive, San Francisco, CA 94118, USA.

References

Aguirre-Villaseñor H, Salas-Singh C, Madrid-Vera J, Martínez-Ortiz J, Didier DA, Ebert DA. New Eastern Pacific records of Hydrolagus melanophasma with annotations of a juvenile female. J Fish Biol. 2013;82:714–24.

Andrade I, Pequeño G. First record of Hydrolagus pallidus Hardy & Stehmann, 1990 (Chondrichthyes: Chimaeridae) in the Pacific Ocean, with comments on Chilean holocephalians. Revista de Biología Marina y Oceanografía, 2006; 41(1):111–115.

Angulo A, López MI, Bussing WA, Murase A. Records of chimaeroid fishes (Holocephali: Chimaeriformes) from the Pacific coast of Costa Rica, with the description of a new species of Chimaera (Chimaeridae) from the eastern Pacific Ocean. Zootaxa. 2014;3861(6):554–574.

Barnett LK, Didier DA, Long DJ, Ebert DA. Hydrolagus mccoskeri sp. nov., a new species of chimaeroid fish from the Galapagos Islands (Holocephali: Chimaeriformes: Chimaeridae). Zootaxa. 2006;1328:27–38.

Bustamante C, Flores H, Concha-Pérez, Y. First record of Hydrolagus melanophasma James, Ebert, Long & Didier, 2009 (Chondrichthyes, Chimaeriformes, Holocephali) from the southeastern Pacific Ocean. 2012. 236–242

Compagno LJV, Dagit DD. Hydrolagus trolli. 2006. The IUCN Red List of Threatened Species, Version 2014.2. <http://www.iucnredlist.org/details/60197/0> Downloaded on 11 August 2014.

Didier DA, Kemper JM, Ebert DA. Phylogeny, biology, and classification of extant Holocephali. In: Carrier JC, Musick JA, and Heithaus M.R. (eds) Biology of Sharks and their Relatives 2nd edition. CRC Press, Boca Raton. 2012.

Didier DA, Séret B. Chimaeroid Fishes of New Caledonia with Description of a New Species of Hydrolagus (Chondrichthyes, Holocephali). Cybium. 2002; 26(3):225–33.

Ebert DA. Sharks, Rays, and Chimaeras of California. Berkeley: University of California Press; 2003. p. 284.

Ebert DA, van Hees KE. Beyond jaws: rediscovering the "Lost Sharks" of southern Africa. In: Ebert, D.A., Huveneers, C., & Dudley, S.F.J. (eds). Advances in shark research. African Journal of Marine Science. 2015.37(2):141–156.

Garman S. The Chismopnea (chimaeroids). Mem Mus Comp Zool. 1911;40:79–101.

Gilbert, C.H. Deep sea fishes of the Hawaiian Islands. In: D.S. Jordan & B.W. Evermann. The Aquatic Resources of the Hawaiian Islands. Bulletin of the U.S. Fish Commission for 1903. 1905. 23:575–713.

James KC, Ebert DA, Long DJ, Didier DA. A new species of chimaera, Hydrolagus melanophasma sp. Nov. (Chondrichthyes: Chimaeriformes: Chimaeridae), from the eastern North Pacific. Zootaxa. 2009;2218:59–68.

Kemper JM, Ebert DA, Naylor G, Didier DA. Chimaera carophila (Chondrichthyes: Chimaeriformes: Chimaeridae), a new species of chimaera from New Zealand. Bull Mar Sci. 2015;91(1):63–81.

Last PR, Stevens JD. Sharks and rays of Australia. Melbourne: CSIRO Publishing; 2009. p. 656.

Lundsten L, McClain CR, Barry JP, Cailliet GM, Clague DA, DeVogelaere AP. Ichthyofauna on Three Seamounts off Southern and Central California, USA. Mar Ecol Prog Ser. 2009;389:223–32.

Schlining B, Jacobsen Stout N. MBARI's video annotation and reference system. Proceedings of the Marine Technology Society/Institute of Electrical and Electronics Engineers Oceans Conference. Boston. 2006, MA. pp. 1–5.

10

First record of the coloured righteye flounder, *Poecilopsetta colorata* (Teleostei: Poecilopsettidae) from the Sakalaves seamounts in the Mozambique Channel

Wei-Jen Chen[1*], Jhen-Nien Chen[1], Eve-Julie Pernet[2] and Karine Olu[2]

Abstract

Background: The coloured righteye flounder, *Poecilopsetta colorata* Günther, 1880 was previously known from the eastern Indian Ocean to the South China Sea and Indonesia. Here, a new record from the western Indian Ocean is reported.

Results: The new record is based on a specimen collected on the Sakalaves seamounts at 375 m in depth in the Mozambique Channel during a recent oceanographic survey. Four other teleost fish species including an uncommon ophidiid species, *Neobythites somaliaensis* Nielsen, 1995 were also collected on the same seamounts.

Conclusions: The presence of *P. colorata* in the Mozambique Channel suggests a broad and Indo-West Pacific wide distribution for this relatively rare deep-sea species. The sequence of the cytochrome oxidase subunit-I for the collected specimen is provided as a genetic reference for further DNA barcoding and systematic studies.

Keywords: Fishes, New record, Distribution, Western Indian Ocean, Mozambique Channel, Sakalaves seamounts, COI, PAMELA-MOZ01 cruise

Background

The bigeye flounders of the genus *Poecilopsetta* Günther, 1880 (Poecilopsettidae) (Sakamoto, 1984; Nelson, 2006) include 15 currently recognized species that inhabit the deep seas of the tropics (Munroe, 2015). Seven species of *Poecliopsetta* occur in the Indian Ocean (*P. albomaculata* Norman, 1939, *P. colorata* Günther, 1880, *P. natalensis* Norman, 1931, *P. macrocophala* Hoshino, Amaoka and Last, 2001, *P. normani* Foroshchuk & Fedorov, 1992, *P. praelonga* Alcock, 1894, *P. vaynei* Quéro et al., 1988, and *P. zanzibarensis* Norman, 1939) (Quéro et al., 1988; Hoshino, 2000; Guibord and Chapleau, 2001, 2002; Hoshino et al., 2001; Evseenko, 2004; Kawai and Amaoka, 2006; Kawai et al., 2010). To date, only three species (*P. natalensis*, *P. vaynei*, and *P. zanzibarensis*) have been recorded from the Mozambique Channel in the western Indian Ocean (Fischer and Bianchi, 1984; Quéro et al., 1988; Foroshchuk and Fedorov, 1992; Evseenko, 2004; Hensley, 1997). The coloured righteye flounder, *P. colorata* Günther, 1880, is a rare bathydemersal species living at depths of 214–800 m (Hensley, 1997; Evseenko, 2004). *P. colorata* is currently known from the eastern Indian Ocean to the South China Sea and Indonesia (Hensley, 1997; Evseenko, 2004).

The authors examined 55 fish specimens collected during the 32-day multi-disciplinary cruise (campaign: PAMELA-MOZ01) in 2014 of the R/V *Atalante* deployed by the French Oceanographic Fleet in the Mozambique Channel in the western Indian Ocean. Among them, one specimen was identified as *P. colorata*.

The purpose of the present work is to record this species in the ichthyofauna of the Mozambique Channel and provide a molecular sequence from a mitochondrial gene as the genetic reference for further DNA barcoding and systematic studies.

* Correspondence: wjchen.actinops@gmail.com
[1]Institute of Oceanography, National Taiwan University, No.1 Sec. 4 Roosevelt Road, Taipei 10617, Taiwan
Full list of author information is available at the end of the article

Methods

The materials described in the present paper were collected during the cruise PAMELA-MOZ01 of the PAMELA project in 2014 conducted by the R/V *Atalante* (Olu, 2014) on the collection sites from three of the explored zones, the slope of the Glorieuses islands, the slope of the Mahajanga basin off Madagascar, and the Sakalaves seamounts on the southern Davie ridge (Fig. 1). The Warén dredge, NIWA seamount sledge and beam trawl were used for sampling at a total of eight sites. The geographic coordinates of the sites, depths, and methods for the deployments are listed in Table 1, and the collected samples are listed in Additional file 1 : Table S1. The specimens examined were deposited in the National Natural History Museum of Paris (MNHN). Muscle tissue samples excised from the specimens for genetic studies were preserved in 95 % ethanol and stored at −20 ° C in the Marine Biodiversity and Phylogenomics laboratory at the Institute of Oceanography, National Taiwan University, Taipei with tissue identification numbers from WIO 001 to WIO 055 (Additional file 1 : Table S1).

Identification and methods for taking counts and measurements of the specimen generally followed Hensley (1997) and Quéro et al. (1988) for the pleuronectiform fishes. All other specimens were identified according to the following taxonomic references: Fischer and Bianchi (1984) (most fishes); Nielsen (1969, 2002) and Nielsen et al. (1999) (Ophidiiformes); Cohen et al. (1990) (Macrouridae).

Whole genomic DNA was extracted from the tissue sample (WIO 041) of the *P. colorata* specimen using an automated extractor: *LabTurbo 48 Compact System* and *LGD 480–500* kits (Taigene Biosciences Corp.) following the manufacturer's protocol. A fragment of the mitochondrial protein-coding gene cytochrome oxidase subunit I (*COI*) was amplified and sequenced for this study. Protocols for collecting molecular data follow those outlined in Ward et al. (2005). Six available *COI* sequences from two congeneric species (*P. natalensis* [$n = 5$] and *P. hawaiiensis* [$n = 1$]) were retrieved from Genbank and compared with our obtained sequence. The sequence alignment was conducted manually using Se-Al v2.0a11 (Rambaut, 2002). The variable nucleotide sites and genetic distance (uncorrected pairwise *p*-distance) among sequences and were calculated using PAUP* (Swofford, 2002).

Results

A total of 55 collected samples were examined. Among them, 49 specimens belong to the 15 following recognized teleost species: *Coloconger raniceps* Alcock, 1889 (Colocongridae) ($n = 6$), *Coloconger scholesi* Chan, 1967 (Colocongridae) ($n = 1$), *Hoplostethus melanopus* (Weber, 1913) (Trachichthyidae) ($n = 3$), *Nezumia semiquincunciata* (Alcock, 1889) (Macrouridae) ($n = 3$), *Ventrifossa johnboborum* Iwamoto, 1982 (Macrouridae) ($n = 2$), *Lophiodes triradiatus* (Lloyd, 1909) (Lophiidae) ($n = 1$), *Neoscopelus macrolepidotus* Johnson, 1863 (Neoscopelidae) ($n = 20$), *Aldrovandia affinis* (Günther, 1877) (Halosauridae) ($n = 1$), *Aldrovandia phalacra* (Vaillant, 1888) (Halosauridae) ($n = 1$), *Barathronus diaphanus* Brauer, 1906 (Aphyonidae) ($n = 1$), *Monomitopus conjugator* (Alcock, 1896) (Ophidiidae) ($n = 2$), *Neobythites somaliaensis* Nielsen, 1995 (Ophidiidae) ($n = 3$), *Pentaceros capensis* Cuvier, 1829 (Pentacerotidae) ($n = 1$), *Poecilopsetta colorata* (Poecilopsettidae) ($n = 1$), and *Paratriacanthodes retrospinis* Fowler, 1934 (Triacanthodidae) ($n = 3$). Six other specimens can only be identified to the genus *Symphurus* (Cynoglossidae) in two morpho-types ($n = 4$ and 2, respectively) based on the available keys of the species identification. Among the identified species, all expect one (described below) have been recorded in the western Indian Ocean.

Poecilopsetta colorata Günther, 1880, new record (Fig. 2)

Material examined

MNHN2016-0180, one specimen, Sakalaves mounts (Fig. 1), 18°0.08589'S 41°46.32208'E, 375 m depth, R/V *Atalante*, NIWA seamount sledge, Station MOZ1_DN5 (Table 1), 100 mm standard length (SL), 44.3 mm body depth, 25.3 mm head length, 7.9 mm upper-jaw length, lateral line scales: ca. 102–105, dorsal soft fin rays: 58,

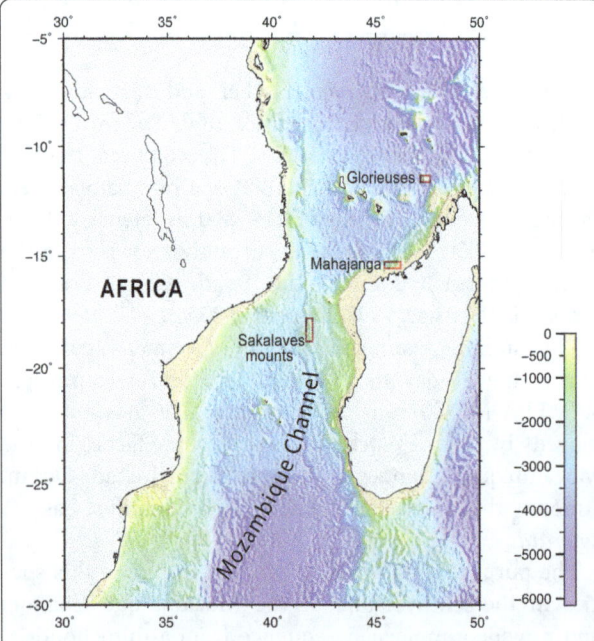

Fig. 1 Bathymetric map of the southwest Indian Ocean region indicating the three explored zones (*red boxes*) where the specimens examined in this study were collected. Depths are in meters

Table 1 Information of the operations during the campaign of PAMELA-MOZ01

Code of operation	Date	Zone	Latitude	Longitude	Type of operation	Depth (m)
MOZ1_DW1	28/09/2014	Glorieuses	11°22.75604'S	47°16.40977'E	Warén dredge	789
MOZ1_CP1	4/10/2014	Mahajanga	15°21.46148'S	45°59.28908'E	Beam trawl	722
MOZ1_CP2	8/10/2014	Mahajanga	15°21.72712'S	45°57.65218'E	Beam trawl	869
MOZ1_CP3	8/10/2014	Mahajanga	15°21.71867'S	45°57.52781'E	Beam trawl	971
MOZ1_CP4	9/10/2014	Mahajanga	15°31.00558'S	45°41.95738'E	Beam trawl	806
MOZ1_DN4	14/10/2014	Sakalaves mounts	18°0.07689'S	41°46.31995'E	NIWA seamount sledge	376
MOZ1_DW4	14/10/2014	Sakalaves mounts	18°0.06847'S	41°46.3343'E	Warén dredge	376
MOZ1_DN5	14/10/2014	Sakalaves mounts	18°0.08589'S	41°46.32208'E	NIWA seamount sledge	375

anal soft fin rays: 48, pectoral fin rays: 10 (8 on blind side), pelvic fin rays: 6, number of nucleotides of the obtained *COI* sequence (Genbank accession number: KX611099): 648.

Diagnosis

The counts of dorsal and anal fin rays and lateral line scales are considered as key features for diagnosing species of *Poecilopsetta* (Hoshino et al., 2001; Kawai et al., 2010). These counts in our examined specimen were 58, 48, and ca. 102–105, which fall into the ranges of the three characters for *P. colorata* (55–62, 46–53, and 90–124) and *P. praelonga* (57–65, 45–55, and 91–113) as being described in Kawai et al. (2010). The former species can be easily distinguished from the latter in having a deeper body (body depth 1.9 to 2.6 times in SL vs. body depth 3.8 to 4 times in SL) and a longer upper-jaw (upper-jaw length 3 to 3.5 times in head length vs. upper-jaw length 3.6 to 3.7 times in head length) (Hensley, 1997). Our specimen was diagnosed with body depth 2.3 times in SL and upper-jaw length 3.2 times in head length. All

these characteristics combined together indicate our specimen is *P. colorata*.

From the aligned *COI* sequences for the samples of the three *Poecilopsetta* species included in this study (see methods), 28 variable sites were observed along the 684-bp long sequenced fragments (Fig. 3). This represents 2.76 % interspecific nucleotide divergence on average (evaluated by uncorrected pairwise p-distance). The nucleotide divergence among five *P. natalensis* samples collected from South Africa ($n = 4$) and from the South China Sea ($n = 1$) was estimated to be 0.64 %. Diagnostic nucleotides of *P. colorata* to *P. natalensis* are site numbers 046, 126, 225, 252, 315, 360, 420, 444, 477, 510, 513, 565, 612, 618, 621, and 639. Diagnostic nucleotides of *P. colorata* to *P. hawaiiensis* are site numbers 210, 225, 313, 315, 369, 390, 477, 505, 510, 513, 537, 549, 561, 594, 603, 618, 621, and 678 (Fig. 3). The available genetic data further confirm that our specimen is distinguishable from the co-occurring species from the region, *P. natalensis*.

Fig. 2 *Poecilopsetta colorata* Günther, 1880, MNHN2016-0180 from the Sakalaves seamount, collected on Oct. 14, 2014. Standard length 100 mm

Species	Nucleotide position																											
Samples, locality, GenBank no.	0	1	2	2	2	3	3	3	3	3	4	4	4	5	5	5	5	5	5	5	5	6	6	6	6	6	6	6
	4	2	1	2	5	1	1	6	6	9	2	4	7	0	1	1	3	4	6	6	9	0	1	1	2	3	7	7
	6	6	0	5	2	3	5	0	9	0	0	4	7	5	0	3	7	9	1	5	4	3	2	8	1	9	5	8
P. natalensis																												
HQ945815, South Africa	?	C	A	T	G	C	A	T	G	C	A	T	G	A	T	A	T	A	C	C	C	C	G	C	A	T	A	T
HQ945811, South Africa	?	.	.	.	.	.	.	.	.	.	.	.	.	.	.	.	.	.	.	.	.	.	.	.	.	.	.	.
GU804926, South Africa	?	.	.	.	.	.	.	.	.	.	.	.	.	.	.	.	.	.	.	.	.	.	.	.	.	.	G	.
GU804911, South Africa	?	.	.	.	.	.	.	.	.	.	.	.	.	.	.	.	.	.	.	.	.	.	.	.	.	.	.	.
JQ700099, South China Sea	C	.	.	.	.	.	.	.	.	.	.	.	.	.	.	.	.	.	.	.	.	.	.	.	.	.	.	.
P. colorata																												
KX611099, Mozambique Channel	T	T	.	C	A	.	G	C	.	.	G	C	A	.	C	G	.	.	.	T	.	.	T	T	G	C	.	.
P. hawaiiensis																												
DQ521023, Hawaii	?	T	G	.	A	T	.	C	A	T	G	C	.	G	.	.	C	G	T	T	T	T	T	.	.	C	.	A

Fig. 3 Polymorphic nucleotide sites at the cytochrome c oxidase subunit-I locus in three *Poecilopsetta* species. Numeration of nucleotide sites starts from first nucleotide of the corresponding gene. Genbank accession numbers of the sequences are given

Remarks

Based on the new record in this study, the distribution of the species extends to the western Indian Ocean from its previously reported area. The four other teleost fish species also collected from the same seamounts are: *Neobythites somaliaensis* (Ophidiidae), *Pentaceros capensis* (Pentacerotidae), *Symphurus* sp. 2 (Cynoglossidae), and *Paratriacanthodes retrospinis* (Triacanthodidae) (Additional file 1: Table S1). *N. somaliaensis* is an uncommon ophidiid species that was described based on specimens collected on the upper continental slope in the Gulf of Aden (Nielsen, 1995). The present record in the Mozambique Channel is new. This species is most similar to the common *Neobythites* species, *N. analis* Barnard 1927, from this region (Nielsen et al. 1999; 2002); it differs from *N. analis* by the distal parts of both dorsal and anal fins being black (Fig. 4).

Discussion

Although some flatfishes including poecilopsettids have large and presumably long-lived larvae that could enhance the probability of achieving long-distance dispersal over large areas (Evseenko, 2000), widespread species crossing two oceans are rare in poecilopsettids (Munroe, 2005). In *Poecilopsetta*, *P. colorata* and *P. praelonga* are the only two species known to have a wide distribution ranging from the eastern Indian Ocean to the western Pacific Ocean (Quéro et al., 1988; Hensley, 1997). The record of *P. colorata* in the western Indian Ocean presented here confirms a wide and an extended distribution for this poorly known deep-sea species. The African righteye flounder, *P. natalensis*, also occurs in the western Indian Ocean. The extension of its distribution into the western Pacific has been suspected because of an unconfirmed record reported in Taiwan (Hensley, 1997). In this study, one of our compared *COI* sequences of *P. natalensis* was from the South China Sea (Fig. 3), confirming the presence of this species in the western Pacific Ocean. It is worthy to mention that the genetic distance between this sample of *P. natalensis* and others from South Africa is very low, from zero to 0.64 % of nucleotide divergence (corresponding to a single nucleotide difference), despite the large distance between the two sampling sites (Fig. 3). A genetic break corresponding to the geology that separates the Indian population from the Pacific one is often present in widespread marine Indo-West Pacific species (Borsa et al., 2016); it was not observed in *P. natalensis*.

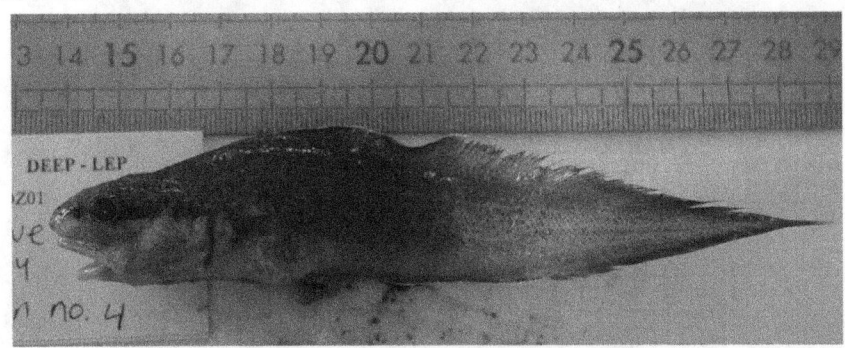

Fig. 4 *Neobythites somaliaensis* Nielsen, 1995, MNHN2016-0176 from the Sakalaves seamount, collected on Oct. 14, 2014. Standard length 140 mm

Conclusions

The first record of *P. colorata* from the Sakalaves seamounts on the southern Davie Ridge in the Mozambique Channel is reported. This record extends the known range of *P. colorata* to the western Indian Ocean. The occurrences of this species elsewhere include the Bay of Bengal to the South China Sea and Indonesia. The individuals are known to live at depths of 214 to 800 m (375 m in this study). The *COI* sequence for the collected *P. colorata* specimen is also provided as a genetic reference. Its availability can permit DNA barcoding work for fish identification purposes and future systematic or other advanced research such as biogeography or studying the evolutionary dynamics of species across the oceans.

Abbreviations

R/V: Research vessel; SL: Standard length; *COI*: Cytochrome c oxidase subunit-l; DNA: Deoxyribonucleic acid; Bp: Base pair

Acknowledgements

Our gratitude goes to the crews of the R/V *Atalante* and participants of the oceanographic cruise (campaign: PAMELA-MOZ01, PI: K. Olu, part of the Mozambique 2014 study) involved in organizing the survey and the capture of the samples. The Mozambique 2014 study is co-funded by TOTAL and Ifremer as part of the PAMELA (Passive Margin Exploration Laboratories) scientific project. The PAMELA project is a scientific project leaded by Ifremer and TOTAL in collaboration with Université de Bretagne Occidentale, Université Rennes 1, Université Pierre and Marie Curie, CNRS and IFPEN. We are grateful to Stephan Jorry as PI of the Mozambique 2014 study and for providing bathymetric data of the sampling sites (Ptolemée cruise: http://dx.doi.org/10.17600/14000900). We thank Anne Basseres (TOTAL, co-PI of Ecosystem studies in Pamela), Philippe Bourges, Jean-Nöel Ferry (TOTAL) and Jean-Fançois Bourillet as PIs of PAMELA for the management of the project. We acknowledge Inge van den Beld for her help onboard for biological samples, Philippe Noël for its help with sampling gears, J.-F. Barazer and M. Clark for the drawings and advices of the Warén dredges and Niwa seamount sledge. We thank Laure Corbari for their involvement in the collaboration between WJC and Ifremer. We are grateful to L.-H. Chen for improving artwork, C.-Y. Chang who helped us to prepare Fig. 1 using Generate Mapping Tool (GMT), and M. A. Campbell for language editing.

Funding

Ministry of Science & Technology, Taiwan (MOST 102-2923-B-002 -001 -MY3) to WJC.

Authors' contributions

WJC contributed to the conception and design of the work, analyzed, interpreted the data, and wrote the paper; JNC collected the data and wrote the paper; JEP collected and managed the samples; KO led the cruise of sample collection and revised the paper. All authors read and approved the final manuscript.

Competing interests

The authors declare that they have no competing interests.

Author details

[1]Institute of Oceanography, National Taiwan University, No.1 Sec. 4 Roosevelt Road, Taipei 10617, Taiwan. [2]Département REM/EEP/Laboratoire Environnement Profond, IFREMER/Centre de Bretagne, Institut Carnot EDROME, 29280 Plouzané, France.

References

Borsa P, Durand JD, Chen W-J, Nicolas H, Muths D, Mou-Tham G, Kulbicki M. Comparative phylogeography of the western Indian Ocean reef fauna. Acta Oecol. 2016;72:72–86.

Cohen DM, Inada T, Iwamoto T, Scialabba N. Gadiform fishes of the world (Order Gadiformes). An annotated and illustrated catalogue of cods, hakes, grenadiers and other gadiform fishes known to date. FAO Fish Synop Rome. 1990;125(10):90–310.

Evseenko SA. Family Achiropsettidae and its position in the taxonomic and ecological classifications of Pleuronectiformes. J Ichthyol. 2000;40:S110–38.

Evseenko SA. Family Pleuronectidae Cuvier 1816 - righteye flounders. Cali Acad Sci Annotated Checklists of Fishes. 2004;37:1–37.

Fischer W and Bianchi G. editors. FAO species identification sheets for fishery purposes. Western Indian Ocean (Fishing Area 51). FAO Rome. 1984; Vol.2-6.

Foroshchuk VP, Fedorov VV. *Poecilopsetta normani* – a new species of flounder (Pleuronectidae) from the Saya de Malha Bank, Indian Ocean. J Ichthyol. 1992;32(7):37–44.

Guibord AC, Chapleau F. *Poecilopsetta dorsialta*: a new species of Poecilopsettidae (Pleuronectiformes) from the Pacific Ocean. Copeia. 2001; 2001(4):1081–6.

Guibord AC, Chapleau F. Poecilopsetta megalepis Fowler 1934, un synonyme junior de *Poecilopsetta plinthus* (Jordan and Starks, 1904) (Pleuronectiformes: Poecilopsettidae). Cybium. 2002;26:135–9.

Hensley DA. Pleuronectidae. Righteye flounders. In: Carpenter KE, Niem V, editors. FAO Identification Guide for Fishery Purposes, vol. 6. Rome: The Western Central Pacific; 1997. p. 3863–77.

Hoshino K. Redescription of a rare flounder, *Poecilopsetta inermis* (Breder) (Pleuronectiformes: Pleuronectidae: Poecilopsettinae), a senior synonym of *P. albomarginata* Reid, from the Caribbean Sea and tropical western Atlantic. Ichthyol Res. 2000;47:95–100.

Hoshino K, Amaoka K, Last P. A new dextral flounder, *Poecilopsetta macrocephala* (Pisces: Pleuronectiformes: Poecilopsettidae), from northwestern Australia. Species Divers. 2001;6:73–81.

Kawai T, Amaoka K. A new righteye flounder, Poecilopsetta pectoralis (Pleuronectiformes: Poecilopsettidae), from New Caledonia. Ichthyol Res. 2006;53:264–8.

Kawai T, Amaoka K, Séret B. A new righteye flounder, *Poecilopsetta multiradiata* (Teleostei: Pleuronectiformes: Poecilopsettidae), from New Zealand and New Caledonia (South-West Pacific). Ichthyol Res. 2010;57(2):193–8.

Munroe TA. Distributions and Biogeography. In: Gibson RN, Nash RDM, Geffen AJ, Van der Veer HW, editors. Flatfishes: Biology and Exploitation. Oxford, UK: Blackwell Science Ltd; 2005. doi:10.1002/9780470995259.ch3.

Munroe TA. Chapter 2. Systematic diversity of the Pleuronectiformes. In: Gibson RN, Nash RM, Geffen AJ, Van der Veer HW, editors. Flatfishes Biology and exploitation 2nd edition. Fish and Aquatic Resources (Series 16). UK: Wiley; 2015. p. 13–51.

Nelson JS. Fishes of the World, fourth ed. New York: Wiley; 2006. p. 442–51.

Nielsen JG. Systematics and biology of the Aphyonides (Pisces, Ophidiodea). Galathea Report. 1969;10:7–90. Pls.1-4.

Nielsen JG. A review of the species of the genus *Neobythites* (Pisces: Ophidiidae) from the western Indian Ocean, with description of seven new species. Ichth Bull. 1995;62:1–19.

Nielsen JG. Revision of the Indo-Pacific species of *Neobythites* (Teleostei, Ophidiidae), with 15 new species. Galathea Report. 2002;19:5–104.

Nielsen JG, Cohen DM, Markle DF, Robins CR. Ophidiiform fishes of the world (Order Ophidiiformes). An annotated and illustrated catalogue of pearlfishes, cusk-eels, brotulas and other ophidiiform fishes known to date. FAO Fish Synop Rome. 1999;125(18):76–87.

Olu K. PAMELA-MOZ01 cruise. RV L'Atalante. 2014. doi: http://dx.doi.org/10.17600/14001000.

Quéro JC, Hensley DA, Maugé AL. Pleuronectidae de l'île de la Réunion et de Madagascar. I. Poecilopsetta. Cybium. 1988;12:2–11.

Rambaut A. Sequence Alignment Editor Version 2.0a11. 2002. http://tree.bio.ed.ac.uk/software/seal/

Sakamoto K. Interrelationships of the family Pleuronectidae (Pisces, Pleuronectiformes). Mem Fac Fish Hokkaido Univ. 1984;31:95–215.

Swofford DL. PAUP*: Phylogenetic Analysis Using Parsimony (* and Other Methods), Version 4. Sunderland, Massachusetts: Sinauer Associates; 2002.

Ward RD, Zemlak TS, Innes BH, Last PR, Hebert PDN. Barcoding Australia's fish species. Phil Trans R Soc B. 2005;360:1847–57.

11

The first definitive record of the giant larvacean, *Bathochordaeus charon,* since its original description in 1900 and a range extension to the northeast Pacific Ocean

R. E. Sherlock[*], K. R. Walz and B. H. Robison

Abstract

Background: Larvaceans in the genus *Bathochordaeus* are large, often abundant filter feeders found throughout much of the world ocean. The first described species, *Bathochordaeus charon*, was reported over 100 years ago by Chun. However in the time since, few specimens have matched Chun's original description, resulting in ambiguity on the validity of *B. charon* as a species.

Methods: Specimens of *Bathochordaeus charon* were identified based on morphological traits, molecular data and observations made on high definition video.

Results: The first records of *Bathochordaeus charon* from the northeast Pacific Ocean off central California and Oregon, USA are reported. Morphology and molecular data clearly distinguish *B. charon* from its congener, *B. stygius.*

Conclusions: This paper establishes the first review of *Bathochordaeus charon* since its original description, extends the range of this species to the northeast Pacific Ocean, and provides the first molecular evidence for two species of *Bathochordaeus.*

Keywords: Appendicularian, Giant larvacean, Larvacean, Molecular, Monterey Bay, Morphology, Oikopleuridae, Taxonomy, Tunicata, Urochordata

Background

Chun (1900) provided the first description of a very large larvacean from specimens collected during the *Valdivia* expedition (1898–1899). He named the new species *Bathochordaeus charon* after the mythical figure who ferries the souls of the dead across the river Styx. During the *Valdivia* expedition two individuals were collected from the South Atlantic and two smaller specimens, from the Indian Ocean. When he studied them later, Lohmann (1914) placed the two largest *Valdivia* specimens in the family Oikopleuridae. Garstang (1936) later referred to them as "veritable giants among Appendicularians, the depressed body being as large as a walnut and the broad tail almost 3 ins. in length" and

his comment may be responsible for them being referred to today simply as "giant larvaceans". The two smaller Indian Ocean specimens were not described until much later (Lohmann 1931). Not only were the Indian Ocean specimens smaller at <20 mm total length, they appeared to differ appreciably from the larger specimens collected in the Benguela Current (Lohmann 1914; Lohmann 1931; Garstang 1936). When Garstang (1936, 1937) collected two specimens of *Bathochordaeus* in very good condition from surface waters near Bermuda, he at once observed that they differed appreciably from Chun's. Garstang's specimens were smaller in size than the two collected by Chun in the South Atlantic but similar in size to the larger of the two collected in the Indian Ocean. However, three features struck Garstang (1936, 1937) as different: the Bermuda specimens lacked the prominent "obconical gill-pouches"

* Correspondence: robs@mbari.org
Monterey Bay Aquarium Research Institute (MBARI), Moss Landing, CA 95039, USA

(spiracles), and the crop-like esophageal expansion of *Bathochordaeus charon* and they possessed an oikoplastic region that was more comparable to other oikopleurids than *B. charon* as described by Chun (1900) and Lohmann (1914).

Consequently, Garstang wondered if the original description was hampered by misinterpretation or poor preservation (Garstang 1936, 1937). He could not fathom a purpose for either the seemingly counter-productive funnel-shape of the spiracles or the improbable and capacious esophageal expansion. Having no access to the original specimens, and unable to account for the absence of these conspicuous features, Garstang somewhat reluctantly described a new species: *Bathochordaeus stygius*, and greatly expanded the slowly growing body of work surrounding these enigmatic animals (Garstang 1937).

Since its description, *Bathochordaeus charon* has appeared in the literature a few times (Thompson 1948; Bückmann and Kapp 1973; Barham 1979; Galt 1979; Castellanos et al. 2009; Lindsay et al. 2015), without specimen collections to accompany them. The lack of specimens combined with Garstang's (1937) concerns about the characteristic features in the original description, have cast doubt on the legitimacy of *B. charon* (Chun) as a species distinct from *B. stygius* (Garstang). The first probable record of *B. charon* since Chun (1900) appeared in 1948, from the Pacific Ocean off Australia, although it was a single, small (trunk 3.2 mm, tail 7.5 mm) specimen (Thompson 1948). Time passed and the lack of specimens caused Fenaux (1966) to synonymize the two species. Subsequently, the second potential specimen(s) of *B. charon* were collected by Galt (1979): five animals acquired during three cruises off southern California, that he called *B. charon*. Unfortunately, those specimens are no longer available and Galt may have used the name *B. charon* for all his specimens *in lieu* of *B. stygius*, since the latter was suggested to him as applying to juvenile specimens (Bückmann and Kapp 1975; Galt 1979 and pers. comm.) and Galt's specimens were on the order of 3–6 cm in total length. In his description of *B. charon* Galt wrote "The present specimens conform generally to published accounts of *B. charon*, detailed descriptions of which were given by Chun (1900), Lohmann (1931)), Garstang (1937 as *B. stygius*)," indicating that Galt likely deferred to Fenaux's synonomy of the species in referring to them all as *B. charon*. More recently, a specimen called *B. charon* was collected by Castellanos et al. (2009)) but no description was provided and, although a photograph was included neither the large spiracles nor esophageal expansion are visible. Lindsay et al, (Lindsay et al. 2015) provide an in situ ROV image of the house of "Bathochordaeus sp. A" observed off the Nansei Island chain of Japan, but the structure of the inner filter differs markedly from the *B. charon* and *B. stygius* we have observed, and that larvacean was not collected.

Giant larvaceans, like other species of larvaceans, use their oikoplastic cells to secrete complex filters or 'houses' that allow them to concentrate and feed on particles (Lohmann 1933; Alldredge 1977; Morris and Deibel 1993; Flood et al. 1998). A house consists of a large, diaphanous outer structure as well as a smaller, more convoluted and bi-lobed inner structure that functions as a filter. Together these serve to concentrate appropriately-sized food particles from the surrounding water. The outer part of the house excludes larger material that would clog the inner filter. Thus, the outer structure often acquires a covering of marine 'snow' that can alter its size and shape (Hamner and Robison 1992; Silver et al. 1998). The inner filter concentrates food particles of ingestible size and are ultimately connected to the animal's mouth via a tube made of the same material as the rest of the structure. However, the inner filter is less diaphanous and more stereotypical in shape, often retaining that shape long after the animal has left its house (Robison et al. 2005).

In situ, larvaceans in the genus *Bathochordaeus* are often visible from several meters away because their houses may span a meter in longest dimension (Hamner and Robison 1992; Robison et al. 2005). When Barham (1979) made the first observations of the occupied houses of large larvaceans he was diving by bathyscaph and saucer in the Pacific Ocean, off southern California and Mexico. He called them "giant" or "large" larvacean houses and inferred they were probably "*Bathochordaeus charon*". However, he recalled, "seeing at least five types of large larvacean houses" and gave "general descriptions of three types". It is not clear he thought all types belonged to *B. charon*. One animal was collected, identified by Donald P. Abbott as a "larvacean" and the specimen was subsequently lost. Intraspecific differences in *Bathochordaeus*' house structure were not known at the time. Resolving the shape of the spiracles and esophagus of *Bathochordaeus* spp. from the portholes of his submersibles seems unlikely, since the larvaceans themselves were often invisible. Barham's dives occurred 10 years prior to his publication and it seems more likely that he used the name *B. charon* because Fenaux (1966) had just synonomized the species and it was the correct name to use.

Almost 30 years after he synonymized the two species, Fenaux apparently acquired specimens collected by manned submersibles that proved to him the validity of both species (Fenaux 1993, 1998). Regrettably, he never published that proof and in the current literature it remains unclear if more than one species of *Bathochordaeus* exists (Hopcroft 2005; Flood 2005).

Using remotely operated vehicles (ROVs) we have carefully observed and collected *B. charon* (Chun) as well as *B. stygius* (Garstang). A combination of morphological features, house structures and molecular evidence clearly distinguish the two species and provide the first records of *B. charon* from Monterey Bay as well as off the coast of Oregon, expanding its range into the eastern North Pacific Ocean (Fig. 1).

Methods

Specimens were collected with MBARI ROVs (remotely operated vehicles) using either detritus samplers or gentle suction (Robison 1995), and preserved in 5 % formalin buffered with sodium tetraborate. The measurements presented here were made on preserved specimens.

Prior to microscopy, specimens were rinsed in seawater then exposed to DAPI (4′,6-diamidino-2-phenylindole) for 2 min, then rinsed again before observation. The oikoplastic cells that generate the mucus feeding filters of *Bathochordaeus* are many times polyploid (Flood 2005). Since DAPI binds strongly to DNA and fluoresces brightly when illuminated with UV light (Russell et al. 1975), the preparation enhances contrast in the oikoplastic region of these animals. The oikoplastic region of *Bathochordaeus* is a monolayer of cells that undulate over the trunk – it is not flat. Wanting to preserve the animal we had for a type-specimen, we chose not to dissect it. Instead, we took images at 12 focal planes, then stacked them together using Adobe Photoshop CS 6. Micrographs were taken with an Olympus DP-71 camera, mounted to a Nikon SMZU dissecting microscope.

For molecular work, animals were frozen in liquid nitrogen and stored at –80 ° C until extraction. DNA samples were extracted from tissue using the DNeasy® Kit (Qiagen, Valencia, CA USA) according to the

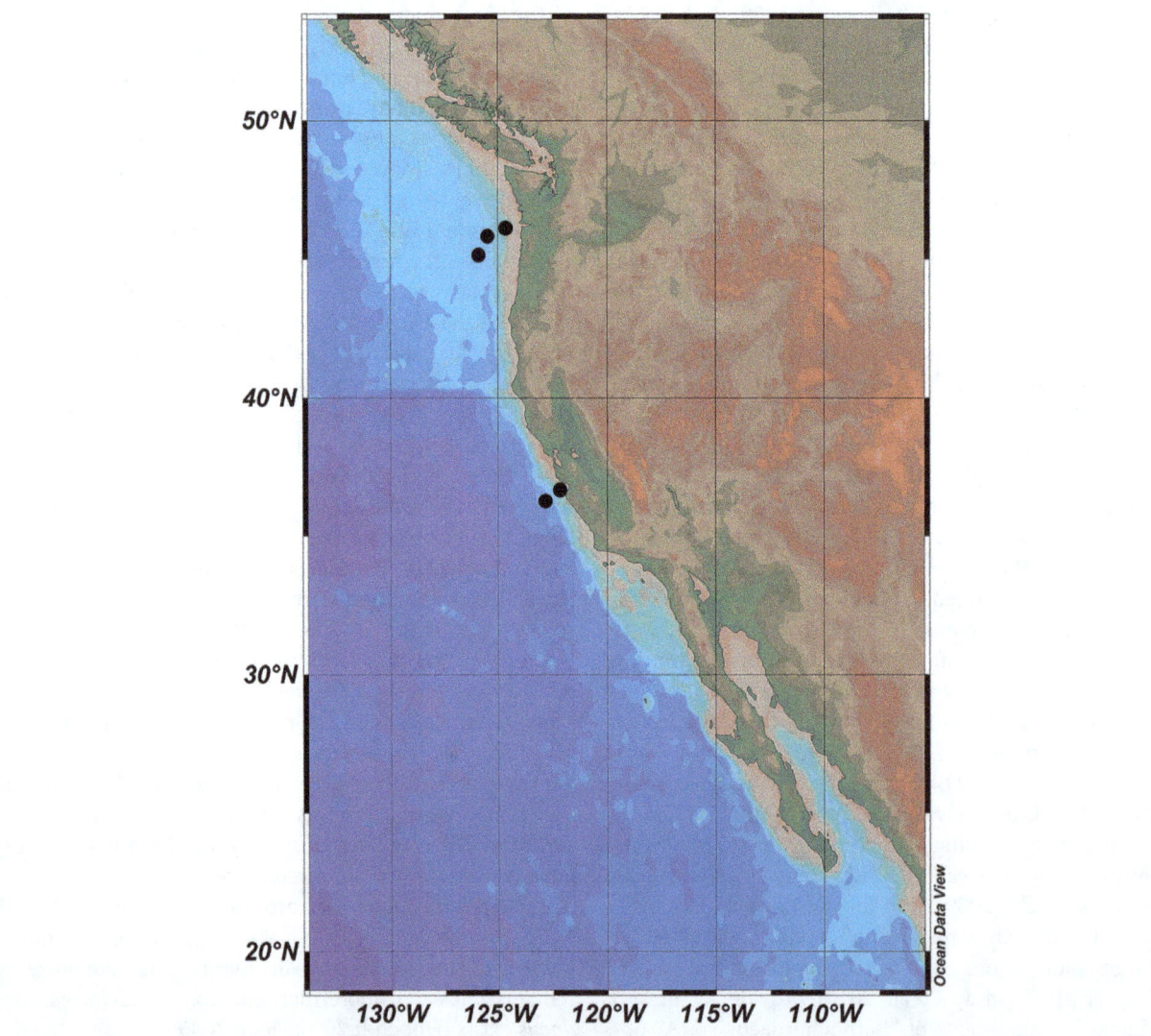

Fig. 1 The eastern North Pacific Ocean, with *black circles* marking the locations where *Bathochordaeus charon* has been observed and/or collected (for waypoints, see Table 1)

manufacturer's instructions. Traditional DNA barcoding primers such as those for cyctochrome c oxidase 1 (Folmer et al. 1994), 12S, cytochrome b, and H3 did not amplify genes for *Bathochordaeus spp*. Instead we selected two primers used by Hirose and Hirose (2009), that were constructed specifically for tunicates, doliolids, salps and larvaceans:

5′ – CATTTWTTTTGATTWTTTRGWCATCC NGA–3′ (UroCox1-244 F)

5′ – GCWCYTATWSWWAAWACATAATGAAAR TG–3′ (UroCox1-387R).

These primers amplified a 400-base pair section of the mitochondrial cytochrome c oxidase subunit I gene (COI) with the following PCR parameters: 35 cycles of 94 ° C for 1 min, 40 ° C for 1 min, 72 ° C for 1 min. Amplification of an 1800-base-pair fragment of small subunit ribosomal DNA (18S) was conducted using the modified universal primers *mitchA* and *mitchB* from Medlin et al. (1988) with the following PCR parameters: 4 cycles of 94 ° C for 1 min, 58 ° C for 1 min stepping 0.1 ° C/second to 72 ° C, 72 ° C for 2 min; followed by 29 cycles of 94 ° C for 1 min, 64 ° C for 1 min, 72 ° C for 1:30 min (Medlin et al. 1988). All products were bi-directionally sequenced using BigDye® Terminator v3.1 Cycle Sequencing Kit on an ABI 3100 or ABI3500 sequencer (Applied Biosystems, Foster City, CA USA). Sequences were edited and aligned using Geneious version 6.0.5 created by Biomatters (Auckland, New Zealand, http://www.geneious.com/), and submitted to GenBank (Accession Numbers: KT881543–KT881545). Genetic distance was calculated using p-distance model in MEGA v6.0 software (Tamura et al. 2013).

High definition (HD) video sequences of larvaceans taken by MBARI ROVs were used to compare the feeding structures ('houses') of *B. charon* and *B. stygius* as well as the morphology of the larvaceans occupying the houses.

Material examined

Physical descriptions of *Bathochordaeus charon* are based on two specimens collected in 2013 from Monterey Bay, CA (D457, D548): at 36.688795 N -122.043871 W and 36.540935 N -122.520577 W; collection depth: 281 m and 598 m (Table 1). These specimens are now at the Smithsonian Institution's National Museum of Natural History (USNIM# 1251907, 1251906). High definition (HD) video from the archive at the Monterey Bay Aquarium Research Institute (MBARI) provided records of 11 more individuals observed from 2006–2013, but not sampled (Table 1). Frozen tissue for molecular work came from two specimens collected in 2013 (D449, D548). A third specimen of *B. charon*, collected from Monterey Bay in 2007 was highly parasitized by ciliates and was not recognized initially as being different from *B. stygius*. This specimen was frozen for subsequent study of its symbionts.

Comparative material examined

No comparative material exists for *Bathochordaeus charon*. *In lieu* of the original specimens described by Chun (1900), figures from *Aus Den Tiefen des Weltmeeres*, the publication from the expedition of the *Valdivia*, are reprinted for comparison (Chun 1900).

Table 1 Specimens of *Bathochordaeus charon* collected or observed

Date	Dive #	ROV	Depth (m)	Lat.	Long.	Location or Smithsonian acquisition #
24-Aug-2006	1024	Tiburon	293	46.15775	−124.790794	N/A
11-Jul-2007	3051	Ventana	236	36.692566	−122.0462	N/A
6-Aug-2007	1112	Tiburon	336	36.34012	−122.90101	MBARI
27-Jul-2009	54	Doc Ricketts	267	45.917487	−125.499868	N/A
29-Jul-2009	56	Doc Ricketts	233	45.151215	−125.91437	N/A
15-Mar-2010	3535	Ventana	255	36.748188	−122.10306	N/A
17-Mar-2010	3538	Ventana	135	36.69935	−122.05209	N/A
15-Apr-2010	146	Doc Ricketts	269	36.702025	−122.047983	N/A
18-Nov-2010	215	Doc Ricketts	286	36.747942	−122.104007	N/A
30-Mar-2011	3614	Ventana	297	36.750654	−122.103026	N/A
31-Mar-2011	3616	Ventana	281	36.732031	−122.040878	N/A
20-Sep-2011	3647	Ventana	261	36.69852	−122.032776	N/A
23-Mar-2013	449	Doc Ricketts	233	36.701236	−122.060411	MBARI
28-Mar-2013	457	Doc Ricketts	281	36.688795	−122.043871	1251907
11-Nov-2013	548	Doc Ricketts	598	36.540935	−122.520577	1251906

Collection records. N/A means that the animal was recorded on high-definition (HD) video. Two specimens were granted to the Smithsonian Institution, National Museum of Natural History

Results and discussion

In March of 2013 a large and fecund specimen was hastily collected by the ROV *Doc Ricketts* as the vehicle was ascending for recovery. At the time, we assumed this large animal was *Bathochordaeus stygius,* a common local species. Since the larvacean spent less than 30 min in the detritus sampler, the individual remained in pristine condition and was preserved immediately, without microscopic observation. Some months later while making measurements, we realized that the larvacean was markedly different from *B. stygius.* Surprisingly, the specimen resembled Chun's (1900) original description of *B. charon* (Fig. 2).

Morphology

Key diagnostic features of *Bathochordaeus charon* are (letters in parentheses refer to structures in Figs. 2, 3 and 4):

- Two funnel-shaped spiracles with the inner, pharyngeal opening many times smaller than the outer, ventral opening (Fig. 3).
- A large expansion of the esophagus (Figs. 1, 2 and 3). The function is unknown, but size and position are reminiscent of a crop to aid in digestion.
- The oikoplastic regions of both *B. charon* and *B. stygius* have paired bands of 12 giant cells (fp), with much smaller trap cells (t) just in front of them (Fig. 4). Garstang (1937) referred to these as "Lohmann's colloplasts" or "glandular crescents" and they are some of the largest and most conspicuous cells in the oikoplastic region. Both species also possess anterior paired bands consisting of eight cells (fa). In *B. stygius* the anterior cells are comparably sized to the posterior band. In *B. charon* the anterior bands of Fol's cells are so much smaller relative to the cells of the posterior bands that, on initial observation, the anterior bands may appear to be missing entirely or, perhaps to be a fourth row of trap cells (Fig. 4).

- The ciliated funnel (cf) is a conspicuous feature, located well to the right of the midline, off the upper wall of the pharynx and posterior to the mouth (Figs. 2, 3 and 4). From above or below, the cf of *B. charon* appears almost circular in shape. The apex of the funnel points left, back toward the brain, which is well-separated from the cf. The cf of *B. charon* is also well back from the dorsal lip and below the opening of the mouth. In contrast, the cf of *B. stygius* is located almost on the midline and only very slightly right of the brain. In *B. stygius,* the cf appears funnel-shaped from above, is obscured from below, its apex is oriented posteriorly/backwards and the opening to the funnel lies directly under the dorsal lip, very close to the mouth (Fig. 4). The granular texture of the cf is apparent in both species (Fig. 5) as are the movements of cilia in living specimens (online supporting video: Bathochordaeus_spp.mov).
- The position of the mouth is terminal, at least in larger specimens. In ventral view, the mouth extends beyond the anterior margin of the trunk, and the ventral oikoplastic region, such that it is visible from underneath. In *B. stygius,* the mouth is located well behind the anterior margin of the trunk (Fig. 4).
- The inner sensory cells of *B. charon,* just below the mouth, are relatively small compared to those of *B. stygius* (Fig. 5).

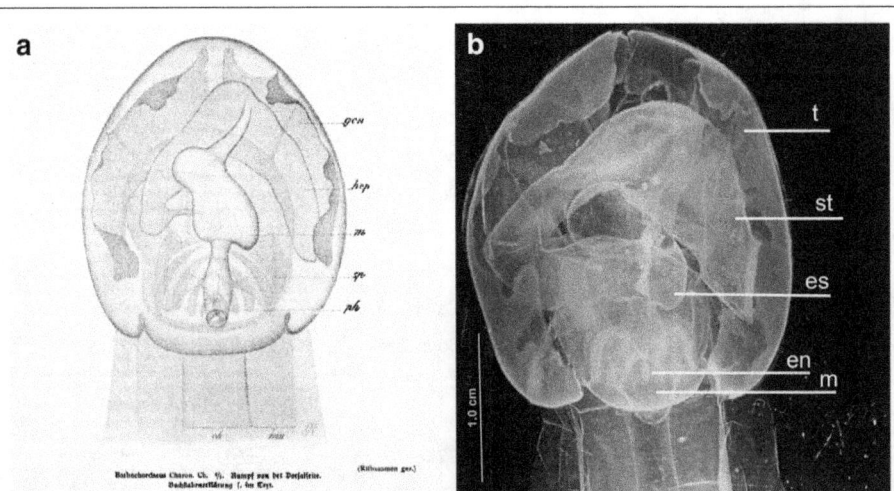

Fig. 2 Dorsal views of the trunk of *Bathochordaeus charon*. **a**. Drawing from *Aus Den Tiefen des Weltmeeres,* Chun's original account of the animal (*left*). **b**. Photo of *B. charon* collected 29 March, 2013 for comparison (m = mouth, en = endostyle, es = esophagus, st = left lobe of stomach, t = testis)

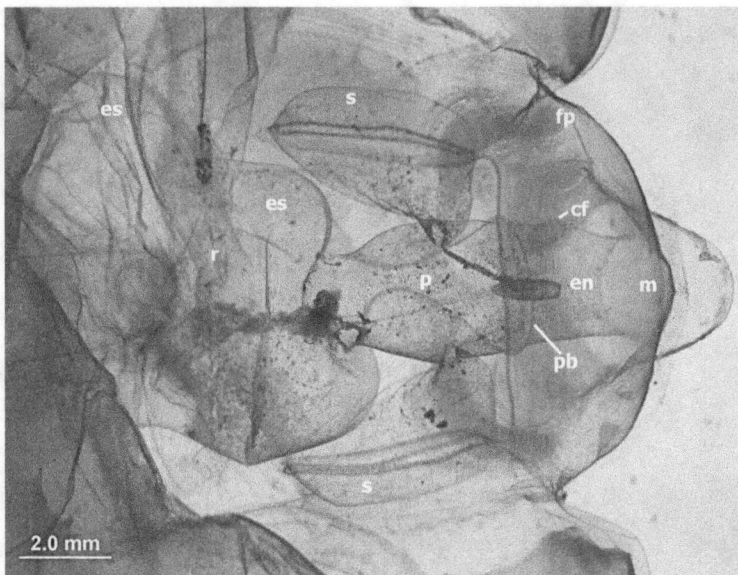

Fig. 3 Ventral view of the trunk of *Bathochodaeus charon*. The size of the external aperture of the spiracles (s = 4.33 mm) is more than 15 times the aperture (0.27 mm) that opens into the pharynx (p). The large, bag-like esophagus extends out the left side of the image (backwards). Note the large, flaccid intestine leading to the rectum (r) and how the mouth, (m), which opens dorsally, protrudes beyond the anterior margin of the trunk (pb = peripharyngeal band, en = endostyle, cf = ciliated funnel, and fp = posterior giant cells of Fol's oikoplast)

- The intestine and rectum of *B. charon* are large and flaccid compared to the same structures in *B. stygius* (Fig. 3).
- The size of a larvacean is not diagnostic of species. We have only collected large *B. charon* (animals >50 mm in total length). However, we have collected *B. stygius* that range from 11 mm to 87 mm in total length and there is no change in the relative size of their spiracles; i.e., the inner pharyngeal openings and outer openings to the spiracles of *B. stygius* are approximately equal in size regardless of the age/size of the specimen.

Inner filter ('House') structure

The pronounced morphological differences in the oikoplastic regions of *B. charon* and *B. stygius* suggest that the structural features of their houses may differ, particularly the inner filter (Fig. 4). The HD video taken by the *Doc Ricketts* immediately prior to the collection of *B. charon* was carefully reviewed and the differences in the inner filter as compared to that of *B. stygius* were readily apparent (Fig. 6).

The inner filter of *B. charon* is only slightly larger than the animals occupying the houses, whereas the inner filter of *B. stygius* is much larger than the occupant. The inner filter of *B. charon* is less convoluted and appears lumpy, more like a cluster of grapes than the many-chambered and accordion-like structure made by *B. stygius*. In short, the relative size and appearance of the inner filters are characteristic for each species, with the inner filter of *B. charon* a distinctly simpler structure than that of *B. stygius* (online supporting video: Bathochordaeus_spp.mov and Fig. 6).

Once the differences in houses had been established by reviewing video from the first *B. charon* captured, all annotated video sequences for the genus *Bathochordaeus* were reviewed. Out of hundreds of larvaceans initially identified as *B. stygius*, eleven were determined to be *B. charon* based on clear, close-up, HD video footage of the houses and/or the spiracles. (Table 1 and Fig. 6).

Molecular analyses

Molecular analyses supported the morphological distinctions observed between *B. charon* and *B. stygius*. Of the 400 base pair region of COI that we amplified, the between-group distance was 16.9 %. For the 18s gene, we compared one *Bathochordaeus charon* sequence to unpublished sequences of *Bathochordaeus stygius* from our studies. The sequences aligned with less than 0.5 % differences among base pairs along the partial (1600-bp) segment of the 18S gene. There were two base pair substitutions and one base pair deletion in *B. charon*. A comparison of the 18S sequence of *B. charon* to sequences in GenBank using NCBI Blast revealed that *B. charon* had 97 % identity to two larvaceans: *Oikopleura dioica* (AB013014) and *Megalocercus huxleyi* (FM244868).

Conclusions

Given that giant larvaceans may contribute up to one third of the vertical carbon flux to the deep seafloor in

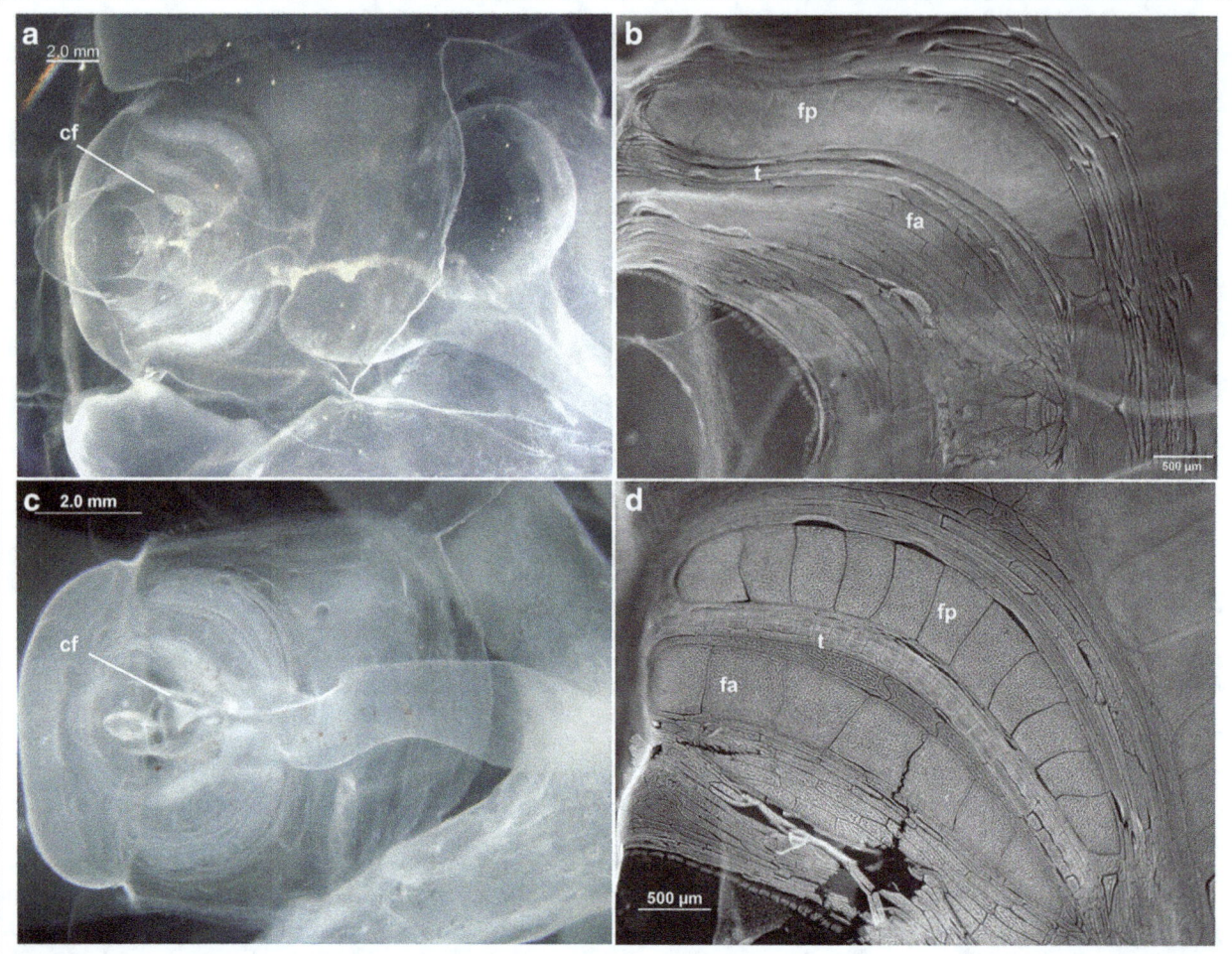

Fig. 4 The dorsal oikoplastic region of both species. *Bathochordaeus charon* (*top*) compared to *B. stygius*. The large esophagus of *B. charon* (**a**) is contrasted with the more slender and tubular esophagus of *B. stygius* (**c**). Plates B and D are magnified views of the right Fol's oikoblast from each specimen. Both species appear to have 12 large posterior Fol's cells (fp) with trap cells (t) just in front of them (**b, d**); however, the anterior row of Fol's cells (fa) is greatly reduced in *B. charon* (**b**)

Monterey Bay (Robison et al. 2005) and that active houses are oases for commensal zooplankton in the mesopelagic habitat (Steinberg et al. 1994), resolving their identities is of ecological as well as taxonomic significance.

Nearly transparent and easily damaged, giant larvaceans are as enigmatic to look at under a microscope as are their descriptions in the literature. Their mouth opens dorsally while their rectum voids ventrally. Many of their structural features are so transparent as to be difficult to see, even under a microscope.

Two of Chun's most contested features of *B. charon* remain its odd, crop-like expansion of the esophagus, and its large, funnel-shaped spiracles constricted where they meet the pharynx. However, Chun's illustration is entirely accurate with regard to both features (Figs. 1, 2 and 3). The simplicity of the inner filter of *B. charon* almost certainly affects their ability to feed on particles. There are clearly more interstices in the inner filter of *B. stygius* (Fig. 6), which seem likely to be a consequence of

B. stygius' much larger bands of anterior Fol's cells (Fig. 5). And the twin supply passages to the inner filter, directly downstream from the tail chamber (tc) of *B. stygius* (Hamner and Robison 1992) appear conspicuously absent from the inner filter of *B. charon* (Fig. 6). If truly absent, it is a mystery how water is pumped through the inner filter. If present, the structure is so diminished as to be invisible in the video sequences we presently have for *B. charon*. Relative to body size, there is decreased surface area of their inner filter compared to that of *B. stygius,* and the seeming lack of split supply passages and the fragile, diaphanous structure suggests that the inner filter of *B. charon* may not generate or be able to accommodate the flow and pressure that the inner filter of *B. stygius* can sustain.

If true, this difference in flow through the house may provide some explanation for the "peculiar" triangular shape of the spiracles that so puzzled Garstang (1937). Larvacean spiracles draw food through the feeding tube

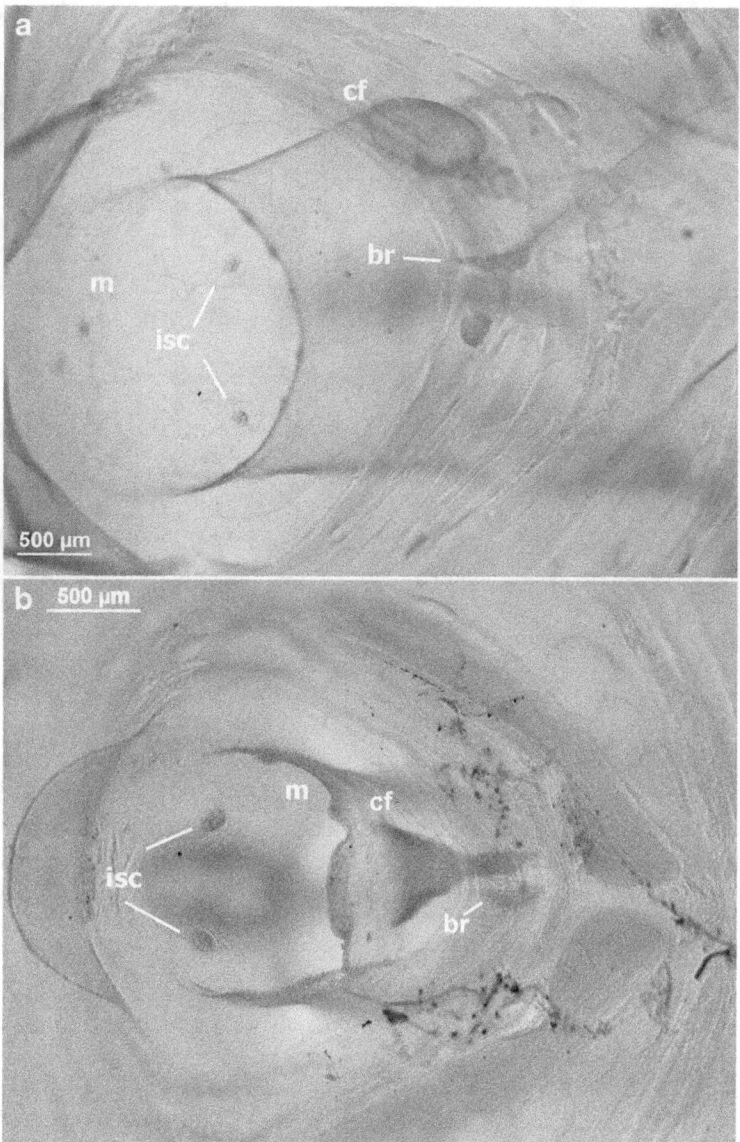

Fig. 5 Detail of the oral (m = mouth) region *Bathochordaeus charon* (**a**), trunk width 225 mm, and *Bathochordaeus stygius* (**b**), trunk width 126 mm. The ciliated funnel (cf) is a conspicuous feature, and in *B. charon* lies well to the right side of the brain (br) while that of *B. stygius* is shifted almost to the midline. The inner sensory cells (isc) of *B. charon* are less conspicuous than those of *B. stygius* although *B. charon* is the larger individual

where the bi-lobed inner filter joins, and into the mouth (Deibel 1998). The large and densely ciliated external opening of the spiracles of *B. charon* could create a venturi that would help to pull water into the mouth (Fig. 3). Because the entry to the spiracles of *B. charon* is close to an order of magnitude smaller than the diameter of their mouth (Fig. 3), evacuating undesirable objects through their spiracles as do other oikopleurids (Alldredge 1977; Lombard et al. 2011), is not an option. If a particle enters the mouth, the only way out seems to be through the gut to the rectum.

A third point of confusion in the early literature came from Chun's assertion that *B. charon* had four pairs of Fol's crescents, when in fact there are two pairs. Lohmann (1914) was also criticized by Garstang (1937) with regard to his description of the large and fairly conspicuous Fol's cells in *B. charon*. Garstang called these "glandular crescents" and because the anterior pair in *B. stygius* consists of eight giant cells, it was clear to him that those were "...identical with Fol's oikoplast in *Oikopluera*, *Stegosoma*, and *Megalocercus* both in structure and in position" (Garstang 1937). However, because Lohmann (1914) counted 12 giant Fol's cells in *B. charon* and asserted that only one pair of "glandular crescents" was visible, Garstang thought Lohmann had mistaken the

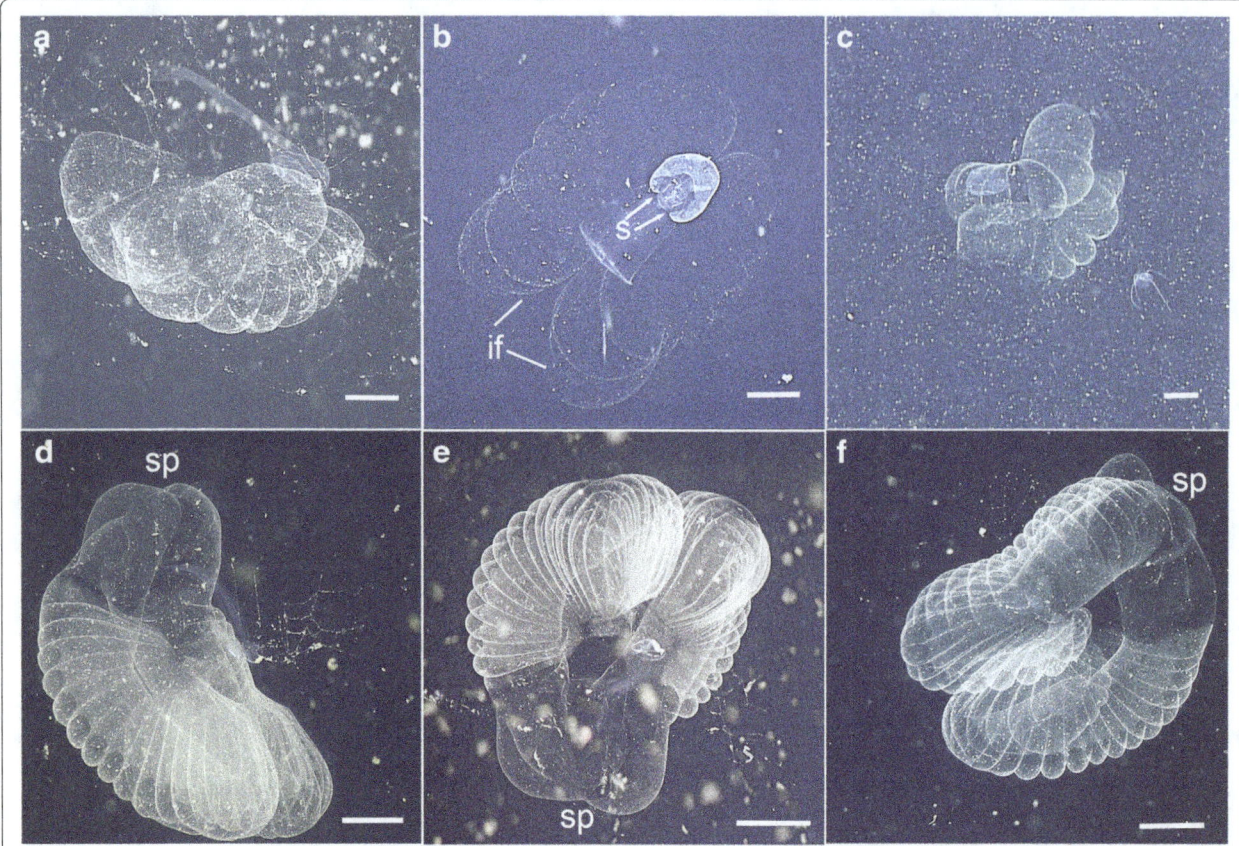

Fig. 6 The inner filter ("house") of *Bathochordaeus charon* (**a–c**) compared to *B. stygius* (**d–f**) are readily distinguishable in high-definition video, taken by MBARI ROVs. *Bathochordaeus charon* is larger in size relative to its inner filter (if) than is *B. stygius*, which also has more conspicuous supply passages (sp) through which water is diverted to either filter. The inner filter of *B. charon* is less convoluted than *B. stygius* and has fewer chambers (compare **c, f**). Plate B demonstrates that the spiracles (s) can be seen on high-definition video. Scale bars ~ 2 cm, **a–f**

posterior giant cells (which were obvious in the specimens of *B. stygius* he collected) for Fol's cells and somehow missed the anterior crescent entirely. Although convinced that Lohmann had erred, Garstang proposed to call the larger, posterior crescent (which he maintained were not Fol's cells) 'Lohmann's colloplasts', in honor of Lohmann's lifetime of work with larvaceans. This terminology is ironic because in this regard, Garstang was incorrect.

Bathochordaeus charon clearly has one band of 12 giant cells on either side of the trunk as does *B. stygius*. But the anterior pair of 8 cells in *B. charon* are so small in comparison to those of *B. stygius* (which are comparable in size to the cells of the posterior band) as to appear missing. This difference in the anterior band of Fol's cells is probably why the inner filter of *B. charon* and *B. stygius* look so different (Fig. 6). Furthermore, we have observed the initial formation of many houses of *B. stygius* and both pairs of "glandular crescents", on either side of the trunk, make the inner filter. Therefore, we regard all of Garstang's "glandular crescents" as Fol's cells and distinguish each band based on its position on the trunk (i.e., anterior or posterior).

In over two decades of exploring the mesopelagic waters of Monterey Bay, CA, remarkably few *Bathochordaeus charon* (*n* = 15) were observed, while thousands of *B. stygius* have been encountered. Both species occupy a similar depth range in Monterey Bay, yet, based on the large differences in COI they are clearly not interbreeding populations. Moving forward we intend to look more closely at the diversity of larvaceans in the mesopelagic habitat, identifying these important organisms responsible for vertical transport of carbon through the water column. As of 2009, the recognized number of appendicularian species was 70, of which 43 are described for the Pacific Ocean (Fenaux et al. 1998; Castellanos et al. 2009), *Bathochordaeus charon* unequivocally among them.

Abbreviations

#, number; br, brain; cf, ciliated funnel; COI, cytochrome oxidase 1; en, endostyle; es, esophagus; et al., et alia, and others; fa, anterior cells of Fol's oikoplast; fp, posterior cells of Fol's oikoplast; HD, high-definition; i.e., id est, that is; isc, inner sensory cells; m, mouth; MBARI, Monterey Bay Aquarium Research Institute; N/A, not applicable; p, pharynx; pb, peripharyngeal band; r, rectum; ROV, remotely operated vehicle; sp, supply passages; st, left lobe of stomach; t, testis; VARS, video annotation and reference system

Acknowledgements
We are grateful for the invaluable skills of our ROV pilots and ships' crews. The annotations and support of MBARI's video lab especially, Kyra Schlining and Susan Von Thun, were essential to this manuscript. Lynne Christianson, Shana Goffredi, and Shannon Johnson Williams helped in the early stages of the molecular work with equipment and advice to get us started. Chuck Galt, Russ Hopcroft, and Marsh Youngbluth provided valuable information and references. Thanks to Hans Jannasch for translating Chun's German to English. Kim Reisenbichler's help at sea and in the lab is indispensible. The comments by reviewers were helpful and improved this manuscript. This work is made possible through the generous support of the David and Lucile Packard Foundation.

Funding
The David and Lucile Packard Foundation.

Authors, contributions
RES identified *B. charon*, made the morphological observations with input from all authors and wrote the first draft of the manuscript. KRW completed all molecular analyses and the majority of video annotations. BHR obtained funding and was responsible for all specimen collections. All authors contributed to editing the manuscript and have read and approved it.

Competing interests
The authors declare that they have no competing interests.

References
Alldredge AL. House morphology and mechanisms of feeding in the Oikopleuridae (Tunicata, Appendicularia). J Zool. 1977;181:175–88.

Barham EG. Giant larvacean houses: observations from deep submersibles. Science. 1979;205:1129–31.

Bückmann A, Kapp H. Untersuchungen am Zooplankton von der Atlantischen Kuppenfahrt der Meteor, März bis Juli 1967. Meteor Forschungsergebnisse D. 1973;13:11–36.

Bückmann A, Kapp H. Taxonomic characters used for the distinction of species of Appendicularia. Mitt Hamb Zool Mus Inst. 1975;72:201–28.

Castellanos IA, Morales-Ramírez A, Suárez-Morales E. Appendicularians (Urochordata). In: Wehrtmann IS, Cortés J, editors. Marine biodiversity of Costa Rica, Central America. Netherlands: Springer; 2009. p. 445–52.

Chun C. Aus den Tiefen des Weltmeeres. Jena: Gustav Fischer; 1900. p. 519–21.

Deibel D. Feeding and metabolism of Appendicularia. In: Bone Q, editor. The biology of pelagic tunicates. Oxford: Oxford University Press; 1998. p. 139–49.

Fenaux R. Synonymie et distribution géographique des appendiculaires. Bull Inst Océanogr Monaco. 1966;66(1363):1–23.

Fenaux R. The classification of Appendicularia (Tunicata): history and current state. Mem Inst Oceranogr: Monaco. 1993;17:1–123.

Fenaux R. Anatomy and functional morphology of the Appendicularia. In: Bone Q, editor. The biology of pelagic tunicates. Oxford: Oxford University Press; 1998. p. 25–34.

Fenaux R, Bone Q, Deibel D. Appendicularian distribution and zoogeography. In: Bone Q, editor. The biology of pelagic tunicates. Oxford: Oxford University Press; 1998. p. 251–64.

Flood PR. Toward a photographic atlas on special taxonomic characters of oikopleurid Appendicularia (Tunicata). In: Gorsky G, Youngbluth M, Deibel D, editors. Response of marine ecosystems to global change: ecological impact of appendicularians. Paris: Contemporary Publishing International; 2005. p. 59–85.

Flood PR, Deibel D, Morris C. The appendicularian house. In: Bone Q, editor. The biology of pelagic tunicates. Oxford: Oxford University Press; 1998. p. 105–24.

Folmer RH, Folkers PJ, Kaan A, Jonker AJ, Aelen J, Konings RN, Hilbers CW. Secondary structure of the single-stranded DNA binding protein encoded by filamentous phage Pf3 as determined by NMR. Eur J Biochem. 1994;224:663–76.

Galt CP. First records of a giant pelagic tunicate, *Bathochordaeus charon* (Urochordata, Larvacea), from the eastern Pacific Ocean, with notes on its biology. Fish Bull. 1979;77:514–9.

Garstang W. On a new Appendicularian, *Bathochordaeus* sp. from Bermuda, with a revision of the genus. P Linn Soc Lond. 1936;148(3):131–2.

Garstang W. On the anatomy and relations of the Appendicularian *Bathochordaeus*, based on a new species from Bermuda (*B. stygius*, sp. n.). J Linn Soc Lond Zoo. 1937;40:283–303.

Hamner WM, Robison BH. In situ observations of giant appendicularians in Monterey Bay. Deep-Sea Res. 1992;39:1299–313.

Hirose M, Hirose E. DNA barcoding in photosymbiotic species of *Diplosoma* (Ascidiacea: Didemnidae), with the description of a new species from the southern Ryukyus, Japan. Zool Sci. 2009;26:564–8.

Hopcroft RR. Diversity in larvaceans: How many species? In: Gorsky G, Youngbluth M, Deibel D, editors. Response of marine ecosystems to global change: ecological impact of appendicularians. Paris: Contemporary Publishing International; 2005. p. 45–57.

Lindsay D, Umetsu M, Grossmann M, Miyake H, Yamamoto H. The gelatinous macroplankton community at the Hatoma knoll hydrothermal vent. In: Ishibashi J, Okino K, Sunamura M, editors. Subseafloor biosphere linked to hydrothermal systems. Japan: Springer; 2015. p. 639–66.

Lohmann H. Die Appendicularien der Valdivia expedition. Verh Dtsch zool Ges. 1914;24:157–92.

Lohmann H. Erste klasse der Tunicaten: Appendiculariae. In: Kükenthal W, Krumbach T, editors. Handbuch der zoologie. De Gruyter. Leipzig: De Gruyter, Berlin; 1933. p. 1–202.

Lohmann H. Die Appendicularien der Deutschen Tiefsee-Expedition. Wiss Ergeb Dtsch. Tiefsee-Exped. 1931;21:1–158.

Lombard F, Selander E, Kiørboe T. Active prey rejection in the filter-feeding appendicularian *Oikopleura dioica*. Limnol Oceanogr. 2011;56:1504–12.

Medlin L, Elwood HJ, Stickel S, Sogin ML. The characterization of enzymatically amplified eukaryotic 16S-like rRNA-coding regions. Gene. 1988;71:491–9.

Morris CC, Deibel D. Flow rate and particle concentration within the house of the pelagic tunicate *Oikopleura vanhoeffeni*. Mar Biol. 1993;115:445–52.

Robison BH. Light in the Ocean's midwaters. Sci Am. 1995;273:60–4.

Robison BH, Reisenbichler KR, Sherlock RE. Giant larvacean houses: Rapid carbon transport to the deep sea floor. Science. 2005;308:1609–11.

Russell C, Newman C, Williamson H. A simple cytochemical technique for demonstration of DNA in cells infected with mycoplasmas and viruses. Nature. 1975;253:461–2.

Silver MW, Coale SL, Pilskaln CH, Steinberg DR. Giant aggregates: Importance as microbial centers and agents of material flux in the mesopelagic zone. Limnol Oceanogr. 1998;43:498–507.

Steinberg DK, Silver MW, Pilskaln CH, Coale SL, Paduan JB. Midwater zooplankton communities on pelagic detritus (giant larvacean houses) in Monterey Bay, California. Limnol and Oceanogr. 1994;39:1606–20.

Tamura K, Stecher G, Peterson D, Filipski A, Kumar S. MEGA6: molecular evolutionary genetics analysis version 6.0. Molecular biology and evolution 2013;30:2725–29.

Thompson H. Bathochordaeus charon Chun 1900. In: Pelagic tunicates of Australia. Australia: Commonwealth Council for Scientific and Industrial Research; 1948. p. 1–196.

12

Cetacean sightings, mixed-species assemblages and the easternmost record of *Indopacetus pacificus* from the northern Indian ocean

Anoukchika D. Ilangakoon[1*] and Abigail K. Alling[2]

Abstract

A visual survey of cetaceans was carried out during a voyage from Singapore to Sri Lanka, through the Straits of Malacca, Andaman Sea and across the Bay of Bengal in the northern Indian Ocean in November/December 2012. Forty sightings of 11 cetacean species were recorded in 19 days of observation. Two mixed-species associations of interest were recorded. One of these contained four species of odontocetes in association with each other. The second group was of Indopacetus pacificus in association with Globicephala macrorhynchus and this while being the easternmost live sighting of I. pacificus in the northern Indian Ocean is also the first such mixed group in the Bay of Bengal.

Keywords: Bay of Bengal, Northern Indian Ocean, Cetaceans, Mixed-species assemblages, *Indopacetus pacificus*

Introduction

There have been few offshore cetacean surveys in Asia and there is a particular dearth of knowledge from the Bay of Bengal area of the northern Indian Ocean. While some countries around the rim of the Indian Ocean have carried out surveys in coastal waters, dedicated offshore cetacean surveys have rarely been undertaken in this area. Therefore, almost all of the knowledge about marine mammal diversity and distribution has come from observations in coastal waters (Alling, 1986; Anderson, 2005; Broker & Ilangakoon,2008; Ilangakoon, 2008; Smith et al., 2008; Smith & Tun, 2008; Ilangakoon, 2009; Ilangakoon & Perera, 2009; Minton et al., 2010; Mansur et al., 2011; Clark et al., 2012; de Vos et al., 2012) and records of dead and stranded animals (Leatherwood & Reeves 1989; Chantrapornsyl et al., 1996; Ilangakoon, 2006, Ilangakoon, 2012a). While dedicated cetacean surveys have been sparse throughout the northern Indian Ocean region (Ballance and Pitman 1998; De Boer, 2000) the only offshore records in the Bay of Bengal have come from observations using platforms of opportunity (Leatherwood et al., 1984; Afsal et al., 2008).

Between 20 November and 12 December 2012, a visual survey for cetaceans was carried out onboard sailing vessel *Mir* while transiting from Singapore to Sri Lanka. The journey began in the Straits of Malacca and continued across the Andaman Sea and Bay of Bengal to Galle, Sri Lanka (Fig. 1). The route was chosen specifically to cross through the Ten Degree Channel (10° N Latitude) instead of the more direct route through the Great Channel (6° N Latitude). It is an unusually deep passage that has strong tides and currents coursing through several 1000 m depth contours running parallel to each other through the channel. Previously no cetacean surveys with published results have taken place in this passage area.

Materials and methods

The vessel *Mir*, a 113′ two-masted ketch with a cruising speed of 3–6 knots was used as the dedicated platform for this survey. One primary observer, stationed on the bow scanning 180° ahead kept a constant watch during daylight hours. The primary observer was assisted by the helmsman who was positioned just off-centre on the port side of the vessel aft of mid-ship. All sightings were

* Correspondence: ai.flukes@gmail.com
[1]Member, Cetacean Specialist Group of IUCN, 215, Grandburg Place, Maharagama, Sri Lanka
Full list of author information is available at the end of the article

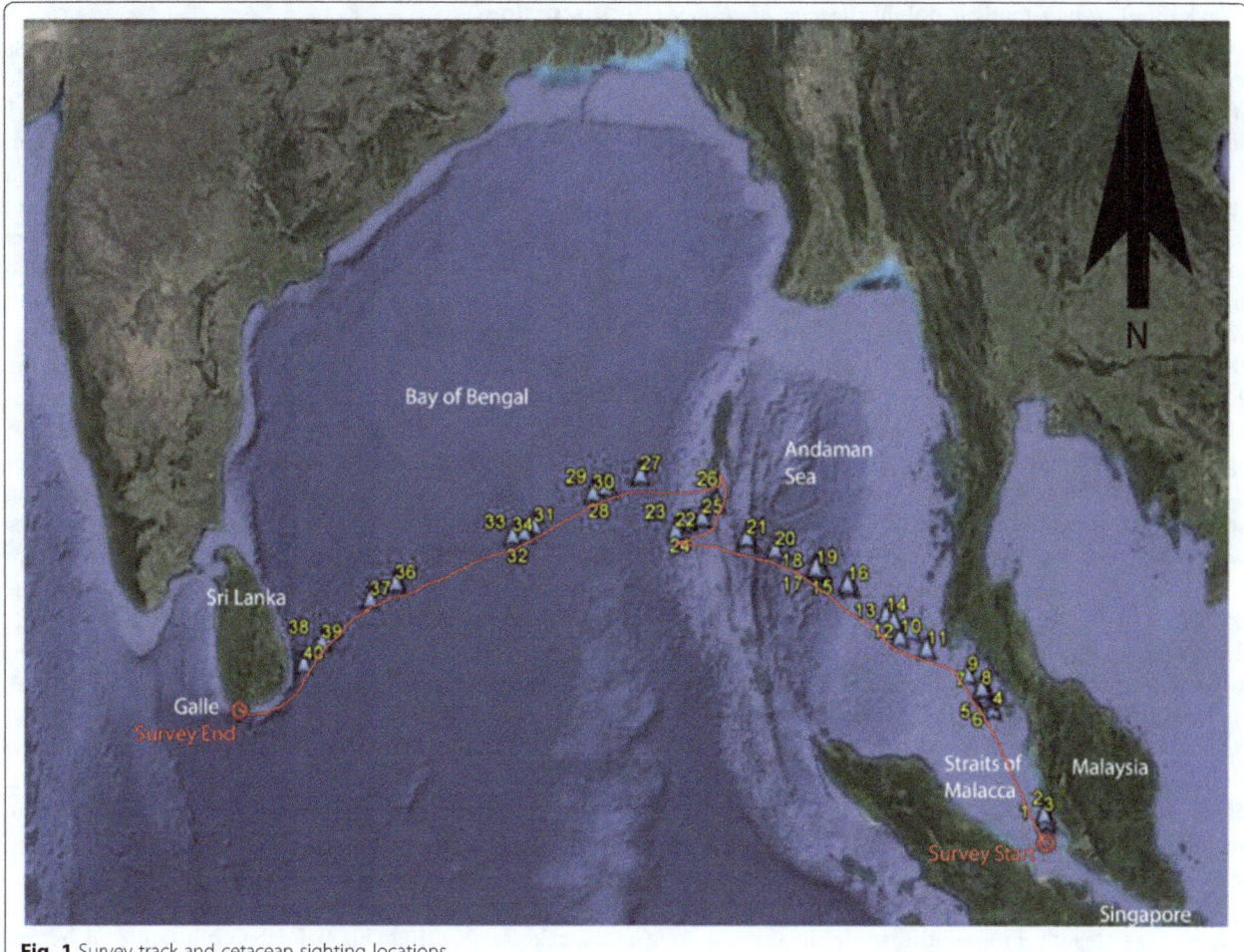

Fig. 1 Survey track and cetacean sighting locations

recorded in passing mode while the vessel maintained its preplanned course and no attempts were made to approach any of the cetaceans sighted.

For each sighting, the date, time, GPS location, species, number of individuals, behaviour and other pertinent information such as group composition, associated organisms and sea state were recorded. Animals were photographed opportunistically to aid identification but as all observations were in passing mode this was possible only when animals approached the vessel. Here we report all cetaceans sighted over the 19 days of observation and discuss some of the significant sightings including mixed-species groups, in terms of expanding knowledge on the cetaceans in the northern Indian Ocean and particularly in the offshore areas of the Bay of Bengal.

Results

The survey commenced at 0800 on 22 November in the Straits of Malacca (3°7′30″N; 100°41′33″E) once *Mir* was outside of the main shipping lane and was completed when reaching the approach to Galle Harbour in Sri Lanka (5°53′48″N; 80°29′6″E) at 0800 on 12 December 2012 (Fig. 1). An estimated distance of 1,660 nm was travelled, with good sighting conditions on all survey days except on 30 November when squalls and an opposing easterly current greater than 4 knots was prevalent just west of the Ten Degree Channel and the Andaman Islands.

Forty cetacean sightings were recorded over the survey period and all sightings were of odontocetes. Species were positively identified in 36 of the 40 sightings due to good sighting conditions and experienced observers while four sightings were recorded as unidentified species. A total of 11 species were positively identified including long-snouted spinner dolphin (*Stenella longirostris* Gray, 1828), pantropical spotted dolphin (*Stenella attenuata* Gray, 1846), striped dolphin (*Stenella coeruleoalba* Meyen, 1833), common bottlenose dolphin (*Tursiops truncatus* Montagu,1821), Indopacific bottlenose dolphin (*Tursiops aduncus* Ehrenberg, 1833), common dolphin (*Delphinus capensis* Gray, 1828), Irrawaddy dolphin (*Orcaella brevirostris*

Gray, 1866), rough-toothed dolphin (*Steno bredanensis* Lesson,1828), short-finned pilot whale (*Globicephala macrorhynchus* Gray, 1846), Longman's beaked whale (*Indopacetus pacificus* Longman, 1926) and sperm whale (*Physeter macrocephalus* Linnaeus, 1758) (Table 1).

Stenella longirostris was the most frequently observed species encountered in 21 of the 40 sightings (Table 1). The only large whale sighting, of *P. macrocephalus* was in deep water (>1000 m) in the Ten Degree Channel off the Andaman Islands. Two mixed-species assemblages were recorded among the sightings and both occurred in the Bay of Bengal. The first of these occurred at location 11° 21' 83" N; 90° 55' 14" E at 1414 on 5 December 2012 and this group of over 100 animals included four species: *S. longirostris, S. coeruleoalba, S. attenuata* and *T. truncatus*. This entire group was moving fast in a southerly direction and flying fish (*Hirundichthys* spp.) were also observed leaping among the dolphins. The second mixed-species assemblage occurred at location 09° 59' 57" N; 88° 01' 18" E at 0825 on 7 December 2012 and comprised of the two species *I. pacificus* and *G. macrorhynchus* in a small group totalling 9–12 animals. When first sighted, three animals of the species *I. pacificus* appeared to be surfacing after a deep dive as they surfaced at a steep angle, with visible blows, which aided species identification along with their bulbous melon, large size and falcate pointed dorsal fin (Dalebout et al., 2003). They were soon followed by 6–9 *G. macrorhynchus* that surfaced beside them, after which they all swam away steadily in a southeasterly direction as one group. Although this group was photographed the animals were moving away, against the light, resulting in low quality photographs, not fit for publication.

Table 1 Cetacean sightings by species and location number as shown in Fig. 1

Species	No. of sightings	Sighting numbers as in Fig. 1
Stenella longirostris	21	3, 4, 5, 6, 8, 9, 11, 12, 13, 14, 16, 19, 25, 26, 27, 30, 32, 34, 35, 38, 39, 40
Stenella coeruleoalba	2	27, 37
Stenella attenuata	3	27, 31, 37
Tursiops aduncus	6	1, 7, 18, 22, 23, 24
Tursiops truncatus	2	27, 29
Steno bredanensis	1	17
Orcaella brevirostris	1	2
Delphinus capensis	1	36
Globicephala macrorhynchus	1	33
Indopacetus pacificus	1	33
Physeter macrocephalus	1	21
Unidentified delphinids	4	10, 15, 20, 28

Discussion

This is the first reported cetacean survey conducted across the Andaman Sea and Bay of Bengal, thus it is not possible to make a comparison with previous studies. While all 11 species recorded have been previously reported from northern Indian Ocean waters (Alling, 1986; Leatherwood and Reeves, 1989; Ballance and Pitman, 1998; Balance et al., 2001; Ilangakoon 2002, Anderson, 2005) some factors of interest that further our knowledge were noted during the sightings of the present survey and particularly so with the mixed-species groups.

Both mixed-species assemblages sighted during the present survey were of significance due to their composition and the species involved. The first was of interest due to four species being in association with each other in a single group. Mixed-species associations involving delphinids including *Stenella* species, *Tursiops* species and *Globicephala* species are not unusual and they have been reported from several areas of the worlds' oceans (Querouil et al., 2008; Rossi-Santos et al., 2009) including areas around the Maldive Islands and Sri Lanka in the tropical Indian Ocean (Ballance & Pitman 1998; Anderson, 2005; Ilangakoon, 2012b). However, most such sightings contained two or three species at most while the present sighting included four species with three *Stenella* species and one *Tursiops* species. The most likely explanation for this association is foraging advantage as previously suggested for similar associations in the Azores and Sri Lanka (Querouil et al., 2008; Ilangakoon, 2012b). This is further substantiated by flying fish being observed with this group. However, it has also been suggested that predation risk can drive sympatric cetacean species to form temporary mixed-species aggregations (Kiszka et al., 2015). Since this sighting was in the deep open waters of the Bay of Bengal this is another possibility to be considered.

The second mixed-species group observed in the present survey is significant firstly as it contained *I. pacificus*, a species that is not commonly sighted anywhere and there is a relative paucity of information on this species in a worldwide context. Sighting records of *I. pacificus* in the northern Indian Ocean are mostly from the western Indian Ocean (Anderson et al., 2006) and this is only the second live sighting to be documented east of Sri Lanka. The only other reported sighting was southeast of Sri Lanka at location 06°18'N; 85° 50' E (Afsal et al., 2009) while the present sighting was in the central Bay of Bengal at location 09° 59' N; 88° 01' E. Therefore, this sighting was over 250 nm northeast of the sighting reported by Afsal et al. (2009), making it the easternmost sighting of this species recorded in the northern Indian Ocean to-date. Secondly, it has been reported that this species occasionally associates with *G. macrorhynchus* (Reeves et al., 2002) but only one such instance of association has been previously recorded in the northern Indian Ocean (Anderson, 2005) in the waters off the Maldive Islands. Therefore the present sighting is important in that it expands our knowledge on the range and behaviour of this little-studies species in the northern Indian Ocean.

The sighting of two *P. macrocephalus* near the Andaman Islands is also noteworthy as there are few recent records of this species from the area. The previously documented sightings of this species from this area are old records from American whaling log books prior to 1920 (Townsend, 1935) and opportunistic observations by British and Dutch merchant seamen (Brown, 1957; Morzer-Bruyns, 1971; de Silva, 1987), that mention sperm whales from the vicinity of the Andaman and Nicobar Islands. Although cetacean stranding and sighting records in Indian waters have been documented in the interim period (Sathasivam, 2004) no sperm whale sightings or strandings have been reported from the area around the Andaman Islands from the 1950's until the present sighting. It is not clear if this dearth of records is due to a lack of dedicated cetacean surveys in the area or because the species is rare in these waters. However, it is also worth mentioning that at the time this sighting was made a French-Indian seismic survey vessel was active in the area and seismic blasts were audible on the hydrophone deployed to attempt recording sperm whale clicks. The sighting was brief as the two whales fluked high and dived rapidly. This behaviour may have been an indication of disturbance due to the proximity of seismic blasts as has been observed elsewhere (Mate et al., 1994).

Of the eight sightings of *Tursiops* spp, observed during the survey, six were of *T. aduncus* in the shallow waters of the Andaman Sea while two were of *T. truncatus* in the deeper waters of the Bay of Bengal beyond the Andaman Islands. Although *T. aduncus* has been commonly reported in coastal waters of southeast Asian countries adjoining the Andaman Sea (Thailand, Myanmar) and in the Swatch of No-Ground off Bangladesh in the northern most regions of the Bay of Bengal, *T. truncatus* predominates around Sri Lanka on the western side of the Bay of Bengal. The present sightings were of interest however, because the clear demarcation of area of occurrence noted in this survey has not been documented before. This observation needs to be treated with caution due to the small number of sightings however, it is a point of interest for future surveys in this area to verify.

The present survey is only a starting point to fill the gaps in knowledge about cetacean occurrence and distribution in the offshore waters of the northeast Indian Ocean, particularly the Bay of Bengal. The data presented here indicates that a lot more work needs to be done in this area in order to gain a proper understanding of the importance of these waters as cetacean habitat

and it is therefore suggested that more systematic surveys are undertaken in the future.

Acknowledgements

This study and voyage were made possible by support from The Ward Family Foundation, Kopcho Family Foundation and Swire Pacific Offshore. We wish to give special thanks to Captain Mark Van Thillo, and crew members Robert Thoren, Ed Baker, Ellie Heywood, Dinouk Perera and Leina Sato for their support in the research. We also thank the two reviewers for their comments.

Authors' contributions

AA and AI conceived the study and and participated in the on-board research. AI prepared the manuscript with input from AA. Both authors read and approved the final manuscript.

Competing interests

The authors declare that they have no competing interests.

Author details

[1]Member, Cetacean Specialist Group of IUCN, 215, Grandburg Place, Maharagama, Sri Lanka. [2]Biosphere Foundation, P.O. Box 112636, Campbell, CA 95011-2636, USA.

References

Afsal VV, Yousuf KSSM, Anoop B, Anoop AK, Kannan P, Rajagopalan M, Vivekanandan B. A note on cetacean distribution in the Indian EEZ and contiguous seas 2003–07. J Cetacean Res Manag. 2008;10(3):209–16.

Afsal VV, Manojkumar PP, Yousuf KSSM, Anoop B, Vivekanandan B. The first sighting of Longman's beaked whale (Indopacetus pacificus) in the southern bay of Bengal. Mar Biodivers Rec. 2009;2:1–3.

Alling A. Records of odontocetes in the Northern Indian ocean (1981–1982) and off the coast of Sri Lanka (1982–1984). J Bombay Nat Hist Soc. 1986;83:376–94.

Anderson RC, Clark R, Madsen P, Johnson C, Kiszka J, Breysse O. Observations of Longman's beaked whale (Indopacetus pacificus) in the Western Indian ocean. Aquat Mamm. 2006;32(2):223–31.

Anderson RC. Observations of cetaceans in the Maldives, 1990–2002. J Cetacean Res Manag. 2005;7(2):119–35.

Ballance LT, Anderson RC, Pitman RL, Stafford K, Shaan A, Waheed Z, Brownell RL. Cetacean sightings around the Republic of the Maldives, April 1998. J Cetacean Res Manag. 2001;3(2):213–8.

Ballance LT, Pitman RL. Cetaceans of the western tropical Indian ocean: distribution, relative abundance, and comparisons with cetacean communities of two other tropical ecosystems. Mar Mammal Sci. 1998;14(3):429–59.

Broker KCA, Ilangakoon A. Occurrence and conservation needs of cetaceans in and around the Bar reef marine sanctuary, Sri Lanka. Oryx. 2008;42(2):286–91.

Brown SG. Whales observed in the Indian ocean. Notes on their distribution. Mar Observations. 1957;27(177):157–65.

Chantrapornsyl S, Adulyanukosol K, Kittiwathanawong K. Records of cetaceans in Thailand. Res Bull Phuket Biol Cent. 1996;61:39–63.

Clark RA, Johnson CM, Johnson G, Payne R, Kerr I, Anderson RC, Sattar SA, Godard CAJ, Madsen PT. Cetacean sightings and acoustic detections in the offshore waters of the Maldives during the northeast monsoon seasons of 2003 and 2004. J Cetacean Res Manag. 2012;12(2):227–34.

Dalebout ML, Ross GJB, Baker CS, Anderson RC, Best PB, Cockroft VG, Hinsz HL, Peddemors V, Pitman RL. Appearance, distribution and genetic distinctiveness of Longman's beaked whale, Indopacetus pacificus. Mar Mammal Sci. 2003;19(3): 421–61.

De Boer MN. A note on cetacean observations in the Indian ocean sanctuary and the South China sea, Mauritius to the Philippines, April 1999. J Cetacean Res Manag. 2000;2:197–200.

De Silva PHDH. Cetaceans (whales, dolphins and porpoises) recorded off Sri Lanka, India, from the Arabian Sea and gulf, gulf of Aden and from the Red Sea. J Bombay Nat Hist Soc. 1987;84(3):505–25.

De Vos A, Clark R, Johnson C, Johnson G, Kerr I, Payne R, Madsen P. Cetacean sightings and acoustic detections in the offshore waters of Sri Lanka: March–June 2003. J Cetacean Res Manag. 2012;12(2):185–93.

Ilangakoon AD. Whales and dolphins Sri Lanka. Colombo: WHT Publications; 2002. p. 99.

Ilangakoon AD. Preliminary analysis of large whale strandings in Sri Lanka 1889–2004. Pak J Oceanography. 2006;2(2):61–8.

Ilangakoon AD. Cetacean species richness and relative abundance around the Bar reef marine sanctuary, Sri Lanka. J Bombay Nat Hist Soc. 2008;105(3):274–8.

Ilangakoon AD. Cetacean survey off southern Sri Lanka, 2008–2009 project completion report. UK: Whale and Dolphin Conservation Society; 2009. p. 25.

Ilangakoon, AD. A review of cetacean research and conservation in Sri Lanka. J Cetacean Res Manag. 2012a;12(2):177–83.

Ilangakoon AD. Cetacean diversity and mixed-species associations off Southern Sri Lanka. In Arai, N. (ed) Proceedings of the 7th International Symposium on SEASTAR2000 and Asian Biologging Science, 8–9 March 2011. Bangkok, Thailand: Kyoto University Press; 2012b 68pp.

Ilangakoon AD, Perera LD. Cetacean and seabird survey off south-west Sri Lanka, 2008–2009. Hong Kong: Ocean Park Conservation Foundation; 2009. p. 37.

Kiszka JJ, Heithaus MR, Wirsing AJ. Behavioural drivers of the ecological roles and importance of marine mammals. Mar Ecol Prog Ser. 2015;523:267–81.

Leatherwood S, Peters R, Santerre R, Santerre M, Clarke JT. Observations of cetaceans in the northern Indian ocean sanctuary, November 1980 – May 1983. Rep Int Whaling Comm. 1984;34:509–20.

Leatherwood S, Reeves RR. Marine mammal research and conservation in Sri Lanka, Marine mammal technical report number 1. Nairobi: United Nations Environment Programme; 1989. p. 138.

Mansur RM, Strindberg S, Smith BD. Mark-resight abundance and survival estimation of indo-pacific bottlenose dolphins, *Tursiops aduncus*, in the swatch-of-no-ground, Bangladesh. Mar Mammal Sci. 2011;28(3):561–78.

Mate BR, Stafford KM, Lungblad DK. A change in sperm whale (*Physeter macrocephalus*) distribution correlated to seismic surveys in the Gulf of Mexico. J Acoustic Soc Am. 1994;965:33268–9.

Minton G, Collins T, Findlay K, Baldwin R. Cetacean distribution in the coastal waters of the sultanate of Oman. J Cetacean Res Manag. 2010;11(3):301–14.

Mörzer-Bruyns W.F.J. (1971) Field Guide of Whales and Dolphins. C. A. Meese, Amsterdam, Netherlands. 258pp.

Quérouil S, Silva MA, Cascão I, Magalhães S, Seabra MI, Machete MA, Santos RS. Why do dolphins form mixed-species associations in the Azores? Ethology. 2008;114:1183–94.

Reeves RR, Stewart BS, Clapham PJ, Powell JA. Guide to marine mammals of the world. New York: National Audubon Society, Alfred A. Knopf Inc; 2002. p. 527.

Rossi-Santos MR, Santos-Neto E, Baracho CG. Interspecific cetacean interactions during the breeding season of humpback whales (*Megaptera noveaengliae*) on the north coast of Bahia State, Brazil. J Mar Biol Assoc UK. 2009;89(5):961–6.

Sathasivam K. Marine mammals of India. Hyderabad: University Press (India) Private; 2004. p. 180.

Smith BD, Ahmed B, Mansur R, Strindberg S. Species occurrence and distributional ecology of nearshore cetaceans in the Bay of Bengal, Bangladesh, with abundance estimates for Irrawaddy dolphins *Orcaella brevirostris* and finless porpoises *Neophocaena phocaenoides*. J Cetacean Res Manag. 2008;10:45–58.

Smith BD, Tun MT. A note on the species occurrence, distributional ecology and fisheries interactions of cetaceans in the Mergui (Myeik) Archipelago, Myanmar. J Cetacean Res Manag. 2008;10:37–44.

Townsend CH. The distribution of certain whales as shown by logbook records of American whaleships. Zoologica (NY). 1935;19(1–2), 1-50 + 6 maps.

13

Records of five bryozoan species from offshore gas platforms rare for the Dutch North Sea

Esther D. Beukhof[1,2,3*], Joop W. P. Coolen[2,3], Babeth E. van der Weide[3], Joël Cuperus[3], Hans de Blauwe[4] and Jerry Lust[3,5]

Abstract

This study reports on bryozoan species collected at three offshore gas platforms in the Dutch part of the North Sea. Four out of thirteen observed species are considered as rare in the Netherlands, whereas *Cribrilina punctata* is a new species for Dutch waters.

Keywords: Bryozoa, North Sea, Netherlands, Offshore, Gas platform, *Cribrilina punctata*, *Arachnidium fibrosum*, *Electra monostachys*, *Scruparia ambigua*, *Scruparia chelata*

Introduction

The Dutch continental shelf of the North Sea largely consists of sandy bottoms. Rocky substrates are only present on the Cleaver Bank (Schrieken et al. 2013), the Borkum Reef Grounds (Coolen et al. 2015) and the Texel Rough (personal observation J.W.P. Coolen). Furthermore, artificial hard substrates are formed by shipwrecks (Lengkeek et al. 2013a), wind farms (Lindeboom et al. 2011; Vanagt et al. 2013) and gas platforms (Van Buuren 1984; Van der Stap et al. 2015).

Bryozoa grow on various hard substrates such as rocks, shells, wood, and plastic material, but also on macroalgae and Hydrozoa (De Blauwe 2009). Previous observations of Bryozoa in the Netherlands concentrated on southern coastal areas (Faasse and De Blauwe 2004). Faasse et al. (2013) recently reviewed the list of known Dutch Bryozoa which now comprises a total of 58 marine and estuarine species. They excluded specimens found on beached material, but included fauna from several recent offshore surveys of the Cleaver Bank (Van Moorsel 2003), the Princess Amalia Wind Farm (PAWF;

Vanagt et al. 2013) and a shipwreck on the sandy Dogger Bank (Schrieken et al. 2013).

This article reports on the finding of 13 bryozoan species on three offshore gas platform in the Dutch part of the North Sea. Of these species, *Cribrilina punctata* is new to the Dutch fauna, and four species are considered rare to the Dutch waters.

Materials and methods

The Bryozoa described here were observed during inventories of the fouling community of three stationary offshore gas platforms. The platforms differed in their distance from the Dutch shore, maximum depth and year of construction, and were sampled at different times during 2014 and 2015 (Table 1). Macrofauna samples were taken from the platform foundation by a commercial diver using a putty knife to detach the organisms and a surface supplied airlift sampler to collect them. In the airlift, all organisms were sieved over a 500 μm mesh. Further details of the airlift sampler and collection methods are described in Coolen et al. (2015). Triplicate samples were taken at 5 m depth intervals between 0 and 25 m and from the scour protection rocks on the bottom. After collection samples were fixed in a 6 % formalin solution buffered with 2 g L^{-1} borax. Following transport to the lab (between three and five days after collection) all organisms were conserved in 70 % ethanol until sorted by major taxonomic units and identified. All

* Correspondence: estb@aqua.dtu.dk
[1]Current address: Centre for Ocean Life, National Institute of Aquatic Resources (DTU Aqua), Technical University of Denmark, Jægersborg Allé 1, 2920 Charlottenlund, Denmark
[2]Wageningen University & Research, Chair Group of Aquatic Ecology and Water Quality Management, PO Box 47, 6700 AA Wageningen, The Netherlands
Full list of author information is available at the end of the article

Table 1 Three offshore gas platforms were visited during several inventories in 2014 and 2015. The platforms differed in their distance from shore, depth at the seabed and year of construction

Platform	Location	Distance from shore (km)	Maximum depth (m)	Year of construction	Sampling date
L10-G	53°29′N, 4°11′E	70	26	1984	June 2014
L10-A	53°24′N, 4°12′E	48	27	1972	April and June 2014, October 2015
L15-A	53°19′N, 4°49′E	5	21	1992	June 2014

shells of *Mytilus edulis*, other biogenic hard substrates and the scour protection rocks were inspected for the presence of Bryozoa. Any detached colonies were identified as well.

For identification Hayward and Ryland (1998), Hayward and Ryland (1999) and De Blauwe (2009) were consulted. The World Register of Marine Species (WoRMS Editorial Board 2015) was used as a taxonomic standard. Specimens were observed using a Zeiss SteREO Discovery.V8 stereomicroscope. When a specimen was covered by a thin layer of organic material which impeded observing the zooids, it was immersed in bleach for half an hour, then rinsed and dried in order to reveal the calcified skeleton.

Results

In 35 samples, a total of 13 species of Bryozoa were observed (Table 2). In ten samples no Bryozoa were encountered. Here, the findings of four species rare to Dutch waters and one new species for the Netherlands are described in more detail.

Five young colonies of the cheilostomatous *Cribrilina punctata* were found on three scour protection rocks of

platform L10-G consisting of 10 to approximately 30 non-ovicellate zooids (Fig. 1). A sub-oral bar with acute median mucro and 3–6 oral spines were present.

The ctenostomatous *Arachnidium fibrosum* was encountered in two samples taken from the foundation of platform L10-G at 10 m depth. Both colonies were attached to *Mytilus edulis*. Another colony of *A. fibrosum* was observed on a scour protection rock collected at the bottom of platform L15-A. Zooids were arranged in rows, and the colonies were sometimes branched.

A colony of *Electra monostachys* comprising several dozens of zooids was found on one of the scour protection rocks from platform L10-G. De Blauwe (2009)

Table 2 Bryozoa species encountered on the platform legs and/or the scour protection rocks at the bottom of the platforms L10-G, L10-A and L15-A

Species	Platform legs		Scour protection rocks	
	L10-G	L10-A	L10-G	L15-A
Alcyonidioides mytili	x	x		x
Arachnidium fibrosum[a]	x	x		x
Aspidelectra melolontha			x	
Callopora dumerilii	x	x	x	
Celleporella hyalina	x	x		
Conopeum reticulum	x	x	x	x
Cribrilina punctata[a]			x	
Electra monostachys[a]			x	
Electra pilosa	x	x	x	x
Microporella ciliata	x			
Schizomavella linearis	x	x		x
Scruparia ambigua[a]	x	x		
Scruparia chelata[a]	x	x		

[a]indicates species discussed in the article

Fig. 1 *Cribrilina punctata* zooids as found on a scour protection rock at the bottom of the platform. Scale bar: 0.2 mm

described a radiating crust as being characteristic for the species; this shape was also observed here. In addition to the relatively long proximal spine and pair of shorter distally located spines, almost all zooids still had their 4–6 pairs of shorter spines located around the frontal membrane.

Two members of the genus *Scruparia* were encountered. *Scruparia ambigua* was observed in one sample attached to *M. edulis* collected at 10 m depth at platform L10-G, but detached specimens were found as well. Several samples from platform L10-A taken between 5 and 15 m depth also contained a number of *S. ambigua* colonies. *Scruparia chelata* was observed at different depths (5–25 m) on platforms L10-G and L10-A, both as detached specimens and attached to *M. edulis*.

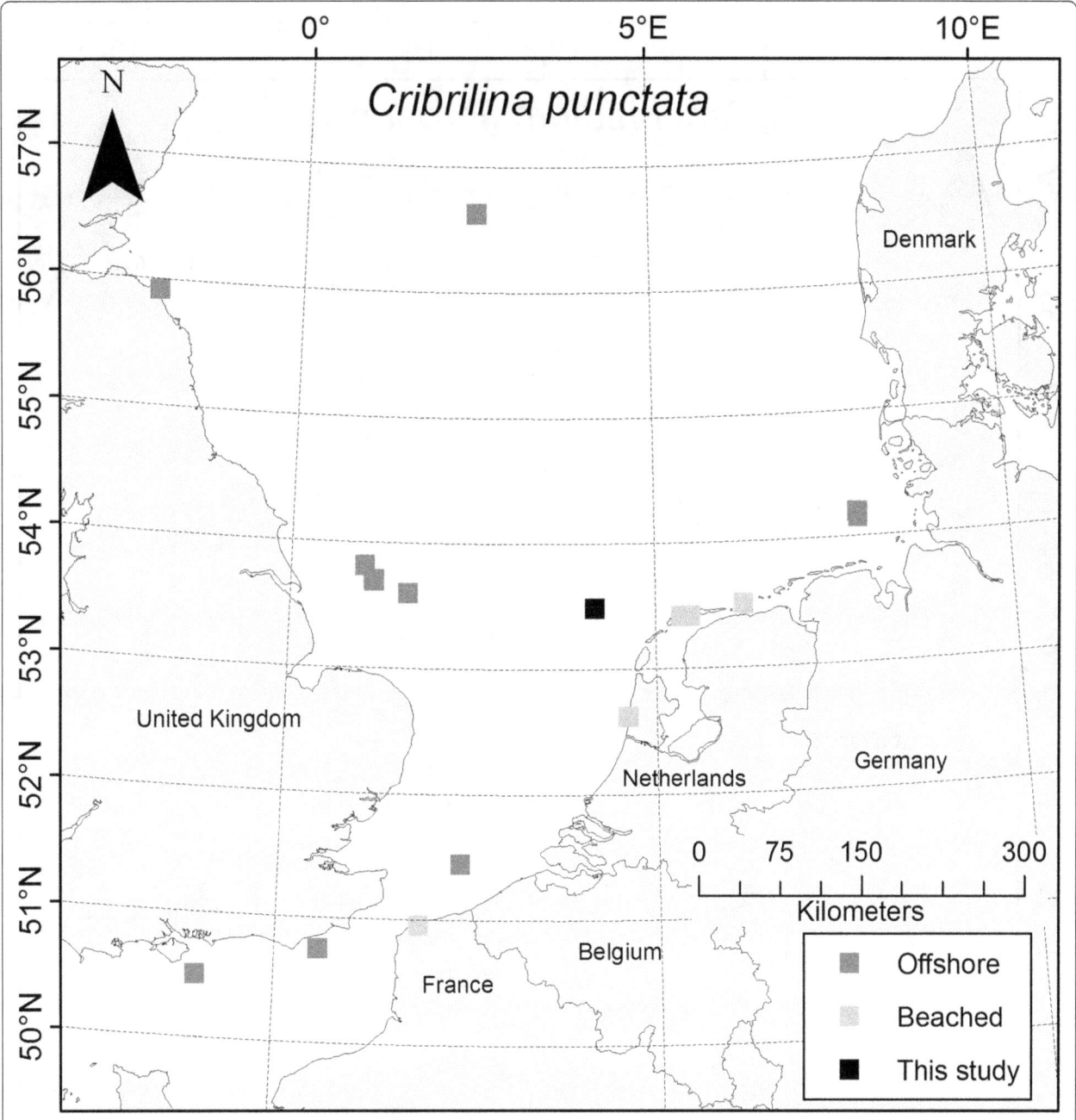

Fig. 2 Observations of *Cribrilina punctata* in the Netherlands (NMNH 1949; NMR 1992; Glorius et al. 2014), Belgium (De Blauwe 2009), France (VLIZ 2007), Germany (Schultze et al. 1990; De Kluijver 1991; Harms 1993; Kuhlenkamp and Kind 2012) and the United Kingdom (Rees et al. 2005; NRM 2010; Joint Nature Conservation Committee, 2005, Joint Nature Conservation Committee 2004; Marine Biological Association 2010)

Discussion

Cribrilina punctata

Cribrilina punctata is considered a rare species for the southern North Sea with only observations of beached specimens in the Netherlands on bivalve shells (NMNH; NMNH 1949), wood (NMR 1992) and plastic (De Ruijter 2014). Figure 2 shows all observations of *C. punctata* in the southern North Sea. The authors here now report the first observation of *C. punctata* attached to a fixed object on the Dutch continental shelf.

Cribrilina punctata is easily confused with *Collarina balzaci*. According to Faasse and De Blauwe (2004) several observations of *C. punctata* had been wrongly identified in the past leading to the exclusion of *C. punctata* from the list of Dutch fauna. Moreover, Faasse et al. (2013) excluded specimens found on beached material.

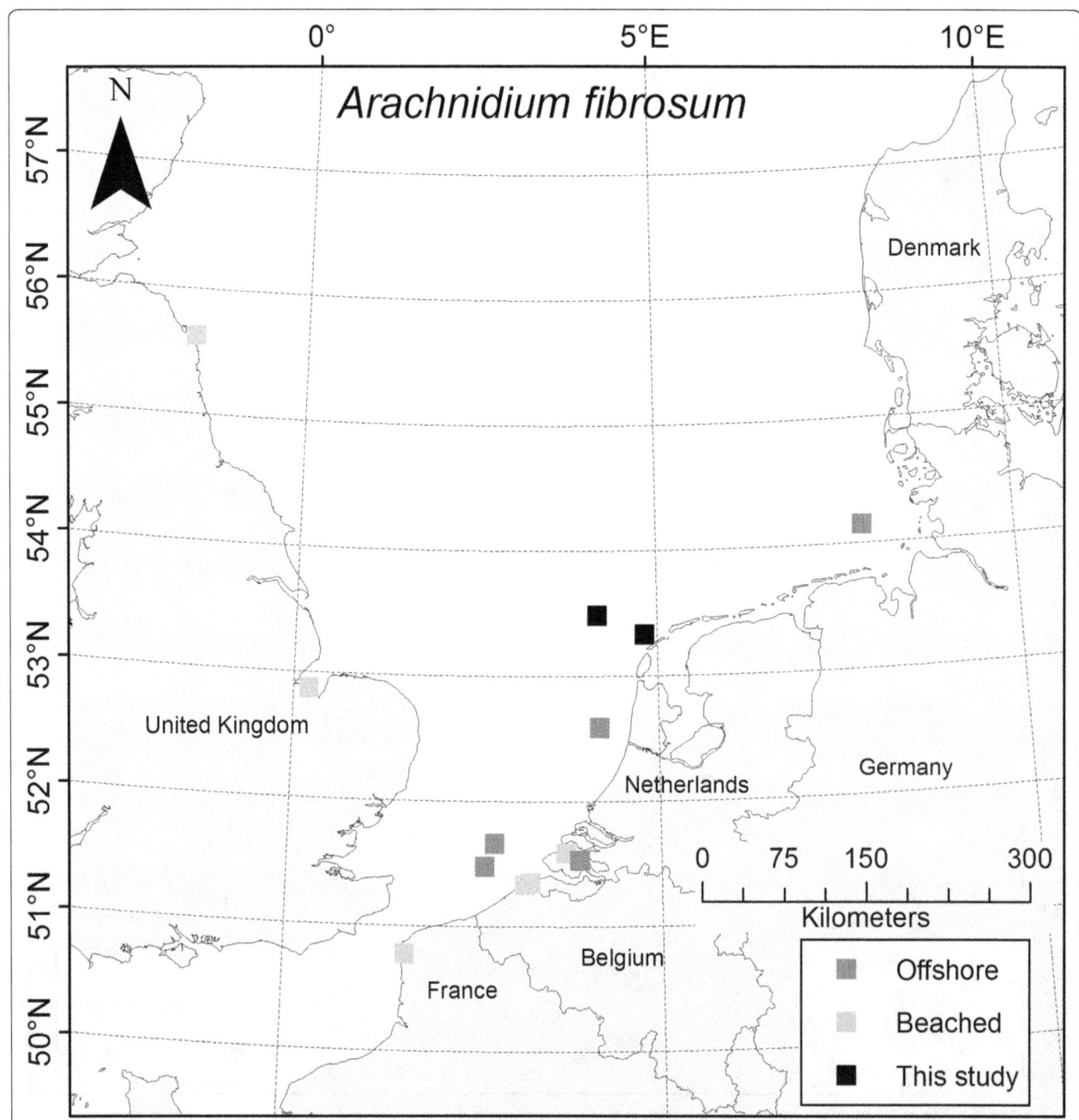

Fig. 3 Observations of *Arachnidium fibrosum* in the Netherlands (De Blauwe 2009; Vanagt et al. 2013), Belgium (Faasse and De Blauwe 2003; VLIZ 2007; Houziaux et al. 2008; De Blauwe 2009; pers. comm. F. Kerckhof), France (VLIZ 2007) and Germany (Kuhlenkamp and Kind 2012)

Arachnidium fibrosum

In the Netherlands *Arachnidium fibrosum* had been observed before on empty shells in southern coastal waters of the North Sea at 5–10 m depth (De Blauwe 2009) and at the PAWF at 5, 10 and 17 m depth (Vanagt et al. 2013). This corresponds to our finding of the species at 10 m depth on *Mytilus edulis*. However, Vanagt et al. (2013) did not encounter the species on scour protection

rocks at the bottom of the wind mill monopiles in contrast to our observation of *A. fibrosum* attached to rocks collected at 21 m and other observations on rocks from the Belgian Hinder Banks (Houziaux et al. 2008; De Blauwe 2009).

Both beached and offshore observations of *A. fibrosum* from the North Sea are relatively uncommon (Fig. 3) (Hayward 1985; De Blauwe 2009). Indeed, Faasse and De

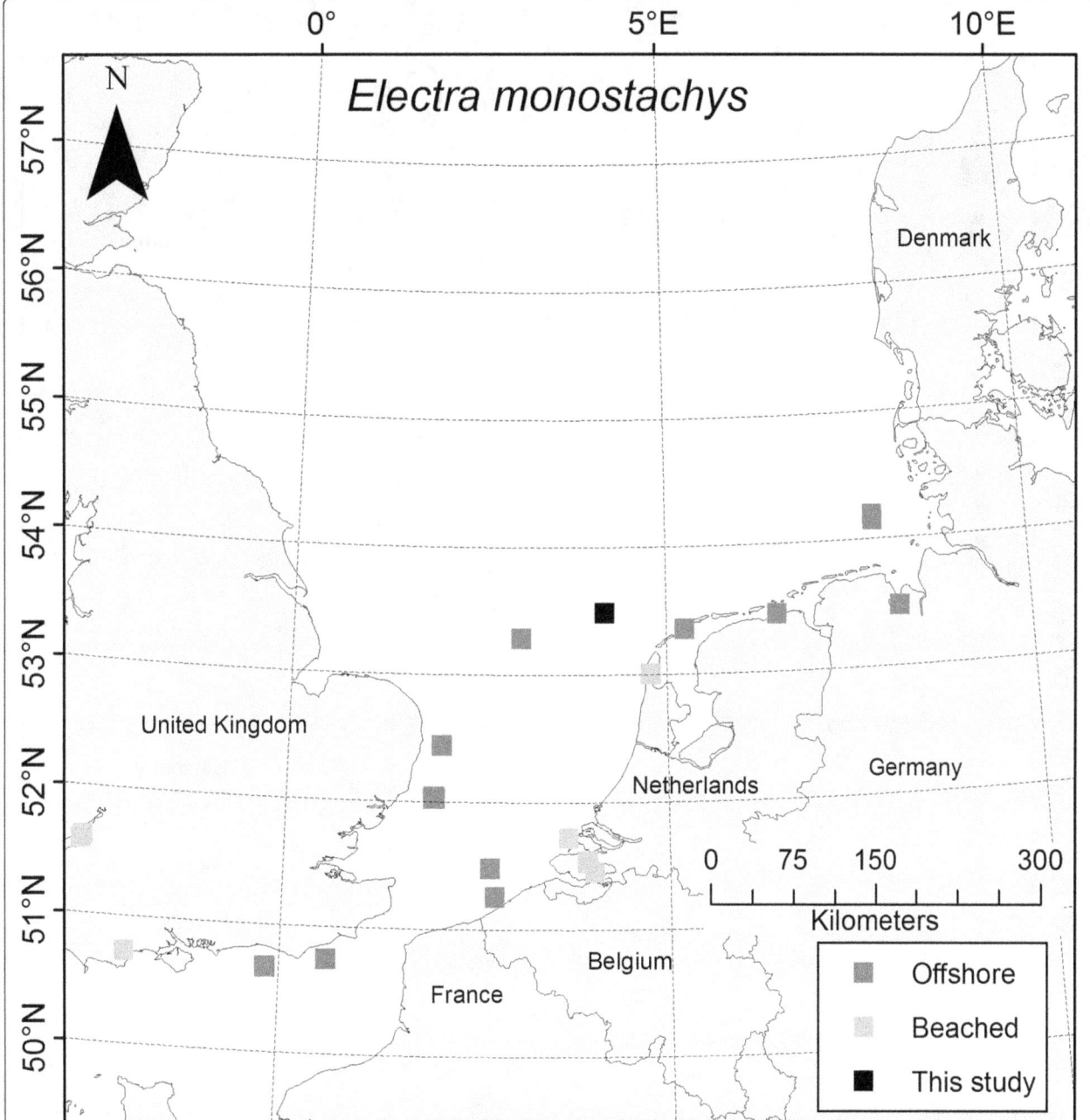

Fig. 4 Observations of *Electra monostachys* in the Netherlands (Faasse and De Blauwe 2004; De Blauwe 2009), Belgium (VLIZ 2007; De Blauwe 2009), Germany (De Kluijver 1991; Harms 1993; Kittelmann and Harder 2005; Kuhlenkamp and Kind 2012) and United Kingdom (Rees et al. 2005, Cooper et al., 1998; Joint Nature Conservation Committee, 2005; UK National Biodiversity Network: Marine Biological Association 2010)

Blauwe (2004) emphasize the capability of *A. fibrosum* zooids to adhere sand and detritus to themselves, making it difficult to observe and identify the species. They suggest this would partly explain the low amount of observations in general. The species often stays unnoticed in preserved material, especially if *Jassa* spp. (Arthropoda, Malacostraca) tubes are present. *Arachnidium fibrosum* was extremely common on reef balls deployed

in 2013 on the Bligh Bank and studied in 2014 (pers. comm. F. Kerckhof). Research on living material would facilitate the discovery and identification of this species in future research.

Electra monostachys

Electra monostachys has been observed several times on beached material along the Dutch coast and on shell

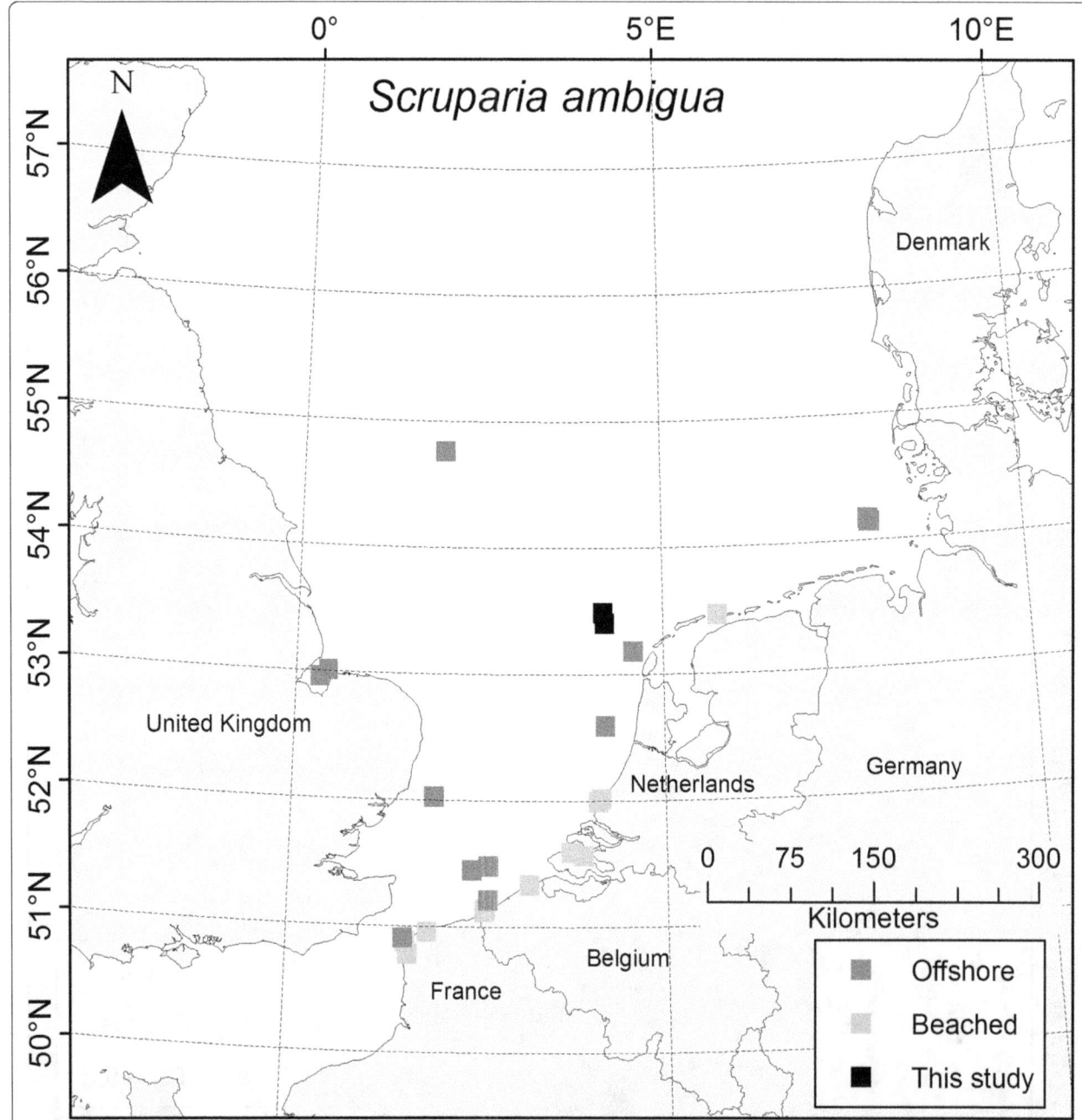

Fig. 5 Observations of *Scruparia ambigua* in the Netherlands (Faasse and De Blauwe 2004; NMR 2013; Vanagt et al. 2013; Coolen et al. 2015), Belgium (RBINS 1908; Vanhaelen et al. 2006; VLIZ 2007; Houziaux et al. 2008; De Blauwe 2009), France (VLIZ 2007), Germany (Senckenberg 2009; Kuhlenkamp and Kind 2012) and United Kingdom (Joint Nature Conservation Committee, 2005; UK National Biodiversity Network: Marine Biological Association 2010)

banks around the Dutch Wadden Sea Islands (De Blauwe 2009). Observations in the entire North Sea reported both a more northern and southern distribution than our specimen (Fig. 4).

Scruparia ambigua

Few observations of *Scruparia ambigua* exist for the Netherlands, and it is considered rare for the southern North Sea (De Blauwe 2009). In the Netherlands the most recent findings were at the PAWF (Vanagt et al. 2013) and the Borkum Reef Grounds (Coolen et al. 2015). In the current study the species was encountered at two platforms between 5 and 10 m depth, though Vanagt et al. (2013) observed the species several times at a slightly greater depth range (2–17 m). Several beached specimens have been reported as well (Fig. 5).

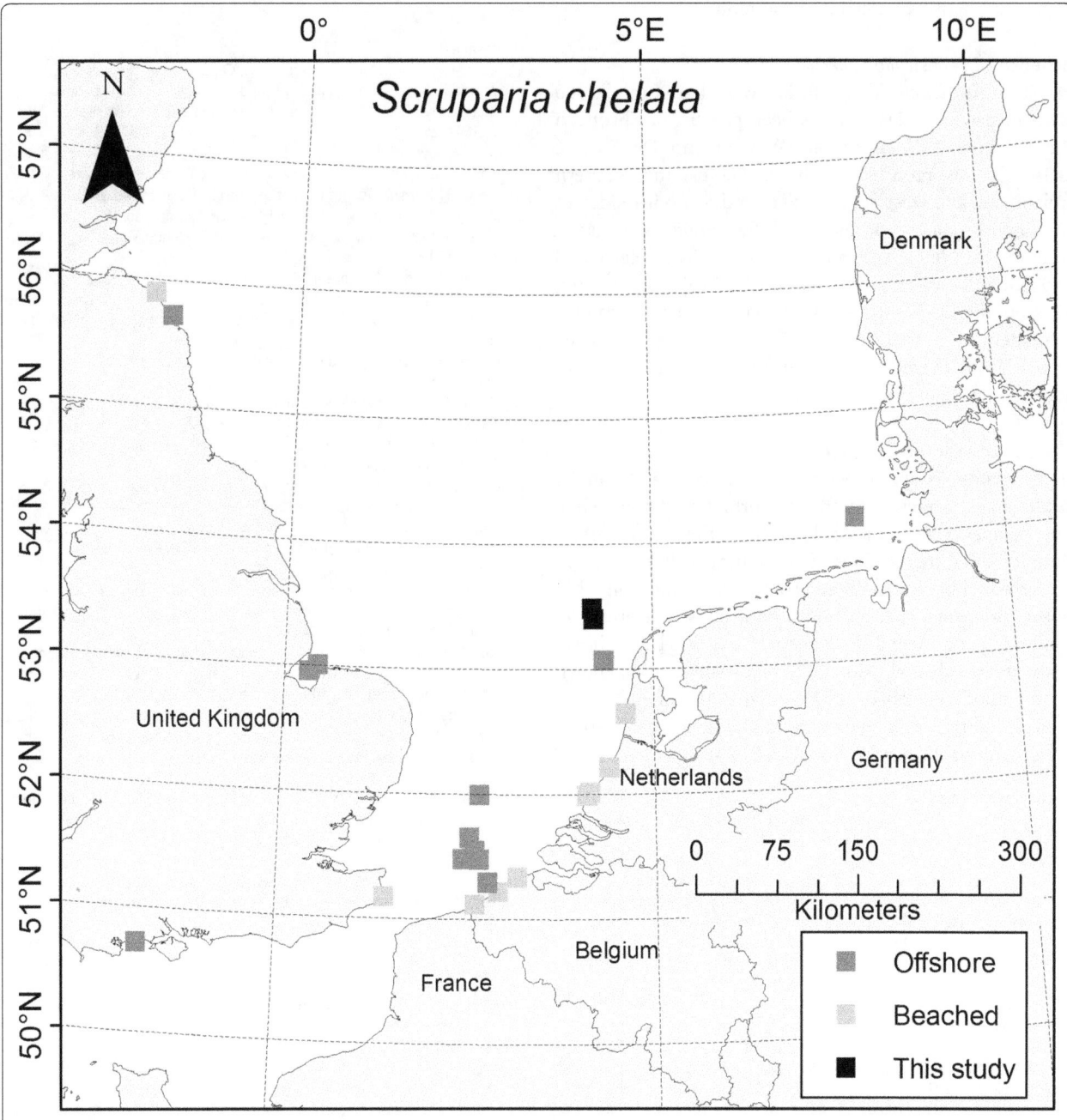

Fig. 6 Observations of *Scruparia chelata* in the Netherlands (Verkuil 1998; De Ruijter 2006; Lengkeek et al. 2013b; NMR 2013), Belgium (De Blauwe 2000, 2009; Vanhaelen et al. 2006; VLIZ 2007; Houziaux et al. 2008; Zintzen and Massin 2010), Germany (Harms 1993; Kuhlenkamp and Kind 2012) and United Kingdom (De Kluijver 1993; Joint Nature Conservation Committee, 2005; Marine Conservation Society, 2010)

Scruparia chelata

Scruparia chelata has only recently been discovered in the Netherlands on a shipwreck (24 km off the coast of Texel) by Lengkeek et al. (2013b) who investigated ten shipwrecks on the Dutch continental shelf. Other observations of the species across the North Sea have been reported from more northerly and southerly locations than the Dutch specimens (Fig. 6). De Blauwe (2009) considered *S. chelata* as rare for the southern North Sea, although beached specimens, most likely originating from the English Channel, are common.

Absence of *Fenestrulina delicia*

Fenestrulina delicia Winston, Hayward and Craig, 2000 is an invasive species that has been present in European waters since 2002 or earlier (Wasson and De Blauwe 2014). It was reported in the Shetlands, in Northern Ireland, on the west coast of Scotland, on both sides of the English channel, in the North Sea along the coast of Belgium, the Netherlands and Germany as far as Helgoland and along the west coast of the UK (De Blauwe et al. 2014; Wasson and De Blauwe 2014). A majority of the locations inhabited by *F. delicia* are wind farms and gas platforms. It is therefore noteworthy that *F. delicia* was absent in our samples.

More species to be expected

In this study six species were encountered mainly on *M. edulis* shells attached to the platform foundation while four species were found exclusively on the scour protection rocks at the seafloor. This indicates differences in preferred substrate and environmental conditions between bryozoan species. Some hard substrate areas on the Dutch continental shelf, such as the Texel Rough, remain uninvestigated. Moreover, bryozoan species known from empty shells on sandbanks on the Belgian and British continental shelf can be expected to be discovered also on Dutch sandbanks nearby (Faasse et al. 2013).

Acknowledgements

This work was funded through the Wageningen UR TripleP@Sea Innovation programme (KB-14-007) and supported by GDF SUEZ E&P Nederland B.V., the Nederlandse Aardolie Maatschappij B.V., Wintershall Holding GmbH and EBN B.V. We are grateful to the staff of GDF SUEZ and the Bluestream dive team for their help during diving and sampling. The authors thank Naturalis Biodiversity Centre and the Natural History Museum Rotterdam for their cooperation in revising their specimens. We thank Britta Kind for her valuable comments on the manuscript and for providing data of several species records.

Authors' contribuions

JCo designed the study, carried out the sampling and created the maps. JCo, EB and JL handled the samples in the lab and prepared them for taxonomic determination. EB, BW, JCu, HB and JL performed the taxonomic determination. EB collected data on observations of species at other locations and drafted the manuscript. All authors read and approved the final manuscript.

Competing interests

The authors declare that they have no competing interests.

Author details

[1]Current address: Centre for Ocean Life, National Institute of Aquatic Resources (DTU Aqua), Technical University of Denmark, Jægersborg Allé 1, 2920 Charlottenlund, Denmark. [2]Wageningen University & Research, Chair Group of Aquatic Ecology and Water Quality Management, PO Box 47, 6700 AA Wageningen, The Netherlands. [3]Maritime Department, Wageningen Marine Reseach, PO Box 57, 1780 AB Den Helder, The Netherlands. [4]Department of Invertebrates, Royal Belgian Institute of Natural Sciences, Vautierstraat 29, 1000 Brussels, Belgium. [5]Van Hall Larenstein, Integrated Coastal Zone Management, 8934 CJ Leeuwarden, The Netherlands.

References

Coolen JWP, Bos OG, Glorius S, Lengkeek W, Cuperus J, Van der Weide BE, Agüera A. Reefs, sand and reef-like sand: a comparison of the benthic biodiversity of habitats in the Dutch Borkum Reef Grounds. J Sea Res. 2015; 103:84–92.

Cooper KM, Boyd SE, Rees HL. Cross Sands broadscale survey 1998, Centre for Environment, Fisheries and Aquaculture. Essex, UK: Burnham laboratory; 1998.

De Blauwe H. Riemwiervoetjes en hun begroeiing. De Strandvlo. 2000;20:89–94.

De Blauwe H. Mosdiertjes van de zuidelijke bocht van de Noordzee - Determinatiewerk voor België en Nederland. Oostende: Vlaams Instituut voor de Zee; 2009.

De Blauwe H, Kind B, Kuhlenkamp R, Cuperus J, Van der Weide B, Kerckhof F. Recent observations of the introduced Fenestrulina delicia Winston, Hayward & Craig, 2000 (Bryozoa) in Western Europe. Studi Trentini di Scienze Naturali. 2014;94:45–51.

De Kluijver MJ. Sublittoral hard substrate communities off Helgoland. Helgoländer Meeresuntersuchungen. 1991;45:317–44.

De Kluijver MJ. Sublittoral hard-substratum communities off Orkney and St Abbs (Scotland). J Mar Biol Assoc UK. 1993;73:733–54.

De Ruijter R. Naar het strand, of toch maar niet? Het Zeepaard. 2006;66:29–30.

De Ruijter R. Cs-verslag. Het Zeepaard. 2014;74:38–45.

Faasse M, De Blauwe H. Het mosdiertje Arachnidium fibrosum Hincks, 1880 nieuw voor België en Nederland. De Strandvlo. 2003;23:47–9.

Faasse M, De Blauwe H. Faunistisch overzicht van de mariene mosdiertjes van Nederland (Bryozoa: Stenolaemata, Gymnolaemata). Nederlandse Faunistische Mededelingen. 2004;21:17–54.

Faasse M, Van Moorsel GWNM, Tempelman D. Moss animals of the Dutch part of the North Sea and coastal waters of the Netherlands (Bryozoa). Nederlandse Faunistische Mededelingen. 2013;41:1–14.

Glorius ST, Wijnhoven S, Kaag NHBM. Benthos community composition along pipeline trajectory A6-A-Ravn. An environmental baseline study. Den Helder: IMARES Wageningen UR; 2014. p. 49. Report Number C116.14.

Harms J. Check list of species (algal, invertebrates and vertebrates) found in the vicinity of the island of Helgoland (North Sea, German Bight) - a review of recent records. Helgoländer Meeresuntersuchungen. 1993;47:1–34.

Hayward PJ. In: Brill EJ, Backhuys W, editors. Ctenostome Bryozoans. Synopsis of the British Fauna (new series). London: The Linnean Society of London; 1985.

Hayward P, Ryland J. Cheilostomatous Bryozoa: 1. Aeteoidea - Cribrilinoidea: notes for the identification of British species. 2nd ed. Shrewsbury: Field Studies Council; 1998.

Hayward P, Ryland J. Cheilostomatous Bryozoa: 2. Hippothooidea - Celleporoidea: notes for the identification of British species. 2nd ed. Shrewsbury: Field Studies Council; 1999.

Houziaux JS, Kerckhof F, Degrendele K, Roche M, Norro A. The Hinder banks: yet an important region for the Belgian marine biodiversity? Brussels: Belgian Science Policy; 2008. p. 249. Report EV/45.

Joint Nature Conservation Committee. Marine benthic dataset (version 1) commissioned by UKOOA. 2004. NBN Gateway. https://data.nbn.org.uk/Datasets/GA000182 [Accessed at 18 Aug 2015].

Joint Nature Conservation Committee. Marine Nature Conservation Review (MNCR) and associated benthic marine data held and managed by JNCC. 2005. NBN Gateway. https://data.nbn.org.uk/Datasets/GA000190 [Accessed at 26 Jan 2016].

Kittelmann S, Harder T. Species- and site-specific bacterial communities associated with four encrusting bryozoans from the North Sea, Germany. J Exp Mar Biol Ecol. 2005;327:201–9.

Kuhlenkamp R, Kind B. Makrozoobenthos Monitoring Helgoland 2011, Maßnahme im Rahmen der WRRL. Report of the State Agency for Agriculture. Germany: Nature and Rural Areas (LLUR) of Schleswig-Holstein; 2012. p. 55.

Lengkeek W, Coolen JWP, Gittenberger A, Schrieken N. Ecological relevance of shipwrecks in the North Sea. Nederlandse Faunistische Mededelingen. 2013a; 40:49–58.

Lengkeek W, Didderen K, Dorenbosch M, Bouma S, Waardenburg H. Bureau Waardenburg bv and Stichting De Noordzee, rapport nr. 13–226, 76 ppx. Culemborg, Nethelands: Bureau Waardenburg bv and Stichting De Noordzee, rapport nr; 2013b. p. 13.

Lindeboom HJ, Kouwenhoven HJ, Bergman MJN, Bouma S, Brasseur S, Fijn RC, De Haan D, Dirksen S, Van Hal R, Hille Ris Lambers R, Ter Hofstede R, Krijgsveld KL, Leopold M, Scheidat M. Short-term ecological effects of an offshore wind farm in the Dutch coastal zone; a compilation. Environ Res Lett. 2011;6:1–13.

Marine Conservation Society. Seasearch Marine Surveys. 2010. NBN Gateway. https://data.nbn.org.uk/Organisations/53 [Accessed at 18 Aug 2015].

NMNH. ZMA.BRYO.95 Cribrilina punctata (Hassell, 1841). the Netherlands: National Museum of Natural History; 1949.

NMNH. ZMA.BRYO.2073 Cribrilina punctata (Hassell, 1841). the Netherlands: National Museum of Natural History.

NMR. NMR992800000029 Cribrilina punctata (Hassell, 1841). the Netherlands: Natural History Museum Rotterdam; 1992.

NMR. Natural history museum Rotterdam, the Netherlands - Invertebrata miscellaneous Collection. 2013. http://www.gbif.org/dataset/7bb2d451-5ffa-4d58-bc7f-19ea7aecb201 [Accessed at 18 Aug 2015].

NRM. Invertebrates Collection of the Swedish Museum of Natural History. 2010. http://www.gbif.org/dataset/56aa0680-0c60-11dd-84cd-b8a03c50a862 [Accessed at 18 Aug 2015].

RBINS. G5004 Scruparia ambigua (d'Orbigny, 1841). Brussels, Belgium: Royal Belgian Institute of Natural Sciences; 1908.

Rees HL, Pendle MA, Waldock R, Limpenny D, Boyd SE. A comparison of benthic biodiversity in the North Sea, English Channel and Celtic Seas - Epifauna, Centre for Environment, Fisheries and Aquaculture Science. Essex, UK: Burnham Laboratory; 2005.

Schrieken N, Gittenberger A, Coolen JWP, Lengkeek W. Marine fauna of hard substrata of the Cleaver Bank and Dogger Bank. Nederlandse Faunistische Mededelingen. 2013;41:69–78.

Schultze K, Janke K, Krüß A, Weidemann W. The macrofauna and macroflora associated with Laminaria digitata and L. hyperborea at the island of Helgoland (German Bight, North Sea). Helgoländer Meeresuntersuchungen. 1990;44:39–51.

Senckenberg. Collection Bryozoa SMF. 2009. http://www.gbif.org/dataset/966c9070-f762-11e1-a439-00145eb45e9a [Accessed at 18 Aug 2015].

UK National Biodiversity Network: Marine Biological Association. Marine Biological Association - Marine survey data (Professional) held by MarLIN. 2010. http://www.gbif.org/dataset/cc2031d1-395d-4052-8ce1-796a08c3dbf2 [Accessed at 18 Aug 2015].

Van Buuren J. Ecological survey of a North Sea gas leak. Mar Pollut Bull. 1984;15:305–7.

Van der Stap T, Coolen JWP, Lindeboom HJ. Marine fouling assemblages on Offshore Gas Platforms in the Dutch part of the North Sea: Effects of depth and distance from shore on biodiversity. PLoS One. 2015;11:e0146324. doi:10.1371/journal.pone.0146324.

Van Moorsel GWNM. Ecologie van de Klaverbank. BiotaSurvey 2002. Doorn: Ecosub; 2003. p. 157.

Vanagt T, Van de Moortel L, Faasse M. Development of hard substrate fauna in the Princess Amalia Wind Farm. Oostende: eCOAST; 2013. p. 42. report 2011036.

Vanhaelen M, Jonckheere I, De Blauwe H. Grote wierenstranding aan de Belgische Westkust tijdens de zomer van 2005. De Strandvlo. 2006;26:5–11.

Verkuil J. CS-verslag. Het Zeepaard. 1998;58:150–7.

VLIZ. Taxonomic Information Sytem for the Belgian coastal area (EurOBIS). 2007. http://www.gbif.org/dataset/83bb72ac-f762-11e1-a439-00145eb45e9a [Accessed at 18 Aug 2015].

Wasson B, De Blauwe H. Two new records of cheilostome Bryozoa from British waters. Marine Biodiversity Records. 2014;7:1–4.

WoRMS Editorial Board. World register of marine species. 2015.

Zintzen V, Massin C. Artificial hard substrata from the Belgian part of the North Sea and their influence on the distributional range of species. Belg J Zool. 2010;140:20–9.

14

Four new records of stranded Kemp's ridley turtle *Lepidochelys kempii* in the NW Iberian Peninsula

Pablo Covelo[1,3*], Lidia Nicolau[2,3] and Alfredo López[1,3]

Abstract

Background: The critically endangered Kemp's ridley turtle (*Lepidochelys kempii*) has a reduced distribution range concentrated in the Gulf of Mexico and only a few pelagic stage juveniles occasionally strand on European coasts. In the study area only three individuals have been recorded previously, most recently in 2001.

Results: Four new records of Kemp's ridley turtle are reported from the northwestern Iberian Peninsula in 2014. All of them were juvenile with a maximum straight carapace length of 26.8 cm. Two of them were found alive, but died on the first two days at the rehabilitation centres. The largest individual stranded in November presented signs of incidental capture in fishing gears.

Conclusions: The four records in just one year are a significant increase compared to previous data for the species in this area, and the data obtained contribute to the local knowledge on species phenology, size, distribution and threats.

Keywords: Kemp's ridley turtle, *Lepidochelys kempii*, Iberian Peninsula, Distribution, Sea turtle, Strandings

Abbreviations: CCL, Curved carapace length; CCW, Curved carapace width; CEMMA, Coordinadora para o Estudo dos Mamíferos Mariños; IUCN, Internation Union for Conservation of Nature; SCL, Straight carapace length; SCW, Straight carapace width; SPVS, Sociedade Portuguesa de Vida Selvagem; TL, Total length

Background

In spite of a recent increase in population (Márquez et al. 2005) the Kemp's ridley turtle *Lepidochelys kempii* (Garman 1880) is considered the most threatened sea turtle species in the world, being categorized as critically endangered by the International Union for Conservation of Nature (IUCN) (Marine Turtle Specialist Group 1996). The Kemp's ridley turtle has a small, restricted geographic range compared to other sea turtles species (Zug et al. 1997). The nesting areas of this species are mainly located in the Mexican coast of the Gulf of Mexico (Márquez, 1990; Bowen et al. 1994) with some secondary isolated rookeries in Texas, Florida, South Carolina and North Carolina (Meylan et al. 1990; Bowen et al. 1994; Johnson et al. 1999).

Adult and immature Kemp's ridley occur year-round in the Gulf of México, and juveniles are found at sea in the northwestern Atlantic Ocean as far as Nova Scotia and Canada (National Marine Fisheries Services US, Service FaW and SEMARNAT 2011). Occasionally some small turtles are also known to drift into European waters during the winter months (Brongersma 1972; Márquez 1994; Witt et al. 2007). There are several documented records showing the presence of Kemp's ridley turtle in the waters around Azores (Brongersma 1972; Bolton and Martins 1990), Madeira (Brongersma 1972), British Islands and France (Brongersma 1972; Witt et al. 2007; Penrose and Gander, 2015) and in the Mediterranean Sea (Brongersma and Carr 1983; Tomás et al. 2003; Oliver and Pigno 2005; Tomás and Raga 2008; Insacco and Spadola 2010; Carreras et al. 2014).

---

* Correspondence: cemmaorganizacion@gmail.com
[1]Coordinadora para o Estudo dos Mamíferos Mariños (CEMMA), P.O. Box 15. 36380, Pontevedra, Gondomar, Spain
[3]Departamento de Biologia & CESAM, Campus de Santiago, Universidade de Aveiro, 3810-193 Aveiro, Portugal
Full list of author information is available at the end of the article

In the NW Iberian Peninsula, previous documented records included only two individuals in the coast of Galicia (Spain), the most recently in 1998 (Fernández 1988; Faraldo and Galán 1999), and one individual in Portugal in 2001 (Dellinger 2008).

The present study provides a new report on the presence of and an increase in the records of Kemp's ridley turtles in the waters around the NW Iberian Peninsula within a short time of ten months. Our work also contributes to the knowledge of the distribution of this species in the north-eastern Atlantic and provide insights to the main threats in the area to contribute to conservation efforts of this species in international waters.

Methods

Study area

The NW Iberian Peninsula coast is about 1,500 km long ranging from Ría de Ribadeo (43°33.20' N; 07°02.00' W Spain) to Peniche (39°21.00' N; 09°22.00' W Portugal) (Fig. 1). The Galician coastline (about 1,200 km in length between Ría de Ribadeo and River Miño estuary) is characterized by a series of large, coastal inlets (rías). The NW Portuguese coast (about 300 km in length between River Minho and Peniche) presents a wider and flat continental shelf (40–70 Km) (Fiúza 1983). This whole area has the influence of winds from south and south-west in autumn and winter, and northerly winds in spring and summer.

Data collection and analysis

In the study area, two stranding networks are established: Coordinadora para o Estudo dos Mamíferos Mariños (CEMMA) in Galicia (NW of Spain) and Sociedade Portuguesa de Vida Selvagem (SPVS) in Portugal, since 1990 and 2000, respectively. These strandings networks record turtles and marine mammals stranded on the beaches, collect samples for Galician and Portuguese Marine Animal Tissue Banks and recover injured individuals for rehabilitation. They are coordinated via a 24 h telephone hotline.

Turtles were moved by the strandings networks mobile units to rehabilitation installations of CEMMA and SPVS to proceed with the veterinary examination and rehabilitation of the two individuals found alive and to carry out the necropsy of the dead ones following detailed protocols (Sociedad Española de Cetáceos 1999; Wyneken 2001). Turtles were measured (total length [TL]; straight carapace length [SCL]; curved carapace length [CCL]; straight carapace width [SCW] and curved carapace width [CCW] according to Bolten 1999), weighed, identified to species level, and the likely cause of death determined based on a complete external and internal examination. In cases when there was evidence

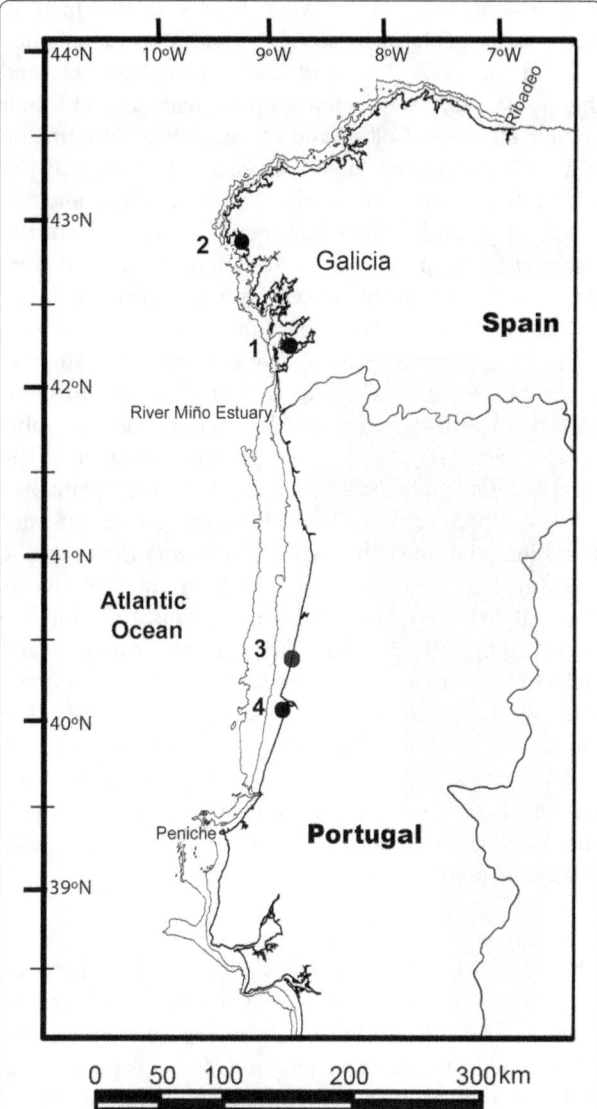

Fig. 1 Location of four new records of stranded Kemp's ridley turtle in the NW Iberian Peninsula between February and November, 2014. Turtles are numbered in chronological order (Table 1)

of multiple lesions, the most severe and recent was assigned as the primary cause of stranding.

The approximated age of the turtles was estimated using the carapace length and applying reference growth models (Chaloupka and Zug, 1997; Zug et al. 1997).

Results

In 2014 four new individuals of Kemp's ridley turtle were recorded stranded on the NW Iberian Peninsula coast (two in Spain and two in Portugal) (Fig. 1). The four turtles were identified as *Lepidochelys kempii* based on their diagnostic features: the presence of a small pore near the rear margin of each of the inframarginal scutes, the coloration of the skin (white bright green), the circular shape of the carapace, the presence of five pairs of coastal scutes, and triangular head with two pairs of frontal scales (Márquez, 1990). The four specimens had SCL of between 19.5 and 26.8 cm (Table 1), and therefore they were considered juveniles in their oceanic life stage (Collard and Ogren, 1990; Witherington et al. 2012). The first three individuals stranded between 11th February and 5th March, within 22 days, and the fourth one stranded on 11th November. Two of the turtles reported in the present study were found alive and were transferred to rehabilitation centres in less than two hours, but died after one and two days (Table 1).

The first individual, found alive in Cangas do Morrazo, Spain, presented small carapace and plastron abrasions, no food in digestive tract and emaciation was noticeable in the neck (Fig. 2), indicating malnourishment. The second individual, found dead on Carnota, Spain, had head, carapace and plastron abrasions, consistent with stranding, and goose barnacles (*Lepas anatifera*) on the plastron. The third individual, found alive in Cantanhede, Portugal, had head and carapace abrasions, probably due to stranding, an ulcerative lesion in the carapace and epibiont crustaceans (goose barnacles, *Lepas anatifera*). Internal observations of its organs indicated sepsis. The fourth individual, found dead in Figueira da Foz, Portugal, had lacerations on its anterior flippers, consistent with incidental capture in nets, hematomas in posterior flippers and evidence of feeding (crustaceans, unidentified crabs) in upper digestive tract.

Discussion

The four turtles were juveniles with a SCL of between 19.5 and 26.8 cm, so according to published growth models derived from skeletonchronological studies in relation with the carapace length (Chaloupka and Zug, 1997; Zug et al. 1997), they had an estimated age of 2 years for the first three individuals and nearly 3 years for the fourth turtle. The time of year in which they stranded is consistent with previous records of the species found stranded on the European continental coasts, that

Fig. 2 Kemp's ridley turtle stranded in Cangas do Morrazo, Spain, 11/02/2014. Emaciation was noticeable in the neck

are more usually found during the autumn and winter (Brongersma 1972; Witt et al. 2007). Goose barnacles found on the first and third turtles are commonly attached to floating objects and when they are found on dead or weak turtles that indicates that they spent a long period floating at the water's surface (De Loreto and Bondioli, 2008), and that is likely associated with a chronic condition such as poor nutrition, chronic infections, parasitic diseases, immune deficiency or a combination of some or all of these causes (Deem et al. 2009).

In the NW Iberian Peninsula, and especially during winter, the seawater temperature is low, sometimes below 10 °C. The average sea surface temperature during the first three stranding events (February and first week in March) was around 12 °C (Feldman and McClain 2014). This low temperature could have been sufficient to affect the sea turtles physiology leading to cold-stunning events, in which turtles become lethargic and float at the water's surface and, as a result, they may be unable to swim or feed (Schwartz 1978). Therefore, lack of feeding due to the low seawater temperature could explain the large weight differences up to 38 % (0.99 kg and 1.37 kg) between the two individuals that had

Table 1 Biometry and geographic data of the four new records of stranded Kemp's ridley turtle in the Iberian Peninsula

	1	2	3	4
Date	11/02/2014	21/02/2014	05/03/2014	11/11/2014
Location	Viñó Beach, Cangas do Morrazo, Spain	Boca do Río, Carnota, Spain	Tocha Beach, Cantanhede, Portugal	Gala Beach, Figueira da Foz, Portugal
Status	Alive (died on first day)	Dead	Alive (died on second day)	Dead
TL (cm)	25.4	30.0	29.5	38.0
SCL (cm)	19.5	21.4	21.4	26.8
CCL (cm)	21.0	23.5	23.9	29.0
SCW (cm)	18.5	20.2	19.3	27.7
CCW (cm)	20.8	24.4	22.6	29.6
Weight (kg)	0.99	1.62	1.37	2.65

TL total length, *SCL* straight carapace length, *CCL* curved carapace length, *SCW* straight carapace width, *CCW* curved carapace width

similar lengths to those found alive in February and March (Table 1). In the case of the turtle stranded in November in Portugal, sea temperature was around 18 °C and therefore cold stunning event is unlikely to explain this turtle stranding. Considering the necropsy findings, this individual was probably a victim of incidental capture by fisheries. Interactions with coastal fisheries, especially interactions with gill or trammel net fisheries, represents an important anthropogenic threat to other sea turtle species (loggerhead *Caretta caretta* and leatherback *Dermochelys coriacea*) in the Portuguese continental coast (Nicolau et al. 2014).

Conclusions

Four new records in one year are a significant increase compared to previous data for the species in this area. In UK and Republic of Ireland an increase in the stranded Kemp's ridley turtle has also been reported in 2014 with 7 stranded turtles, while in the previous ten year period 2004–2013 the average rate was just 1 stranded individual (Penrose and Gander, 2015). It will be important to pay attention to the occurrence of the species in the next years in the European coasts to find out whether the 2014 numbers are isolated findings or whether they indicate a trend.

Acknowledgements
We thank the staff of Dirección Xeral de Conservación da Natureza that found the two animals at the Galician beaches and took care of the alive one in the first hours. We also thank Juan Ignacio Díaz for helping with the field work and Ángela Llavona and Marisa Ferreira for their comments and suggestions during the writing of this note. The Quiaios Marine Animal Rehabilitation Centre provided technical support for the turtle found in Portugal. We are very grateful to the anonymous reviewers for their very useful comments that improved this note.

Funding
PC is supported by a research grant BI/UI88/7056/2014, AL is supported by a postdoctoral grant SFRH/BPD/82407/2011 and LN is supported by PhD grant SFRH/BD/51416/2011 both from the Portuguese Foundation for Science and Technology of the Portuguese Ministry of Science and Education. The Galician stranding network is supported by the regional government Xunta de Galicia, cofinanced with European Regional Development Funds (ERDF/FEDER).

Authors' contributions
PC, LN and AL wrote the manuscript. PC and LN participated in the field data collection and AL prepared the figures. All authors approved the final manuscript.

Competing interests
The authors declare that they have no competing interests.

Author details
[1]Coordinadora para o Estudo dos Mamíferos Mariños (CEMMA), P.O. Box 15. 36380, Pontevedra, Gondomar, Spain. [2]Sociedade Portuguesa de Vida Selvagem (SPVS). Departamento de Biologia, Campus de Gualtar, Universidade do Minho, 4710-057 Braga, Portugal. [3]Departamento de Biologia & CESAM, Campus de Santiago, Universidade de Aveiro, 3810-193 Aveiro, Portugal.

References

Bolten AB. Techniques for measuring sea turtles. In: Eckert KL, Bjorndal KA, Abreu-Grobois FA, Donnelly M, editors. Research and management techniques for the conservation of sea turtles. Washington: IUCN/SSC Marine Turtle Specialist Group; 1999. p. 110–4.

Bolton AB, Martins HR. Kemp's ridley captured in the Azores. Mar Turt Newsl. 1990;48:23.

Bowen BW, Conant TA, Hopkins-Murphy SR. Where are they now? The Kemp's Ridley Headstart Project. Conserv Biol. 1994;8(3):853–6.

Brongersma LD. European Atlantic turtles. Zoologsische Verhandelingen. 1972;121. 318 pp.

Brongersma, L.D. and Carr, A.F. (1983).*Lepidochelys kempii* (Garman) from Malta. Proceedings of the Koninklijke Nederlandse Akademie van Wetenschappen, Series C 86, 445–454

Carreras C, Monzón-Argüello C, López-Jurado LF, Calabuig P, Bellido JJ, Castillo JJ, Sánchez P, Medina P, Tomás J, Gozalbes P, Fernández G, Marco A, Cardona L. Origin and dispersal routes of foreign green and Kemp's ridley turtles in Spanish Atlantic and Mediterranean waters. Amphibia-Reptilia. 2014;35:73–86.

Chaloupka M, Zug GR. A polyphasic growth function for the endangered Kemp's ridley sea turtle, *Lepidochelys kempii*. Fish Bull. 1997;95(4):849–56.

Collard SB, Ogren LH. Dispersal scenarios for pelagic post-hatchlings sea turtles. Bull Mar Sci. 1990;47:233–43.

De Loreto BO, Bondioli ACV. Epibionts associated with green sea turtles (*Chelonia mydas*) from Cananéia, Southeast Brazil. Mar Turt Newsl. 2008;12:5–8.

Deem SL, Norton TM, Mitchell M, Segars A, Alleman AR, Cray C, Poppenga RH, Dodd M, Karesh WB. Comparison of blood values in foraging, nesting, and stranded loggerhead turtles (*Caretta caretta*) along the coast of Georgia, USA. J Wildl Dis. 2009;45(1):41–56.

Dellinger, T. (2008). Tartarugas marinhas. In: Loureiro, A. Ferrand de Almeida,N. Carretero, M.A. e Paulo, O.S. (eds.) Atlas dos Anfíbios e Repteis de Portugal. Lisboa: Instituto da Conservação da Natureza e da Biodiversidade; 257 pp

Faraldo, R. and Galán, P. (1999). *Lepidochelys kempii* (Tortuga golfina) en la costa de A Coruña. Boletín de la Asociación Herpetológica Española, 10:17

Feldman, G. C. and McClain, C.R. (2014), http://oceancolor.gsfc.nasa.gov, Ocean Color Web, Eds. Kuring, N., Bailey, S. W., Franz, B. A., Meister, G., Werdell, P. J., Eplee, R. E.. 05-09-2014. NASA Goddard Space Flight Center. 11-12-2014. http://oceancolor.gsfc.nasa.gov

Fernández E. Islas Cíes, Serie Naturaleza Gallega. Vigo: Asociación Gallega para la Cultura y la Ecología; 1988. 232pp.

Fiúza, A.F., 1983. Upwelling patterns off Portugal. In: Suess, E., Thiede, J. (Eds.), Coastal Upwelling. New York: Plenum Publishers; pp. 85–87

Garman, S. On certain species of Chelonidae. Bulletin of the Museum of Comparative Zoology, Harvard University. 1880;6:123–126.

Insacco G, Spadola F. First record of Kemp's ridley sea turtle, *Lepidochelys kempii* (Garman 1880) (Cheloniidae), from the italian waters (Mediterranean Sea). Acta Herpetologica. 2010;5(1):113–7.

Johnson SA, Bass AL, Libert B, Marshall M, Fulk D. Kemp's ridley (*Lepidochelys kempii*) nesting in Florida. Florida Scientist. 1999;62:3–4.

Marine Turtle Specialist Group (1996). *Lepidochelys kempii*. In: IUCN 2014. IUCN Red List of Threatened Species. Version 2014.3. <www.iucnredlist.org>. Downloaded on December 30th, 2014.

Márquez, M.R. (1990). FAO species catalogue Vol.11 Sea turtles of the world. An annotated and illustrated catalogue of sea turtle species known to date. FAO Fisheries Synopsis No. 125. Vol. 11. Rome. FAO. 81 pp.

Márquez, M.R. (1994). Synopsis of the biological data on the Kemp's Ridley turtle *Lepidochelys kempii* (Garman 1880). NOAA Technical Memorandum. NMFS-SEFCSC-343.

Márquez MR, Burchfield PM, Diaz J, Sanchez M, Carrasco M, Jimenez C, Leo A, Bravo R, Pena J. Status of Kemp's Ridley sea turtle, Lepidochelys kempii. Chelonian Conserv Biol. 2005;4:761–6.

Meylan A, Castaneda P, Coogan C, Lozon T, Fletemeyer J. First recording nesting by Kemp's ridley in Florida, USA. Mar Turt Newsl. 1990;48:8–9.

National Marine Fisheries Services US, Service FaW and SEMARNAT. Bi-National Recovery Plan for the Kemp's Ridley sea turtle (Lepidochelys kempii). Silver Spring, Mariland: National Marine Fisheries Service; 2011.

Nicolau, L., Marçalo, A., Ferreira, M., Sequeira, M., Vingada, J., Eira, C. (2014). Sea turtle strandings along the Portuguese continental coast: distribution, patterns and insights of bycatch. In: Proceedings of XIII Iberian Congress of

Herpetology, University of Aveiro, Aveiro, 30th September – 04th October 2014, pp. 160–161.

Oliver G, Pigno A. Première observation d'une Tortue de Kemp, *Lepidochelys kempii* (Garman, 1880), (Reptilia, Chelonii, Cheloniidae) sur les côtes françaises de Méditerranée. Bulletin de la Société Herpétologique de France. 2005;116:31–8.

Penrose, R.S. and Gander, L.R. (2015). British Isles & Republic of Ireland marine turtle strandings and sightings annual report 2014. Marine Environmental Monitoring http://www.strandings.com/Graphics%20active/2014%20 Turtle%20Annual%20Strandings%20Report.pdf

Schwartz, F.J. (1978). Behavioural and tolerance responses to cold water temperature by three species of sea turtles (Reptilia, Cheloniidae) in North Carolina. Florida Marine Research Publications 33, 16–18

Sociedad Española de Cetáceos. Recopilación, análisis, valoración y elaboración de protocolos sobre las labores de observación, asistencia a varamientos y recuperación de mamíferos y tortugas marinas de las aguas españolas, Dirección General de Conservación de la Naturaleza. Madrid: Ministerio de Medio Ambiente; 1999. p. 268.

Tomás, J. and Raga, J.A. (2008). Occurrence of Kemp's ridley sea turtle (*Lepidochelys kempii*) in the Mediterranean. *Marine Biodiversity Records* Vol1, e58, 1–2

Tomás J, Formia A, Fernández M, Raga JA. Occurrence and genetic analyses of a Kemp's ridley sea turtle (*Lepidochelys kempii*) in the Mediterranean Sea. Sci Mar. 2003;67:367–9.

Witherington B, Hirama S, Hardy R. Young sea turtles of the pelagic *Sargassum*-dominated drift community: habitat use, population density and threats. Mar Ecol Prog Ser. 2012;463:1–22.

Witt MJ, Penrose R, Godley BJ. Spatio-temporal patterns of juvenile marine turtle occurrence in waters of the European continental shelf. Mar Biol. 2007;151:873–85.

Wyneken, J. (2001). The Anatomy of Sea Turtles. U.S. Department of Commerce NOAA Technical Memorandum NMFS-SEFSC-470, 172 pp.

Zug GR, Kalb HJ, Luzard SJ. Age and growth in wild Kemp's ridley seaturtles *Lepidochelys kempii* from skeletonchronological data. Biol Conserv. 1997;80:261–8.

15

New distributional records of three soldier fishes (Pisces: Holocentridae: *Myripristis*) from Indian waters

Rekha J. Nair* and S. Dineshkumar

Abstract

Three species of *Myripristid* fishes identified as *Myripristis seychellensis*, *M. formosa* and *M. greenfieldi* were collected from Indian waters. This adds to the already existing six species of soldierfishes *Myripristis adusta*, *M. berndti*, *M. botche*, *M. hexagona*, *M. murdjan* and *M. violacea* recorded from Indian and Andaman waters. Since these three fishes are new distributional records for Indian waters, meristic counts, body measurements and descriptions of the specimens are presented.

Keywords: Seychelles soldier, Holocentridae, Range extension, Indian waters

Introduction

Order Beryciformes with two sub families Myripristinae and Holocentrinae (Nelson, 2006) occur mainly on reefs or rocky substrate at shallow to moderate depths of tropical and sub-tropical waters of Indo-Pacific. Subfamily Holocentrinae consists of 3 genera *Holocentrus* (2 Atlantic species), *Sargocentron* (5 Atlantic and about 21 Indo-Pacific species) and *Neoniphon* (1 Atlantic and 4 Indo-Pacific species); subfamily Myripristinae consists of 5 genera *Corniger* Agassiz, *Pristilepis* Randall, Shimizu and Yamakawa, *Plectrypops* Gill, *Ostichthys* Cuvier and *Myripristis* Cuvier. Of these five genera, *Myripristis* (soldierfishes) is the largest with about 28 species recorded from all over the world (Randall and Greenfield 1996; Froese and Pauly 2010 online). Six species of soldierfishes *Myripristis adusta*, *M. berndti*, *M. botche*, *M. hexagona*, *M. murdjan* and *M. violacea* are recorded from Indian and Andaman waters (Randall and Greenfield 1996; Rajan et al., 2011). Myripristid fishes are characterized by deep ovate moderately compressed body with the last dorsal spine distinctly longer than the penultimate; snout very short, blunt; eye very large with slightly concave interorbital; coarsely ctenoid body scales. These fishes are reported to occur on reefs or rocky strata from shallow to moderate depths. They are nocturnal and characterized by

their large eyes hiding in crevices or caves in aggregations during day and dispersing at night to feed on zooplankton at night. During the course of study on the reef fishes of India, three soldier fishes which are new distributional records for Indian waters have been recorded. The fishes were identified as *Myripristis seychellensis*, *M. formosa* and *M. greenfieldi*. *M. seychellensis* was earlier recorded from Seychelles, Red Sea from the Western Indian Ocean; *M. greenfieldi* from Ogasawara Islands, southern Japan and is reported to be endemic to the region while *M. formosa* was first reported from Taiwan. The three fishes were hitherto not reported from the Indian waters and the present account is the first report of the species from these waters.

Materials and methods

Nine specimens of *Myripristis seychellensis* in the length range 168.01–217.01 mm and weight 95–190 g respectively were collected on 25 February 2012 during routine fish survey at Cochin Fisheries Harbour, from the commercial trawler vessels operating off west coast of India at depth range between 50 and 200 m. Later in February 2012 one specimen of *M. formosa* was also collected from Cochin Fisheries Harbour. *M. greenfieldi* was from a previous 2006 collection from Cochin Fisheries Harbour. Methods of taking counts and measurements follow Hubbs and Lagler (1949). Gut content and sex of the collected specimens were also studied. The fishes were identified as *Myripristis seychellensis* (Fig. 1) *M. formosa*

* Correspondence: rekhacmfri@gmail.com
Central Marine Fisheries Research Institute, Ernakulam North PO, Kochi 682018, Kerala, India

Fig. 1 *Myripristis seychellensis* collected from Cochin Fisheries Harbour

Fig. 3 *Myripristis greenfieldi* collected from Cochin Fisheries Harbour

(Fig. 2) and *M. greenfieldii* (Fig. 3) respectively based on available literature (Randall and Greenfield 1996). The specimens were preserved and deposited in the reef fish collection in the Demersal Fisheries Division, CMFRI, Kochi, India as DFD-1-2012(1/9), DFD-2-2012 and DFD-3-2012 respectively.

Results

Systematics

Order BERYCIFORMES
    Family BERYCIDAE Lowe 1839
    Subfamily Myripristinae Nelson 1955
    Genus *Myripristis* Cuvier 1829
    Myripristis seychellensis Cuvier 1829
    (Fig. 1)

Materials examined

Myripristis seychellensis Cuvier 1829 DFD-1-2012(1/9) 9 specimens with total length range 168.01–217.01 mm and weight 95–190 g collected from Cochin Fisheries Harbour, Kerala, India.

Diagnosis

Dorsal fin rays X+I, 14; anal fin rays IV, 13–14; pectoral fin rays 15–16; lateral-line scales 28–30; mouth terminal with lower jaw slightly inferior when closed; posterior border of teeth on vomer rounded.

Fig. 2 *Myripristis formosa* collected from Cochin Fisheries Harbour

Description

Body deep, ovate moderately compressed with large rough ctenoid scales on the body; gill rakers 11–13 on upper limb and 22–23 on lower limb, with a total of 33–36. Body oblong, moderately compressed, depth 2.3–2.7 times in standard length; head length 2.6–2.7 times in standard length; eyes large, its diameter 2.1–2.7 times in head length; mouth moderate, terminal, slightly inferior when closed; maxilla vertically reaching the posterior edge of orbit; pre-operculum finely serrated. Fourth dorsal spine longest, its length 1.3 times in longest dorsal ray; third anal spine length 1.6 times in longest anal ray; caudal peduncle depth less than eye diameter, its depth 1.3–1.8 times in eye diameter; caudal fin forked, lobes rounded.

Colour: Body and head reddish pink to white dorsally and pinkish silvery ventrally; edges of the scales reddish. Head light red, scales on operculum pinkish silver with edges tipped light red. A prominent black blotch seen on opercular membrane, diameter equal to the length of eye diameter; blotch fades or becomes dusky ventrally; axil of pectoral, dorsal, pectoral base blackish; iris dark reddish with a black blotch dorsally. Dorsal fin spines reddish to pink; upper half of inter spine membranes orangish – red, lower part whitish; soft dorsal, anal and caudal rays reddish yellow with a whitish leading edge, followed by a black streak. Pelvic fin rays pinkish white with a whitish outer edge, followed by reddish band. Pectoral fin pale red to transparent.

Colour in formalin: Body colour fades to creamy white - brownish except the black blotches on the opercular membrane and the black pectoral axil.

Order BERYCIFORMES
    Family BERYCIDAE Lowe 1839
    Subfamily Myripristinae Nelson 1955
    Genus *Myripristis* Cuvier 1829
    Myripristis formosa (Fig. 2)

Material examined

Myripristis formosa (DFD-2-2012), 1 specimen with total length 177.41 mm and weight 120 g collected from Cochin Fisheries Harbour, Kerala, India.

Table 1 Morphometric comparison of *Myripristis seychellensis* (in SL and HL) from India with the earlier records

Morphometric measurements	Randall and Gueze 1981 Red Sea	Chen et al. 1990 Taiwan	Randall and Greenfield 1996 Seychelles	Present specimens 2012 Kochi
Dorsal fin	X+I, 14–15	X+I, 14	X+I, 14–15	X+I, 14–15
Anal fin	IV, 13	IV, 11	IV, 13	IV, 13–14
Pectoral Fin	14–15	15	14–15	15–16
Lateral line	28	28	28	28–30
Gill rakers	(12–15) + (25–29)	11 + 22	(12–15) + (25–29)	(11–13) + (22–23)
In standard length				
Body depth	2.19–2.38	2.20	2.2–2.4	2.3–2.7
Head length	2.58–3.02	2.94	2.6–3	2.6–2.7
In head length				
Eye diameter	*	2.17	*	2.1–2.7
Snout length	4.75–5.28	7.28	*	7.6–8.8
Inter orbital	4.19–4.50	4.88	4.1–4.5	5–5.8
3rd dorsal spine	2.29–2.49	2.11	*	2.7
2nd dorsal ray	*	1.54	*	1.8–2
4th anal spine	*	2.44	2.1–2.45	2.9
2 or 3rd anal ray	1.31–1.58	*	*	1.6–1.9
Depth of caudal peduncle	3.39–3.83	3.27	*	3.7–3.9
Pectoral fin length	1.47–1.56	*	*	1.4–1.6
Ventral fin Length	1.46–1.66	*	*	1.5–1.9
Caudal fin length	1.16–1.29	*	*	1.3

*Not available

Diagnosis

D XI, 14; A IV, 12; P 16; V I, 7; Ll 33; Gr 11+22. Deep bodied, oblong soliderfish with deeply forked caudal fin and serrated pre-operculum.

Description

Body deep, oblong, depth 2.4 times in SL; head length 2.9 times in SL; snout short; eyes big, diameter 2.2 times in HL. Mouth oblique, terminal; lower jaw protruding slightly and fitting into a notch in the upper jaw. Very small villiform teeth on the vomer in a triangular a patch. Preoperculum serrated; operculum with single spine. Dorsal fin origin on a vertical straight above pelvic fin origin; above second to third lateral line scale. Fins with hard, sharp spines; caudal peduncle small, narrow; caudal fin deeply forked. Scales strongly ctenoid with ridges on surface.

Colour in fresh condition: Body reddish to silvery white; dorsal, anal and paired fins reddish, interfin membranes slightly pale red to white; tips of soft dorsal and anal fins with a blackish spot; the leading edges of all fins white. Axil of pectoral fins blackish except the white scaled part. Outer free central end of opercular membrane is blackish; a black patch extends vertically across the eye.

Order BERYCIFORMES
Family BERYCIDAE Lowe 1839
Subfamily Myripristinae Nelson 1955
Genus *Myripristis* Cuvier 1829
Myripristis greenfieldi Randall & Yamakawa 1996
(Fig. 3)

Diagnosis

Myripristis greenfieldi Randall & Yamakawa 1996 (DFD-3-2012), 1 specimen with total length 215 mm and standard length 182 mm collected from Cochin Fisheries Harbour, Kerala, India in 2006

Material examined

D X+I, 14; A IV, 12; pectoral fin rays 14, pelvic I, 5; lateral line scales 29; gill rakers 43. Small soldier fish with oblong body, broad interorbital space, forked caudal fin with pointed lobes. Scales absent in the axil of the pelvic fins; pelvic, second, dorsal and caudal fin with white leading lines. Body covered with thick osseous scales.

Description

Body compressed, depth 2.4 times in SL; head length 2.9 times in SL; eyes large, diameter 2.1 times in HL; broad inter-orbital space; mouth moderate, terminal; narrow

Table 2 Morphometric comparison of *Myripristis formosa* from India with its holotype and paratype

Morphometric measurements	Chen et al. 1990 HOLOTYPE Taiwan	Chen et al. 1990 PARATYPE Taiwan	Present Specimen 2012 Kochi
Dorsal fin	X+I, 14	X+I, 14	X+I, 14
Anal fin	IV, 12	IV, 12	IV, 12
Pectoral Fin	14–15	15	15
Lateral line	28	28	28
Gill rakers	10–11 + 22–23	10–11 + 22–23	23
As % of standard length			
Body depth	47.5	45.2	28.55
Body width	21.8	21.1	39.64
Head length	35.8	34.8	34.36
Snout length	7.5	7.3	5.15
Orbit diameter	15.1	13.6	15.24
Inter-orbital width	7.6	8.1	7.77
Upper jaw length	20.3	19.4	19.53
Caudal peduncle depth	11	10.7	9.62
Caudal peduncle length	12.3	13	11.98
Pre-dorsal length	40.2	38.3	40.38
Pre-anal length	74.5	73.0	70.41
Pre-pelvic length	42.9	39.7	39.64
First dorsal spine	10.4	9.4	9.5
Second dorsal spine	13.7	12.9	12.5
Longest dorsal spine	15.1	14.2	15.46
Tenth dorsal spine	4.9	5.1	5.2
Eleventh dorsal spine	10.1	10.3	*
Longest dorsal ray	22.2	21.4	20.04
First anal spine	2.8	1.9	*
Third anal spine	13.2	13.2	*
Fourth anal spine	13.9	13.3	*
Longest anal ray	22.4	21.6	18.42
Caudal fin length	27.3	24.5	25.74
Caudal concavity	14.2	13.3	*
Pectoral fin length	24.9	23.7	26.33
Pelvic spine length	15.9	15.1	*
Pelvic fin length	22.7	22.8	17.83

*not available

band of villiform teeth on palatine. Scales absent in the axil of the pectoral fin; pre-operculum finely serrated. Dorsal rays longer than spines, length 1.6 times in HL; third anal spine length 2.7 times in HL

Colour in fresh condition: Body scales reddish dorsally, silvery white ventrally with a brassy sheen. Upper half of dorsal fin reddish and basal region white. Prominent black blotches on pectoral axil and opercular membrane. Head reddish in colour; soft dorsal, anal, caudal and pelvic fins light red.

Colour in formalin: Body colour fades to creamy white - brownish except the black blotches on the opercular membrane and the black pectoral axil.

Remarks

M. seychellensis gets its name from its unique locality from where it was first described. *M. seychellensis* is identified by its unique characters, terminal mouth, slightly inferior lower jaw when mouth is closed, posterior border of vomerine tooth patch rounded and the

Table 3 Morphometric comparison of *Myripristis greenfieldi* from India with earlier records

Morphometric measurements	Randall and Greenfield 1996	Present specimen
Dorsal fin	X+I, 14–15	X+I, 14
Anal fin	IV, 11–12	IV, 12
Pectoral fin	14–16	14
Lateral line scales	28–29	29
Gill rakers	(15–17) + (28–31)	-
In standard length		
Body depth	2.25–2.5	2.4
Head length	2.75–2.9	2.9
Inter orbital space	3.6–4.1	3.8
Third anal spine	1.85–2.1	2.7

orange red upper half of inter spine membrane of dorsal fin. The morphometric and meristic ratios were compared with previous records from different waters, agrees well except that of interorbital values (Table 1). The sex and gut content of the collected 10 species were studied. The majority (8 specimens) were female at the stage F4 and F5 with fully digested material in the gut. One big specimen of length 166 mm had a small trigger fish in its gut. This species was previously only reported from Seychelles, Reunion, St. Brandon's Shoals and Madagascar of Western Indian Ocean and Taiwan of Western Pacific. But now with the present study, the distribution range of this species has extended to Indian waters. *M.seychellensis* differs from its close relative *M.murdjan* in lesser number of lateral line scales (27–32 in *M. murdjan*) and more number of pectoral fin rays (14–16 in *M. murdjan*). Randall's observation of the largest specimen was only 175 mm while the present specimens were in the length range of 168–217 mm.

Myripristis formosa was previously identified as *M. seychellensis* a species known only from Western Indian Ocean by Chen et al. (1990). However, it differs from the latter in fewer number of gill rakers (33) compared to 37–43 in latter, shorter anal spines compared to latter, 14 dorsal rays compared to 15 in 15 in latter and 12 anal rays in former and 13 in latter. The figure given by Chen et al. (1990: fig. 9) is not of *M. seychellensis* but of *M. formosa* as can be seen by the close resemblance in the picture with the present specimen. The morphometric and meristic measurements of the present specimen match well with that of the holotype and paratype, except slight variations in the case of pectoral and pelvic fin length as can be seen in Table 2. *Myripristis formosa* was available only from Taiwan (Randall and Greenfield 1996), with the present collection from Kochi, Kerala, the fish has shown distributional extension to the southwest coast of India to the Arabian sea. The fish comes

under the Data Deficient (DD) category of the IUCN as per IUCN 2013 online (McEachran et al. 2010). This confirms the presence of this specimen in Indian waters (Table 3).

In the *case of Myripristis greenfieldi* only one sample was obtained. This fish was previously reported only from Japan, Ogasawara and Ryukyu Islands (Randall and Heemstra 1985). With the collection from Kochi waters, Kerala, India the fish has shown a distributional extension to Indian waters.

Acknowledgement
The authors wish to thank Director and Head, Division of Demersal Fisheries Central Marine Fisheries Research Institute, Kochi, India for the encouragement and support. The financial support from Ministry of Environment and Forest is gratefully acknowledged.

Authors' contributions
RJN prepared the design of the study, RJN and SD were both involved in the collection of fish samples, its meristic and morphometric measurements and taxonomic identification. RJN prepared and evaluated the manuscript. Both authors thoroughly read the manuscript approved and prepared the final version of it.

Competing interests
The authors declare that they have no competing interests.

References
Chen JP, Shao KT, Mok H-K. A review of the myripristin fishes from Taiwan with description of a new species. Bull Inst Zool Acad Sinica. 1990;29(4):249–64.

Froese R, Pauly D, editors. FishBase. 2010. Available at http://www.fishbase.org. (Accessed 11/2010).

Hubbs CL, Lagler KF. Fishes of the Great Lakes region. Bull Cranbrook Inst Sci. 1949;26:1–186.

McEachran J, Moore JA, Williams J. Myripristis formosa. The IUCN Red List of Threatened Species. 2010. e.T155249A4757203. http://dx.doi.org/10.2305/IUCN. UK.2010-4.RLTS.T155249A4757203.en.

Nelson JS. Fishes of the World. 4th ed. Hoboken: John Wiley & Sons; 2006; i-xix + 1–601.

Rajan PT, Sreeraj CR, Immanuel T. Fish fauna of coral reef, mangrove, freshwater, offshore and seagrass beds of Andaman and Nicobar Islands. Zoological Survey of India, Andaman and Nicobar Regional Centre, Haddo, Port Blair. 2011.

Randall JE, Greenfield DW. Revision of the Indo-Pacific holocentrid fishes of the genus Myripristis, with descriptions of three new species. Indo-Pacific Fish. 1996;25:61.

Randall JE, Gueze P. The holocentrid fishes of the genus Myripristis of the Red Sea, with clarification of the murdjan and hexagonus complexes. Cent Sci Los Angeles. 1981;334:1–16.

Randall JE, Heemstra PC. A review of the squirrel fishes of the subfamily Holocentrinae from the western Indian Ocean and the Red Sea. Ichth Bull Smith Inst Ichth. 1985;49:1–27.

Randall JE, Yamakawa T. Two new soldierfishes (Beryciformes: Holocentridae: Myripristis) from Japan. Ichthyol Res. 1996;43(3):211–22.

16

First record of *Siganus randalli* (Teleost, Siganidae) in New Caledonia, and comments on its diet

Thibaud Moleana[1,3,4*], Luc Della Patrona[2], Tarik Meziane[3] and Yves Letourneur[1]

Abstract

Background: Most of the 29 Siganidae species are widely distributed through the Indo-Pacific area. In New Caledonia, these family was represented by 12 species. The present report is the first record of *Siganus randalli* in New Caledonian waters and provide information on its diet.

Methods: Three specimens of *Siganus randalli* were caught in shallow mangrove waters of the southern part of New Caledonia. Their stomach contents and isotopic signatures (carbon and nitrogen) were analyzed and compared to others siganids species.

Results and conclusion: This note provides the most southerly record of the rabbitfish *Siganus randalli*, which extends its distribution range by 1200 km southward and 1300 km southwest. The data on its diet, when compared with other co-occurring or more reef-associated siganid species, provide information on feeding processes and ecological functions associated with its mangrove habitat.

Keywords: Variegated rabbitfish, Diet, Stomach content, Isotope, SW pacific

Abbreviations: ADECAL, Agence de développement économique de la Nouvelle-Calédonie; ARR, Arrêté; C, Carbon; DENV, Direction de l'environnement; G, grams; IRI, Index of relative importance; LIVE, Laboratoire insulaire du vivant et de l'environnement; MM, millimeter; N, Nitrogen; SD, standard deviation; TL, Total length; UNC, Université de la Nouvelle-Calédonie

Background

Most of the 29 Siganidae species are widely distributed through the Indo-Pacific area (Woodland 1990; Randall and Kulbicki 2005). Their habitats range from estuaries and mangroves to outer reefs. This local distribution is closely related to their ecological function as browsers, croppers, mixed-feeders and spongivores (Hoey et al. 2013). In New Caledonia, 12 species have been recorded, though one of them needs further verification (*S. vermiculatus*; Fricke et al. 2011). Siganid species are currently identified by color and morphology characteristics such as snout shape, color patterns, body depth and caudal fin shape (Woodland 1990; Randall and Kulbicki 2005;

Borsa et al. 2007), but similarities in these characteristics between closely related species may lead to misidentification (Randall and Kulbicki 2005).

The variegated rabbitfish *Siganus randalli* (Woodland 1990) is a little known species and the information available mostly relates to aquaculture (Collins and Nelson 1993; Brown et al. 1994; Nelson and Wilkins 1994). It lives in brackish waters of mangroves (Collins and Nelson 1993; Nelson and Wilkins 1994), as well as above sandy bottoms of coral reef flats (Woodland 1990). The species is distributed widely from Guam (Kamikawa et al. 2015) and the Philippines archipelago (Galenzoga and Quiñones 2014) to Fiji (Blaber et al. 1993).

The aim of this paper is, firstly, to record the presence of *Siganus randalli* in New Caledonia and, secondly, to provide informations on its diet through an analysis of stomach contents and stable isotope signatures. The diet

* Correspondence: t.moleana@gmail.com
[1]Laboratoire LIVE et LABEX « Corail », Université de la Nouvelle-Calédonie, BP R4, Nouméa 98851, New Caledonia
[3]UMR BOREA 7208, MNHN/CNRS/UPMC/IRD, 61 rue Buffon, Paris Cedex 5 CP 53, 75231, France
Full list of author information is available at the end of the article

Fig. 1 Specimen (UNC-Y1002) of *Siganus randalli* caught during the experimental fishing session. Photo T. Moleana

data allow a better understanding of the functional role and habitat use of this species.

Materials and methods

On 5 March 2014, three specimens of *Siganus randalli* were collected (Fig. 1) at Goro (22° 17′ S; 167° 1′ E), southeast of New Caledonia. All specimens were caught with a gillnet in shallow mangrove waters (0.5–1 m depth). Individual weights were measured to the nearest gram and total, fork and standard lengths and body depth were recorded to the nearest millimeter. In addition, three specimens of *Siganus lineatus* (Valenciennes, 1835) and two specimens of *Siganus fuscescens* (Houttuyn, 1782) were collected along with the three *S. randalli*, and were also analyzed for a comparison of diets.

Stomach contents were collected, weighed and preserved in a 4 % formaldehyde solution. Stomach fullness was determined using the ratio between total stomach content weight and total fish weight. Each item found in the stomachs was then identified at the lowest taxonomic level and weighed. Occurrence (F%) of items and proportions by number (N%) and by weight (W%) were monitored to estimate the Index of Relative Importance (IRI; Pinkas 1971) and express it as a percentage (%IRI) of the total IRI of the stomach content. To ascertain feeding patterns, standardized niche breadth (B; Hurlbert 1978) and niche overlap (O; Pianka 1974) were determined using %IRI of each item to take into account their

occurrence and proportion by number and weight. Following Pinnegar and Polunin (1999), carbon and nitrogen stable isotope ratios (δ^{13}C and δ^{15}N) were measured on dorsal white muscle of the jointly collected three *S. randalli*, two *S. fuscescens* and three *S. lineatus*. Other δ^{13}C and δ^{15}N values obtained from siganids collected at reef sites off Noumea (22° 14′ S; 166° 28′ E) were also used for species and habitat comparisons (Letourneur, unpublished data for 10 *Siganus lineatus* 186–209 mm TL; 5 *S. punctatus* 200–261 mm TL; 6 *S. puellus* 170–212 mm TL).

Results

The three *Siganus randalli* specimens exhibited the same meristic characters, body shape and color pattern as those described by Woodland (1990). Morphometric data are given in Table 1.

The stomach fullness values of *S. randalli* and *S. lineatus* caught in the mangrove was respectively 3.35 ± 0.67 and 3.42 ± 0.57. By comparison, low stomach fullness values of the two *S. fuscescens* (0.43 ± 0.03) showed lower feeding activity in the mangrove.

The diet of the *S. randalli* comprised 15 food items (Table 2). Rhodophyta and Phaeophyta were dominant in the stomach contents, which consisted mainly of *Bostrychia* sp. (75.6 %), *Dictyota* sp. (10.7 %) and *Lomentaria* sp. (4.2 %). This trend was also apparent in terms of %W (67.2, 17.7 and 6.9 %) and hence of

Table 1 Morphometric data of *S. randalli*, *S. lineatus* and *S. fuscescens* collected in Goro, New Caledonia

N° UNC	S. randalli			S. lineatus			S. fuscescens	
	Y1001	Y1002	Y1003	Y1008	Y1014	Y1019	Y1004	Y1005
Weight (g)	247.6	308.4	223.1	228.2	250.9	171.2	378.5	245
Total length (mm)	240	253	231	235	237	212	306	274
Fork length (mm)	220	235	213	217	220	197	278	246
Standard length (mm)	178	189	170	174	177	186	232	201
Body depth (mm)	95	101	90	92	92	85	100	85

Table 2 Diet composition of *S. randalli* (*n* = 3), *S. fuscescens* (*n* = 2) and *S. lineatus* (*n* = 3) caught in Goro, New Caledonia, and expressed in frequency of occurrence (%F), percent by number (%N) and percent by weight (%W). %IRI: Percent of Index of Relative Importance

Items prey	S. randalli				S. lineatus				S. fuscescens			
	F%	N%	W%	%IRI	F%	N%	W%	%IRI	F%	N%	W%	%IRI
Magnoliophyta												
Halophila spp.	-	-	-	-	-	-	-	-	50.0	1.0	2.8	1.5
Thalassia hemprichii	-	-	-	-	-	-	-	-	100.0	5.7	8.3	11.5
Rhodophyta												
Amphiroa sp.	-	-	-	-	-	-	-	-	50.0	42.9	55.5	40.2
Bostrychia sp.	100.0	75.6	67.2	74.2	100.0	49.0	44.6	62.0	-	-	-	-
Centroceras sp.	-	-	-	-	25.0	0.1	<0.1	<0.1	-	-	-	-
Ceramiaceae spp.	100.0	0.4	0.1	0.3	50.0	3.0	0.1	1.0	100.0	18.1	1.4	15.9
Ceramium spp.	-	-	-	-	50.0	0.9	0.1	0.3	-	-	-	-
Gelidiella sp.	-	-	-	-	-	-	-	-	50.0	17.1	18.5	14.6
Gelidium sp.	66.7	3.7	3.8	2.6	50.0	0.7	0.4	0.4	-	-	-	-
Gracilaria sp.	-	-	-	-	25.0	0.1	0.1	<0.1	-	-	-	-
Hypnea spp.	-	-	-	-	25.0	<0.1	0.1	<0.1	50.0	1.0	2.3	1.3
Hypoglossum sp.	33.3	<0.1	0.1	<0.1	-	-	-	-	-	-	-	-
Lomentaria sp.	100.0	4.2	6.9	5.7	100.0	4.6	6.4	7.3	-	-	-	-
Polysiphonia sp.	66.7	1.1	1.8	1.0	50.0	9.0	16.1	8.3	-	-	-	-
Tolypiocladia sp.	-	-	-	-	25.0	0.3	0.1	0.1	-	-	-	-
Chlorophyta												
Caulerpa verticillata	66.7	0.7	0.4	0.4	75.0	3.3	1.7	2.5	-	-	-	-
Chlorodesmis sp.	-	-	-	-	25.0	6.0	10.7	2.8	-	-	-	-
Ulva sp.	-	-	-	-	-	-	-	-	50.0	4.8	0.9	2.3
Phaeophyta												
Padina sp.	33.3	1.5	2.0	0.6	25.0	4.2	2.3	1.1	-	-	-	-
Dictyota sp.	100.0	10.7	17.7	14.8	50.0	9.9	16.7	8.8	-	-	-	-
Invertebrates												
Amphipods	66.7	0.2	0.1	0.1	75.0	0.4	0.3	0.4	-	-	-	-
Ascidians	33.3	0.3	<0.1	0.1	25.0	0.1	0.1	<0.1	50.0	2.9	3.8	2.7
Copepods	-	-	-	-	75.0	0.5	<0.1	0.2	-	-	-	-
Foraminiferans	66.7	0.3	<0.1	0.1	100.0	3.1	<0.1	2.2	50.0	1.9	<0.1	0.8
Halacarids	33.3	0.5	<0.1	0.1	25.0	0.2	<0.1	<0.1	-	-	-	-
Hydrozoans	-	-	-	-	-	-	-	-	100.0	4.8	6.6	9.2
Isopods	-	-	-	-	25.0	0.1	<0.1	<0.1	-	-	-	-
Diptera larvae	-	-	-	-	25.0	0.4	<0.1	0.1	-	-	-	-
Nematods	33.3	0.5	<0.1	0.1	100.0	3.3	<0.1	2.2	-	-	-	-
Polychaetes	33.3	0.3	<0.1	0.1	75.0	0.8	<0.1	0.4	-	-	-	-

%IRI (74.2, 14.8 and 5.7 %). Consequently, niche breadth had a very low value (*B* = 0.05).

S. lineatus consumed more food items than *S. randalli*, but a similar trend occurred with regard to dominance of Rhodophyta and Phaeophyta. In accordance with %IRI, the diet of *S. lineatus* was dominated by *Bostrychia* sp., *Dictyota* sp., *Lomentaria* sp. and *Polysiphonia* sp.

The niche breadth of this species also had a very low value (*B* = 0.07).

By contrast, only 10 items were found in the diet of *S. fuscescens*, again dominated by Rhodophyta, but also by Magnoliophyta and, to a lesser extent, Chlorophyta and invertebrates. The niche breadth of this species had a higher value than *S. randalli* and *S. lineatus* (*B* = 0.37).

Table 3 Mean stable isotope ratios (±sd) δ13C and δ15N of *S. randalli*, *S. lineatus* and *S. fuscescens* from mangrove, and of *S. lineatus*, *S. punctatus* and *S. puellus* from coral reef, New Caledonia

Species	Mangrove			Reef		
	N	δ^{13}C (‰)	δ^{15}N (‰)	N	δ^{13}C (‰)	δ^{15}N (‰)
S. fuscescens	2	−14.10 ± 0.29	7.00 ± 0.34	-	-	-
S. randalli	3	−27.95 ± 0.68	5.67 ± 0.21	-	-	-
S. lineatus	3	−26.13 ± 1.50	5.57 ± 0.23	10	−25.83 ± 2.22	5.06 ± 1.03
S. punctatus	-	-	-	5	−14.91 ± 1.10	5.72 ± 0.35
S. puellus	-	-	-	6	−16.42 ± 1.97	6.70 ± 1.23

Strong niche overlap was found between *S. randalli* and *S. lineatus* (O = 0.99), but not between *S. randalli* and *S. fuscescens* (O = 0.01) or between *S. fuscescens* and *S. lineatus* (O = 0.01).

S. fuscescens has the most ^{13}C-enriched and ^{15}N-enriched values, whereas *S. randalli* and *S. lineatus* from both habitats (mangrove and coral reef) exhibited quite similar values for δ^{13}C and δ^{15}N (Table 3). *Siganus punctatus* and *S. puellus* had values similar to *S. fuscescens* for δ^{13}C, but their nitrogen isotopic signatures occupied a more intermediate position between *S. fuscescens* and *S. lineatus* from mangrove.

Discussion

The record of *Siganus randalli* from New Caledonian waters greatly extends its distribution, by approximately 1200 km southward and 1300 km southwest, compared to the closest previously recorded locations (i.e. the Solomon Islands and Fiji; Woodland 1990; Blaber et al. 1993). In view of the geographical distribution overlay (Woodland 1990) and the close phylogenetic similarities between *S. lineatus* and *S. randalli* (Borsa et al. 2007), this range extension is not surprising.

Apart from similar morphology characteristics, siganids can be separated into functional groups according to their feeding patterns. *Siganus randalli* revealed considerable differences from *S. fuscescens* in stomach content composition, a finding that explains the very low niche overlap for the two species. The close similarity found with *S. lineatus* in terms of stomach fullness, niche breadth and niche overlap suggests that *S. randalli* consumes the food resources of the mangrove in a comparable manner. Finally, the dominance of *Bostrychia* sp., a mangrove-associated algae (Zuccarello et al. 2006; Zuccarello and West 2008) in the diet of *S. randalli* highlights the importance of mangrove habitats for this species. Isotopic signatures support this view. Whereas *S. fuscescens* and *S. puellus* differ from *S. randalli* in their diets, functional groups (Debenay et al. 2011; Hoey et al. 2013) and hence their δ^{13}C and δ^{15}N, the mixed feeder *S. punctatus* revealed higher δ^{13}C, suggesting a diet of different origin. *Siganus lineatus* exhibits similar δ^{13}C and δ^{15}N for both locations, suggesting that its diet

has a single habitat origin. Most notably, similarities in the isotopic ratios between *S. randalli* and *S. lineatus* indicated a comparable resource origin, utilization and assimilation by these two species. This low or even negligible food partitioning between *S. lineatus* and *S. randalli* may indicate trophic competition in the mangrove habitat. Our observations need further confirmation because of the small number of individuals examined.

Conclusion

This note provides the most southerly record of the rabbitfish *Siganus randalli*. The data on its diet, when compared with other co-occurring or more reef-associated siganid species, provide information on feeding processes and ecological functions associated with its mangrove habitat. Several other specimens of *S. randalli* have been reported by fishermen, always in mangroves, in the northern part of New Caledonia, but individuals need to be captured to confirm these reports.

Acknowledgements
We warmly thank Germain Navarre Vama for allowing us to collect these specimens. We also thank Philippe Borsa and Gérard Mou-Tham for their valuable help.

Funding
This work was financially supported by Operation SIGA-NC, Zonéco Program, ADECAL Technopôle, New Caledonia.

Authors' contribution
TMo, LDP and YL carried out fieldwork; TMo and LDP analyzed the stomach contents; TMo and YL analyzed the isotopic signatures; TMo analyzed the data and wrote the manuscript; LDP, YL and TMe coordinated the study and corrected the draft manuscript. All authors read and approved the final manuscript.

Competing interests
The authors declare that they have no competing interests.

Author details
[1]Laboratoire LIVE et LABEX « Corail », Université de la Nouvelle-Calédonie, BP R4, Nouméa 98851, New Caledonia. [2]Ifremer LEAD-NC, 101 Promenade Roger Laroque, Nouméa BP 2059, 98846, New Caledonia. [3]UMR BOREA 7208, MNHN/CNRS/UPMC/IRD, 61 rue Buffon, Paris Cedex 5 CP 53, 75231, France. [4]Aqualagon SARL, BP 2525 Mont-Dore, Nouméa 98800, New Caledonia.

References

Blaber SJM, Milton DA, Rawlinson NJF, Sesewa A. A checklist of fishes recorded by the Baitfish Research Project in Fiji from 1991 to 1993. Pages 102–110. In: Blaber SJM, Milton DA, Rawlinson NJF, Sesewa A, editors. Tuna Baitfish in Fiji and Solomon Islands: Proceedings of a workshop, Nadi, Fiji 17–18 August 1993, ACIAR Proceedings, 52. Canberra: Australian Centre for International Agricultural Research; 1993. http://trove.nla.gov.au/work/30776616?selectedversion=NBD11098819

Borsa P, Lemer S, Aurelle D. Patterns of lineage diversification in rabbitfishes. Mol Phylogenet Evol. 2007;44:427–35.

Brown JW, Chiricheti P, Crisostomo D. A cage culture trial of *Siganus randalli* in Guam. Asian Fish Sci. 1994;7:53–6.

Collins LA, Nelson SG. Effects of temperature on oxygen consumption, growth, and development of embryos and yolk-sac larvae of Siganus randalli (Pisces: Siganidae). Mar Biol. 1993;117:195–204.

Debenay JP, Sigura A, Justine JL. Foraminifera in the diet of coral reef fish from the lagoon of New Caledonia: Predation, digestion, dispersion. Rev Micropaleontol. 2011;54(2):87–103.

Fricke R, Kulbicki M, Wantiez L. Checklist of the fi shes of New Caledonia, and their distribution in the Southwest Pacific Ocean (Pisces). Stuttg Beitr Natkd Ser A. 2011;4:341–463. http://www.naturkundemuseum-bw.de/sites/default/files/publikationen/serie-a/ns04-19frickekulbickiwantiez.pdf.

Galenzoga D, Quiñones G. Species Composition and Abundance of Marine Fishes in Selected Landing Areas of Northern Samar, Philippines. Pages 81–87. In: Rahman A, Ahmadi R, editors. International conference on Chemical, Environnement & Biological Sciences Sept. 17–18, 2014, (CEBS-2014). Kuala Lumpur; 2014.

Hoey AS, Brandl SJ, Bellwood DR. Diet and cross-shelf distribution of rabbitfishes (f. Siganidae) on the northern Great Barrier Reef: Implications for ecosystem function. Coral Reefs. 2013;32:973–84.

Hurlbert SH. The measurement of niche overlap and some relatives. Ecology. 1978;59(1):67–77. http://onlinelibrary.wiley.com/doi/10.2307/1936632/abstract;jsessionid=59F6A092D7D0075EA2F5CDE71AEEE07D.f03t02

Kamikawa KT, Cruz E, Essington TE, Hospital J, Brodziak JKT, Branch TA. Length–weight relationships for 85 fish species from Guam. J Appl Ichthyol. 2015;31:1171–4.

Nelson SG, Wilkins S. Growth and respiration of embryos and larvae of the rabbittish *Siganus randalli* (Pisces, Siganidae). J Fish Biol. 1994;44(3):513–25.

Pianka ER. Niche overlap and diffuse competition. Proc Natl Acad Sci. 1974;71(5):2141–5.

Pinkas L. Food habits study. Fish Bull. 1971;152:5–10.

Pinnegar JK, Polunin NVC. Differential fractionation of δ13C and δ15N among fish tissues: implications for the study of trophic interactions. Funct Ecol. 1999;13(2):225–31.

Randall JE, Kulbicki M. *Siganus woodlandi*, new species of rabbitfish (Siganidae) from New Caledonia. Cybium. 2005;29(1):185–9.

Woodland DJ. Revision of the fish family Siganidae with descriptions of two new species and comments on distribution and biology. Indo-Pac Fish. 1990;19:1–136.

Zuccarello GC, West JA. *Bostrychia* (Rhodomelaceae, Rhodophyta) species of New Zealand, and relationships in the Southern Hemisphere. N Z J Mar Freshw Res. 2008;42(3):315–24.

Zuccarello GC, West JA, Loiseaux-de-Goër S. Diversity of the Bostrychia radicans/Bostrychia moritziana species complex (Rhodomelaceae, Rhodophyta) in the mangroves of New Caledonia. Cryptogam Algol. 2006;27(3):245–54.

17

First record of the Berber ponyfish *Leiognathus berbis* Valenciennes, 1835 (Osteichthyes: Leiognathidae) from Syrian marine waters (Eastern Mediterranean)

Firas Alshawy[*], Murhaf Lahlah and Chirine Hussein

Abstract

Background: Climatic changes and human activities have worked to pave the way for alien species to invade new areas far from their native habitat. The Mediterranean sea has received many invasive species (Eissa and Zaki, Procedia Environmental Sciences 4:251-259, 2011; Occhipinti-Ambrogi, Marine Pollution Bulletin 55(7):342-352, 2007), and some of these species had been recorded in the Syria coastal (Saad, Turkish Journal of Fisheries and Aquatic Sciences 5:99-106, 2005).

Method: One specimen of the Berber ponyfish *Leiognathus berbis,* with a total length of 78 mm, was caught by gillnet at a depth of 35 m, where the bottom is sandy soft, on 05 May 2016, in Syrian marine waters at Ibn Hani area (The Eastern Mediterranean Sea).

Results: This study reports that Berber ponyfish *Leiognathus berbis*, a member of Lessepsian species, was found in Syrian marine waters and recorded for the first time there.

Conclusion: This is the first record for *Leiognathus berbis* in the Syrian costal waters, and observations for the first time from the fishermen, There are several factors helped this specimen to arrive to this area of Mediterraean; one of these factors is ballast water.

Keywords: *Leiognathus berbis*, Lessepsian, Mediterranean, Berber ponyfish, Syrian costal

Background

Climatic changes and human activities have worked to pave the way for alien species to invade new areas far from their native habitat. The climatic changes have made the environmental conditions suitable for these species and similar to their original habitat in terms of temperature, salinity and food. While leading human activity, the opening of the Suez Canal, and the movement of ships across the world, is an important factor for making the road, which was impassable, for the fish species to move into new marine environments (Occhipinti-Ambrogi 2007, Eissa and Zaki 2011). The Mediterranean sea has received many invasive species coming from the Atlantic, Pacific and Red Sea. Many species have invaded the Mediterranean species, and settled in, because the marine environment has become suitable for their growth and reproduction (Golani 1998a, Golani 1998b, Oral 2010). To this day, new marine organisms still reach the Mediterranean; of these organisms, those belonging to the family Leiognathidae: Small to medium-sized fish (rarely exceeding 16 cm); body oblong or rounded, moderately to markedly compressed laterally. Eyes moderate to large. Mouth highly protractible, when extended forming a tube directed either upwards (Secutor spp.), forward (Gazza spp.), or forward or downward (Leiognathus spp.). Color :silvery, with characteristic markings on the upper half of sides which are useful for identification (Capenter and Niem 2001, Abraham et al. 2011). In reference to the ocean biogeographic information system (http://www.iobis.org/) and the encyclopedia of life (http://www.eol.org/), the family Leiognathidae exists in water's temperature range (18.528–28.954)°C, and salinity (32.183–35.468) PPS.

* Correspondence: falshawy@gmail.com
Department of Marine Biology, High Institute of Marine Research, Tishreen University, Lattakia, Syria

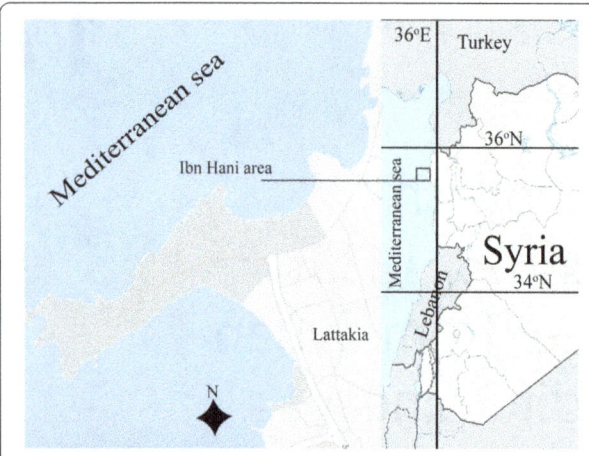

Fig. 1 A map showing the collection site of the specimen from the Syrian marine water (According Google earth)

Methods

The fish sample was collected during May 2016 from Ibn–Hani, Lattakia, Syria (Latitude:35.591632°, Longitude:35.732343°) (Fig. 1). On 15/05/2016, one specimen of *Leiognathus berbis* was carried out by using gillnet, with a mesh size is 15 mm, at a depth of 35 m; the bottom of the fishing zone is sandy soft mixed with some little stones; the net had been deployed in the coastal water for five hours (from 1 am to 6 am). The morphometric measurements and meristic details were recorded for this fish, and conserved at the fish biology lab of the Higher Institute of marine Research, Tishreen University (lenght to the nearest mm, weight to the nearest gram). This sample was identified according to (Carpenter and Niem 2001), depending on the morphological characters. The head

length, the caudal fin length and the eye diameter were measured by vernier caliper as in the Fig. 2. The specimen was preserved in 4% formaldehyde.

Results

In the current study, the specimen *Leiognathus berbis* (Fig. 3a) has the following properties: compressed body and elongated body more than the depth of the body, dorsal and ventral sides are convex, and mouth tapering and downward when protracted (Fig. 3b); the dorsal side is greenish with light gray and contains dark irregular vermiculations. The ventral side is coloured with belly gray; the base of anal and caudal fin are light yellowish. The morphometric measurements are shown in Table 1. The meristic data were: D VIII + 16; P,16; V, I + 5; A, III + 15, the features of *Leiognathus berbis* are in agreement with (Chakrabarty et al. 2010). The sex of the fish is male. The bottom of fishing zone is soft sandy; it is similar to the bottom in the native region of this species (Carpenter and Niem 2001), and is convenient for feeding on benthic invertebrates. The temperature of fishing area is (27.5) °C, and the salinity (38.2) PPS on 15/05/2016; this parameters are close to that are found in the native habitat.

Discussion

The specimen of *Leiognathus berbis* lives from Madagascar to the Red Sea and the Gulf of Aden, along the Indian coasts and off Sri Lanka, eastward to Malaysia, Indonesia, and the Philippines; north to Taiwan and Fukien provinces of China; prefers coastal inshore water as a habitat; at a depth of about 40 m (Carpenter and Niem 2001); and has never been recorded in the Syrian coast before (Saad 2005, Ulman et al. 2015); this fish has been registered in the Suze gulf in 2005 (El-Ganainy et al. 2005). This led to arrival of

Fig. 2 The head length, the caudal fin length and the eye diameter of the fish

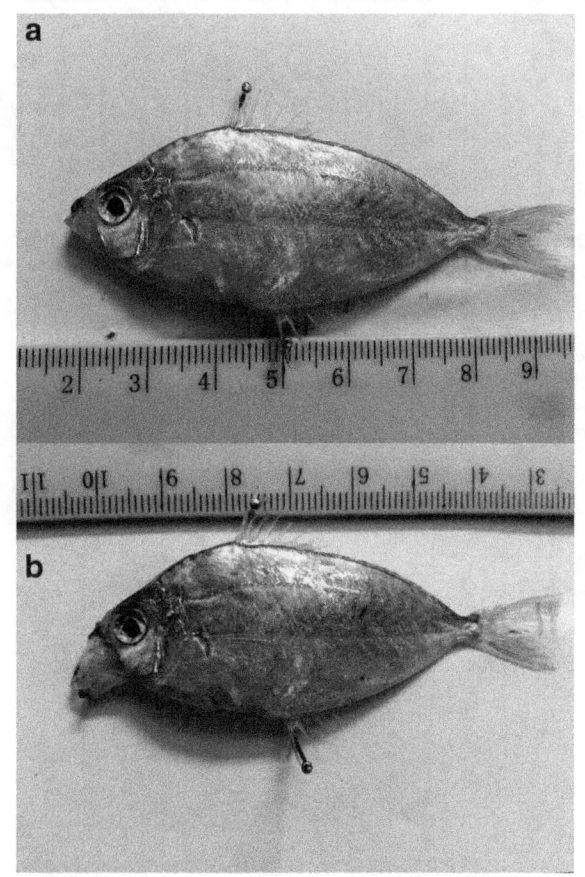

Fig. 3 *Leiognathus berbis* (Valenciennes, 1835) (**a**: general view , **b**: the shap of fish's mouth when feed) with78 mm total f3:2 length, was carried out during May 2016, from Ibn_Hani area (Lattakia–Syria)

the Syrian coast is that the small size of *Leiognathus berbis* allows it to move through the ballast water. This is the first record for *Leiognathus berbis* which is spreading from the Indian Ocean to the Red Sea (Carpenter and Niem 2001); this result shows that the waters of the Syrian coastal water has become more convenient than before to the invasive species which they will compete with the native species; it is possible that the invasive species would become useful by entering into the food chains of other marine organisms, allowing the increase of species and diversity in the new area, particularly in the eastern basin of the Mediterranean (Dial and Roughgarden 1998).

Conclusion

This is the first record for *Leiognathus berbis* in the Syrian costal waters, and the first time they are observed by fishermen; this indicates that there are several factors helped this specimen to arrive to this area of Mediterraean such as ballast water.

Acknowledgements
The authors thank Tishreen University and the High Institute of Marine Research, Lattakia who provided the financial and logistic supports to this work.

Funding
The University of Tishreen, Syria.

Authors' contributions
All authors have equal participation in this work. All authors read and approved the final manuscript.

Competing interests
The authors declare that they have no competing interests.

the specimen of *Leiognathus berbis* from the Gulf of Suez, where the hydrological factors are very close to those found in the Mediterranean Sea; they moved then through the Suez Canal to reach the Syrian coastal. A new Suez Canal had been opened on 9 August 2015, which has made a big chance for fish to move into the Mediterranean sea. On other hand, the climatic changes in the world, especially in the eastern Mediterranean, are making the environment very suitable for invasive species in terms of the temperature, food, and the place for reproduction (Sorte et al. 2010). One point of view explaining the arriaval of this types of fish to

Table 1 Morphometric measurements of *Leiognathus berbis* was caputuer from syrian coastal water during May 2016

Morphometric Measurements	
Total length	78 mm
Stander length	64 mm (82.05% Total Length)
Head length	16 mm (20.51% Total Length)
Caudal fin length	13 mm (16.66% Total Length)
Eye diameter	4.7 mm (6.02% Total Length)
Total weight	6.12 g

References
Abraham K, Joshi K, Murty VS. Taxonomy of the fishes of the family Leiognathidae (Pisces, Teleostei) from the West coast of India. Zootaxa. 2011; 2886:1–18.

Capenter K, Niem V. The living marine resources of the Western Central Pacific: vol 5 bony fishes part 3 (Menidae to Pomacentridae). Roma: Food and Agriculture Organization of the United Nations; 2001.

Carpenter KE, Niem VH. FAO species identification guide for fishery purposes. The living marine resources of the Western Central Pacific. Volume 5. Bony fishes part 3 (Menidae to Pomacentridae). FAO Library; 2001.

Chakrabarty P, Chu J, Nahar L, Sparks JS. Geometric morphometrics uncovers a new species of ponyfish (Teleostei: Leiognathidae: Equulites), with comments on the taxonomic status of Equula berbis Valenciennes. Zootaxa. 2010; 2427(1):15–24.

Dial R, Roughgarden J. Theory of marine communities: the intermediate disturbance hypothesis. Ecology. 1998;79(4):1412–24.

Eissa AE, Zaki MM. The impact of global climatic changes on the aquatic environment. Procedia Environmental Sciences. 2011;4:251–9.

El-Ganainy AA, Yassien MH, Ibrahim EA. Bottom trawl discards in the Gulf of Suez, Egypt. Egypt J Aquat Res. 2005;31:240–55.

Golani D. Distribution of Lessepsian migrant fish in the Mediterranean. Italian Journal of Zoology. 1998a;65(sup1):95–99

Golani D. Impact of Red Sea fish migrants through the Suez Canal on the aquatic environment of the Eastern Mediterranean. Bulletin Series Yale School of Forestry and Environmental Studies. 1998b;103:375–87.

Occhipinti-Ambrogi A. Global change and marine communities: alien species and climate change. Mar Pollut Bull. 2007;55(7):342–52.

Oral M. Alien fish species in the Mediterranean-Black Sea Basin. Journal of the Black Sea/Mediterranean Environment. 2010;16(1):87–132.

Saad A. Check – list of Bony Fish Collected from the Coast of Syria. Turk J Fish Aquat Sci. 2005;5:99–106.

Sorte CJ, Williams SL, Zerebecki RA. Ocean warming increases threat of invasive species in a marine fouling community. Ecology. 2010;91(8):2198–204.

Ulman A, Saad A, Zylich K, Pauly D, Zeller D. Reconstruction of syria's fisheries c atches from 1950–2010: signs of overexploitation. Acta Ichthyol Piscat. 2015; 45:3–259.

18

First record of genus *Paramphitrite* (Polychaeta: Terebellidae) in Mediterranean Sea

Marco Loia[1*], Luisa Nicoletti[2] and Barbara La Porta[1]

Abstract

Background: The presence of species belonging to genus *Paramphitrite* (Terebellidae) has been recorded in the Temperate Northern Atlantic and Arctic regions (Iberian Atlantic, Norwegian and Russian seas). This paper describes the first occurrence of *Paramphitrite birulai* (Ssolowiew, 1899) in the Mediterranean Sea.

Methods: Sampling surveys were carried out in the North Adriatic Sea about 30 nm offshore Chioggia (Italy) on soft seabed at depths ranging from 29 to 32 meters. The sampling plan provided 18 stations.

Results: Seventy-four *Paramphitrite birulai* specimens were examined from a morphological point of view and described in comparison with the existing literature. The species was collected in sandy sediments.

Conclusions: The record of *P. birulai* in the North Adriatic Sea represents the first report of the genus *Paramphitrite* in the Mediterranean Sea extending the distribution range of this species into different ecological and environmental conditions if compared with those where it was previously recorded.

Keywords: Terebellomorpha, *Paramphitrite birulai*, Non-indigenous species, Sandy sediment, North Adriatic Sea

Background

Family Terebellidae Johnston 1846 was originally revised by Malmgren (1866) and the structure of his classification has been accepted with small modifications until today. Further analyses of this family were argued by Saint Joseph (1894), Hessle (1917), Chamberlin (1919) and Fauvel (1927). The most recent revisions of Terebellomorpha were carried out by Nogueira et al. (2010) and Jirkov & Leontovitch (2013) based on previous works by Jirkov (2001) and Holthe (1986).

Genus *Paramphitrite*, erected by Holthe (1976), was originally composed by one species, *P. tetrabranchia* found in Norwegian Sea, and subsequently reported along Atlantic Iberian coast by Parapar et al. (1991). Holthe (1986) included in the genus *Paramphitrite* the species *Amphitrite birulai* Ssolowiew 1899 and in addition remarked the personal comment of Jirkov

(Holthe 1986) suggesting that *P. tetrabranchia* was a junior synonym of this species. Jirkov (2001) re-describing *P. birulai* confirmed that *P. tetrabranchia* is a junior synonym of *P. birulai*.

Genus *Paramphitrite* is characterized by the presence of lateral lobes on segments 2, 3 and 4, not much developed, and by the presence of 13 chaetigerous thoracic segments with notochaetae having finely serrated brim on one side. These characteristics distinctly differ from genus *Amphitrites* Augener (1922). The presence of two pairs of dorsal dichotomous branchiae is the main difference from genera *Neoamphitrite* Hessle (1917) and *Amphitrite* (Müller 1771 *sensu* Hessle 1917). The body shape and the notochaetae characters resemble to *Lanassa* Malmgren (1866) but this genus (not present in Mediterranean Sea) differs from *Paramphitrite* in the absence of branchiae.

Several specimens of the genus *Paramphitrite*, never signaled in the Mediterranean Sea, were found in the North Adriatic Sea (Italy) during an environmental characterization of relict sand deposits, funded by the Regione Veneto local authority and carried out by the Italian National Institute for Environmental Protection

---

* Correspondence: marco.loia@tiscali.it
[1]Laboratory of Benthic Ecology - ISPRA, Italian National Institute for Environmental Protection and Research, Via di Castel Romano 100, Rome 00128, Italy
Full list of author information is available at the end of the article

and Research (ISPRA) of Rome (Italy), in order to evaluate the environmental compatibility of dredging activities for beach nourishment. In this work the main diagnostic characteristics of the Adriatic species *Paramphitrite birulai* are provided and some ecological information about its habitat and distribution is enhanced.

Methods

Sampling surveys were carried out in the North Adriatic Sea about 30 nm offshore Chioggia (Italy) (Fig. 1) on soft seabed at depths ranging from 29 to 32 meters, in October 2013. The sampling plan included 18 stations, strictly set for the project aimed at the environmental characterization of relict sands deposits. In each station macrozoobenthos samples were collected in two replicates using a Van Veen grab (0.1 m² covering area) and sieved using a 1 mm mesh according to methods proposed by Castelli et al. (2004). The collected material was preserved in seawater adding 4% CaCO3-buffered formalin. Surface sediment samples were collected at each station with a box-corer. At ISPRA Laboratory of Benthic Ecology (LEB) in Rome, the morphological

features of the specimens belonging to *Paramphitrite birulai* were examined mainly following Holthe (1976, 1986), Parapar et al. (1991), Jirkov (2001) and Jirkov & Leontovitch (2013).

For the description of the specimens, total length and number of abdominal segments of complete specimens were considered; thorax length and width at the 9th thoracic segment were measured for all the collected material. The body measures were obtained by image analysis of photographs taken using a Zeiss SteREO Discovery V20 microscope equipped with a Zeiss Axiocam ERc5s digital camera and Zen 2011 (blue edition) software. Moreover, for each sampled station surface sediments were analysed and classified according to sediment texture percentage, following Shepard (1954) classification. Depth and geographical position (UTM projection, datum WGS84 system) were recorded at sampled stations (Table 1). The examined material, preserved in 70% ethanol for long-term storage, has been deposited in the reference collection of benthic marine organisms available at ISPRA Laboratory of Benthic Ecology in Rome.

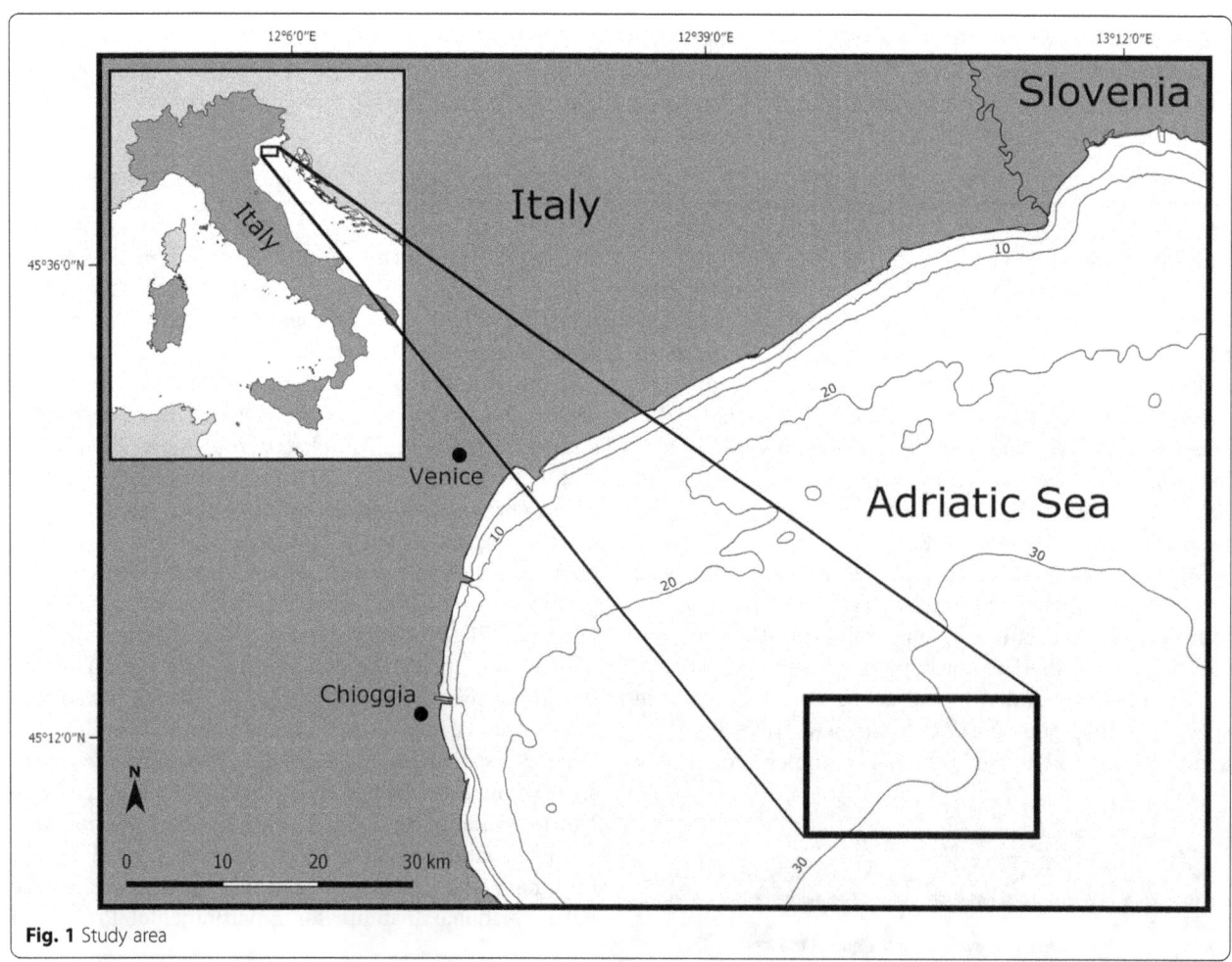

Fig. 1 Study area

Table 1 Records of *Paramphitrite birulai* specimens collected in 15 sampling stations in the North Adriatic Sea (Chioggia, Italy)

Station	Station coordinates		Depth	Specimens collected	Reference collection code
	Latitude N	Longitude E			
RVH-C8	45°10.410	12°54.890	30.0	14	ISPRA-26
RVH-E17	45°09.880	12°55.733	30.8	12	ISPRA-27
RVH-C6	45°10.505	12°54.019	29.8	10	ISPRA-28
RVH-D14	45°10.305	12°55.345	30.5	7	ISPRA-29
RVH-C9	45°10.562	12°55.330	30.3	6	ISPRA-30
RVH-C7	45°10.471	12°54.697	29.9	5	ISPRA-31
RVH-D15	45°10.211	12°55.858	30.7	5	ISPRA-32
RVH-B2	45°10.976	12°54.751	29.9	3	ISPRA-33
RVH-B4	45°10.959	12°56.676	31.2	3	ISPRA-34
RVH-C12	45°10.458	12°56.081	30.8	3	ISPRA-35
RVH-C11	45°10.647	12°55.724	30.6	2	ISPRA-36
RVH-A1	45°11.618	12°55.807	31.0	1	ISPRA-37
RVH-B3	45°10.905	12°55.503	30.6	1	ISPRA-38
RVH-C10	45°10.597	12°55.547	30.5	1	ISPRA-39
RVH-E18	45°09.950	12°56.584	31.3	1	ISPRA-40

Results
Material examined

A total of 74 specimens of *Paramphitrite birulai* were examined. The number of specimens collected in each station is reported in Table 1, together with the reference collection code. The granulometric composition of the 18 sampling stations is reported in Table 2. Surface sediments are classified as sand except for the stations RVH-B4 and RVH-E16.

Table 2 Granulometric composition of the 18 sampling stations

Sites	Granules(%)	Sand(%)	Silt(%)	Clay(%)	Shapard classification
RVH-A1	0.0	77.2	12.5	10.3	Sand
RVH-B2	0.0	78.9	11.2	9.9	Sand
RVH-B3	0.0	83.8	8.5	7.7	Sand
RVH-B4	0.0	74.8	12.9	12.3	Silty sand
RVH-B5	0.0	77.3	12.0	10.7	Sand
RVH-C6	0.0	82.5	9.0	8.5	Sand
RVH-C7	0.0	84.1	8.3	7.6	Sand
RVH-C8	0.0	80.5	10.2	9.3	Sand
RVH-C9	0.0	84.7	7.7	7.6	Sand
RVH-C10	0.0	83.2	8.9	7.9	Sand
RVH-C11	0.0	81.9	9.5	8.6	Sand
RVH-C12	0.0	79.4	10.3	10.3	Sand
RVH-C13	0.0	75.6	12.9	11.5	Sand
RVH-D14	0.0	83.3	8.8	7.9	Sand
RVH-D15	0.0	81.8	9.5	8.7	Sand
RVH-E16	0.0	68.2	16.5	15.3	Silty sand
RVH-E17	0.0	79.9	10.7	9.4	Sand
RVH-E18	0.0	83.4	8.8	7.8	Sand

Morphological description of *Paramphitrite birulai* found in the North Adriatic Sea

Among the 74 specimens examined, only 8 are complete. These last specimens range from 15.2 to 46.2 mm in total length, 4.2 to 7.8 mm in thorax length and 0.7 to 1.9 mm in width. Table 3 shows the values of measurements performed on the 8 complete specimens and the number of total segments counted for each specimen.

For incomplete or damaged specimens at the abdominal level, length measurements have always been made for the thoracic segments. These specimens range from 2.6 to 13.6 mm in thorax length and from 3.2 to 0.5 mm in width.

Table 3 Values of measurements performed on the 8 complete specimens

Specimens	total length	thorax length	width (9°CH)	no. of segments
1	46.277	7.800	1.986	65
2	32.022	6.280	1.492	69
3	28.015	6.820	1.083	75
4	24.185	4.581	0.883	70
5	22.880	5.582	0.959	64
6	20.260	5.139	1.160	67
7	19.805	4.355	0.945	45
8	15.215	4.239	0.722	42

The body is elongated, composed of a thorax with 16 segments and a long abdomen with numerous segments (Fig. 2a). The prostomium shows a crenulated edge and many eyespots (Fig. 2f, Fig. 3b). A few short buccal tentacles are present (Fig. 2b,c,d, Fig. 3a,b), clearly grooved in their full length. Ventrally, the prostomial upper lip is broad, rounded and triangular. The lower lip is broad and thin (Fig. 2b, d, Fig. 3a). Two pairs of dorsal

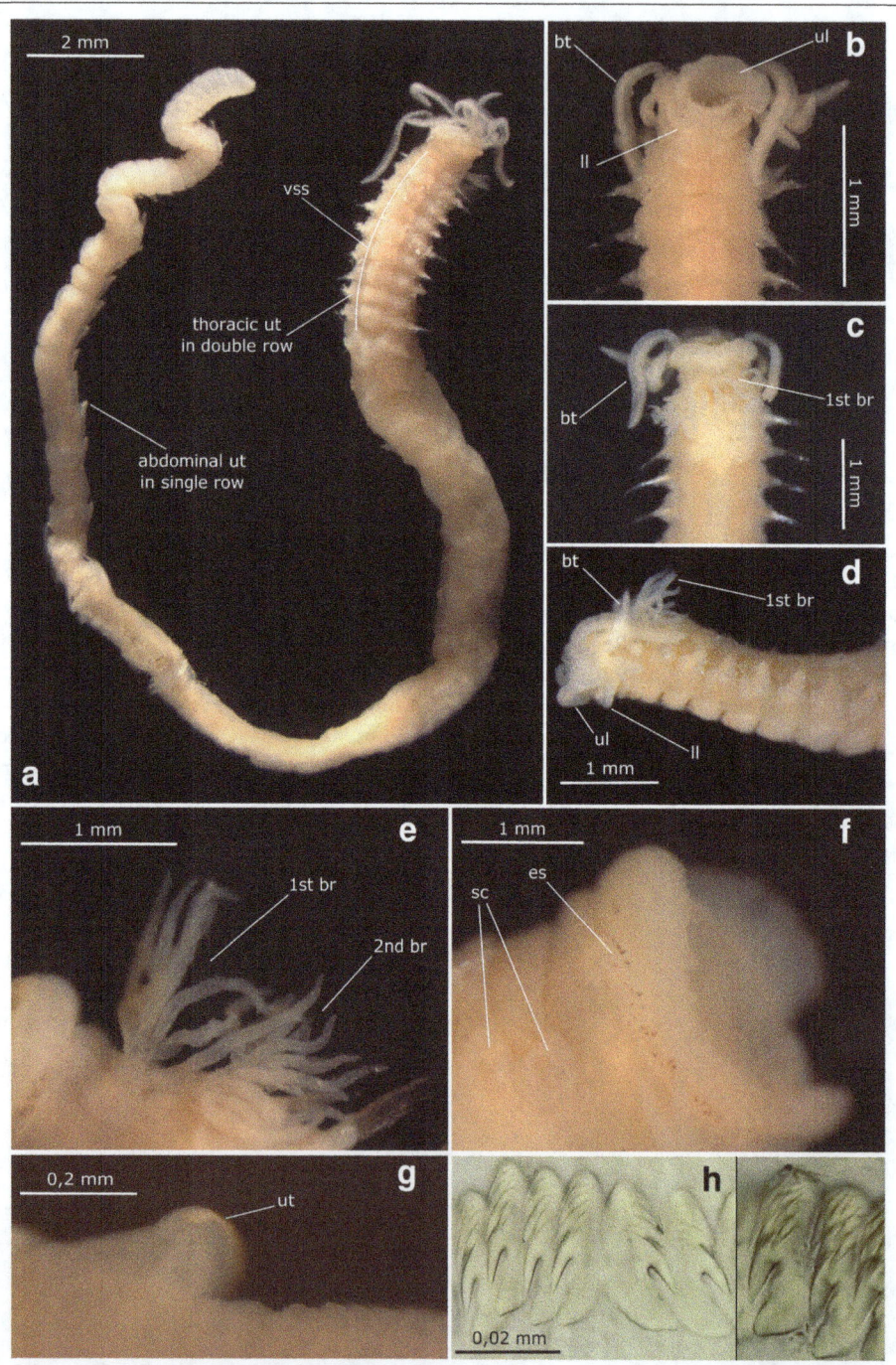

Fig. 2 *Paramphitrite birulai* of North Adriatic Sea: (**a**) entire worm, ventral view, indicating thoracic uncigerous tori in double rows, abdominal uncigerous tori in single row and ventral shields; (**b**) anterior end, ventral view, indicating buccal tentacles, upper lip and lower lip; (**c**) anterior end, dorsal view, indicating buccal tentacles and first pair of branchiae; (**d**) anterior end, lateral view, indicating buccal tentacles, upper lip, lower lip and first pair of branchiae; (**e**) anterior end, lateral view, indicating first and second pair of branchiae; (**f**) anterior end, dorsal view, indicating eyespots and scars of branchiae; (**g**) 20th abdominal segment, ventral view, uncigerous tori in single row; (**h**) thoracic and abdominal uncini. Abbreviations: br, branchiae; bt, buccal tentacles; es, eyespots; ll, lower lip; sc, scars; ul, upper lip; ut, uncigerous tori; vss, ventral shields

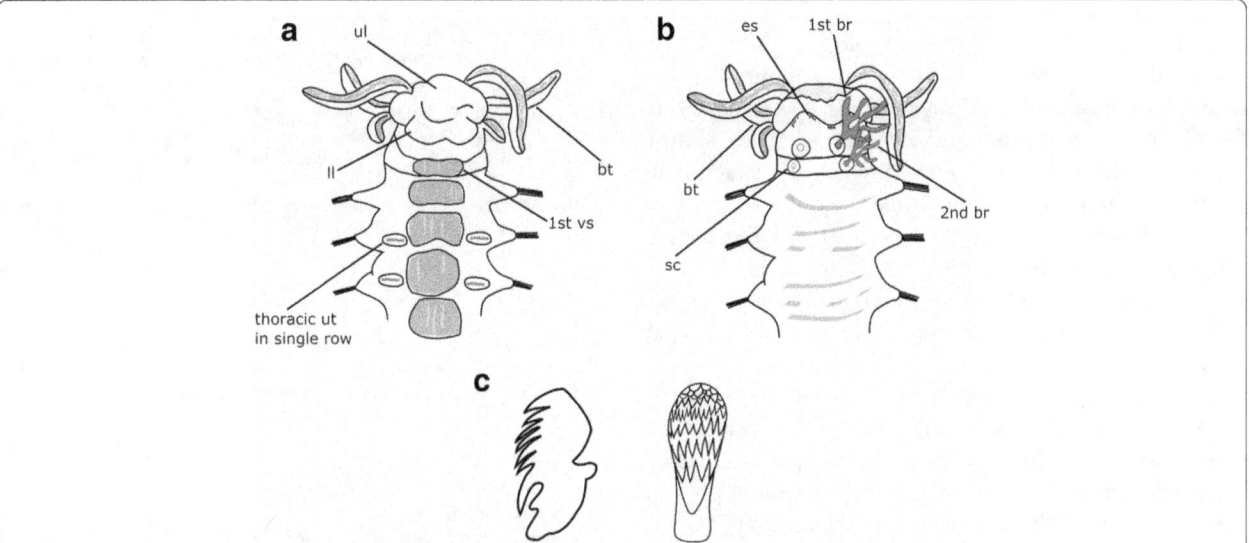

Fig. 3 *Paramphitrite birulai* of North Adriatic Sea: (**a**) anterior end, frontal view,1-6 segments indicating buccal tentacles, upper and lower lip, uncigerous tori, first ventral shield; (**b**) anterior end, dorsal view, 1–6 segments, indicating buccal tentacles, first and second pair of branchiae, eyespots, scars of branchiae; (**c**) abdominal uncini. Abbreviations: br, branchiae; bt, buccal tentacles; es, eyespots; ll, lower lip; sc, scars; ul, upper lip; vs, ventral shield

dichotomous branchiae are arranged on segments 2 and 3, with short stems and short tips. Anterior branchiae (on segment 2) are always longer than posterior ones and are nearer to the middorsum (Fig. 2c, e, Fig. 3b). On segments 2, 3 and 4 lateral lobes are present, not much developed but distinctly visible. Ten distinct ventral shields are visible on thorax (Fig. 2a) with comparable dimension and shape, except for the first one (on the second segment) thinner than the others (Fig. 3a). Notochaetae start from segment 4 and are present on 13 thoracic chaetigers. Prominent uncinigerous tori are present on thorax from segment 5, moderately long and arranged in single row (Fig. 3a). Uncini in double rows start on segment 11 (Fig. 2a) to segment 20 (4th abdominal segment). On the rest of abdomen uncini are again arranged in a single row (Fig. 2a, g). Uncini are small and avicular and appear to have six rows of teeth above the main fang (MF). Dental formula *sensu* Day (1967) MF:4–5:ca6:ca8:∞:∞:∞.

The pygidium is simple with crenulated edge and no appendages. The color in alcohol is yellowish pink. Tube is composed by a thin and transparent layer of secretion incrusted with sand grains, present in just a few of the specimens examined. Tube grain size is classifiable as medium sand, using Shepard (1954) classification.

Habitat and Distribution

White Sea (Ssolowiew 1899; Norwegian Sea, on mixed sediments of clay and silt with stones, 55–138 meters (Holthe 1976; Atlantic Iberian coast, on mixed sediments of sand, clay and shell fragments, 10–15 meters Parapar et al. (1991). North Adriatic Sea specimens here described were found on sandy sediments at depths ranging from about 29 to 31 meters.

Discussion

The examination of North Adriatic specimens of *Paramphitrite birulai* collected in Italian coasts reveals that they conform morphologically to the specimens described by Ssolowiew (1899), Holthe (1976, 1986), Parapar et al. (1991), Jirkov (2001) and Jirkov & Leontovitch (2013). No consistent intraspecific morphological variations have been observed in the specimens analyzed. In the studied specimens 13 thoracic segments have been always counted, comparatively to Jirkov (2001) and Jirkov & Leontovitch (2013) that accounted 13, rarely 14–15 chaetigerous segments. Ten ventral shields are clearly visible in all specimens examined, starting on segment 2 up to the 11th. The first shield, on segment 2, appears to be thinner than the others, which have comparable dimensions. This last aspect differs from Holthe (1976) description of Norwegian Sea specimens, which indicated the last two shields shorter than the previous ones. In all Adriatic specimens, ventral uncini always begins from the fifth segment (2nd chaetigerous segment).

Exclusively 10 segments (from 8th chaetigerous segment up to 4th abdominal segment) have uncini in double rows. From the 5th abdominal segment the uncini are in a single row again, supported by well visible neuropodia.

The biggest *P. birulai* collected in the North Adriatic result to be longer and with a larger number of segments (46,2 mm long and 65 segments) if compared with Norwegian Sea specimens described by Holthe (1986) (40 mm and 57 segments) and the Iberian specimens described by Parapar et al. (1991) (32 mm and 31 segments).

According to Holthe (1976) the tube grain size was mainly composed of sand, even if his specimens were found on mixed bottoms of clay and silt with stones, suggesting the ability of *P. birulai* to select a specific dimension of grains to build the tube.

The Adriatic records of *P. birulai* extend the habitat range of this species into different environmental conditions (grain size and depth) if compared with those where it was previously recorded (Holthe 1976; Parapar et al. 1991.

Given the reported presence of *P. birulai* exclusively in Artic, Norwegian and Iberian Atlantic Sea waters (Ssolowiew 1899; Holthe 1976, 1986; Parapar et al. 1991, the specimens collected in the North Adriatic represent the first report of this species for the Mediterranean Sea, updating the checklist of polychaete in Adriatic Sea Mikac (2015) and the checklist of flora and fauna of the Italian Seas Castelli et al. (2008). The finding of this species in the Northern Adriatic Sea supports the strong boreal affinity of this region, as well as its ecological and biogeographical similarities with the North Atlantic, documented by Bianchi et al. (2004) and Boero & Bonsdorff (2007). The Adriatic Sea is the northernmost Mediterranean region and, together with the Gulf of Lions and the Northern Aegean Sea, the coldest sector of the Mediterranean. This sector is a geomorphologically, hydrographically and biogeographically peculiar Mediterranean region, with lower average temperatures that allow the occurrence of a specific flora and fauna with cold water affinities (Mikac & Musco 2010; Parapar et al. 2013). Beside these aspects, considering that *P. birulai* has not intra-specific morphological variations, as often occur in polychaetes cryptic species, and that the presence of this species in Mediterranean has been never signaled, *P. birulai* might be a non-indigenous species (NIS) whose translocation from separate biogeographical regions in the North Adriatic Sea waters has been man-mediated (either intentionally or accidentally) (Çinar & Dagli 2012; Streftaris et al. 2005). Further molecular analyses could possibly assist in clarifying this hypothesis.

Conclusions

The first occurrence of *Paramphitrite birulai* Ssolowiew 1899 in the North Adriatic Sea represents the first record of the genus *Paramphitrite* in the Mediterranean Sea, previously known only from Temperate Northern Atlantic and Arctic regions (Iberian Atlantic, Norwegian and Russian seas).

Acknowledgements
The authors' acknowledgements go to their ISPRA workgroup of researchers for their assistance in the project planning, field and laboratory activities; Dr. Paola La Valle, Dr. Loretta Lattanzi, Dr. Daniela Paganelli, Dr. Alfredo Pazzini, Dr. Raffaele Proietti and Dr. Monica Targusi. We are grateful to Dr. Fabio Bertasi for his support with the image analysis. Moreover, we would like to thanks ISPRA's Laboratory of sedimentology and environmental micropaleontology for sediments data analyses. Furthermore, we want to express our gratitude to Prof. Adriana Giangrande (University of Salento, Lecce, Italy), Dr Maria Cristina Gambi (Zoological Station of Naples "Anton Dohrn", Ischia Island, Italy), Prof. Alberto Castelli (University of Pisa, Italy), Dr. Maria Flavia Gravina (University of Rome "Tor Vergata", Italy) and Dr Marco Lezzi (University of Salento, Lecce, Italy) for their invaluable support on species identification and contribution on taxonomy of Polychaetes.

Funding
This research has been carried out as a part of an environmental characterization project funded by the Regione Veneto local authority and carried out by the Italian National Institute for Environmental Protection and Research (ISPRA) of Rome (Italy).

Authors' contributions
The authors conceived the study, participated in the field data collection, data analysis, background analysis, preparation of the manuscript, and read and approved the final manuscript.

Competing interest
The authors declare that they have no competing interests.

Author details
[1]Laboratory of Benthic Ecology - ISPRA, Italian National Institute for Environmental Protection and Research, Via di Castel Romano 100, Rome 00128, Italy. [2]ISPRA, Italian National Institute for Environmental Protection and Research, Via Vitaliano Brancati 60, Rome 00144, Italy.

References
Augener H. Über littorale Polychaeten von Westindien. Sitzungsberichte der Gesellshaft naturforschender Freunde zu Berlin. 1922;3–5:38–53.

Bianchi CN, Boero F, Fraschetti S, Morri C. The wildlife of the Mediterranean. In: Minelli A, Chemini C, Argano R, Ruffo S, editors. Wildlife in Italy. Rome: Touring Editore, Milan - Italian Ministry for the Environment and Territory; 2004. p. 248–335.

Boero F, Bonsdorff E. A conceptual framework for marine biodiversity and ecosystem functioning. Mar Ecol. 2007;28 Suppl 1:134–45. http://dx.doi.org/10.1111/j.1439-0485.2007.00171.x.

Castelli A, Lardicci C, Tagliapietra D. Soft-bottom macrobenthos. Mediterranea marine benthos: a manual of methods for its sampling and study. Gambi MC, Dappiano M, editors. Biologia Marina Mediterranea 2004. Suppl. 11:99–131

Castelli A, Bianchi CN, Cantone G, Çinar ME, Gambi MC, Giangrande A, Iraci Sareri D, Lanera P, Licciano M, Musco R, Simonini Sanfilippo R. Annelida Polychaeta. In: Relini G, editor. Checklist of flora and fauna of the Italian seas. Biologia Marina Mediterranea 2008; 15 Suppl. 1: 385 pp.

Chamberlin RV. The Annelida Polychaeta [Albatross Expeditions]. Memoirs of the Museum of Comparative Zoology at Harvard College. 1919;48:1–514. http://dx.doi.org/10.5962/bhl.title.49195.

Çinar ME, Dagli E. New records of alien polychaete species for the coasts of Turkey. Mediterr Mar Sci. 2012;13(1):103–7. http://dx.doi.org/10.12681/mms.26.

Day JH. A monograph on the Polychaeta of Southern Africa. London: British Museum (Natural History); 1967. http://dx.doi.org/10.5962/bhl.title.8596.

Fauvel P. Polychètes sédentaires. Addenda aux errantes, Archiannélides, Myzostomaires. Faune de France. 1927;16:1–494. http://dx.doi.org/10.1038/113528b0.

Hessle C. Zur Kenntnis der terebellomorphen Polychaeten. Zoologiska bidrag från Uppsala. 1917;5:39–258.

Holthe T. Paramphitrite tetrabranchia gen. et sp. nov. A new terebellid polychaete from western Norway. Sarsia. 1976;61:59–62. http://dx.doi.org/10.1080/00364827.1976.10411303.

Holthe T. Polychaeta Terebellomorpha. In: Marine Invertebrates of Scandinavia No.7. Oslo: Norwegian University Press; 1986. p. 1–194. http://dx.doi.org/10.1017/s0025315400026783.

Jirkov IA. Polychaeta of the Arctic Ocean, Moscow, Yanus-K Press; 2001. 632 pp.

Jirkov IA, Leontovitch MK. Identification keys for Terebellomorpha (Polychaeta) of the eastern Atlantic and the North Polar basin. Invertebrate Zoology [aka Zoologiya Bespozvonochnykh]. 2013;10(2):217–43.

Johnston G. An index to the British Annelides. Ann Mag Nat Hist. 1846;series 1(16):433–62. http://dx.doi.org/10.1080/037454809495980.

Malmgren AJ. Nordiska Hafs-Annulater. Öfversigt af Kongiliga Veteskaps-Akademiens Förhandlingar. 1866;22:355–410.

Mikac B. A sea of worms: polychaete checklist of the Adriatic Sea. Zootaxa. 2015; 3943(1):1–172. http://dx.doi.org/10.11646/zootaxa.3943.1.1.

Mikac B, Musco L. Faunal and biogeographic analysis of Syllidae (Polychaeta) from Rovinj (Croatia, northern Adriatic Sea). Sci Mar. 2010;74(2):353–70. http://dx.doi.org/10.3989/scimar.2010.74n2353.

Müller OF. Von Würmern des süssen und salzigen Wassers. H. Mumme and Faber, Copenhagen; 1771. http://dx.doi.org/10.5962/bhl.title.14428.

Nogueira JMM, Hutchings PA, Fukuda MV. Morphology of terebelliform polychaetes (Annelida: Polychaeta: Terebelliformia), with a focus on Terebellidae. Zootaxa. 2010;2460:1–185.

Parapar J, Besteiro C, Urgorri V. Primera cita en el litoral ibérico de Paramphitrite tetrabranchia Holte, 1976 (Polychaeta, Terebellidae). Miscellània Zoològica. 1991;15:63–8.

Parapar J, Mikac B, Fiege D. Diversity of the genus Terebellides (Polychaeta: Trichobranchidae) in the Adriatic Sea with the description of a new species. Zootaxa. 2013;3691(3):333–50. http://dx.doi.org/10.11646/zootaxa.3691.3.3.

Saint-Joseph A. Les Annélides polychètes des côtes de Dinard. Troisième partie Annales des Sciences Naturelles (Zoologie et Paléontologie). 1894;Series 7(17):1–395.

Shepard FP. Nomenclature based on sand, silt, clay ratios. Journal Sedimentary Petrology. 1954;24:151–8. http://dx.doi.org/10.1306/D4269774-2B26-11D7-8648000102C1865D.

Ssolowiew M. Polychaeten-Studien. Die Terebelliden des Weissen Meeres. Annales du Museé Zoologique. Académie Impérial des Sciences St. Pétersbourg. 1899;4(2):179–220.

Streftaris N, Zenetos A, Papathanassiou E. Globalisation in marine ecosystems: The story of non-indigenous marine species across European seas. Annu Rev Oceanogr Mar Biol. 2005;43:419–53. https://doi.org/10.1201/9781420037449.ch8.

19

First record of the chimaera *Neoharriota carri* (Bullis and Carpenter 1966) in the Caribbean of Guatemala

Francisco Polanco-Vásquez[1], Ana Hacohen-Domené[1*], Thalya Méndez[1], Alerick Pacay[1] and Rachel T. Graham[2]

Abstract

Background: A new record of *Neoharriota carri* is here reported for the Caribbean of Guatemala.

Results: Two chimaeras, a male and female *Neoharriota carri*, were caught with a single panel trammel net off the coast of El Quetzalito, Guatemala in February 2015 and January 2016 respectively. Details concerning the identification and measurement of these species are presented.

Conclusions: These records represent the first records in Guatemalan waters and the northernmost records in the western Atlantic for the distribution of *N. carri*.

Keywords: Rhincomeridae, First record, Range extension, Caribbean

Background

The family Rhincomeridae, belongs to the subclass Holocephali, order Chimaeriformes, commonly known as the longnose chimaera. Longnose chimaeras are small to medium chondrichthyans with a broad head and elongated spear-like snout. Currently, Rhincomeridae is represented by three genera: *Harriota* Goode and Bean 1895, *Neoharriotta* Bigelow and Schroeder 1950, and *Rhinochimaera* Garman 1901. Species of *Neoharriota* are distinguished from *Harriota* by the possession of a prominent anal fin. *Neoharriota* species are represented by: *N. pinnata*, Schnakenbeck 1931, which appear to be restricted to the eastern Atlantic, off the western coast of Africa, *N. pumila*, Didier and Stehmann 1996, presently known only from the northwestern Indian Ocean and *N. carri* with known occurrence in the upper and mid continental slopes in the Southern Caribbean (219–458 m depth range) (Bullis and Carpenter 1966).

Globally, Chimaeroids are captured incidentally in commercial, recreational and artisanal fisheries (Barnett et al. 2012). Despite captures, few data exist regarding their population status; the International Union for Conservation of Nature lists 16 out of 35 chimaera species as data deficient (IUCN 2011). During February 2015 and January 2016, while conducting landings verification of elasmobranchs in Quetzalito, a fishing village on Guatemala's Caribbean coast, two specimens of the chimaera *Neoharriota carri* were collected, with the 2016 specimen was being preserved for further examination.

Methods

On 14th February 2015 and 30th January 2016, two chimaera specimens were captured by artisanal fishermen of El Quetzalito, Izabal (Fig. 1). According to the fisherman, both specimens were captured approx. 16 Km from El Quetzalito, Izabal, Guatemala (15° 52.374 N, 88° 18.712 W), approximately 240 m depth, with a 1000 m bottom trammel net of 3.5 in. mesh size and one panel. Images of the 2015 specimen as well as total length (TL) and sex were recorded for the captured chimaera in 2015. The specimen captured in 2016 was kept on ice and later preserved in formaldehyde (10%) for 3 weeks before finally being transferred to ethyl alcohol (70%) and donated to the Laboratory of Biological Science and Oceanography, Centro de Estudios del Mar y Acuicultura (CEMA) of the Universidad San Carlos de Guatemala (USAC). The specimen is part of the

* Correspondence: ahacohen@fundacionmundoazul.com
[1]Fundación Mundo Azul, Blvd. Rafael Landivar 10-05 Paseo Cayala Zona 16, Edificio D1 Oficina 212, Guatemala City, Guatemala
Full list of author information is available at the end of the article

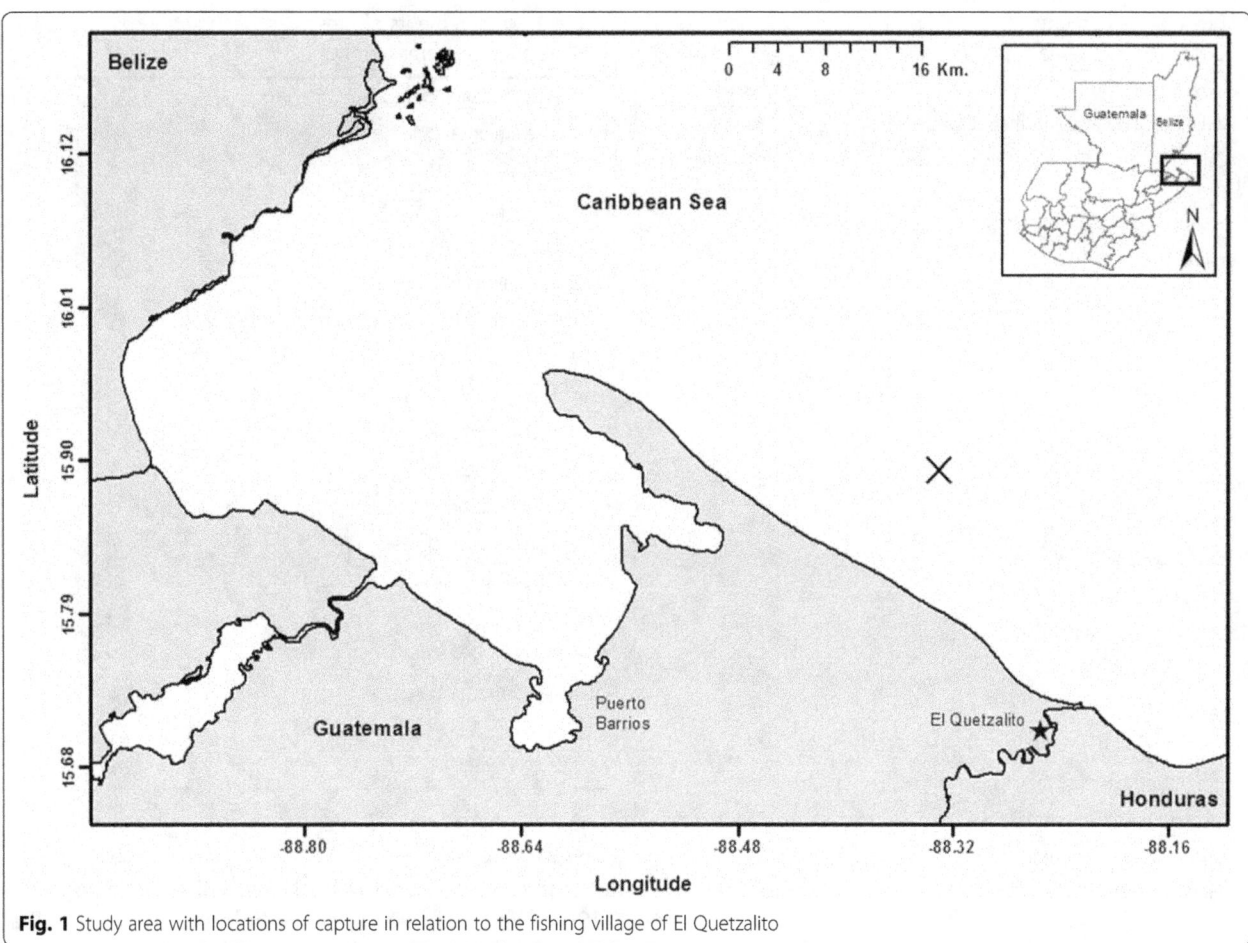

Fig. 1 Study area with locations of capture in relation to the fishing village of El Quetzalito

collection registered to the Consejo Nacional de Áreas Protegidas (CONAP) under the reference number 162.

The 2016 specimen was measured using a ruler and measuring tape. A total of 46 measurements were taken (Table 1) as proposed by Compagno et al. (1990) and Bullis and Carpenter (1966). Specimen examination and species confirmation were based on Didier (2002) and Bullis and Carpenter (1966).

Results

Systematic account
   Family: Rhinochimaeridae Garman, 1901
   Genus: *Neoharriotta* Bigelow and Schroeder, 1950
   Neoharriotta carri Bullis and Carpenter 1966
   Common name: Dwarf sicklefin chimaera; Quimera pálida con hocico largo (Spanish), tiburón elefante (local name).

Material examined

2015 specimen, male 730 mm TL (Fig. 2). 2016 specimen, female, 880 mm TL, 349 mm body length (BDL) (Table 1, Fig. 3).

Description

2016 specimen: Medium to large. Snout elongated and pointed. Caudal fin axis weakly raised, prominent anal fin present and is separated from ventral caudal lobe. Caudal filament broken off. Pectoral and pelvic fins are triangular in shape, darker in color than the body. First dorsal fin is preceded by a spine. Second dorsal fin base terminates immediately above and slightly anterior to anal fin origin.

For the 2015 specimens, photographs were taken by the fishermen for evidence (Fig. 2), but no specimen was kept for preservation or further identification. Based on the fisherman's report and on images taken with known reference lengths, this specimen was male with a 730 mm TL, presumed mature due to clasper formation and size (Dagit 2006).

A year later, in January 2016, the same fisherman collected a new specimen: a female chimaera, 880 mm TL (Fig. 3). This specimen was identified as *Neoharriota carri* and according to size at sexual maturity (Dagit 2006), this organism was also presumed to be sexually mature.

Table 1 Measurements of preserved *Neoharriota carri* landed in El Quetzalito, Guatemala on January 2016

Measurements (mm)	Female (*n* = 1)
Total length (TL- caudal filament broken off)	880
Precaudal length (PCL), snout tip to posterior end of anal fin base	646
Body length (BDL), gill opening to upper caudal origin	349
Prenarial length (PRN)	142
Preorbital length (POB)	179
Head length (HDL)	249
Head height (HDH)	81
Head width (HDW)	53
Prepectoral length (PP1)	277
Prepelvic length (PP2)	420
Trunk height (TRH)	85
Trunk weight (TRW)	36
Caudal peduncle height (CPH)	21
Caudal peduncle width (CPW)	11
Pectoral length (P1L)	86
Pectoral anterior margin (P1A)	175
Pectoral base (P1B)	38
Pre first dorsal length (PD1)	264
Pre second dorsal length (PD2)	382
Snout vent length, snout tip to front of anus (SVL)	442
Snout greatest width (SWF)	19
Snout basal width (SWB)	41
Snout basal height (SHB)	34
Mouth length (MOL)	16
Mouth width (MOW)	30
Upper labial furrow length (ULA)	27
Lower labial furrow length (LLA)	4
Upper labial furrow height (ULH)	10
Nostril width (NOW)	7
Internarial space (INW)	34
Outer internarial space (IOW)	37
Anterior nasal flap length (ANF)	11
Eye length (EYL)	31
Eye height (EYH)	22
Eye mouth space (EMO)	12
Interorbital space (INO)	21
Trunk length (TRL)	194
First dorsal insertion to second dorsal origin (IDS)	28
Dorsal caudal space (DCS)	61
First dorsal anterior margin (D1A)	105
Length dorsal spine (DSA)	94
Second dorsal fin length (D2L)	164
Second dorsal fin base (D2B)	162

Table 1 Measurements of preserved *Neoharriota carri* landed in El Quetzalito, Guatemala on January 2016 *(Continued)*

Second dorsal fin height (D2H)	31
First dorsal height (D1H)	101
First dorsal base (D1B)	91
Second dorsal length (D2L)	4
Total caudal length (CTL)	248
Dorsal caudal margin (CDM)	198
Caudal filament length (CFI-broken off)	51
Maximum caudal height (CHI)	44
Ventral caudal margin (CVM)	215
Pelvic caudal space (PCA)	180
Caudal lower ray length (CLR)	35
Caudal upper ray length (CUR)	6
Gill opening (gill split) (GS1)	34
Prenarial length (PRN)	162
Preoral length (POR)	165
Pectoral pelvic space (PPS)	149
Intergill width (IG1)	40
Anal fin length (AL)	27
Anal fin base (AB)	11

Discussion

N. carri was first described by Bullis and Carpenter (1966), who described a female holotype of 428 mm TL, collected in Panama. In Colombia, Acero (1998) reported the occurrence of two individuals, 1 female 640 mm TL and 1 male 820 mm TL. *N. carri* has also been reported in Venezuela (Dagit 2006). Benavides et al. (2014), recorded *N. carri* while using bottom

Fig. 2 Chimaera specimen collected near El Quetzalito, in Guatemala (February 2015), male, 730 mm TL (**a**. lateral view, **b**. dorsal view)

Fig. 3 *Neoharriota carri* (Bullis and Carpenter 1966) collected near El Quetzalito, in Guatemala (January 2016), female, 860 mm TL, 349 mm BDL, (A- lateral view; B-dorsal view)

Authors' contributions
FP, AH participated in the identification of the species, supporting literatures and contributed to draft the manuscript. TM, AP, RTG participated in the identification of the species and contributed to draft the manuscript. All authors read and approved the final manuscript.

Competing interests
The authors declare that they have no competing interests.

Author details
[1]Fundación Mundo Azul, Blvd. Rafael Landivar 10-05 Paseo Cayala Zona 16, Edificio D1 Oficina 212, Guatemala City, Guatemala. [2]MarAlliance, 32 Coconut Drive, Po Box 283, San Pedro, Belize.

References
Acero A. Registros nuevos de peces cartilaginosos para el Caribe Colombiano. Actu biol. 1998;17(63):36–39.
Barnett LAK, Ebert DA, Cailliet GM. Evidence of stability in a chondrichthyan population: case study of the spotted ratfish *Hydrolagus colliei* (Chondrichthyes: Chimaeridae). J Fish Biol. 2012;80:1765–88.
Benavides R, Brenes CL, Márquez A. Análisis de la población de condrictios (Vertebrata: Chondrichthyes) de aguas demersales y profundas del Caribe centroamericano, a partir de faenas de prospección pesquera con redes de arrastre. Rev Cienc Mar Cost. 2014;6:9–27.
Bullis Jr HR, Carpenter JS. Neoharriotta carri: A New Species of Rhinochimaeridae from the Southern Caribbean Sea. Copeia. 1966;3:443–50.
Compagno LJV, Stehmann M, Ebert DA. *Rhinochimaera africana*, a new longnose chimaera from southern Africa, with comments on the systematics and distribution of the genus *Rhinochimaera* Garman, 1901 (Chondrichthyes, Chimaeriformes, Rhinochimaeridae). S Afr J Marine Sci. 1990;9:201–22.
Dagit DD. *Neoharriotta carri*. In: The IUCN Red list of threatened species. 2006. e. T60141A12312391. http://dx.doi.org/10.2305/IUCN.UK.2006.RLTS. T60141A12312391.en. Accessed date 30 March 2016.
Didier DA. Chimaeras. In: Carpenter KE, editor. FAO species identification guide for fishery purposes. The living marine resources of the western central Atlantic, Vol. 1: Introduction, molluscs, crustaceans, hagfishes, sharks, batoid fishes, and chimaeras. 2002. p. 591–6.
IUCN. IUCN Red List of Threatened Species: Version 2011.1http://www.iucnredlist.org. Accessed date 30 March 2016.

trawling nets in Costa Rica. During these surveys in Costa Rica, the authors collected a large number of specimens, 31 males (285–545 mm TL) and 31 females (285–545 mm TL), all presumed sexually immature per Dagit (2006). According to Acero (1998), the maximum size reported for this species is 820 mm TL. The 2016 specimen described here is therefore the largest specimen on record (880 mm TL).

To date fishers report catching at least 10 additional *N. carri* specimens near the actual coordinates at which the specimens were found. There is no reported seasonal variance in the captures of this species. Captured *N. carri* are generally released at sea, but if landed, are neither corned and/or consumed as is customary with other chondrichthyan species.

Conclusion
The importance of the present records resides in the fact that they represent the first record of *N. carri* for Guatemala and a significant range extension and northernmost report in the Western Atlantic. No targeted fisheries exist for this species, and captures represent bycatch from traditional small-scale finfish and elasmobranch fisheries.

Abbreviations
BDL: Body length; Km: Kilometer; m: Meter; TL: Total length

Acknowledgements
FUNMZ would like to thank the community of El Quetzalito for their constant support to the research program in the area. Also, we would like to thank Josué Ayala for reporting the collection of the specimen.

Funding
Fundación Mundo Azul and MarAlliance provided funding for the elasmobranch monitoring project during 2015.

20

Nephroselmis viridis (Nephroselmidophyceae, Chlorophyta), a new record for the Atlantic Ocean based on molecular phylogeny and ultrastructure

Karoline Magalhães Ferreira Lubiana[1*], Sônia Maria Flores Gianesella[2], Flávia Marisa Prado Saldanha-Corrêa[2] and Mariana Cabral Oliveira[1]

Abstract

Nephroselmis is composed by unicellular nanoplanktonic organisms, occurring predominantly in marine environments. Currently, 14 species are taxonomically accepted. *Nephroselmis viridis* was described in 2011 and strains were isolated from Indic and Pacific Oceans. Since then, it was not recorded in other places. A strain was isolated from coastal waters of Brazil by micropipetting and washing, and cultivated in f/2 medium for morphological observations (light, confocal, SEM and TEM) and molecular phylogeny inferences (maximum likelihood and Bayesian). The cells are asymmetrical, have two unequal flagella, one cup-shaped chloroplast with an eyespot, and a large starch covered pyrenoid. Chloroplast thylakoids intrude into the pyrenoid and organic scales cover all cell body and flagella. Molecular phylogeny (18S rRNA) clustered the isolated strain with other *Nephroselmis viridis* sequences, and the species is the sister of the *N. olivacea*, the type species of the genus. Morphology and molecular phylogeny corroborate the strain identification, and it is the first time this species is recorded in Brazil and in the Atlantic Ocean.

Keywords: Brazilian coast, 18S rRNA, Strain isolation, Morphology, Biodiversity

Background

Nephroselmis was described in 1879 by the typification of *Nephroselmis olivacea* Stein, and initially was allocated into Cryptophyceae (Parke and Rayns 1964). Further studies moved it to Chlorophyta, and Bourelly, in 1970, classified it as Prasinophyceae (Norris 1980). In the last decades, many studies taking into account molecular phylogeny have shown that Prasinophyceae is not monophyletic (Marin and Melkonian 2010; Marin and Melkonian 1994; Nakayama et al. 1998; Steinkotter et al. 1994). Hence, the class Nephroselmidophyceae (Nephrophyceae) was proposed to accommodate the genus (Cavalier-Smith 1993; Nakayama et al. 2007). This class seems to be an early derived clade of the core Chlorophyta (Daugbjerg et al. 1995; Nakayama et al. 1998; Steinkotter et al. 1994; Turmel et al. 2009; Turmel et al. 1999), keeping a high number of ancestral characters.

Currently, 14 species of *Nephroselmis* are taxonomically accepted (Guiry and Guiry 2016), and except for *N. olivacea* which is freshwater, all other species are brackish or marine. The nuclear gene coding for the ribosomal small subunit RNA (18S rDNA) is the most widely used molecular marker for this group (Bell 2008; Faria et al. 2012; Faria et al. 2011; Nakayama et al. 2007; Nakayama et al. 1998; Yamaguchi et al. 2011). However, sequences for just nine species of this molecular marker are available in Genbank, representing less than 65% of the genus biodiversity.

Nephroselmis viridis Inouye, Pienaar, Suda & Chihara was described in 2011 and strains were isolated from marine waters of Fiji, Japan and South Africa, in the Pacific and Indic Oceans (Yamaguchi et al. 2011). In the Atlantic Ocean, just five *Nephroselmis* species were

* Correspondence: karolinemfl@usp.br
[1]Laboratório de Algas Marinhas "Édson José de Paula", Departamento de Botânica, Instituto de Biociências, Universidade de São Paulo, Rua do Matão 277, São Paulo, SP CEP 05508-090, Brazil
Full list of author information is available at the end of the article

recorded previously, vis., *N. discoidea* Skuja (Menezes and Bicudo 2008), *N. fissa* (Lackey 1940), *N. minuta* (N.Carter) Butcher (Butcher 1959; Domingos and Menezes 1998), *N. pyriformis* (N.Carter) Ettl (Bergesch et al. 2008; Moestrup 1983; Steinkotter et al. 1994), and *N. rotunda* (N.Carter) Fott (Bell 2008; Butcher 1959). Therefore, here we report for the first time the occurrence of *N. viridis* in Atlantic Ocean, isolated from the coast of Brazil, identified by molecular and microscopy tools.

Methods

Strain isolation and culturing conditions

The strain was isolated from a water sample collected in coastal area of Ubatuba, São Paulo, Brazil, close to Anchieta Island (23° 35.847′ S, 45° 01.70′ W), at a depth of 40 m. In the laboratory, a drop of the water sample was used to select the cell, which was transferred successively to sterile sea water drops until just the desired cell was present. Then, the cell was placed in 3 ml of medium, and after 1 month transferred to higher volume. The isolated strain is being maintained in f/2 medium (without Si stock solution) (Guillard and Ryther 1962), salinity 32–35, temperature 20 °C (±1), photoperiod of 12 h light/12 h dark, and 80 µE m^{-2}.s^{-1} radiation. The strain is deposited in the Microorganisms Collection Aidar & Kutner from Oceanographic Institute, University of São Paulo (strain number BMAK193).

Morphological characterization

Cultures of 1–3 weeks old were used for morphological observations. Living and fixed cells (2% glutaraldehyde) were observed under light microcopy Leica DM 4000 B (Leica Microsystems, Wentzler, Germany), and confocal microscopy Zeiss LSM 440 Axiovent 100 (lp870/543 nm) (Carl Zeiss, Jena, Germany). For SEM and TEM, cultures were harvested by centrifugation (3 min, 100–150 g), and transferred for 90 min to a fixative solution (2% glutaraldehyde plus sodium cacodylate trihydrate 0.1 M, and sucrose 0.8 M buffer). For the SEM preparation, cells were washed in cacodylate plus-sucrose buffer, and then post-fixed in osmium tetroxide (1%) for 60 min. After that, the cells were washed again in buffer, and dehydrated in an ethanol series (70, 90, 95 and 100%). Finally, the sample was dried to critical point (Balzers CPD 030, Bal-Tec, Vaduz, Liechtenstein) and gold-coated (Balzers SCD 050) for visualization in Zeiss Sigma VP. For TEM, cells were dehydrated in an acetone series (50, 70, 95 and 100%), and after embed in Spurr resin. Lastly, thin sectioned, stained, and observed in Zeiss EM 900.

DNA extraction, amplification, sequencing and molecular phylogeny

Genomic DNA was extracted using *NucleoSpin® Plant* II kit (Macherey-Nagel, Düren, Germany), according to the manufactures instructions. PCRs of 18S (small ribosomal subunit), ITS 1 and 2 (internal transcribed spacers), 5.8S and partial 28S (large ribosomal subunit) rRNA were amplified with Platinum® *Taq* DNA polymerase kit (Invitrogen™, Carlsbad, USA) and purified with the GFX Illustra kit (GE Healthcare Life Sciences, Little Chalfont, Buckinghamshire, UK), both done in accordance with the manufactures instructions. PCRs programs and primers are available as Additional file 1. Terminator Cycle Sequencing Ready Reaction kit (Applied Biosystems™, Hammonton, NJ, USA) was used for sequencing reactions, and samples were sequenced using a 3730 DNA Analyzer (Applied Biosystems™, Hammonton, NJ, USA).

Sequences were assembled with *Sequencher* 4.7 software (Gene Codes Corporation, Ann Arbor, Michigan, USA), and were used to seek for other sequences in GenBank database. Thirty-four sequences were used in the matrix data (see Additional file 2). Four sequences of phylogenetically close species were used to root the tree (*Pyramimonas aurea, Pseudoscourfieldia marina*, and *Pycnococcus provasolii*). These sequences were chosen based on previous studies of *Nephroselmis* phylogeny (Faria et al. 2012; Faria et al. 2011; Nakayama et al. 2007; Yamaguchi et al. 2011). Introns were removed from the data. Dataset alignment was performed in AliView (Larsson 2014), using the Muscle algorithm (Edgar 2004). The appropriate evolution method was selected according to JModelTest 2.1.7 analysis (Darriba et al. 2012). Maximum likelihood (ML) phylogeny inference was performed in Garli (Bazinet et al. 2014) using 1000 bootstrap replicates (Felsenstein 1985), and two searches per run. MrBayes (Ronquist et al. 2012) was used to perform Bayesian analysis, with nodes confidence supported by posterior probability. Two runs were done consecutively, each one with 4×10^6 generations, four chains, and sampling at 100 generations. MrBayes generated 8×10^4 trees, whereas 6×10^4 were used to build the consensus tree (burn-in 2×10^4).

Results

SYSTEMATICS

Order NEPHROSELMIDALES
Family NEPHROSELMIDACEAE
Genus *Nephroselmis* Stein 1878
Nephroselmis viridis Inouye, Pienaar, Suda & Chihara, 2011 (Fig. 1 and Yamaguchi et al. 2011)

Description

The cells decant on the flasks bottom and the color of the culture is green in exponential phase and become olive in stationary and senescent phases. Cells are flattened when observed in ventral view and almost symmetrical in lateral

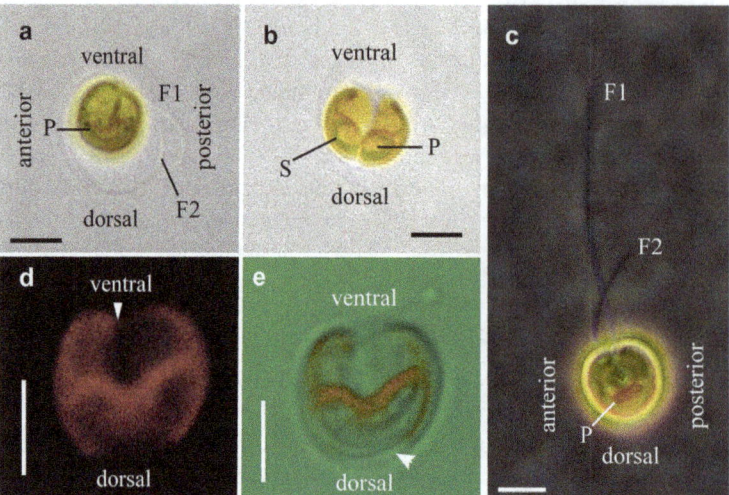

Fig. 1 *Nephroselmis viridis* morphology by light and confocal microscopy. Scale bars represents 5 μm. **a** Living cell in bright field coiling the flagella around the body; **b** Living cell in duplication observed in bright field. **c** Fixed cell in phase contrast evidencing the flagella length and pyrenoid. **d** Chloroplast natural fluorescence evidencing the chloroplast sinus (*arrow*) and pyrenoid. **e** Chloroplast fluorescence and cell morphology showing disk-like structure (*arrow*). (*F1*) longer flagellum, (*F2*) shorter flagellum, (*P*) pyrenoid. (*S*) starch sheath

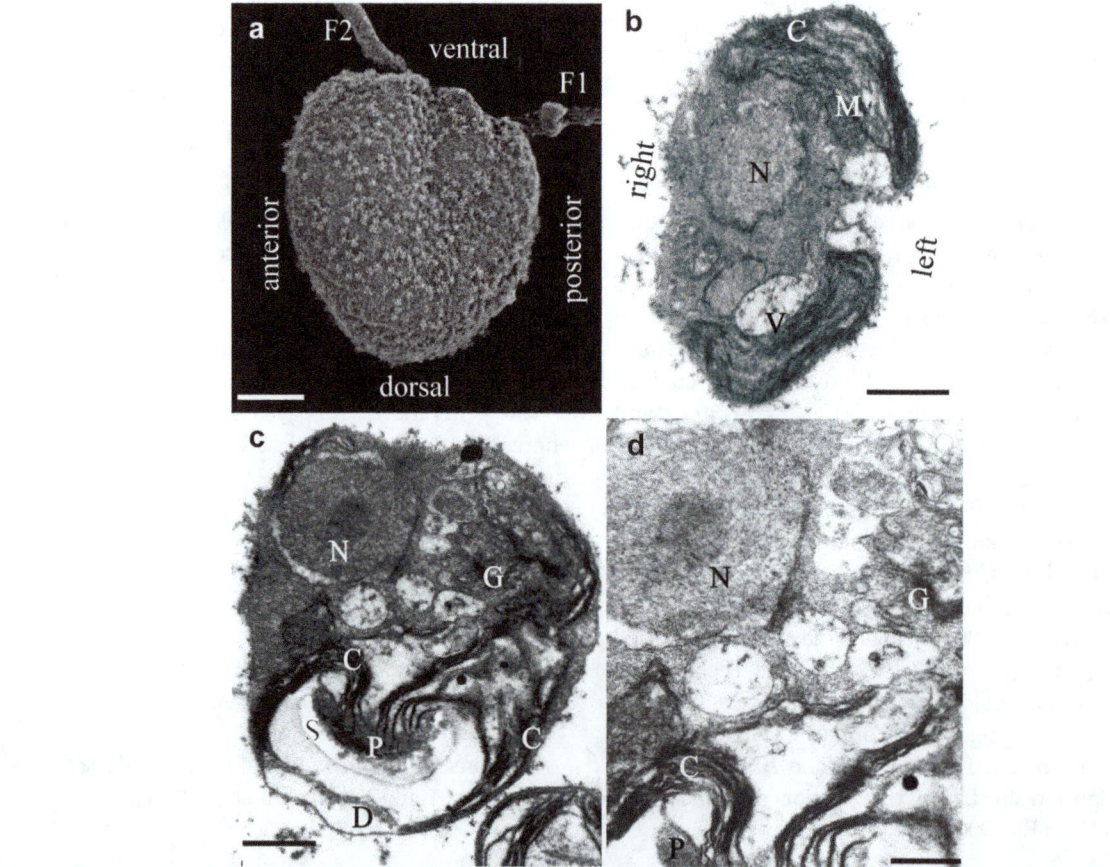

Fig. 2 *Nephroselmis viridis* morphology by electron microscopy. **a** SEM image showing cell surface and organic scales (Scale bar 1 μm). **b** TEM image of the ventral view a cell showing the right nucleus (Scale bar 1 μm). **c** TEM image form the right- anterior view evidencing the organellar placement (Scale bar 1 μm). **d** More detailed view of organellar arrangement (Scale bar 0.5 μm). (*C*) chloroplast, (*D*) disk-like structure, (*F1*) longer flagellum, (*F2*) shorter flagellum, (*G*) Golgi body, (*M*) mitochondria, (*N*) nucleus, (*P*) pyrenoid. (*S*) starch sheath, and (*V*) vacuoles

view, bean-shaped, ranging 5 to 7.5 μm in length and 5.5 to 9 μm width (Figs. 1 and 2). During the cellular cycle, cells enlarge becoming more rounded, and the first noticeable feature is the expansion of the pyrenoid. The cells reproduce by bisection in the longitudinal axis (Fig. 1b), and sexual reproduction was not observed. Two unequal heterodynamic flagella emerge from a frontal groove, ventrally located (Figs. 1a, c and 2a). The bigger flagellum (F1), ranged from 20 to 27 μm (3–4×), and the smaller flagellum (F2), ranged between 8.5 and 11.5 μm (1–1.5×) (Fig. 1d). The cells commonly coil both flagella around the body when resting (Fig. 1a). An unique green parietal cup-shaped chloroplast was located at cells dorsal face (Figs. 1a, c, d, e and 2c) which has an eyespot in the anterior/ventral face (not show in figures). The chloroplast has a large sinus in

the ventral portion (Fig. 1d), and a big cup-shaped pyrenoid starch sheath is at the dorsal region (Figs. 1b, c and 2c). Thylakoids sheets penetrate the pyrenoid (Fig. 2c). A disc-like structure is located at the dorsal part of the cell (Figs. 1e and 2c). The nucleus is located in the right position, near the ventral face (Fig. 2b and c). A single reticulate mitochondrion (Fig. 2b) is situated in the inner part of the chloroplast cavity, and a high number of Golgi vesicles are visible (Fig. 2c and d).

Molecular phylogeny

The 18S rDNA of BMAK193 does not have introns. The sequences of ITS 1 and 2, 5.8S, and partial 28S rRNA obtained in this work were not used to infer phylogeny, due to few sequences of these markers in the Genbank

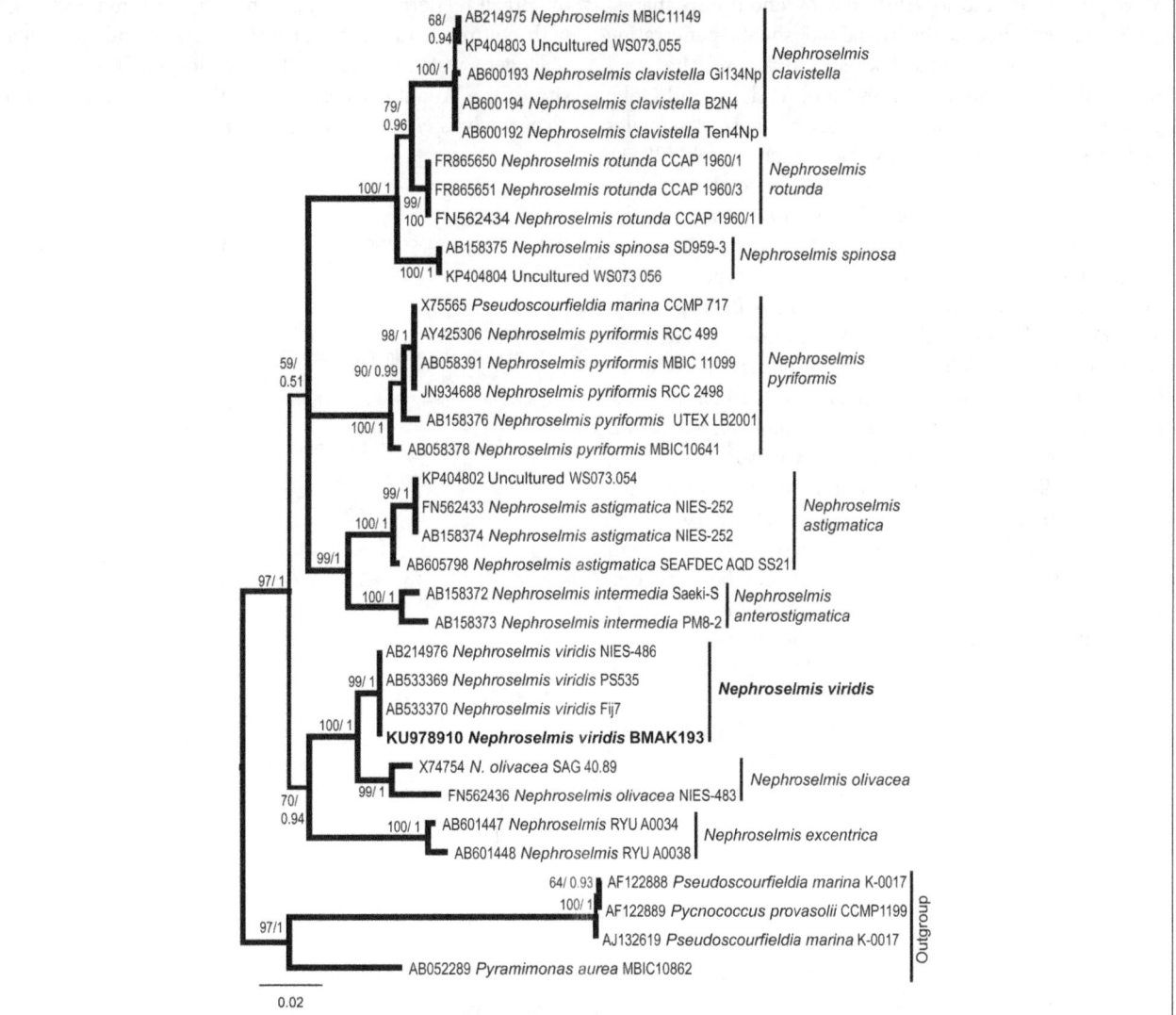

Fig. 3 *Nephroselmis* maximum likelihood phylogeny tree inferred by 18S rRNA performed with 1000 bootstraps replicates in two consecutive runs. General time reversible with invariant sites (*I* = 0.6902) and gamma distribution rate (α = 0.6101) was the evolutionary substitution nucleotide model used (GTR + G + I, lnL −5465.48501). Nodes supports are bootstrap/posterior probability. Branch width represents the node bootstrap support. Scale bar is the rate of nucleotide substitution per site

for the genus and the absence for the species. In the alignment matrix of 18S rDNA, *Nephroselmis viridis* strains sequences are 100% identical (DNA matrix is available upon request). Maximum likelihood and Bayesian phylogenetic analysis clustered BMAK193 into *Nephroselmis viridis* strains (Fig. 3). It also pointed out that *Nephroselmis* is a monophyletic genus, and *N. viridis* is the sister group of the freshwater species *N. olivacea*.

Discussion

The cell measures of *Nephroselmis viridis* from the Atlantic Ocean, such as width and length of cell body and flagella, are exactly the same of the type described in Yamaguchi et al. (2011). Ultrastructural features observed also endorse the identification, such as the chloroplast form and location, the pyrenoid cup shaped and its starch sheath, the thylakoid sheets penetrating into the pyrenoid, the disk like structure, and the positions of the reticulate mitochondrion, nucleus, and Golgi apparatus. The shape and location of these organelles are the same as observed by Yamaguchi et al. (2011). However, the color of the cells and culture are different from the species description. The isolated strain is olive when in stationary and senescent physiological culture stage, different from the green color of the type.

The most common cell shape in *Nephroselmis* species is bean-shaped or semicircular, and symmetrical in anterior/ posterior and right/left axis, as in *N viridis* (Faria et al. 2012; Faria et al. 2011; Yamaguchi et al. 2011). The cell and flagella size are overlapping in some *Nephroselmis* species. Another common feature widespread in this genus is coiling the flagella around the cell body when cells are resting (Faria et al. 2012; Faria et al. 2011; Suda 2003; Yamaguchi et al. 2011). For these reasons, *N. viridis* could be easily mistaken with *N. rotunda* in light microscopy investigation. Therefore, for reliable morphological identification ultrastructural information is need.

Molecular markers are more suitable for identification of species, once are less affected by erroneous or incomplete observations and morphological plasticity. The clustering of *Nephroselmis viridis* isolated from coastal waters of Brazil with other *N. virids* strains from Japan, Fiji and South Africa give a clear evidence that they are the same species. The 18S rDNA pointed out that *Nephroselmis* is a monophyletic genus, and *N. viridis* is the sister group of the freshwater species *N. olivacea*, as observed in previous studies (Faria et al. 2012; Faria et al. 2011; Marin and Melkonian 2010; Yamaguchi et al. 2011; Yoshii et al. 2005). The 18S rDNA of BMAK193 does not have introns, such as the strain isolated in Fiji (Fiji7). But, introns in the 18S rDNA are present in other two strains of *N. viridis*, NIES-486 and PS537, isolated from Japan (Yamaguchi et al. 2011).

Other species of *Nephroselmis* were detected in Brazilian coastal waters, such as *N discoidea* (Menezes and Bicudo 2008), *N. minuta*, (Domingos and Menezes 1998), and *N. pyriformis* (Bergesch et al. 2008). However, most of the studies performed in South Atlantic Ocean investigate the composition of diatoms and dinoflagellates (Garcia and Odebrecht 2012; Jardim and Cardoso 2013; Lubiana and Dias Júnior 2016, and others). Consequently, the biodiversity status is rarely updated for other groups, such as the chlorophytes.

The small cell size, the challenge of morphological identification, and the species tendency to reside in sediments makes *Nephroselmis viridis* detection difficult. However, the species geographic distribution ought to be worldwide, especially in tropical and temperate marine regions (Yamaguchi et al. 2011), such as coastal waters of Brazil. Therefore, using an integrate methodology with culturing, morphological description, and molecular phylogeny we contribute to the knowledge of the biodiversity of the Atlantic Ocean, presenting the first record of *Nephroselmis viridis* in coastal waters of Brazil.

Abbreviations

SEM: Scanning electron microscopy; TEM: Transmission electron microscopy

Acknowledgments

We wish to thanks, André Nakasato, Willian da Silva Oliveira, and Rosario Petti for technical support. Also, Irwandro Pires and Waldir Caldeira from Electron Microscopy Laboratory of IB USP for the help with electron and confocal microscopes. Financial support was obtained from FAPESP (2010/50187-1), and CNPq scholarships to K. M. F. Lubiana (163070/2013-0), and to M.C. Oliveira (301491/2013-5).

Funding

FAPESP (2010/50 187-1), and CNPq scholarships to KMFL (163070/2013-0), and to MCO (301491/2013-5).

Authors' contributions

KL isolated the strain, obtained morphological and molecular data, and wrote the manuscript. SG assisted article composing. FSC did strain culturing for experiments. MCO draw the experiment, did the phylogenetic analysis, and assisted article composing. All authors read and approved the final manuscript.

Competing interests

The authors declare that they have no competing interests.

Author details

[1]Laboratório de Algas Marinhas "Édson José de Paula", Departamento de Botânica, Instituto de Biociências, Universidade de São Paulo, Rua do Matão 277, São Paulo, SP CEP 05508-090, Brazil. [2]Departamento de Oceanografia Biológica, Instituto Oceanográfico, Universidade de São Paulo, Praça do Oceanográfico, 191, São Paulo, SP CEP 05508-120, Brazil.

References

Bazinet AL, Zwickl DJ, Cummings MP. A gateway for phylogenetic analysis powered by grid computing featuring GARLI 2.0. Syst Biol. 2014;63(5):812–8.

Bell TG. A taxonomic and phylogenetic study of *Nephroselmis* Stein. 2008.

Bergesch M, Odebrecht C, Moestrup Ø. Nanoflagellates from coastal waters of southern Brazil (32°S). Bot Mar. 2008;51(1):35–50.

Butcher RW. An introductory account of the smaller algae of British coastal waters. Part I: Introduction and Chlorophyceae. London; Fish Investig. 1959;4:1–74

Cavalier-Smith T. The origin, losses and gains of chloroplast. In: Lewin RE, editor. Orig Plast Symbiogenes prochlorophytes Orig chloroplast. New York: Chapman and Hall; 1993. p. 291–48.

Darriba D, Taboada GL, Doallo R, Posada D. jModelTest 2: more models, new heuristics and parallel computing. Nat Methods. 2012;9(8):772.

Daugbjerg N, Moestrup Ø, Arctander P. Phylogeny of genera of Prasinophyceae and Pedinophyceae (Chlorophyta) deduced from molecular analysis of the rbcL gene. Phycol Res. 1995;43:203–13.

Domingos P, Menezes M. Taxonomic remarks on planktonic phytoflagellates in a hypertrophic tropical lagoon (Brazil). Hydrobiologia. 1998;369/370:297–313.

Edgar RC. MUSCLE: multiple sequence alignment with high accuracy and high throughput. Nucleic Acids Res. 2004;32(5):1792–7.

Faria DG, Kato A, de la Peña MR, Suda S. Taxonomy and phylogeny of *Nephroselmis clavistella* sp. nov. (Nephroselmidophyceae, Chlorophyta). J Phycol. 2011;47(6):1388–96.

Faria DG, Kato A, Suda S. *Nephroselmis excentrica* sp. nov. (Nephroselmidophyceae, Chlorophyta) from Okinawa-jima, Japan. Phycologia. 2012;51(3):271–82.

Felsenstein J. Confidence limits on phylogenies: an approach using bootstrap. Evolution. 1985;39(4):783–91.

Garcia M, Odebrecht C. Remarks on the morphology and distribution of some rare centric diatoms in southern Brazilian continental shelf and slope waters. Braz J Oceanogr. 2012;60(4):415–27.

Guillard RR, Ryther JH. Studies of marine planktonic diatoms. I. *Cyclotella nana* Hustedt, and *Detonula confervacea* (Cleve) Gran. Can J Microbiol. 1962;8:229–39.

Guiry MD, Guiry GM. AlgaeBase. World-wide eletronic Publ Natl Univ Ireland, Galw. 2016. http://www.algaebase.org. Accessed 2 Oct 2016.

Jardim PFG, de Cardoso LS. New distribution records of Dinophyta in Brazilian waters. Check List. 2013;9(3):631–9.

Lackey JB. Some new flagellates from the Woods Hole Area. Am Midl Nat. 1940; 23(2):463–71.

Larsson A. AliView: a fast and lightweight alignment viewer and editor for large datasets. Bioinformatics. 2014;30(22):3276–8.

Lubiana KMF, Dias Júnior C. The composition and new records of micro- and mesophytoplankton near the Vitória-Trindade Seamount Chain. Biota Neotrop. 2016;16(3):e20160164.

Marin B, Melkonian M. Flagellar hairs in Prasinophytes (Chlorophyta): ultrastructure and distribution on the flagellar surface. J Phycol. 1994; 30(4):659–78.

Marin B, Melkonian M. Molecular phylogeny and classification of the Mamiellophyceae class. nov. (Chlorophyta) based on sequence comparisons of the nuclear- and plastid-encoded rrna operons. Protist. 2010;161(2):304–36.

Menezes M, Bicudo CEM. Flagellate green algae from four water bodies in the state of Rio de Janeiro. Southeast Brazil. Hoehnea. 2008;35(3):435–68.

Moestrup Ø. Further studies on *Nephroselmis* and its allies (Prasinophyceae). I. The question of the genus *Bipedinomonas*. Nord J Bot. 1983;3(1979):609–27.

Nakayama T, Marin B, Kranz HD, Surek B, Huss VAR. The basal position of scaly green flagellates among the green algae (Chlorophyta) is revealed by analyses of nuclear-encoded SSU rRNA sequences. Protist. 1998;149:367–80.

Nakayama T, Suda S, Kawachi M, Inouye I. Phylogeny and ultrastructure of *Nephroselmis* and *Pseudoscourfieldia* (Chlorophyta), including the description of *Nephroselmis anterostigmatica* sp. nov. and a proposal for the Nephroselmidales ord. nov. Phycologia. 2007;46:680–97.

Norris RE. Prasinophytes. In: Cox ER, editor. Phytoflagellates. New York: Elsevier; 1980. p. 85–146.

Parke M, Rayns DG. Studies on marine flagellates: VII *Nephroselmis gilva* sp nov. and some allied forms. J Mar Biol Ass UK. 1964;44:209–17.

Ronquist F, Teslenko M, Van Der Mark P, Ayres DL, Darling A, Höhna S, et al. Mrbayes 3.2: Efficient bayesian phylogenetic inference and model choice across a large model space. Syst Biol. 2012;61(3):539–42.

Steinkotter J, Bhattacharya D, Semmelroth I, Bibeau C, Melkonian M. Prasinophytes form independent lineages within the Chlorophyta: evidence from ribosomal RNA sequence comparisons. J Phycol. 1994;30(2):340–5.

Suda S. Light microscopy and electron microscopy of Nephroselmis spinosa sp. nov. (Prasinophyceae, Chlorophyta). J Phycol. 2003;39(3):590–9.

Turmel M, Gagnon MC, O'Kelly CJ, Otis C, Lemieux C. The chloroplast genomes of the green algae *Pyramimonas*, *Monomastix*, and *Pycnococcus* shed new light on the evolutionary history of prasinophytes and the origin of the secondary chloroplasts of euglenids. Mol Biol Evol. 2009;26(3):631–48.

Turmel M, Otis C, Lemieux C. The complete chloroplast DNA sequence of the green alga *Nephroselmis olivacea*: insights into the architecture of ancestral chloroplast genomes. Proc Natl Acad Sci U S A. 1999;96(18):10248–53.

Yamaguchi H, Suda S, Nakayama T, Pienaar RN, Chihara M, Inouye I. Taxonomy of *Nephroselmis viridis* sp. nov. (Nephroselmidophyceae, Chlorophyta), a sister marine species to freshwater *N. olivacea*. J Plant Res. 2011;124(1):49–62.

Yoshii Y, Takaichi S, Maoka T, Suda S, Sekiguchi H, Nakayama T, et al. Variation of siphonaxanthin series among the genus *Nephroselmis* (Prasinophyceae, Chlorophyta), including a novel primary methoxy carotenoid. J Phycol. 2005;41(4):827–34.

21

Medusae and ctenophores from the Bahía Blanca Estuary and neighboring inner shelf (Southwest Atlantic Ocean, Argentina)

M. Sofía Dutto[1,2*], Gabriel N. Genzano[1,3,4], Agustín Schiariti[1,3,5], Julieta Lecanda[6,7], Mónica S. Hoffmeyer[1,2,8] and Paula D. Pratolongo[1,2,9]

Abstract

An updated checklist of medusae and ctenophores is presented for the first time for the area comprised by the Bahía Blanca Estuary, the adjacent shelf El Rincón and Monte Hermoso beach, on the southwest coast of Buenos Aires province (Argentina). The area is highly productive and provides several ecosystem services including fishing and tourism. Updated information on the biodiversity of medusae and ctenophores species is essential for the study area, given that these species can affect ecosystem services. The list includes 23 hydromedusae, 3 scyphomedusae, and 3 ctenophores. Five hydromedusae (Halitiara formosa, Amphinema dinema, Aequorea forskalea, Clytia lomae and Halopsis ocellata) were firstly observed in this area. Three species of medusae, 2 hydromedusae (Olindias sambaquiensis and Liriope tetraphylla) and 1 scyphomedusae (Chrysaora lactea) pose a potential health risk, due to their toxicity to humans. Considering the size of the study area, the Bahía Blanca region has a comparatively high species richness of hydromedusae, higher than larger zones previously studied along the temperate SW Atlantic Ocean. The present report provides the baseline knowledge of gelatinous species for the Bahía Blanca region.

Keywords: Gelatinous species, Composition, Richness, Coastal ecosystem, South America

Introduction

Our knowledge of gelatinous fauna (medusae and ctenophores particularly) in the Argentine Sea has been much enhanced since major contributions by Ramírez and Zamponi (1981), Bouillon (1999), Mianzan (1999), Mianzan and Cornelius (1999), Genzano et al. (2008a) and Rodriguez (2012). However, the Bahía Blanca Estuary and its neighboring inner shelf is a still largely understudied area, in spite of the widely recognized ecosystem services that these coastal waters provide, such as wildlife support (Delhey and Petracci 2004; Hoffmeyer et al. 2009; Guinder et al. 2013) and fisheries production (Carozza and Fernández Aráoz 2009), as well as nutrient cycling and amelioration of heavy metal pollution (Negrin et al. 2016). These coastal waters are intensely used for fishing, recreational

* Correspondence: msdutto@criba.edu.ar
[1]Consejo Nacional de Investigaciones Científicas y Técnicas (CONICET), Av. Rivadavia 1917, C1033AAJ Ciudad Autónoma de Buenos Aires, Argentina
[2]Instituto Argentino de Oceanografía (IADO-CONICET/UNS), Área Oceanografía Biológica, La Carrindanga km 7.5, B8000FWB Bahía Blanca, Argentina
Full list of author information is available at the end of the article

purposes, and provide space for industrial, port and commercial activities (Acha et al. 2004; Pizarro and Piccolo 2008). Several species of medusae and ctenophores have been reported in the region (Ramírez and Zamponi 1980; Hoffmeyer and Mianzan 2004), including species which may exert considerable predation pressure on fishing resources, mainly through consumption of larvae and juveniles, as well as competition for food (Mianzan and Sabatini 1985; Mianzan 1986a; Hoffmeyer 1990). Medusae species of public health concern such as *Olindias sambaquiensis* and *Liriope tetraphylla* have also been reported in high concentrations (Mianzan and Ramírez 1996; Mianzan et al. 2000, 2001). Species composition and ecological studies on medusae and ctenophores in the Bahía Blanca region were mainly conducted in the 1980's (Mianzan and Sabatini 1985; Zamponi and Mianzan 1985; Mianzan 1986a, 1986b, 1989a, 1989b, 1989c; Mianzan and Zamponi 1988; Hoffmeyer 1990), but discontinued afterwards.

Given the potential impacts of gelatinous species on valuable ecosystem services, updated information on their biodiversity in the study area is essential. This work aims at compiling information on the species composition of medusae and ctenophores in the Bahía Blanca Estuary and its neighboring inner shelf. To do so, we considered earlier faunal lists and own unpublished data from plankton surveys.

Materials and methods
Study area
The study area is located in a temperate, semiarid zone in the southwestern Atlantic, Argentina (38°30′ - 41°S, ~60°W) (Fig. 1). According to the biogeographic division of the Argentine Sea based on its marine fauna (Balech and Ehrlich 2008), the study area represents the transition between the Rionegrine and Uruguayan Districts of the Argentine Province. In this work, we considered information on the biodiversity of medusa and ctenophore species from the Bahía Blanca Estuary (BBE), the associated sandy beaches of Monte Hermoso (MH) located north of the estuary, and the adjacent shelf El Rincón (ER) (Fig. 1). The BBE is a shallow, funnel shaped system, oriented NW/SE. The estuarine area comprises a dense arrangement of meandering channels and islands, surrounded by extensive intertidal mudflats and marshes (Pratolongo et al. 2013). The inner zone of the estuary is characterized by a shallow (~7 m depth) and vertically

homogeneous water column, due to the strong tidal mixing and wind forcing (Popovich and Marcovecchio 2008). Turbidity and nutrient contents are typically high, with a strong seasonal pattern (Popovich and Marcovecchio 2008). ER, in the adjacent shelf, extends to the 50 m isobath (Acha et al. 2004). Within the ER area, low-salinity waters and sediments from the continent (Río Negro, Río Colorado and BBE), highly saline waters from the south (San Matías Gulf) and shelf waters of typically intermediate salinity combine to create oceanic fronts and circulation cells that favor retention mechanisms, creating an appropriate environment for successful larval fish development (Hoffmeyer et al. 2009; Acha et al. 2012; Delgado et al. 2015). Sandy beaches of MH and Pehuen Có are one of the most visited touristic destinations of Argentina (Vaquero et al. 2007). These beaches are transitional environments between estuarine shorelines and open beaches of Buenos Aires Province (Carcedo et al. 2015; Menéndez et al. 2016), and commonly receive significant amounts of estuarine sediments (Delgado et al. 2016).

Own data collection and literature review
We present own data from 89 plankton samples, collected in 33 sampling campaigns. Plankton sampling

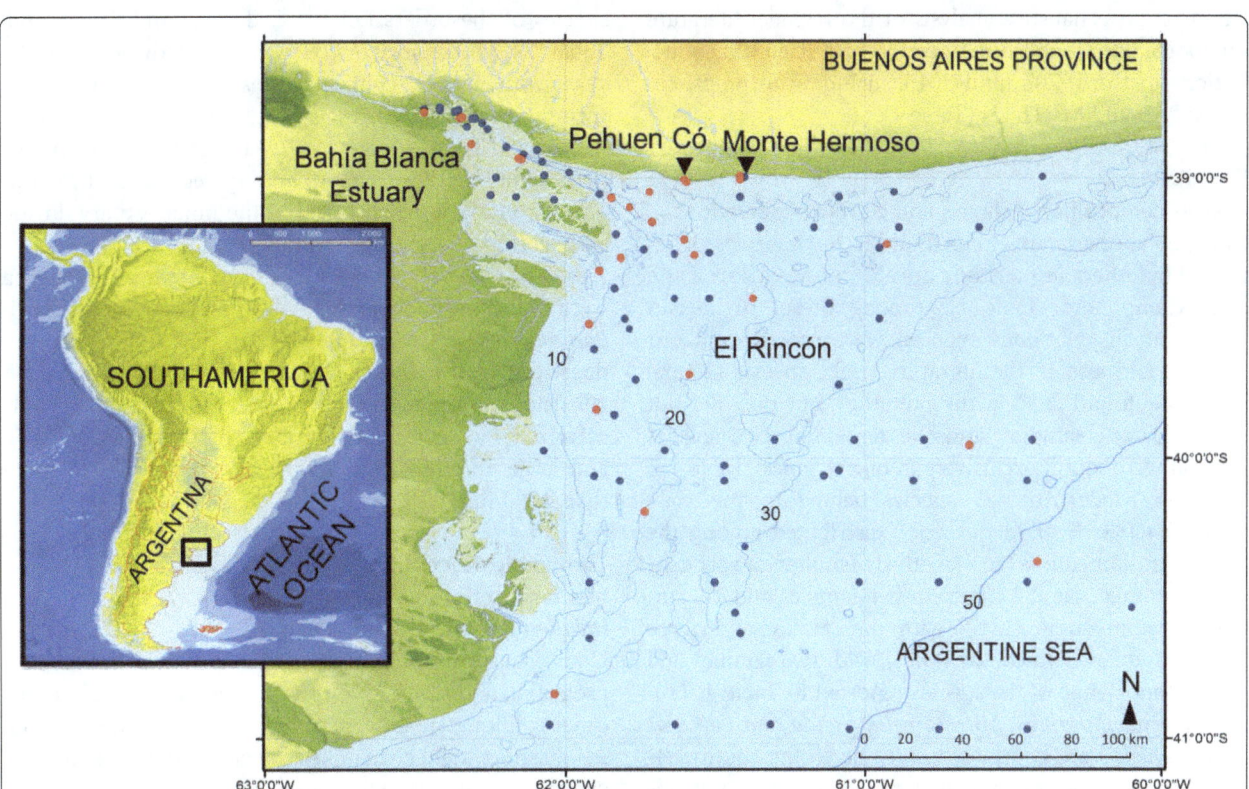

Fig. 1 Study area located in the southwestern Atlantic, Argentina. The area comprised the Bahía Blanca Estuary, the adjacent shelf El Rincón and the surf beach of Monte Hermoso to the north. Numbers represent bathymetry expressed in meters. Samplings sites included in the checklist are shown (*red circles* = own sampling surveys, blue circles = literature sampling points)

within the study area was performed in December 2012, and recurring monthly campaigns were carried out from April 2013 to May 2014, and from December 2014 to February 2015 (Fig. 1). To collect the samples, Hensen-like zooplankton nets were used (mouth diameters 30 and 40 cm, and mesh sizes 67, 200 and 500 μm) and also a modified RMT (Rectangular Midwater Trawl; mouth opening 2.25 m^2 and mesh size 1000), designed to capture a wider size range of gelatinous organisms. Hensen nets were used in oblique tows (bottom to surface) from a motorboat or ship moving at ~2 knots, during 7 min. The RMT net was deployed against the ebb tide current, during 20 min. Records of stranded individuals were also considered.

Gelatinous species were fixed using a solution of 4% formaldehyde in seawater and later identified following Mayer (1910), Kramp (1961), Bouillon (1999), Mianzan (1999), Mianzan and Cornelius (1999) and Bouillon et al. (2006). When necessary, the hydrozoan collection stored at the J. J. Nágera Biological Station (Universidad Nacional de Mar del Plata, Argentina) was consulted. Species which do not tolerate fixation (as the ctenophore *Mnemiopsis leidyi*) were immediately identified after capture. Photographic records of the collected species were obtained with a digital camera attached to a stereomicroscope, before their deposit in the Instituto Argentino de Oceanografia (IADO-CONICET). The bibliographical review was based on an exhaustive analysis of the regional literature on zooplankton published since the first scientific contributions in the 1970's until 2015, including technical reports, PhD and Ms Thesis (Table 1).

Results
Species composition, richness and current observations
In all, 29 species were either found in our samples or reported by others in the study area (26 of Medusozoa and 3 of Ctenophora; Table 1). Among these, 16 species (55.2% of all gelatinous species) were recorded in our samples and also in the literature, eight species (27.6%) were only found cited in the literature, but they did not appear in our samples, and the remaining five species (17.2%) were found exclusively in our samples (Table 1).

Within Medusozoa, 23 species belong to the class Hydrozoa (88.5% of Medusozoa), distributed among the orders Anthomedusae (26%), with six families and six genera, Leptomedusae (52.2%), with five families and 12 genera, Limnomedusae (13%), with one family and three genera, and Trachymedusae (8.7%), with two families and two genera. Three of the species reported in Table 1, *Proboscidactyla mutabilis*, *Mitrocomella frigida* and *Olindias sambaquiensis* are considered endemic to the southwestern Atlantic. Species identified to genus level only included *Obelia* spp. (Hydrozoa, Leptothecata), for which medusa specimens cannot be reliably determined down to species level (Bouillon et al. 2006). However, when

hydroids were found at the time of medusae, the nominal species *O. longissima*, *O. dichotoma* and *O. bidentata* were cited for the study area. Five hydrozoan species (2 Anthomedusae and 3 Leptomedusae), collected in our samples, were reported for the first time in the study area: *Halitiara formosa* (Fewkes 1882), *Amphinema dinema* (Péron and Lesueur, 1810), *Aequorea forskalea* (Péron and Lesueur 1810), *Clytia lomae* (Torrey 1909) and *Halopsis ocellata* (Agassiz, 1865) (Fig. 2). Their taxonomic descriptions are provided below. The record of *H. ocellata* in MH is, as far as we know, the northern record for the Southwestern Atlantic. With respect to *A. forskalea*, large numbers of adult and juvenile specimens were commonly observed in MH, either stranded on the beach or in coastal waters, during January 2014, and the massive occurrence repeated in February 2016.

Scyphozoa was represented by three species of the order Semaeostomeae, families Pelagiidae, Ulmaridae and Cyaneidae. *Chrysaora lactea*, *Aurelia aurita* and *Drymonema gorgo* were the only scyphozoans recorded in the area (Table 1). We did not find *A. aurita* ephyrae in our samples, although the use of this coastal area as a reproduction site had been suggested (Mianzan 1986a). Regarding *C. lactea*, ephyrae of this species were frequently found in our samples, mainly in BBE and ER.

The Phylum Ctenophora was represented by three species of the orders Lobata, Cydippida and Beroida (cf. Table 1). *Mnemiopsis leidyi* was found throughout the year, with higher concentrations during autumn and spring, mainly in channels within the BBE connected with island zones. *Beroe ovata* was also found in the estuary almost all year round, and aggregations of *Pleurobrachia pileus* were observed in the inner estuary during early spring.

Finally, three stinging species of public health concern were found in the area: the hydromedusae *Olindias sambaquiensis* and *Liriope tetraphylla*, and the scyphomedusa *Chrysaora lactea*. Regarding stinging species, a consistent summer trend was observed since 2013, characterized by decreasing numbers of *O. sambaquiensis*, and large amounts of *L. tetraphylla* (from 600 to more than 1000 ind.m^{-3}).

Taxonomic descriptions of the species observed for the first time in the study area
Halitiara formosa Fewkes 1882
Umbrella about 3 mm high, pear-shaped, with solid apical projection about half as long as bell cavity, 4 straight radial canals; 4 long, hollow perradial marginal tentacles and 24–35 short, solid cirrus-like tentacles; mouth simple, cruciform; with or without mesenteries; "gonads" interradial, smooth, sometimes extending over mesenteries; without ocelli; cnidome, when known, with merotrichous isorhizae (Bouillon 1999; Bouillon et al. 2006) (Fig. 2 A).

Table 1 Review of medusae and ctenophores from the Bahía Blanca region (Argentina) based on own data and literature data. In each case, data related to study area and type of collection from own samples are highlighted in bold while those from literature are not highlighted; when both sources overlap, data are italic

Taxa	Sampling date	Study area	Source	Type of collection
Phylum CNIDARIA				
Class HYDROZOA				
Subclass HYDROIDOLINA				
Order ANTHOATHECATA				
Suborder FILIFERA				
Family OCEANIIDAE				
Turritopsis nutricula	1979–1981; 1983–2008; 2010; 2013–2014	*BBE-ER-***MH**	Hoffmeyer 1983; Ramírez and Zamponi 1980; Genzano et al. 2008a; Rodriguez 2012; **This study**	**J**, *A*
Family PANDEIDAE				
Amphinema dinema[a]	2013; 2015	**ER-MH**	**This study**	**A**
Family PROBOSCIDACTYLIDAE				
Proboscidactyla mutabilis	1970; 1983–2008	ER	Ramírez and Zamponi 1980; Genzano et al. 2008a; Rodriguez 2012	A
Family PROTIARIDAE				
Halitiara formosa[a]	2013–2014	**ER-MH**	**This study**	**J, A**
Suborder APLANULATA				
Family CORYMORPHIDAE				
Corymorpha januarii	1982; 1983–2008; 2010; 2013–2014	*BBE*-ER	Genzano et al. 2009a; Rodriguez 2012; **This study**	*H*[c], J, A
Family TUBULARIIDAE				
Hybocodon chilensis	1983–2008; 2010	BBE-ER	Genzano et al. 2008a; Rodriguez 2012; Rodriguez et al. 2012	J, A
Order LEPTOTHECATA				
Family AEQUOREIDAE				
Aequorea forskalea[a]	2014	**MH**	**This study**	**J, A**
Family CAMPANULARIIDAE				
Clytia gracilis	1971–1972; 1997; 1998; 2003; 2013–2014	BBE-*ER-***MH**	Bastida and Torti 1971; Bastida et al. 1977; Genzano et al. 2009b; Rodriguez 2012; **This study**	H, A
Clytia hemisphaerica	1983–2008; 2010; 2014	ER-**MH**	Genzano et al. 2009b; Rodriguez 2012; **This study**	H, *A*
Clytia lomae[a]	2014	**MH**	**This study**	**A**
Clytia simplex	1970	ER	Ramírez and Zamponi 1980	A
Obelia spp.	1979–1981; 1993–2006; 2013–2014	*BBE-ER-MH*	Hoffmeyer 1983; Blanco 1994; Hoffmeyer and Barría de Cao 2007; Genzano et al. 2008b; Genzano et al. 2009b; Rodriguez 2012; **This study**	H, J, A
Family EIRENIDAE				
Eutonina scintillans	2006	ER	Rodriguez et al. 2007	A
Family LOVENELLIDAE				
Eucheilota ventricularis	1983–2008;2010; 2013–2014	**BBE**-*ER-MH*	Rodriguez 2006; Rodriguez 2012; **This study**	J, A
Family MITROCOMIDAE				
Cosmetirella davisi	1983–2008; 2010	ER	Ramírez and Zamponi 1980; Rodriguez 2012	A
Mitrocomella frigida	1970; 1983–2002	ER	Ramírez and Zamponi 1980; Rodriguez 2006	A
Mitrocomella brownei	1983–2008; 2010	ER	Genzano et al. 2008a; Rodriguez 2012	J, A

Table 1 Review of medusae and ctenophores from the Bahía Blanca region (Argentina) based on own data and literature data. In each case, data related to study area and type of collection from own samples are highlighted in bold while those from literature are not highlighted; when both sources overlap, data are italic *(Continued)*

Halopsis ocellata[a]	2015	**MH**	**This study**	A
Subclass TRACHYLINAE				
Order LIMNOMEDUSAE				
Family OLINDIIDAE				
Aglauropsis kawari	1983–2008; 2015	ER-**MH**	Genzano et al. 2008a; **This study**	A, **J**
Gossea brachymera	1982–1984; 2013–2014	*BBE*-**ER-MH**	Mianzan 1986a; **This study**	na; **J**, **A**
Olindias sambaquiensis	1981–1984; 2013–2014	*BBE*-ER-*MH*	Mianzan 1986a; Zamponi and Mianzan 1985; Mianzan 1989c; Macchi et al. 1995; Chiaverano et al. 2004; Chiaverano & Mianzan 2001; Rodriguez 2012; **This study**	J, A
Order TRACHYMEDUSAE				
Family GERYONIIDAE				
Liriope tetraphylla	1983; 1987–1988; 1992–2002; 2013–2014	**BBE**-*ER*-**MH**	Gaitán 2004; Rodriguez 2012; **This study**	J, A
Family HALICREATIDAE				
Halitrephes maasi	1970	ER	Ramírez and Zamponi 1980	A
Class SCYPHOZOA				
Subclass DISCOMEDUSAE				
Order SEMAEOSTOMEAE				
Family CYANEIDAE				
Drymonema gorgo	1982–1984; 2008; 2012	*BBE*-**MH**	Mianzan 1986a; Mianzan 1989a, b; **This study**	J[b]
Family PELAGIIDAE				
Chrysaora lactea	1982–1984; 2013–2014	*BBE*-**ER**-*MH*	Mianzan 1986a; Mianzan 1989a, b; **This study**	E, J, A
Family ULMARIDAE				
Aurelia aurita	1982–1984; 2008; 2013	*BBE*-**MH**	Mianzan 1986a; Mianzan 1989a, b; **This study**	E, J, A
Phylum CTENOPHORA				
Class TENTACULATA				
Subclass CYCLOCOELA				
Order LOBATA				
Family BOLINOPSIDAE				
Mnemiopsis leidyi	1982–1984; 1990; 2013–2014	*BBE*-ER-**MH**	Mianzan and Sabatini 1985; Mianzan 1986a; Hoffmeyer 1990 (as *Mnemiopsis mccrady*); **This study**	J, A
Subclass TYPHLOCOELA				
Order CYDIPPIDA				
Family PLEUROBRACHIIDAE				
Pleurobrachia pileus	1979–1981; 2013–2014	*BBE*-**ER-MH**	Hoffmeyer 1983; **This study**	na; **J**, **A**
Class NUDA				
Order BEROIDA				
Family BEROIDAE				
Beroe ovata	1982–1984; 2013–2014	*BBE*-*ER*-**MH**	Mianzan 1986a, b; **This study**	J, A

Abbreviations: *BBE* Bahía Blanca Estuary, *MH* Monte Hermoso, *ER* El Rincón
E ephyrae, *H* hydroids, *J* juveniles, *A* adults, *Ac* actinula larvae
na not available
[a] first record for the region
[b] material provided by Prof. Verónica Arias and storaged at Cátedra de Zoología de Invertebrados I, Universidad Nacional del Sur, Bahía Blanca, Argentina
[c] material provided by Alberto Conte and storaged at the Instituto Argentino de Oceanografía, Bahía Blanca, Argentina

Amphinema dinema Péron and Lesueur 1810

Umbrella up to 4 mm wide and 6 mm high with considerable conical, solid, apical projection, jelly of uniform thickness around top. Four undivided radial canals. Manubrium with broad base, cross-like in section, flask-shaped, almost as long as bell cavity. Mouth with four prominent, recurved lips. With eight simple adradial gonads, smooth, on manubrium wall only. Two perradial, hollow, marginal tentacles with large elongated conical basal bulbs; bulbs without ocelli. With 14–24 small marginal warts (Bouillon 1999) (Fig. 2 B1-2).

Aequorea forskalea Péron and Lesueur 1810

Flat umbrella, 14–32 mm in diameter. Short manubrium, mouth large, about half the diameter of umbrella. Numerous radial canals (usually 60–80, sometimes fewer, and up to 160). Tentacles with elongate conical bulbs, generally less numerous than radial canals but varying from half to the same number of them; 5–10 statocysts between successive radial canals. Gonads along almost the entire length of the radial canals (Nagata et al. 2014) (Fig. 2 C1-3).

Clytia lomae Torrey 1909

Umbrella 9–12 mm in diameter, about 4 times broader than high, thin. Gonads narrow, elongated along less than 1/2 of the distal part of radial canals; about 32 tentacles and some young bulbs. Manubrium short, cruciform; mouth with 4 slightly frilled lips; bulbs elongated; 1 (rarely 2) statocysts between successive tentacles (Bouillon 1999). Smaller medusae (umbrella 3–5 mm wide) have been observed in Argentine waters (Rodriguez 2012; this study) (Fig. 2 D1-2).

Halopsis ocellata Agassiz 1865

Umbrella 50–65 mm in diameter, about 4 times as wide as high, watch-glass-shaped; mesoglea thick toward centre; manubrium broad, flat, 1/5 of bell diameter, circular to star-shaped in outline; mouth with 4 fairly short lips; 12–16 radial canals in 4 groups branching usually within outline of manubrium; gonads linear, about 2/3 of radial canals; up to 450 marginal tentacles; 1 marginal cirrus between successive tentacles; about 80 statocysts (Bouillon 1999). Smaller medusae (umbrella up to 28 mm in diameter) have been observed in Argentine waters (Rodriguez 2012; this study) (Fig. 2 E1-2).

Discussion

Our study provides the first compiled list of medusae and ctenophores species of the Bahía Blanca region, adding five hydromedusae species that had not been previously reported for the area. We found a high richness of hydromedusae species compared to values reported along the temperate Southwestern Atlantic platform (Genzano et al. 2008a; Rodriguez 2012). Based on an exhaustive sampling carried out across the neritic region from 33° to 55°S, over 20 years, Genzano et al. (2008a) recognized 71 hydromedusae species. Our study area covers less than 3% of the area covered by Genzano et al. (2008a), but we found 32.4% of the total number of hydromedusae species detected by these authors, a

disproportionally large richness for the small area considered.

Taking into account the transitional location of our study area and the hydromedusae faunal list by Rodriguez (2012), we found species that equally represent both the Rionegrin and Uruguayan biogeographic districts. According to Balech and Ehrlich (2008), the Argentine Province is essentially neritic, characterized by a marked biological heterogeneity due to the mix of subtropical and subantarctic waters. This combination of subtropical and subantartic elements also determines a low level of endemism for organisms in this region. The fundamentally neritic character of the Argentine Province is further reflected in the dominance of meroplanktonic species in the Bahía Blanca region (see Bouillon et al. 2006). Among hydromedusae, the meroplanktonic Leptomedusae showed the highest genus diversity followed by Anthomedusae, and only three species of hydromedusae considered endemic to the southwestern Atlantic were observed (Mianzan 1989c; Genzano et al. 2008a).

The species observed for the first time in the study area also belong to the orders Leptomedusae and Anthomedusae. *Halitiara formosa*, *A. dinema* and *C. lomae* were previously reported in the Argentine Sea south and north of our study area, while *H. ocellata* and *A. forskalea* were reported in austral waters (from 51° to 54°50′S and from 43° to 53°S, respectively). This later species was reported in Patagonian waters and there is only one record northward from our study area within the Argentine Sea (37°40′S-56°02′W) (Genzano et al. 2008a). The underlying causes of the massive occurrences of *A. forskalea* observed in MH during January 2014 and February 2016 are still unresolved. We hypothesize that changes in currents and wind patterns might have produced a recurrent advection of large numbers of individuals, but further studies are required to understand the origin of these mass occurrences. However it has to be also considered that jellyfish research in the study area has been neglected over the past 20 years, and the increasing sampling effort on gelatinous zooplankton throughout the last 5 years increased the proportion of findings related to gelatinous species.

Chrysaora lactea, *A. aurita* and *D. gorgo* were the only scyphozoans found in the study area (Mianzan 1989a, b; Schiariti et al. 2016; this study). Although we did not find *A. aurita* ephyrae in our samples, recently released ephyrae of this species were reported by Mianzan (1989a) who suggested the use of this coastal area as a reproduction site. Regarding *C. lactea* even though medusae are rather common and widespread, ephyrae have been rarely observed in plankton samples elsewhere (Mianzan 1989a, 1989b; Tronolone et al. 2002). In our samples, however, ephyrae of *C. lactea* were frequently found, which supports the suggestion by Mianzan

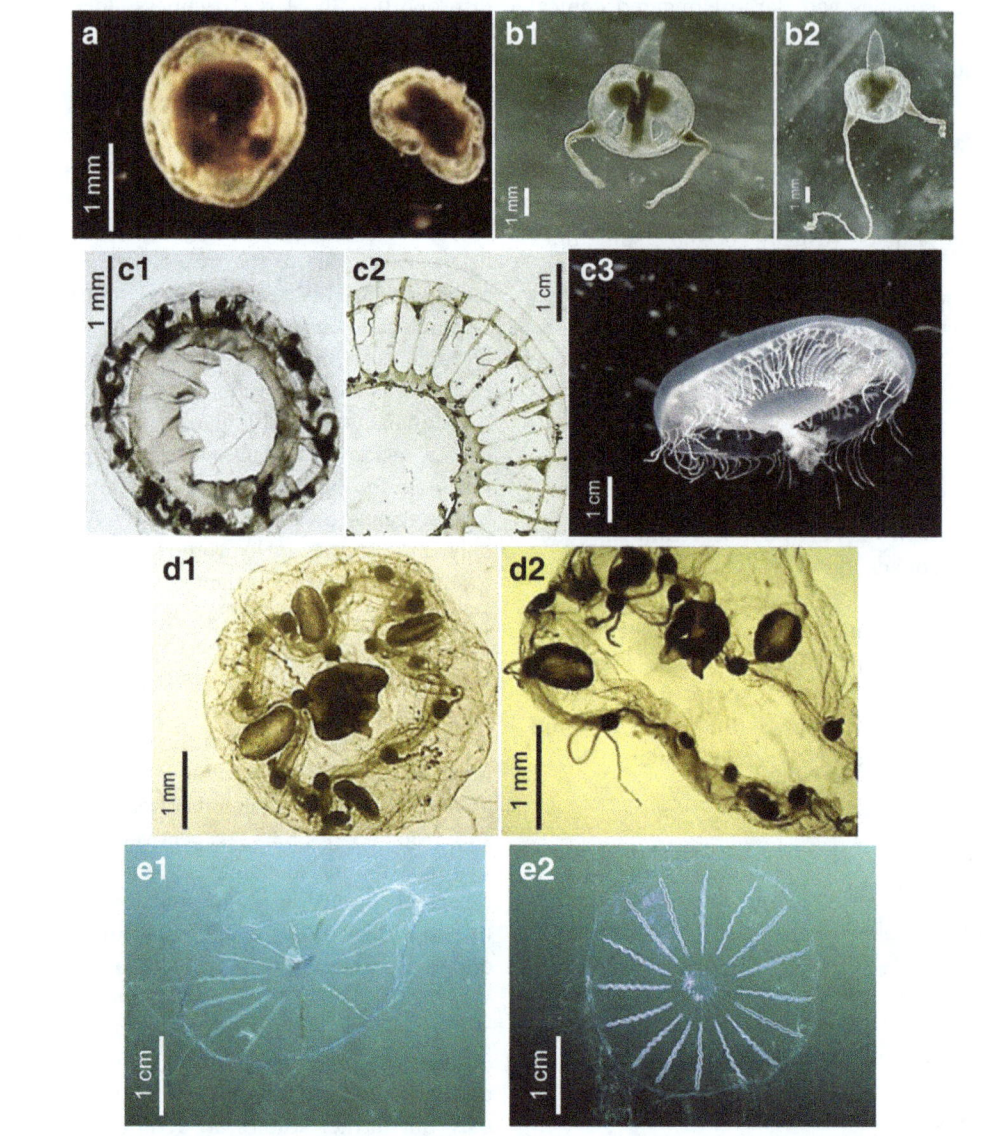

Fig. 2 Photographs of the five hydromedusae species firstly observed in the study area. A) Adults of *H. formosa*, B1-2) Adults of *A. dinema*, C1) juvenile and C2-3) adults of *A. forskalea*, D1-2) adults of *C. lomae*, E1-2) adults of *H. ocellata*. C3 and E1-2 reproduced with the permission of Kåre Telnes (www.seawater.no)

(1989a) about the reproduction area for these scyphozoans. Polyps of *C. lactea* have never been found in nature (Morandini et al. 2004). Potential substrata for polyp settlement include docks, harbors, support structures of industries, dredged material storage piles, buoys, fouling fauna, native vegetation, rocky, muddy and sandy bottoms (Miyake et al. 2002; Morandini et al. 2004; Lucas et al. 2012), all of them available in the study area. Future research should include benthic surveys to explore the presence of benthic stages, and their association with natural and human-made substrates. Finally, the occurrence of *D. gorgo* is a rare event that reconfirms its geographical distributional range for these latitudes (Mianzan 1989a, 1989b). Information on its

ecology and distribution is very scarce due to its sporadic occurrence and the few specimens available for study (Williams et al. 2001; Bayha and Dawson 2010).

Coastal ctenophores are a major macroplanktonic group in the Southwestern Atlantic (Mianzan 1999), that may dominate zooplankton abundance and biomass (Mianzan et al. 1996; Mianzan and Guerrero 2000). In our study area, aggregations of ctenophores were observed at different times and sites (Hoffmeyer 1983; Mianzan and Sabatini 1985; Mianzan 1986a; Hoffmeyer 1990, this study). The occurrences of *M. leidyi*, *B. ovata*, and *P. pileus* in our samples are in agreement with findings by Mianzan and Sabatini (1985) and Mianzan (1986a).

Regarding the three species of public health concern, *O. sambaquiensis* has been long considered the most problematic species in terms of its health consequences, as well as its detrimental effects on touristic development (Mianzan et al. 2001). It causes severe skin damage and pain (Kokelj et al. 1993). Adults range in size from 6 to 10 cm (Nagata et al. 2014) although specimens up to 21 cm were observed in our study area (Mianzan 1986a). This species has a clear seasonal pattern of high-density aggregations during the warmest months (Macchi et al. 1995; Mianzan and Ramírez 1996), and were reported in the area from October to April (Mianzan 1986a). In spite of its presence year after year, the asexual polyp phase of *O. sambaquinsis* remains unidentified, and little is known about its population dynamic and reproduction (Macchi et al. 1995; Chiaverano et al. 2004). Regarding our samples, immature stages were expected to be found in late spring, as well as adults in summer. Nevertheless, juveniles did not appear and observations of adults reduced to sporadic occurrence in the area. This is in accordance with the unusual trend observed since 2013, characterized by a disappearance of the high-density aggregations usually observed (Brendel AS, Dutto MS, Menéndez MC, Huamantinco Cisneros MA, Piccolo MC: Wind pattern variation in a SW Atlantic beach: An explanation for changes in the coastal occurrence of the medusa Olindias sambaquiensis, submitted). The large amounts of *L. tetraphylla* observed in summer, since 2014, have also raised concern. These aggregations which can cause severe pruritus and strong itching sensation in sensitive areas of human skin conform a locally-known phenomenon called "tapioca", which was well documented on northern beaches in Argentina and Uruguay (Mianzan et al. 2000), but never reported in this geographic area or further south. Finally, *Chrysaora lactea* was the last stinging species found in the study area. It is one of the most common blooming scyphozoan along the entire South Western Atlantic coast (Mianzan and Cornelius 1999; Migotto et al. 2002; Schiariti et al. 2016). This species can cause mild to moderate local pain and burning sensation. Although less common, erythema and edema forming lesions were also reported (Marques et al. 2014).

The background list provided lays the foundation for the development of further investigations on gelatinous zooplankton in this highly relevant economic and biological coastal area. Benthic surveys are required to confirm the occurrence of polyps and to provide potential valuable information on the biology of the gelatinous species (e.g. life cycle) inhabiting this geographic region.

Acknowledgements

This work was supported by Agencia Nacional de Promoción Científica y Tecnológica (PICT 2011-2096 to M.S. Hoffmeyer, PICT 2013-1773 to A. Schiariti, and PICT 2012-2071 to P.D. Pratolongo), Secretaría General de Ciencia y Tecnología, Universidad Nacional del Sur (PGI 24 B/236), Universidad Nacional de Mar del Plata (EXA 734/15 to G.N. Genzano), and Consejo Nacional de Ciencia y Tecnología of Argentina (PIP 0152 to G.N. Genzano and PIP 2013-00615 to A. Schiariti). We thank greatly to A. Conte, E. Redondo, E. Contardi, C. Bernárdez, and J. Albrizio, to Cámara de Pescadores de Monte Hermoso and Pehuen Có, especially to E. Flores, and to all the colleagues which help and assist during samplings and laboratory activities: A. Berasategui, E. Nahuelhual, R. Uibrig, D. Muro Schenfelt, M. Garcia, J. Chazarreta, C. López Abbate, C. Menéndez, A. Delgado, V. Guinder, F. Thomsen, R. Elicer, M. Tártara, E. Dos Santos, L. Diaz Briz, C. Rodriguez, and A. Puente Tapia. The collaboration with material of A. Conte, V. Arias and D. Tanzola is also much appreciated. Thanks are also due to A. Migotto for providing photographs and W. Melo for drawing the map. We appreciate the thorough review of the manuscript by Dr. M. Thiel and two anonymous referees.

Authors' contributions

GNG, AS, MSH and PDP contributed to draft the manuscript. MSD wrote the manuscript. GNG and AS participated in the identification of the species. JL contributed with biological material. All authors read and approved the final manuscript.

Competing interests

The authors declare that they have no competing interests.

Author details

[1]Consejo Nacional de Investigaciones Científicas y Técnicas (CONICET), Av. Rivadavia 1917, C1033AAJ Ciudad Autónoma de Buenos Aires, Argentina. [2]Instituto Argentino de Oceanografía (IADO-CONICET/UNS), Área Oceanografía Biológica, La Carrindanga km 7.5, B8000FWB Bahía Blanca, Argentina. [3]Instituto de Investigaciones Marinas y Costeras (IIMyC), Funes 3350, B7602AYL Mar del Plata, Argentina. [4]Departamento de Ciencias Marinas, Facultad de Ciencias Exactas y Naturales, Universidad Nacional de Mar del Plata (UNMdP), Funes 3350, B7602AYL Mar del Plata, Argentina. [5]Instituto Nacional de Investigación y Desarrollo Pesquero (INIDEP), Paseo Victoria Ocampo N° 1, Escollera Norte, B7602HSA Mar del Plata, Argentina. [6]Museo Municipal de Ciencias Naturales de Monte Hermoso, N. Fossatty (ex Rio Paraná) N° 250, 8153 Balneario Monte Hermoso, Buenos Aires, Argentina. [7]Universidad Nacional del Sur (UNS), Bahía Blanca, Argentina. [8]Facultad Regional Bahía Blanca, Universidad Tecnológica Nacional (UTN), 11 de Abril 461, B8000LMI Bahía Blanca, Argentina. [9]Departamento de Biología, Bioquímica y Farmacia, UNS, San Juan 670, B8000DIC Bahía Blanca, Argentina.

References

Acha EM, Mianzan HW, Guerrero RA, Favero M, Bava J. Marine fronts at the continental shelves of austral South America: Physical and ecological processes. J Mar Syst. 2004;44:83–105.

Acha EM, Orduna M, Rodrigues K, Militelli MI, Braverman M. Caracterización de la zona de "El Rincón" (provincia de Buenos Aires) como área de reproducción de peces costeros. Revista de Investigación y Desarrollo Pesquero. 2012;21:31–43.

Agassiz A. Halopsis ocellata. Proc Boston Soc Nat Hist. 1865;9:219–20.

Balech E, Ehrlich MD. Esquema Biogeográfico del Mar Argentino. Revista de Investigación y Desarrollo Pesquero. 2008;19:45–75.

Bastida RO, Torti MR. Estudio preliminar sobre las incrustaciones biológicas de Puerto Belgrano. LEMIT-Anales. 1971;3:47–75.

Bastida RO, L'Hoste S, Spivak E, Adabbo H. Las incrustaciones biológicas de Puerto Belgrano. I. Estudio de la fijación sobre paneles mensuales, período 1971/72. Corrosión y Protección. 1977;8:1–23.

Bayha KM, Dawson MN. New Family of Allomorphic Jellyfishes, Drymonematidae (Scyphozoa, Discomedusae), Emphasizes Evolution in the Functional Morphology and Trophic Ecology of Gelatinous Zooplankton. Biol Bull. 2010; 219:249–67.

Blanco O. Enumeración sistemática y distribución geográfica preliminar de los Hidroides de la República Argentina. Suborden Athecata (Gymnoblastea, Anthomedusae), Thecata (Calyptoblastea, Leptomedusae) y Limnomedusae. Revista del Museo de La Plata. 1994;14:181–216.

Bouillon J. Hydromedusae. In: Boltovskoy D, editor. South Atlantic Zooplankton. Leiden: Backhuys Publishers; 1999. p. 385–465.

Bouillon J, Gravili C, Pagès F, Gili JM, Boero F. An introduction to Hydrozoa. *Memoires du Museum National d'Histoire Naturelle. Publications Scientifiques*, vol. 194. 2006. p. 591.

Carcedo MC, Fiori SM, Piccolo MC, López Abbate MC, Bremec CS. Variations in macrobenthic community structure in relation to changing environmental conditions in sandy beaches of Argentina. Estuar Coast Shelf Sci. 2015;166:56–64.

Carozza C, Fernández Aráoz NC. Análisis de la actividad de la flota en el área de "El Rincón" dirigida al variado costero durante el período 2000-2008 y situación de los principales recursos pesqueros. *INIDEP Informe técnico oficial* N° 23/09. 2009.

Chiaverano L, Mianzan H. Dinámica y estructura poblacional de *Olindias sambaquiensis*, Muller, 1861 (Limnomedusae, Olindiidae) durante su fase sexual en la zona de Bahía Blanca (Buenos Aires, Argentina). San Andrés Isla: IX Congreso Latinoamericano sobre Ciencias del Mar; 2001.

Chiaverano L, Mianzan H, Ramírez F. Gonad development and somatic growth patterns of *Olindias sambaquiensis* (Limnomedusae, Olindiidae). Hydrobiologia. 2004;530–531:373–81.

Delgado AL, Loisel H, Jamet C, Vantrepotte V, Perillo GME, Piccolo MC. Seasonal and inter-annual analysis of chlorophyll-a and inherent optical properties from satellite observations in the inner and mid-shelves of the south of Buenos Aires Province (Argentina). Remote Sens. 2015;7:11821–47.

Delgado AL, Menéndez MC, Piccolo MC, and Perillo GME. Hydrography of the inner continental shelf 'along the southwest Buenos Aires Province, Argentina: Influence of an estuarine plume on coastal waters. J Coastal Res. 2016. doi: http://dx.doi.org/10.2112/JCOASTRES-D-16-00064.1.

Delhey K, Petracci PF. Aves marinas y costeras. In: Piccolo MC, Hoffmeyer MS, editors. Ecosistema del Estuario de Bahía Blanca. Bahía Blanca: EdiUNS; 2004. p. 203–20.

Fewkes JW. Notes on acalephs from the Tortugas, with a description of new genera and species. In: Agassiz A, editor. Explorations of the surface fauna of the Gulf Stream, under the auspices of the U.S. Coast Survey. Bulletin of the Museum of comparative Zoölogy of Harvard College, vol. 9. 1882. p. 251–89.

Gaitán EN. *Distribución, abundancia y estacionalidad de Liriope tetraphylla (Hidromedusa, Trachymedusae) en el Océano Atlántico Sudoccidental y su rol ecológico en el estuario del Río de la Plata*. MSc thesis. Mar del Plata: Universidad Nacional de Mar del Plata; 2004.

Genzano G, Mianzan H, Bouillon J. Hydromedusae (Cnidaria: Hydrozoa) from the temperate southwestern Atlantic Ocean: a review. Zootaxa. 2008a;1750:1–18.

Genzano GN, Mianzan HW, Diaz BL, Rodriguez CS. On the occurrence of *Obelia* medusa bloom and empirical evidence of an unusual *Obelia* and *Amphisbetia* hydroids shoreline massive accumulations. Lat Am J Aquat Res. 2008b;36:301–7.

Genzano GN, Giberto D, Schejter L, Bremec C, Meretta P. Hydroid assemblages from the Southwestern Atlantic Ocean (34–42° S). Mar Ecol. 2009a;30:33–46.

Genzano GN, Rodriguez C, Pastorino G, Mianzan HW. The hydroid and medusa of *Corymorpha januarii* in temperate waters of the Southwestern Atlantic Ocean. Bull Mar Sci. 2009b;84:229–35.

Guinder VA, Popovich CA, Molinero JC, Marcovecchio J. Phytoplankton summer bloom dynamics in the Bahía Blanca Estuary in relation to changing environmental conditions. Cont Shelf Res. 2013;52:150–8.

Hoffmeyer MS. Zooplancton del área interna de Bahía Blanca (Buenos Aires - Argentina). I- Composición faunística. Hist Nat. 1983;3:73–94.

Hoffmeyer MS. Algunas observaciones sobre la alimentación de *Mnemiopsis mccradyi* Mayer (Ctenophora-Lobata). Iheringia Serie Zoologia. 1990;70:55–65.

Hoffmeyer MS, Barría de Cao MS. Zooplankton assemblage from a tidal channel in the Bahía Blanca Estuary. Braz J Oceanogr. 2007;55:97–107.

Hoffmeyer MS, Mianzan HW. Macrozooplancton del estuario de Bahía Blanca y aguas adyacentes. In: Piccolo MC, Hoffmeyer MS, editors. Ecosistema del Estuario de Bahía Blanca. Bahía Blanca: EdiUNS; 2004. p. 143–51.

Hoffmeyer MS, Menéndez MC, Biancalana F, Nizovoy AM, Torres ER. Ichthyoplankton spatial pattern in the inner shelf off Bahía Blanca Estuary, SW Atlantic Ocean. Estuar Coast Shelf Sci. 2009;84:383–92.

Kokelj F, Mianzan H, Avian M, Burnett JW. Dermatitits due to *Olindias sambaquiensis*: a case report. Cutis. 1993;51:339–42.

Kramp PL. Synopsis of the medusa of the world. J Mar Biol Assoc U K. 1961;40:1–469.

Lucas CH, Graham WM, Widmer C. Jellyfish life histories: role of polyps in forming and maintaining scyphomedusa populations. Adv Mar Biol. 2012;63:133–96.

Macchi G, Mianzan H, Cristiansen H, Ramírez F. Histology of the gonadal cycle of the stinging hydromedusa *Olindias sambaquiensis*, Muller, 1861 at Blanca Bay, Argentina. Bolletino della Societa Adriatica di Scienze. 1995;76:59–68.

Marques AC, Haddad Jr V, Rodrigo L, Marques-da-Silva E, Morandini AC. Jellyfish (*Chrysaora lactea*, Cnidaria, Semaeostomeae) aggregations in southern Brazil and consequences of stings in humans. Lat Am J Aquat Res. 2014;42:1192–9.

Mayer A. Medusae of the world. Hydromedusae. 1910;1:1–230.

Menéndez MC, Fernández Severini MD, Buzzi NS, Piccolo MC, Perillo GME. Assessment of surf zone environmental variables in a southwestern Atlantic sandy beach (Monte Hermoso, Argentina). Environ Monit Assess. 2016;188:495–507. doi:10.1007/s10661-016-5495-9.

Mianzan HW. *Estudio sistemático y bioecológico de algunas medusas Scyphozoa de la región subantártica*. Phd thesis. La Plata: Universidad Nacional de La Plata; 1986a.

Mianzan HW. *Beroe ovata* en aguas de la Bahía Blanca, Argentina (Ctenophora). Spheniscus. 1986b;2:29–32.

Mianzan HW. Las medusas Scyphozoa de la Bahía Blanca. Boletim do Instituto Oceanografico São Paulo. 1989a;37:29–32.

Mianzan HW. Sistemática y zoogregrafía de Scyphomedusae en aguas neríticas argentinas. Investigaciones Marinas CICIMAR. 1989b;4:15–34.

Mianzan HW. Distribución de *Olindias sambaquiesis* Müller, 1861 (Hydrozoa, Limnomedusae) en el Atlántico Sudoccidental. Iheringia Série Zoologia. 1989c;69:155–7.

Mianzan HW. Ctenophora. In: Boltovskoy D, editor. South Atlantic Zooplankton. Leiden: Blackhuys Publishers; 1999. p. 561–73.

Mianzan HW, Cornelius PFS. Cubomedusae and Scyphomedusae. In: Boltovskoy D, editor. South Atlantic Zooplankton. Leiden: Backhuys Publishers; 1999. p. 513–59.

Mianzan HW, Guerrero RA. Environmental patterns and biomass distribution of gelatinous macrozooplankton. Three study cases in the South-Western Atlantic Ocean. Sci Mar. 2000;64:215–24.

Mianzan HW, Ramírez FC. *Olindias sambaquiensis* stings in the South West Atlantic. In: Williamson JAH, Fenner PJ, Burnett JW, Rifkin JF, editors. Venomous and poisonous marine animals: a medical and biological handbook. Brisbane: University of New South Wales Press; 1996. p. 206–8.

Mianzan HW, Sabatini M. Estudio preliminar sobre distribución y abundancia de *Mnemiopsis maccradyi* en el estuario de Bahía Blanca (Ctenophora). Spheniscus. 1985;1:53–68.

Mianzan HW, Zamponi MO. Estudio bioecológico de *Olindias sambaquiensis* Müller, 1861 (Limnomedusae, Olindiidae) en el área de Monte Hermoso. II. Factores meteorológicos que influyen en su aparición. Iheringia Série Miscelanea. 1988;2:63–8.

Mianzan HW, Mari N, Prenski B, Sanchez F. Fish predation on neritic ctenophores from the Argentine continental shelf: A neglected food resource? Fish Res. 1996;27:69–79.

Mianzan HW, Sorarrain D, Burnett JW, Lutz LL. Mucocutaneous junctional and flexural paresthesias caused by the holoplanktonic trachymedusae *Liriope tetraphylla*. Dermatology. 2000;201:46–8.

Mianzan HW, Fenner PJ, Cornelius PFS, Ramírez C. Vinegar as disarming agent to prevent further discharge of the nematocysts of the stinging hydromedusa *Olindias sambaquiensis*. Cutis. 2001;6:45–8.

Migotto AE, Marques AC, Morandini AC, da Silveira FL. Checklist of the Cnidaria Medusozoa of Brazil. Biota Neotropica. 2002;2:1–31.

Miyake H, Terazaki M, Kakinuma Y. On the polyps of the common jellyfish *Aurelia aurita* in Kagoshima Bay. J Oceanogr. 2002;58:451–9.

Morandini AC, da Silveira FL, Jarms G. The life cycle of *Chrysaora lactea* Eschscholtz, 1829 (Cnidaria, Scyphozoa) with notes on the scyphistoma stage of three other species. Hydrobiologia. 2004;530:347–54.

Nagata RM, Nogueira M, Haddad MA. Faunistic survey of Hydromedusae (Cnidaria, Medusozoa) from the coast of Paraná State, Southern Brazil. Zootaxa. 2014;3768:291–326.

Negrin VL, Botté SE, Pratolongo PD, González TG, Marcovecchio JE. Ecological processes and biogeochemical cycling in salt marshes: synthesis of studies in the Bahía Blanca estuary (Argentina). Hydrobiologia. 2016;774:217–35.

Péron F, Lesueur CA. Tableau des caractères génériques et spécifiques de toutes les espèces de méduses connues jusqu'à ce jour. Annales du Muséum National d'histoire Naturelle de Paris. 1810;14:325–66.

Pizarro N, Piccolo MC. Socio-economic issues in the Bahía Blanca Estuary. In: Neves R, Baretta J, Mateus M, editors. Perspectives on integrated coastal zone management in South America. Lisbon: IST Press; 2008. p. 287–300.

Popovich CE, Marcovecchio JE. Spatial and temporal variability of phytoplankton and environmental factors in a temperate estuary of South America (Atlantic coast, Argentina). Cont Shelf Res. 2008;28:236–44.

Pratolongo P, Mazzon C, Zapperi G, Piovan MJ, Brinson MM. Land cover changes in tidal salt marshes of the Bahía Blanca estuary (Argentina) during the past 40 years. Estuar Coast Shelf Sci. 2013;133:23–31.

Medusae and ctenophores from the Bahía Blanca Estuary and neighboring inner shelf...

131

Ramírez FC, Zamponi MO. Medusas de la plataforma bonaerense y sectores adyacentes. Physis. 1980;39:33–48.

Ramírez FC, Zamponi MO. Hydromedusae. In: Boltovskoy D, editor. Atlas del Zooplancton del Atlántico Sudoccidental y Métodos de Trabajo con el Zooplancton Marino. Mar del Plata: Publicaciones Especiales de INIDEP; 1981. p. 443–69.

Rodriguez CS. *Distribución, abundancia y estacionalidad de* Mitrocomella frigida *y* Eucheilota ventricularis *(Hydrozoa, Leptomedusae) en el Atlántico Sudoccidental (33-55° S).* MSc thesis. Mar del Plata: Universidad Nacional de Mar del Plata; 2006.

Rodriguez C. *Hidromedusas del Atlántico Sudoccidental: biodiversidad y patrones de distribución.* PhD thesis. Mar del Plata: Universidad Nacional de Mar del Plata; 2012.

Rodriguez C, Genzano G, Mianzan H. First record of *Eutonina scintillans* Bigelow, 1909 (Hydrozoa: Leptomedusae: Eirenidae) in temperate waters of the southwestern Atlantic Ocean. Investig Mar. 2007;35:135–8.

Rodriguez C, Miranda TP, Marques AC, Mianzan H, Genzano G. The genus *Hybocodon* (Cnidaria, Hydrozoa) in the southwestern Atlantic Ocean, with a revision of *Hybocodon* species recorded in the area. Zootaxa. 2012;3523:39–48.

Schiariti A, Dutto MS, Morandini AC. Diversity and spatial distribution of Scyphomedusae and Cubomedusae from Argentina and Uruguay. Barcelona: 5th International Jellyfish Bloom Symposium, May 30-June 3 2016; 2016. p. 152.

Torrey HB. The Leptomedusae of the San Diego region. Univ Calif Publ Zool. 1909;6:11–31.

Tronolone VB, Morandini AC, Migotto AE. On the occurrence of scyphozoan ephyrae (Cnidaria, Scyphozoa, Semaeostomeae and Rhizostomeae) in the southeastern brazilian coast. Biota Neotropica. 2002;2:1–18.

Vaquero M del C, Rodríguez C, Trellini M, de Bulnes Cernadas M, Marcilese J. El turismo residenciado en Monte Hermoso. In: Proceedings of IV Jornadas interdisciplinarias del sudoeste bonaerense, Universidad Nacional del Sur, 7-9 September 2006. Bahía Blanca: Cuestiones políticas, socioculturales y económicas en el sudoeste bonaerense; 2007. p. 201–6.

Williams Jr EH, Bunkley-Williams L, Lilyestrom CG, Larson RJ, Engstrom NA, Ortiz-Corps EAR, Timber JH. A population explosion of the rare tropical/subtropical purple sea mane, *Drymonema dalmatinum*, around Puerto Rico in the summer and fall of 1999. Caribb J Sci. 2001;37:127–30.

Zamponi MO, Mianzan HW. La mecánica de captura y alimentación de *Olindias sambaquiensis* Müller, 1861 (Limnomedusae) en el medio natural y en condiciones experimentales. Hist Nat. 1985;5:269–78.

22

Morphological and genetic analyses of the first record of the Niger Hind, *Cephalopholis nigri* (Perciformes: Serranidae), in the Mediterranean Sea and of the African Hind, *Cephalopholis taeniops*, in Malta

Noel Vella*, Adriana Vella and Sandra Agius Darmanin

Abstract

Background: Non-native marine species, including tropical eastern Atlantic fish species are on the increase in Malta, with shipping activities being the main vector for the movement of these alien species from the Atlantic into the Mediterranean Sea. This calls for cooperation and collaboration between various sea-users and researchers to ensure continuous monitoring of coastal biodiversity.

Methods: Research methods involving local fishermen cooperation in monitoring efforts to identify and track populations of alien species in the Central Mediterranean has led to new records for the genus *Cephalopholis* (Perciformes: Serranidae) in Malta. Morphological characteristics, meristic counts and mitochondrial DNA sequences from specimens of both species sampled from Maltese waters were analysed to confirm their species identify accurately, essential for tracking their respective population expansions in the Mediterranean.

Results and conclusion: Results from this study have led to confirmation of the first record of the Niger Hind, *Cephalopholis nigri* (Günther, 1859), in the Mediterranean Sea and of the establishment of the African Hind, *Cephalopholis taeniops* (Valenciennes, 1828) in Maltese waters.

Keywords: *Cephalopholis nigri*, *Cephalopholis taeniops*, Non-native, New record, Mediterranean

Background

Groupers are economically important species and are caught by commercial, artisanal and recreational fishermen. In the Mediterranean, the subfamily Epinephelinae is represented by six native species, *Epinephelus aeneus*, *E. caninus*, *E. costae*, *E. marginatus*, *Hyporthodus haifensis* and *Mycteroperca rubra* (Heemstra and Randall 1993; Froese and Pauly 2016). In addition to these, there are non-native tropical Epinephelinae species, including six Indo-Pacific *Epinephelus* species, *E. malabaricus*, *E. coioides*, *E. merra*, *E. fasciatus*, *E. geoffroyi* and *E. areolatus*, that were first recorded in 1966 (Heemstra and Randall 1993), 1969 (Ben-Tuvia and Lourie 1969;

Heemstra and Golani 1993), 2004 (Lelong 2005), 2011 (Bariche and Heemstra 2012), 2015 (Golani et al., 2015) and 2015 (Rothman et al., 2016) respectively. Additionally, another two Atlantic species, *Cephalopholis taeniops*, first noted in 2002 (Ben Abdallah et al., 2007), and *Mycteroperca fusca*, reported in 2010 (Heemstra et al., 2010) were also found in the Mediterranean Sea. The increase in new records of non-native Epinephelinae species in this region follows the trend noted for a number of other tropical fish groups (Golani 2010; Golani 2013; Vella et al., 2015a, b & 2016a).

The genus *Cephalopholis* Bloch & Schneider, 1801 is composed of 24 species, 19 of which occur in the Red Sea and the Indo-Pacific region, one is from the eastern Pacific, two from the western Atlantic and two from the eastern Atlantic (Heemstra and Randall 1993; Froese and

* Correspondence: noel.vella@um.edu.mt
Conservation Biology Research Group, Department of Biology, University of Malta, Msida MSD2080, Malta

Pauly 2016). These species are primarily tropical and sub-tropical species, none of which are native to the Mediterranean Sea. The two species analysed in this study are *C. nigri* and *C. taeniops*, both of which are of tropical eastern Atlantic Ocean origin, with their native range extending from Angola to the Canary Islands and Western Sahara respectively (Heemstra and Randall 1993; Froese and Pauly 2016). While specimens of *C. taeniops* have been morphologically analysed in Libya (Ben Abdallah et al., 2007) and Israel (Salameh et al., 2009), prior to this study there have been no records of *C. nigri* in the Mediterranean Sea.

Results and discussion

Morphometrics

The voucher specimen of *C. nigri* sampled weighed 36.43 grams and had a total length of 140.0 mm (Fig. 1) while the voucher specimen of *C. taeniops* weighed 402.42 grams and had a total length of 288.0 mm (Fig. 2). The appearance, morphology and meristics are presented in Table 1 and Figs. 1 and 2. These match the descriptions of *C. nigri* and *C. taeniops* given by Heemstra and Randall (1993) and Froese and Pauly (2016). The *C. nigri* specimen had a meristic formula with a dorsal fin count of IX + 14; pectoral fin count of 14; pelvic fin count of I + 5; and anal fin count of III + 8. The pectoral fin length was 59% the head length, the pelvic fins reached the anus and were 53% the head length, while the caudal fin was rounded. The specimen had 45 scales on the lateral line and 22 gill rakers. The colour of the fish was brown, with reddish orange reticulated spots. Four darker brown bars were noted over the body extending over the dorsal fin, and another two bars on the caudal peduncle. The margin of the distinctly indented membrane on the dorsal spines was orange. Unlike the rest of the Epinephelinae species, the continuous dorsal fin of *Cephalopholis* species has 9 hard dorsal spines, a feature that is important especially in identifying *C. nigri*, given that there are other groupers such as *E. coioides* that have similar orange spots and banding patterns (Froese and Pauly 2016). The *C. taeniops* specimen collected had a meristic formula with a dorsal fin count of IX + 15; pectoral fin count of 18; pelvic fin count of I + 5; and anal fin count of III + 9. The pectoral fins were longer than the pelvic fin, and their length was

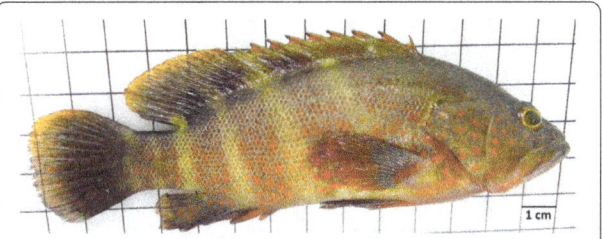

Fig. 1 Photograph of the first record of *Cephalopholis nigri* from the Mediterranean Sea (collected in July, 2016)

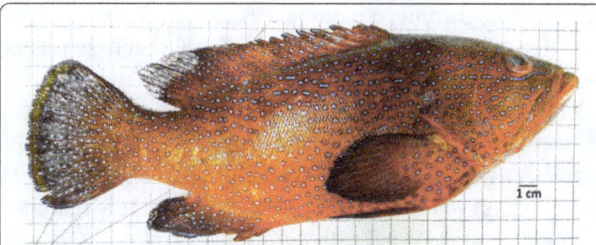

Fig. 2 Photograph of *Cephalopholis taeniops* voucher specimen analysed in this study (collected in April 2016)

64% the head length. The specimen had 72 scales on the lateral line and 23 gill rakers. The colour of the fish was reddish orange and its body, including the head, was covered in small blue spots. The fins had a darker blue colouration.

Genetic analyses

A total of 2026 bp were sequenced from the mtDNA of both specimens. The sequence lengths obtained were 412 bp, 604 bp, 585 bp and 425 bp for cytochrome b

Table 1 Measurements and meristic counts of the first record of *Cephalopholis nigri* in the Mediterranean Sea and of a *Cephalopholis taeniops* both specimens caught from Maltese waters

	Cephalopholis nigri		Cephalopholis taeniops	
Parameter	Measurements (mm)	% SL	Measurements (mm)	% SL
Total length	140.0		288.0	
Standard length	116.0		247.0	
Maximum body depth	36.9	31.8	81.3	32.9
Length of dorsal fin base	14.6	12.6	113.8	46.1
Pectoral fin base	8.9	7.7	17.8	7.2
Anal fin base	19.1	16.5	43.5	17.6
Pre-pelvic length	41.0	35.3	86.9	35.2
Pre-anal length	77.0	66.4	143.4	58.1
Head length	43.1	37.2	88.1	35.7
Pre-orbital length	10.6	9.1	29.3	11.9
Eye diameter	9.1	7.8	12.7	5.1
	Counts		Counts	
Dorsal fin spines	9		9	
Dorsal fin soft rays	14		15	
Pectoral fin soft rays	14		18	
Pelvic fin spines	1		1	
Pelvic fin soft rays	5		5	
Anal fin spines	3		3	
Anal fin soft rays	8		9	
Scales on lateral line	45		72	
Gill rakers	22		23	

(Cytb), cytochrome c oxidase I (COI), 16S rRNA (16S) and 12S rRNA (12S) genes respectively. Each sequence was run via BLASTn to identify sequence matches.

The two studied specimens were genetically confirmed to species level at the 12S and the 16S genes at >99.3% matches with Craig and Hastings (2007). The 12S matched AY949451 and AY949387, while 16S matched AY947604 and AY947589 for *C. nigri* and *C. taeniops* respectively. Cytochrome B gene of *C. taeniops* also confirmed the species with a 100% match to EF455990-1 specimens from Mauritania (Gonzalez-Sevilla et al., unpublished). It was not possible to compare CytB in *C. nigri* and COI for both species with any other sequences as there is no publically available data for them. This study presents the first sequences for these alien species collected from the Mediterranean Sea.

Other specimens
In December, 2015, an additional three specimens of *C. taeniops* have been caught in close proximity to each other (Fig. 4). The three recorded individuals included one specimen with a colouration similar to the one described in Fig. 2, and another two specimens of the less common darker variety (Seret, 1981), one of which can be seen in Fig. 4.

Discussion
The number of alien species in the Mediterranean Sea are on the increase (Golani 2010; 2013), including species of both Atlantic and Indo-Pacific origin. In fact the list of non-native species extends across several taxa, including groupers from the family Epinephelinae. *Cephalopholis nigri* is the latest record of a non-native grouper in the region. The species is not known to be a natural migrant as there are no records of this species elsewhere in the Mediterranean Sea. Moreover, given that Epinephelinae species are highly prized catches, it is very unlikely that records of this species would have gone unnoticed by divers and fishermen. This first Mediterranean record of *C. nigri* is from a busy Maltese harbour which caters for dockyards, oil platform servicing, transhipment activities and berthing of large marine vessels including cruise-liners and super-yachts. Like elsewhere in the region (Galil, 2006; Katsanevakis et al., 2014), such maritime activity can be considered as the main vector in the introduction of alien species. This new record of alien species, follows others from areas that are characterized by intensive marine activity in Malta, such as the first Mediterranean records of *Stegastes variabilis*, *Lutjanus fulviflamma* and *Abudefduf hoefleri* (Vella et al., 2015a, b & 2016a). Nonetheless, one cannot exclude the possibility that *C. nigri* was an aquarium release given that members of the genus *Cephalopholis* are exported as ornamental fish (Monteiro-Neto et al., 2003) and in recent years this industry has led to an increase in alien species within the Mediterranean Sea (Guidetti et al., 2016; Zenetos et al., 2016).

Prior to this study there have only been isolated sighting reports for *C. taeniops* in Maltese waters without the

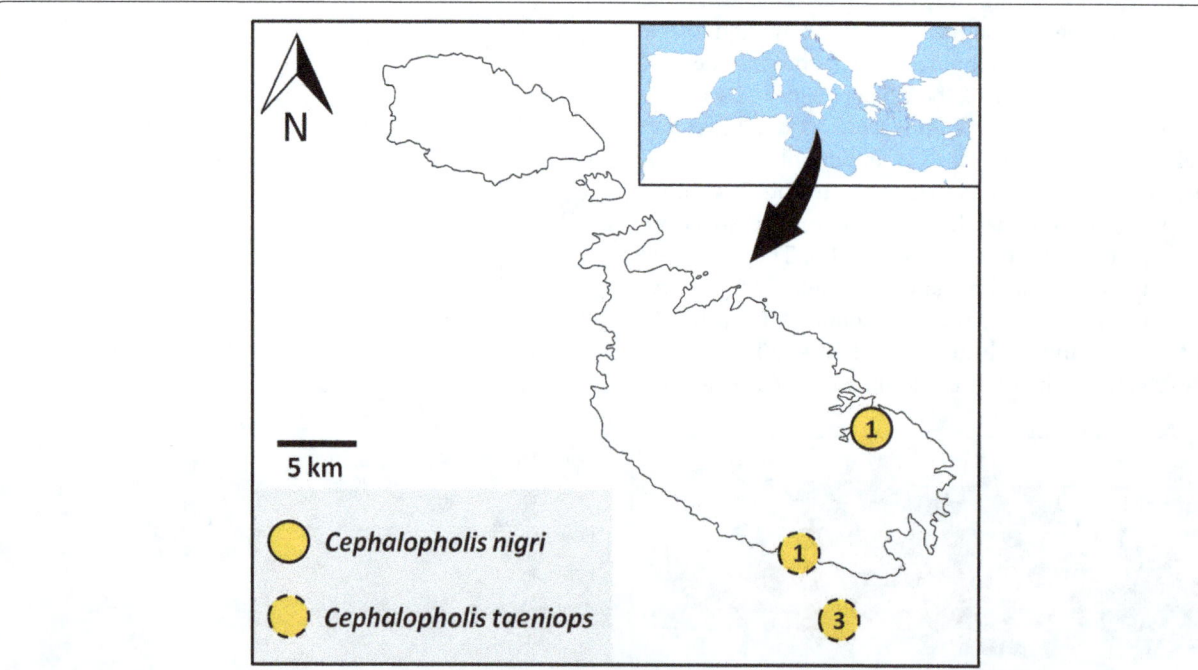

Fig. 3 A map of Malta showing the locations where *Cephalopholis nigri* and *Cephalopholis taeniops* were recorded during this study. The numbers in the circles indicate the number of specimens collected per location

analyses of any voucher specimen for this alien grouper species. This study has therefore presented the first scientific morphometric and genetic analyses of this species from Maltese waters, while reporting an increasing number of sightings and photographic records. The four new records of *C. taeniops* indicate the establishment of a population in Maltese waters. Ongoing scientific monitoring would be required to further study the expanding range of this species, although the permanence of its population is threatened by the local fishing industry that is looking at *C. taeniops* as another grouper that may be exploited for commercial means (pers. comm. with fishermen). This latter activity can prove to be beneficial in controlling the spread of this non-native species.

Fig. 4 Photograph showing the different colourations of *Cephalopholis taeniops* (collected in December, 2015)

Conclusion

The occurrence and proliferation of these carnivorous species need monitoring as one cannot exclude the possibility that expanding populations of these species might lead to interspecific competition for resources with other already vulnerable native Epinephelinae species in the Mediterranean and in Maltese waters. However, since groupers are species under pressure from over-exploitation these alien species are also being exploited for local consumption, keeping their numbers low.

Methods

As part of ongoing research with fishermen to study species caught in Maltese waters, voucher specimens of two alien grouper species were collected and analysed morphologically and genetically. On 10th July 2016 a voucher specimen of *C. nigri* (Fig. 1) was caught at 8 m depth from Senglea, Malta [GPS: 35°53′14.26″N, 14°30′55.90″E] (Fig. 3). On 9th April 2016 a voucher specimen of *C. taeniops* (Fig. 2) was caught from Żurrieq [GPS: 35°49′8.88″N, 14°27′8.66″E] (Fig. 3). Various sea-users provided sightings and photographic records of *C. taeniops* around Malta (Figs. 2, 3 and 4) since 2015. These two voucher specimens have been deposited in the ichthyological collection of the Conservation Biology Research Group laboratory at the University of Malta with reference code number CBRG/F.160710/CN001 and CBRG/F.160409/CT001 respectively.

The diagnostic features used in the morphological identification of both specimens followed Heemstra and Randall (1993) and Froese and Pauly (2016). All length measurements were taken to the nearest 0.1 mm using electronic calipers and mass was recorded to the nearest 0.01 g.

DNA was extracted using GF-1 Tissue DNA Extraction Kit (Vivantis Technologies). PCR amplifications were carried out for Cytb, COI and 16S using the primers sets as described in Vella et al., (2016a, b), and for the 12S using H1478 and L1091 primers (Kocher et al., 1989). PCR amplifications were carried out following the amplification protocols described in Vella et al., (2016a). PCR products were purified and sequenced via ABI3730XL sequencer using both the forward and reverse primers. The sequences, at both nucleotide and amino acid level, were analyzed using Geneious v6 (http://www.geneious.com, Kearse *et al.*, 2012). The sequences obtained were deposited in GenBank, with accession numbers of *C. taeniops* KX758563-6 and *C. nigri* KX758567-70 for Cytb, COI, 16S and 12S respectively. These sequences were compared to other sequences available in genomic databases using BLASTn.

Abbreviations

12S: 12S rRNA gene; 16S: 16S rRNA gene; COI: Cytochrome c oxidase I gene; Cytb: Cytochrome b gene

Acknowledgements

Thanks are due to sport fishermen, in particular to Mr. R. Farrugia, Hooked on Fishing sports club and Mr. J. Tanti who have assisted in this research.

Funding

The genetic research disclosed in this publication is funded by the REACH HIGH Scholars Programme-Post Doctoral Grants. The grant may be part-financed by the European Union, Operational Programme II - Cohesion Policy 2014–2020 "Investing in human capital to create more opportunities and promote the well being of society" - European Social Fund.

Authors' contributions

NV and AV have contributed to all aspects of the research work presented here including the conception and design of the molecular genetics research, analyses and interpretation of both genetic and morphological data and were involved in finalizing the manuscript. SAD contributed to the morphological research work of the specimens collected. All three authors were involved in the drafting of the manuscript and gave approval for publication.

Authors' information

All three authors are researchers of the Conservation Biology Research Group, Department of Biology, University of Malta.

Competing interests

None of the authors of this paper have financial or non-financial competing interests associated with this research work.

References

Bariche M, Heemstra PC. First record of the blacktip grouper *Epinephelus fasciatus* (Teleostei: Serranidae) in the Mediterranean Sea. Mar Biodiversity Rec. 2012;5:e1.

Ben Abdallah A, Ben Soussi J, Méjri H, Canapé C, Golani D. First record of *Cephalopholis taeniops* (Valenciennes) in the Mediterranean Sea. J Fish Biol. 2007;71:610–4.

Ben-Tuvia A, Lourie A. A Red Sea grouper *Epinephelus tauvina* caught on the Mediterranean coast of Israel. Isr J Zool. 1969;18:245–7.

Craig MT, Hastings PA. A molecular phylogeny of the groupers of the subfamily Epinephelinae (Serranidae) with a revised classification of the Epinephelini. Ichthyol Res. 2007;54:1–17.

Froese R, Pauly D. FishBase. 2016. URL http://www.fishbase.org. Accessed 10 Aug 2016.

Galil BS. Shipwrecked - Shipping impacts on the biota of the Mediterranean Sea. Chapter 3. In: Davenport J, Davenport JL, editors. The Ecology of Transportation: Managing Mobility for the Environment. The Netherlands: Springer publishers; 2006. p. 392.

Golani D. Colonization of the Mediterranean by Red Sea fishes via the Suez Canal - Lessepsian migration. In: Golani D, Appelbaum-Golani B, editors. Fish Invasions of the Mediterranean Sea: Change and Renewal. Sofia: Pensoft; 2010. p. 145–88.

Golani D, Orsi-Relini L, Massuti E, Quignard JP, Dulčić J, Azzurro E. CIESM - Atlas of Exotic Fishes - List [WWW Document]. 2013. URL http://www.ciesm.org/atlas/appendix1.html. Accessed 10 Aug 2016.

Golani D, Askarov G, Dashevsky Y. First record of the Red Sea spotted grouper, *Epinephelus geoffroyi* (Klunzinger, 1870) (Serranidae) in the Mediterranean. BioInvasions Records. 2015;4(2):143–5.

Guidetti P, Magnani L, Navone A. First record of the acanthurid fish *Zebrasoma xanthurum* (Blyth, 1852) in the Mediterranean Sea, with some considerations on the risk associated with aquarium trade. Mediterr Mar Sci. 2016;17(1):147–51.

Heemstra PC, Golani D. Clarification of the Indo-Pacific groupers (Pisces: Serranidae) in the Mediterranean Sea. Isr J Zool. 1993;39:381–90.

Heemstra PC, Randall JE, Groupers of the world (Family Serranidae, Subfamily Epinephelinae). An annotated and illustrated catalogue of the grouper, rockcod, hind, coral grouper and lyretail species known to date. Rome: FAO Fisheries Synopsis No. 125 Vol. 16. Food and Agriculture Organization of the United Nations; 1993. p. 382.

Heemstra P, Aronov A, Goren M. First record of the Atlantic island grouper *Mycteroperca fusca* in the Mediterranean Sea. Mar Biodiversity Rec. 2010;3, e92.

Katsanevakis S, Coll M, Piroddi C, Steenbeek J, Ben Rais Lasram F, Zenetos A, Cardoso AC. Invading the Mediterranean Sea: biodiversity patterns shaped by human activities. Frontiers in Marine Science. 2014. doi:10.3389/fmars.2014.00032.

Kearse M, Moir R, Wilson A, Stones-Havas S, Cheung M, Sturrock S, Buxton S, Cooper A, Markowitz S, Duran C, Thierer T, Ashton B, Mentjies P, Drummond A. Geneious Basic: an integrated and extendable desktop software platform for the organization and analysis of sequence data. Bioinformatics. 2012;28:1647–9.

Kocher TD, Thomas WK, Meyer A, Edwards SV, Pääbo S, Villablanca FX, Wilson AC. Dynamics of mitochondrial DNA evolution in animals: amplification and sequencing with conserved primers. Proc Natl Acad Sci U S A. 1989;86:6196–200.

Lelong P. Capture d'un macabit, *Epinephelus merra* Bloch, 1793 (Poisson, Serranidae), en Méditerranée nord- occidentale. Mar Life. 2005;15:63–6.

Monteiro-Neto C, Cunha FED, Nottingham MC, Araujo ME, Rosa IL, Barros GML. Analysis of the marine ornamental fish trade at Ceara State, northeast Brazil. Biodivers Conserv. 2003;12(6):1287–95.

Rothman SB, Stern N, Goren M. First record of the Indo-Pacific areolate grouper *Epinephelus areolatus* (Forsskål, 1775) (Perciformes: Epinephelidae) in the Mediterranean Sea. Zootaxa. 2016;4067(4):479–83.

Salameh P, Sonin O, Golani D. A first record of the African hind (*Cephalopholis taeniops*) (Pisces: Serranidae) in the Levant. Annales, Series Historia Naturalis. 2009;19(2):151–4.

Séret B. Poissons de Mer de l'ouest Africain Tropical. Paris: ORSTOM; 1981. p. 416.

Vella A, Agius Darmanin S, Vella N. Morphological and genetic barcoding study confirming the first *Stegastes variabilis* (Castelnau, 1855) report in the Mediterranean Sea. Mediterr Mar Sci. 2015a;16(3):609–12.

Vella A, Vella N, Agius DS. First record of *Lutjanus fulviflamma* (Osteichthyes: Lutjanidae) in the Mediterranean Sea. J Black Sea/Mediterr Environ. 2015b;21(3):307–15.

Vella A, Vella N, Agius DS. The first record of the African sergeant, *Abudefduf hoefleri* (Perciformes: Pomacentridae), in the Mediterranean Sea. Mar Biodiversity Rec. 2016a;9(1):1–5.

Vella N, Vella A, Agius DS. The first record of the lowfin chub *Kyphosus vaigiensis* (Quoy & Gaimard, 1825) from Malta. J Black Sea/Mediterr Environ. 2016b;22(2):175–81.

Zenetos A, Apostolopoulos G, Crocetta F. Aquaria kept marine fish species possibly released in the Mediterranean Sea: First confirmation of intentional release in the wild. Acta Ichthyol Piscat. 2016;46(3):255–62.

23

Ampelisca lusitanica (Crustacea: Amphipoda): new species for the Atlantic coast of Morocco

Z. Belattmania[1], A. Chaouti[1], M. Machado[2], A. Engelen[2], E. A. Serrão[2], A. Reani[1] and B. Sabour[1*] ⓘ

Abstract

Background: This study reports for the first time the presence of the Lusitanian ampeliscid amphipod *Ampelisca lusitanica* Bellan-Santini & Marques, 1986 in the northwestern Atlantic coast of Morocco.

Methods: Specimens were collected in January 2015 from intertidal rock pools along the El Jadida shoreline associated with the brown algae *Bifurcaria bifurcata* and *Sargassum muticum*.

Results: Systematic description of the species is presented, as well as a discussion of its ecological and geographical distribution.

Conclusion: This new finding extends the geographical distribution from the Lusitanian (Europe) to the Mauritanian (Africa) region and increases knowledge of the ecology and the global distribution of *A. lusitanica* found, previously, only on Portuguese and Spanish coasts.

Keywords: *Ampelisca lusitanica*, Amphipoda, epifauna, Atlantic coast, Morocco

Introduction

Crustacean amphipods are distributed in all ecosystems worldwide including terrestrial, freshwater and marine aquatic environments (Mosbahi et al. 2015). They often play a critical role in aquatic food webs, acting as conduits of nutrients and energy to higher trophic levels (Vainola et al. 2008). The Ampeliscidae is one of the most diversified amphipod families in the ocean (Barnard and Karaman 1991). This family is commonly found in a variety of habitats from shallow to deep waters although some species are restricted to limited bathymetric ranges and sediment types (Dauvin et al. 2012). Ampeliscid communities are generally composed of several species belonging to three dominant genera: *Ampelisca*, *Byblis*, and *Haploops*. The genus *Ampelisca* Krøyer, 1842 is a diverse benthic genus of marine amphipods, containing approximately 236 species (WORMS 2017) mainly reported from shallow waters with a worldwide distribution (Dauvin 1988; Dauvin and Bellan-Santini 1988; Filhoa et al. 2009; Dauvin et al. 2012). In the northeastern Atlantic region between the northern coast of Norway and the Gulf of Guinea, *Ampelisca* is represented by more than 52 valid species (Dauvin and Bellan-Santini 1988). *Ampelisca* species are found from the intertidal zone to abyssal depths, but most of them live on the continental shelf (Bellan-Santini and Dauvin 1988a; Dauvin 1988; Dauvin and Bellan-Santini 1988). Among *Ampelisca* species, the Lusitanian marine amphipod *Ampelisca lusitanica* Bellan-Santini and Marques, 1986 was reported first from the Atlantic coast of Portugal (Bellan-Santini and Marques 1986) and later from the northeastern Atlantic coast of Spain (Martínez et al. 2007).

The present paper reports the first discovery of *A. lusitanica* in North African Atlantic marine waters (littoral of El Jadida, NW Morocco). Some data on the external morphology, ecology and distribution of the species are provided.

Material and methods

Samples were collected as a part of a study on macroalgae-associated epifauna in the northwestern Atlantic coast of Morocco. Seaweed species were carefully removed from the intertidal substratum, and

* Correspondence: sabour.b@ucd.ac.ma
[1]Phycology Research Unit, Department of Biology, Faculty of Sciences, Chouaib Doukkali University, P.O. Box 20, 24000, El Jadida, Morocco
Full list of author information is available at the end of the article

placed in plastic bags with 5% formalin. In the laboratory, macroalgae were washed in fresh water to collect the majority of the associated fauna. Specimens were sieved (0.5 mm mesh size), sorted and preserved (70% ethanol) for later identification and counting. *Ampelisca lusitanica* specimens were conserved separately, examined and identified according to descriptions provided by Bellan-Santini and Marques (1986), and Dauvin and Bellan-Santini (1988).

Results

Systematics

Order AMPHIPODA Latreille, 1816
 Suborder GAMMARIDEA Latreille, 1802
 Family AMPELISCIDAE Krøyer, 1842
 Genus *Ampelisca* Krøyer, 1842
 Ampelisca lusitanica Bellan-Santini and Marques, 1986

Material examined

A total of 6 individuals of the amphipod *A. lusitanica* with size ranging between 1 and 4 mm, were found associated to the brown algae *Bifurcaria bifurcata* and *Sargassum muticum* collected from the intertidal areas of El Jadida shorelines (33° 10' 50.2" N - 8° 36' 56.5" W) in January 2015 (Fig. 1). Among the collected individuals, two specimens were juveniles and four were adults.

Description

A. lusitanica can be distinguished from other *Ampelisca* species by the following characters (Fig. 2): The head is obliquely truncated, shorter than the three first segments of the pereon. The lower pair of eyes set at lower front corners of the head. The first antenna is shorter than A2 reaching the middle of its flagellum. The latter is about half the length of the body (Fig. 2a, b). The third epimeral plate is rounded-quadrate at the postero-distal corner. Pereopod 7 has the basis distally rounded; the merus (without a large posterior lobe) is shorter than the carpus with an anterior peg-like process projecting beyond the proximal end of the carpus but not reaching its distal end; the dactylus is rounded and slightly curved in its terminal part (Fig. 2c). The first and second uropods have spinous

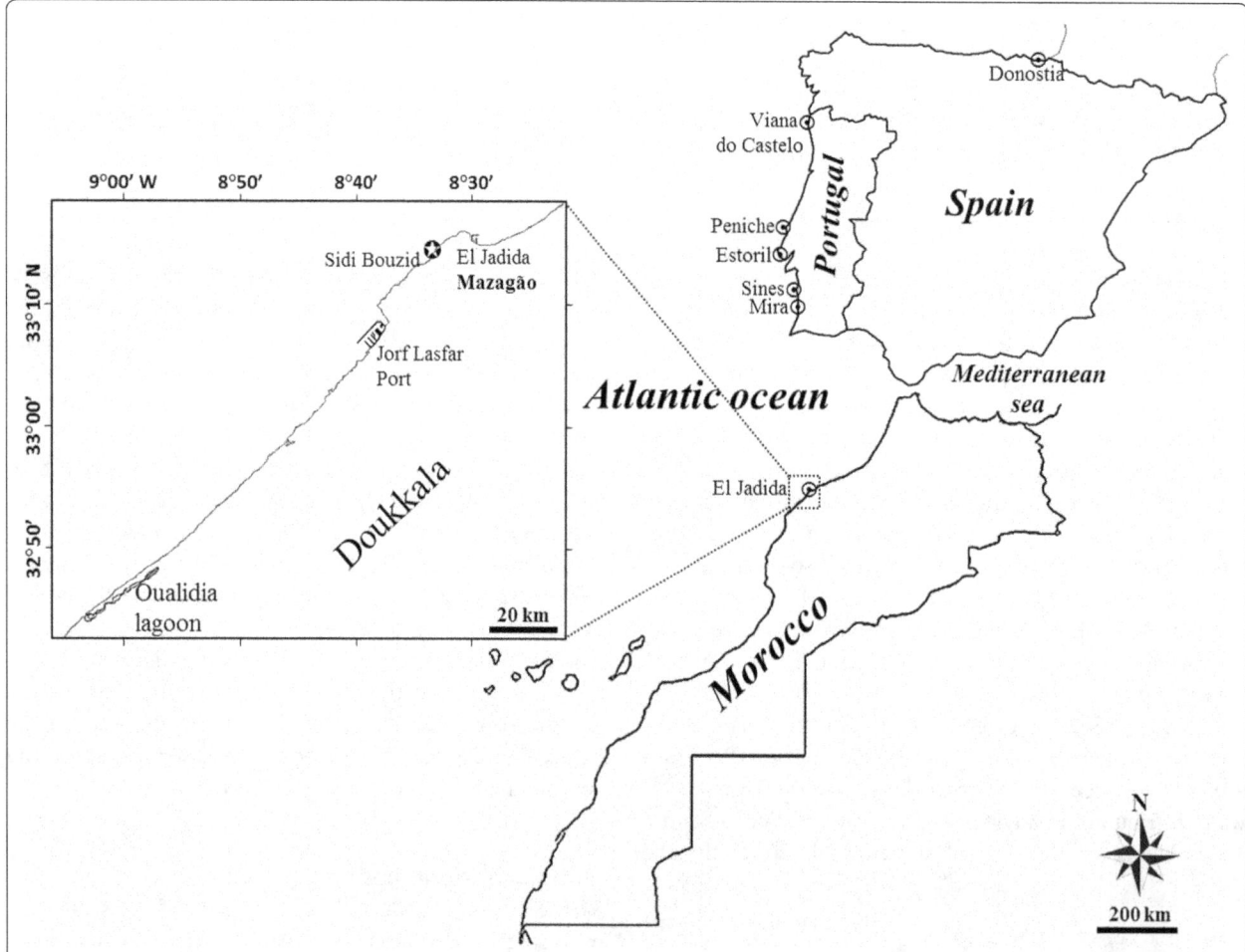

Fig. 1 Sampling site (⊘) and geographical distribution (⊙) of *Ampelisca lusitanica*, indicating new record reported from the Moroccan shores and previous literature records along the Atlantic coast of the Iberian Peninsula

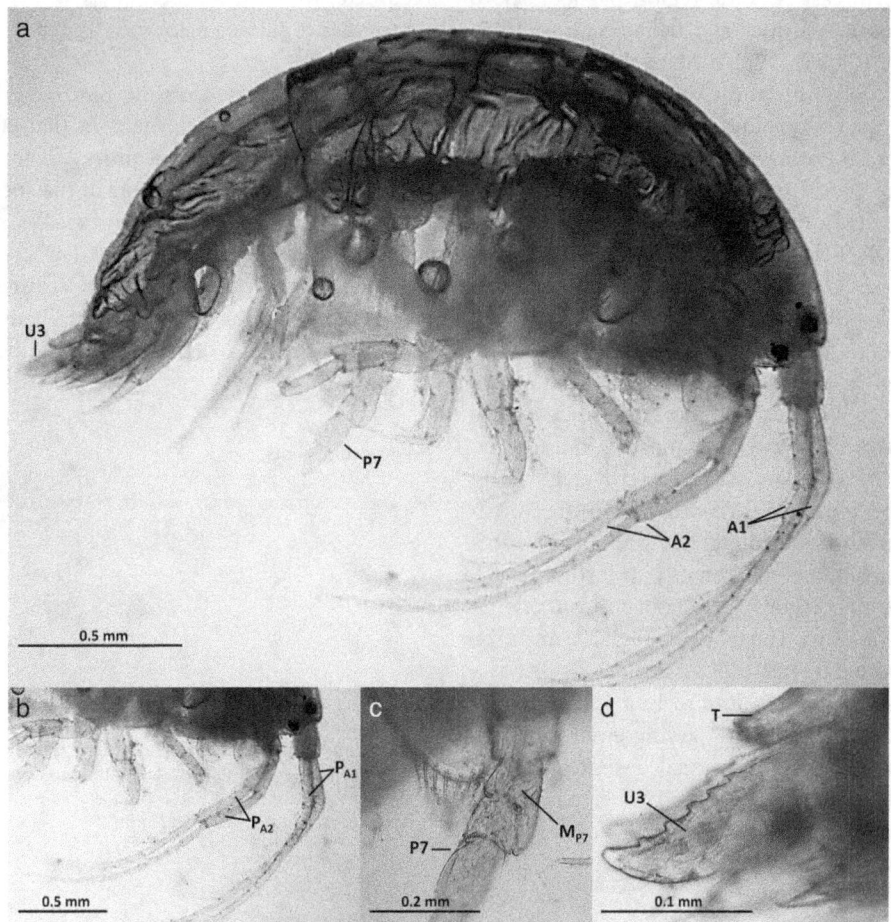

Fig. 2 Specimens of *Ampelisca lusitanica* from northwestern Atlantic coast of Morocco collected during winter 2015. **a** Body shape with indication of the main morphological characters; **b** Antenna 1 (A1) shorter than antenna 2 (A2) showing antennal peduncle 2 (P$_{A2}$) < A1; **c** Pereopod 7 (P7) showing the merus (M$_{P7}$) with an anterior peg-like process projected beyond the proximal end of the carpus; and **d** Uropod 3 (U3) with inner ramus denticulate

inner and outer rami. The inner ramus of the uropod 3 is longer than the outer ramus, with its inner margin denticulate (10–12 denticulations). *A. lusitanica* can be confused with *A. unidentata* Schellenberg, 1936 (Bellan-Santini and Marques 1986) but the latter differs from the former by antenna formula (Al sub-equal to A2) and the epimeral plate 3 which has postero-distal corner quadrate.

Discussion

The amphipod *A. lusitanica* was considered as an uncommon to rare species for Portuguese coasts and the Lusitanian region (Bellan-Santini and Dauvin 1988a,b; Marques and Bellan-Santini 1990), and southeast of the Bay of Biscay, Spain (Martínez et al. 2007). This species is known as typical of coastal warm temperate waters and occurs in various habitats. It was found on deep rocky bottoms (between 8 and 37 m) (Bellan-Santini and Marques 1986; Bellan-Santini and Dauvin 1988a,b) covered with sand or silt and on intertidal rocky substrates with algal communities such as *Corallina elongata*, *Jania*

rubens and *Cystoseira* sp. (Bellan-Santini and Marques 1986). The species was also reported from muddy bottoms at 7.5 km upstream from the mouth river of the Mira estuary where the salinity is lower than in open coastal waters (Marques and Bellan-Santini 1987) and from sandy sediments of continental shelf of Guipúzcoa at 41 m depth from Basque Country, Spain (Martínez et al. 2007). However, in the present investigation, *A. lusitanica* was found associated to the brown seaweeds *Bifurcaria bifurcata* and *Sargassum muticum* on protected rocky intertidal pools of 1–4 m depth where surface seawater temperature ranged between 16 and 20 °C and salinity varied from 33 to 37‰. According to Bellan-Santini and Marques (1986), the ecology of *A. lusitanica* seems to be similar to that of *A. rubra* Chevreux, 1925 (accepted name *Ampelisca heterodactyla* Schellenberg, 1925), *A. serraticaudata* Chevreux, 1888 and especially *A. unidentata* Schellenberg, 1936.

Based on the literature data, the geographical distribution of this species (Fig. 1) would seem to be currently limited to the North-east Atlantic region particularly to the Portuguese coast (Marques and Bellan-Santini 1987; Bellan-Santini and Dauvin 1988a; Marques and Bellan-Santini 1990; Bellan-Santini and Costello 2001) and the Basque Country, Spain (Martínez et al. 2007). Like most *Ampelisca* species, *A. lusitanica* is restricted to small geographical areas. It had previously only been reported thrice inside the Lusitanian region; twice from coastal waters of Portugal (Bellan-Santini and Marques 1986; Marques and Bellan-Santini 1987) and once from southeast of the Bay of Biscay, Spain (Martínez et al. 2007). The species was not found in the Azores archipelago (Portugal), located in the middle north Atlantic region (Rosa Lopes et al. 1993) nor in the eastern side of the Iberian Peninsula, western Mediterranean (Conradi and López-González 1999; de-la-Ossa-Carretero et al. 2010, 2016). Also the species was not reported from the Portuguese continental shelf during a comprehensive sampling of the soft-bottom macrofauna (Martins et al. 2013). This is the first record of the species outside its native geographical area (Lusitanian region) and the third one at the regional scale (NE Atlantic).

Due to the presence of *A. lusitanica* in a wide range of habitats and its ability to support varying environmental conditions in intertidal and in subtidal deeper areas, we assume that the species could occur in other coastal localities away from its previously recorded biogeographic region (Lusitanian). Wherever encountered it is always a rare species with low abundances.

The lack of previous records in Morocco and North Africa might be due to the small size of individuals, the low abundance of local populations, or its misidentification (including a possible confusion with its congeneric species *Ampelisca unidentata*). Also, the limited distribution of the species might be due to an insufficient investigation effort.

Significant numbers of individuals of *A. lusitanica* were found with *B. bifurcata* (5 individuals), whereas only one individual was found on *S. muticum*. It can be suggested that the complexity of the algal habitat was not an important factor affecting the abundance of *A. lusitanica*. Previous studies showed that the structural complexity of algae was not a consistent predictor of the number of individuals and species of amphipods (Russo 1990; Schreider et al. 2003; Engelen et al. 2013). Conversely, Taylor and Cole (1994) and Wernberg et al. (2004) reported a higher abundance of small crustaceans in more structurally complex algal habitats. The number of individuals found associated with intertidal algal samples in the present investigation corroborates results obtained by Bellan-Santini and Marques (1986) on the Portuguese coast, where the lower number of specimens ranging from 2 to 5 individuals, was reported from intertidal habitats and the higher number (up to 28 individuals) was recorded from the deeper areas (37 m depth) where a more stable environment accounted for higher species abundances.

This study reported on the first record of *A. lusitanica* in Moroccan waters (i) suggests that *A. lusitanica* may also be present in other localities of Moroccan shores, (ii) extends the geographical range of the species *A. lusitanica* into the northeastern Atlantic (species shifting from the Lusitanian region), (iii) adds a new contribution to the macrofauna diversity thriving in seaweed beds and (iv) yields a contribution to the growing body of knowledge of the Moroccan and North African Atlantic biodiversity.

Acknowledgements
Z. Belattmania acknowledges her doctoral fellowship from the Ministry of higher education and scientific research of Morocco.

Funding
Not applicable

Authors' contributions
BS, ZB and AE conceived the study and carried out the field work. MM, ES and AR provided logistic support during the study and did the taxonomic analysis of the specimens. BS, ZB and AC drafted the manuscript. All authors read and approved the final manuscript.

Competing interests
The authors declare that they have no competing interests.

Author details
[1]Phycology Research Unit, Department of Biology, Faculty of Sciences, Chouaib Doukkali University, P.O. Box 20, 24000, El Jadida, Morocco. [2]CCMAR – Centre of Marine Sciences, University of Algarve, Faro, Portugal.

References
Barnard JL, Karaman G. The families and genera of marine gammaridean Amphipoda (except marine gammaroids). Rec Aust Mus. 1991;13:1–866.
Bellan-Santini D, Costello MJ. Amphipoda. In: Costello MJ, Emblow CS, White R, editors. European Register of Marine Species. A check-list of the marine species in Europe and a bibliography of guides to their identification, Collection Patrimoines Naturels, vol. 50. Paris: Muséum National d'Histoire Naturelle; 2001. p. 295–308.
Bellan-Santini D, Dauvin JC. Éléments de synthèse sur les *Ampelisca* du Nord-Est Atlantique. Crustaceana. 1988a;13:20–60. Suppl.
Bellan-Santini D, Dauvin JC. Actualisation des données sur l'écologie, la biogéographie et la phylogénie des Ampeliscidae (Crustacés - Amphipodes) atlantiques après la révision des collections d'E. In: IFREMER, editor. Chevreux, Aspects récents de la biologie des crustacés, Actes de Colloques, vol. 8. 1988b. p. 207–16.
Bellan-Santini D, Marques JC. Une nouvelle espèce d'*Ampelisca* (Crustacea-Amphipoda) des côtes du Portugal (Atlantique Nord-Est): *Ampelisca lusitanica* n.sp. Cah Biol Mar. 1986;27(2):153–62.
Conradi M, López-González PJ. The benthic Gammaridea (Crustacea, Amphipoda) fauna of Algeciras Bay (Strait of Gibraltar): distributional ecology and some biogeographical considerations. Helgol Mar Res. 1999;53:2–8.

Dauvin JC. Biologie, dynamique, et production de populations de crustacés amphipodes de la Manche occidentale. 1. *Ampelisca tenuicornis* Liljeborg. J Exp Mar Biol Ecol. 1988;118:55–84.

Dauvin JC, Bellan-Santini D. Illustrated key to *Ampelisca* species from the north-Eastern Atlantic. J Mar Biol. 1988;68:659–76.

Dauvin JC, Alizier S, Weppe A, Gujmundsson G. Diversity and zoogeography of Icelandic deep-sea Ampeliscidae (Crustacea: Amphipoda). Deep-Sea Res I. 2012;68:2–23.

de-la-Ossa-Carretero JA, Dauvin JC, Del-Pilar-Ruso Y, Giménez-Casalduero F, Sànchez-Lizaso JL. Inventory of benthic amphipods from fine sand community of the Iberian Peninsula east coast (Spain), western Mediterranean, with new records. Mar Biodivers Rec. 2010;3:1–10.

de-la-Ossa-Carretero JA, Del-Pilar-Ruso Y, Giménez-Casalduero F, Sànchez-Lizaso JL. Amphipoda assemblages in a disturbed area (Alicante, Spain, Western Mediterranean). Mar Ecol. 2016;37:503–17.

Engelen AH, Primo AL, Cruz T, Santos R. Faunal differences between the invasive brown macroalga *Sargassum muticum* and competing native macroalgae. Biol Invasions. 2013;15:171–83.

Filhoa JFS, Souzab AMT, Valério-Berardo MT. Description of four new species of the genus *Ampelisca* (Amphipoda, Ampeliscidae) from the northeastern and southeastern coasts of Brazil and designation of a neotype for *Ampelisca soleata* Oliveira, 1954. J Nat Hist. 2009;43:2391–423.

Marques JC, Bellan-Santini D. Crustacés Amphipodes des côtes du Portugal: faune de l'estuaire du Mira (Alentejo, côte sud-ouest). Cah Biol Mar. 1987; 28(3):465–80.

Marques JC, Bellan-Santini D. Benthic amphipod fauna (Crustacea) of the Portuguese coast: Biogeographical considerations. Mar Nat. 1990;3:43–51.

Martínez J, Adarraga I, Ruiz JM. Tipificación de poblaciones bentónicas de los fondos blandos de la plataforma continental de Guipúzcoa (Sureste del golfo de Vizcaya). Bol Inst Esp Oceanogr. 2007;23(1–4):85–110.

Martins R, Quintino V, Rodrigues AM. Diversity and spatial distribution patterns of the soft-bottom macrofauna communities on the Portuguese continental shelf. J Sea Res. 2013;83:173–86.

Mosbahi N, Dauvin JC, Neifar L. First record of the amphipods *Leucothoe incisa* (Robertson, 1892) and *Lysianassa pilicornis* (Heller, 1866) from Tunisian waters (central Mediterranean Sea). Vie Milieu. 2015;65(3):175–9.

Rosa Lopes MF, Marques JC, Bellan-Santini D. The benthic amphipod fauna of the Azores (Portugal): an up-to-date annotated list of species, and some biogeographic considerations. Proceedings of the First European Crustacean Conference, 1992. Crustaceana. 1993;65(2):204–17.

Russo AR. The role of seaweed complexity in structuring Hawaiian epiphytal amphipod communities. Hydrobiologia. 1990;194:1–12.

Schreider MJ, Glasby TM, Underwood A. Effects of height on the shore and complexity of habitat on abundances of amphipods on rocky shores in New South Wales, Australia. J Exp Mar Biol Ecol. 2003;293:7–71.

Taylor RB, Cole RG. Mobile epifauna on subtidal brown seaweeds in northeastern New Zealand. Mar Ecol Prog Ser. 1994;115:271–82.

Vainola R, Witt JDS, Grabowski M, Bradbury JH, Jazdzewski K, Sket B. Global diversity of amphipods (Amphipoda; Crustacea) in freshwater. Hydrobiologia. 2008;595:241–55.

Wernberg T, Thomsen MS, Staehr PA, Pedersen MF. Epibiota communities of the introduced and indigenous macroalgal relatives *Sargassum muticum* and *Halidrys siliquosa* in Limfjorden (Denmark). Helgol Mar Res. 2004;58:154–61.

World Register of Marine Species (WoRMS). Available at: http://www.marinespecies.org. Accessed 24 Jan 2017.

24

Range extension of a vulnerable Sea horse *Hippocampus fuscus* (Actinopterygii: Syngnathidae) on the north-eastern Bay of Bengal coast

Debasish Mahapatro[1], R. K. Mishra[2*] and S. Panda[3]

Abstract

The study describes the range extension of the sea horse Hippocampus fuscus from the south to north east coastal waters of the India, Bay of Bengal. After 99 years since initial discovery, the Hippocampus fuscus was reported within the southern sector of the Chilika Lake. The extension range may be due to the East India Coastal Current of the Bay of Bengal and the predominance of extensive sea grass meadows within the southern sector of Lake.

Keywords: Sea horse, *Hippocampus fuscus*, Gopalpur coast, Chilika lake, Range extension

Introduction

Sea horses are predominantly found in Indo-Pacific regions, covering approximately 45°S to 45°N (Froese & Pauly, 2014; Sreepada et al., 2002). The International Union of Conservation of Nature (IUCN, 2014) reported 38 species of sea horse worldwide, out of which about 50% of the total dominate the coastal region of the Indian sub-continent. The habitats of sea horses pertain to various coastal ecosystems, such as seagrass meadows, mangrove, estuaries, lagoons, and coral reefs (Froese & Pauly, 2014; Kendrick & Hyndes, 2003; Sreepada et al., 2002; Lourie et al., 1999). Sea horses are also found in association with other animals, like gorgonians, sponges and sea quirts, exhibiting wide ranging adaptations to different environmental conditions and locations. Sea horses were observed ubiquitously as little as 10 years ago (Lourie et al., 1999; Sreepada et al., 2002; Froese & Pauly, 2014). However, in recent times the scenario has changed dramatically, due to illegal poaching and hunting for Chinese medicines, as well as Korean and Japanese recipes. As a result, the global sea horse population has undergone significant decline, causing the IUCN to give sea horses as endangered, threatened

vulnerable statuses (Baillie & Groombridge, 1996; Vincent, 1996; Sreepada et al., 2002; IUCN, 2014). In India, the Ministry of Environment and Forests banned the exportation of Syngnathids in 2001, also conserving them under Schedule I of the Indian Wildlife Protection Act (1972) (Sreepada et al., 2002). Sea horses mostly belong to the genus *Hippocampus*, within the family "Syngnathidae" (Froese & Pauly, 2014; Lourie et al., 1999). *Hippocampus* species have very peculiar body shapes, consisting of five different features:(i) head and neck resembles a horse- hence the name "Sea horse", (ii) the middle body portion looks like a fish (because of the presence of dorsal and pectoral fin rays), (iii) the tail portion looks similar to the tail of a monkey, (iv) body colour changes according to the surrounding environment, like a chameleon and (v) existence of brood pouches, like kangaroos.

For protection from predators, sea horses are able to change body colour rapidly and frequently, bringing about short and long term changes in body colour. Thus, these organisms are regarded as the "masters of camouflage" in the marine environment (Project Sea horse, 2014). Rapid changes in body colour are also noticed during mating, which depends upon age and the surrounding oceanic environmental conditions. Therefore, the proper identification of species of sea horse is considered difficult. The male sea horse brood pouch

* Correspondence: rajanimishra@yahoo.com
[2]National Centre for Antarctic and Ocean Research, Ministry of Earth Sciences, Government of India, Goa, India
Full list of author information is available at the end of the article

helps to incubate fertilized eggs laid by the female until the young are able to swim. This form of paternal care is a rare phenomenon amongst marine organisms (Project Sea horse, 2014; Froese & Pauly, 2014; Lourie et al., 1999). Considering future climatic changes, resultant environmental changes could have significant impacts on the sources of food such as phytoplankton biomasses and other marine organisms (Fabry et al., 2008; Jena et al., 2013; Mishra et al., 2015). This may trigger cascading effects on sea horse populations, other marine communities and ecosystem dynamics.

India is a hotspot of sea horse abundance and diversity; thus reflected in the past research activities (Lipton and Thangaraj 2002; Sreepada et al., 2002; Thangaraj & Lipton 2007; Lipton & Thangaraj 2013). However, detailed taxonomic research on sea horses within India is sparse. The first reliable documentation on sea horses within an Indian context is by Choudhury (1916) from the Chilika Lake. Subsequently, major research programmes have not been carried out in this particular region to substantiate the distribution, diversity, biology, and population size of this vulnerable species. The sea horse species documented by Choudhury (1916) from Chilika Lake was known as *Hippocampus brachyrhynchus* Duncker 1914, later on identified as *H. fuscus* Rüppell 1838 (Bailly, 2015). Furthermore, Bailly (2015) states that this particular species only breeds in the lake, but is mostly found within the lagoon's southern sector (i.e. mouth of the Rambha bay).

Marichamy et al.(1993) report two new records of sea horses (*H. kuda* and *H. trimaculatus*) from Palk Bay. Since, Lipton and Thangaraj (2002) have added *H. fuscus* to the Palk Bay record. Simultaneously, Balasubramanian (2002) reported another species *H. kellogi* from the southeast coast of India. Sreepada et al. (2002) communicate a detailed review on the sea horse status in India, highlighting the key issues of sea horse use in medicine. This basic information pertained to their threats and conservation, trade and commerce in India and the world.

Thangaraj & Lipton (2007) note the occurrence of a Japanese sea horse *Hippocampus mohnikei* from the Palk Bay. Murugan et al., (2008) describe the presence of five species of sea horses along the south-eastern coast of India. Significant work on morphological characterization was carried out by Thangaraj & Lipton (2011) on species such as *Hippocampus fuscus, H. kelloggi, H.kuda* and *H. trimaculatus* within the Gulf of Mannar. A similar study was also made by Lipton & Thangaraj (2013) at Tamilnadu and Kerala coasts of India, observing six sea horse species named *Hippocampus fuscus, H. kelloggi, H. kuda, H. histrix, H. mohnikei* and *H. trimaculatus*. Among them, *H. fuscus, H. kuda* and *H. trimaculatus* are common and widely distributed across the south-eastern coast of

India. However, Gopalpur of India's north-eastern coast is still being deficient in the incidence of sea horses. Therefore, the present study is carried out on the evidence of any re-occurrence of *H. Fuscus* and its range extension from the south to the northeast coast of India, in Chilika Lake and the adjoining of Bay of Bengal (BoB).

Materials and methods
Study area
The Gopalpur coast is situated between 19^0 256′ 381″N and 84^0 909′ 366″E on north-eastern coast of India, approximately 160 km south of Paradip and 260 km north of Visakhapatnam. It's 4 km coastline extends with sandy beaches, whilst the adjacent areas are covered with casuarinas vegetation. The beach is covered with sand dune vegetations. On the northern side, a 7 km long backwater creek exists, known as Haripur creek. The climate is tropical, receiving southwest and northeast rainfall with an average of 8.6 mm per month. The East India Coastal Current (EICC) of the BoB travels from the south to the northeast coast of India during January to October touching to it. The region is highly susceptible to tropical cyclones (Mahapatro et al., 2015b) (Fig. 1a).

Lake Chilika is situated between 19^0 28′N and 19^0 54′N and 85^0 05′E and 85^0 38′E on the east coast of India (Fig. 1b). Chilika is a pear shaped brackish water lagoon, with length of 64 km and width varying between 2 and 20 km. The lagoon is divided into four ecological sectors, namely the northern-fresh water sector, central-brackish water sector, southern-brackish to marine sector and outer channel, which is solely marine. The outer channel is 32 km long, orientated parallel to the coastline, and terminating in the BoB through two inlets. The tidal amplitude is semi-diurnal. The salinity of the southern sector is marine throughout the year, in contrast to the lagoon average. During the southwest monsoon, Chilika's average surface salinity approaches to zero because of the influx of fresh water from the rivers and channels connected to the Lake from the northern and western catchments (Panda et al., 2008). The southern sector connects to the Palur Canal (16 km long channel), which transports marine waters during high tide periods, from Bay of Bengal along with the transport sof and or sandy-clay substratum. This condition supports the growth of sea grass meadows such as *Halophila ovalis, Halophila ovata, Halophila beccari, Halodule uninervis* and *Halodule pinnifolia* (Panda et al., 2008).

Sample collection
A trawl net was hauled at a depth of 4 m on 9th March 2009, within in the Gopalpur coastal waters for fishing purposes. During these activities, a sea horse was caught and kept for later analysis. The sample was identified

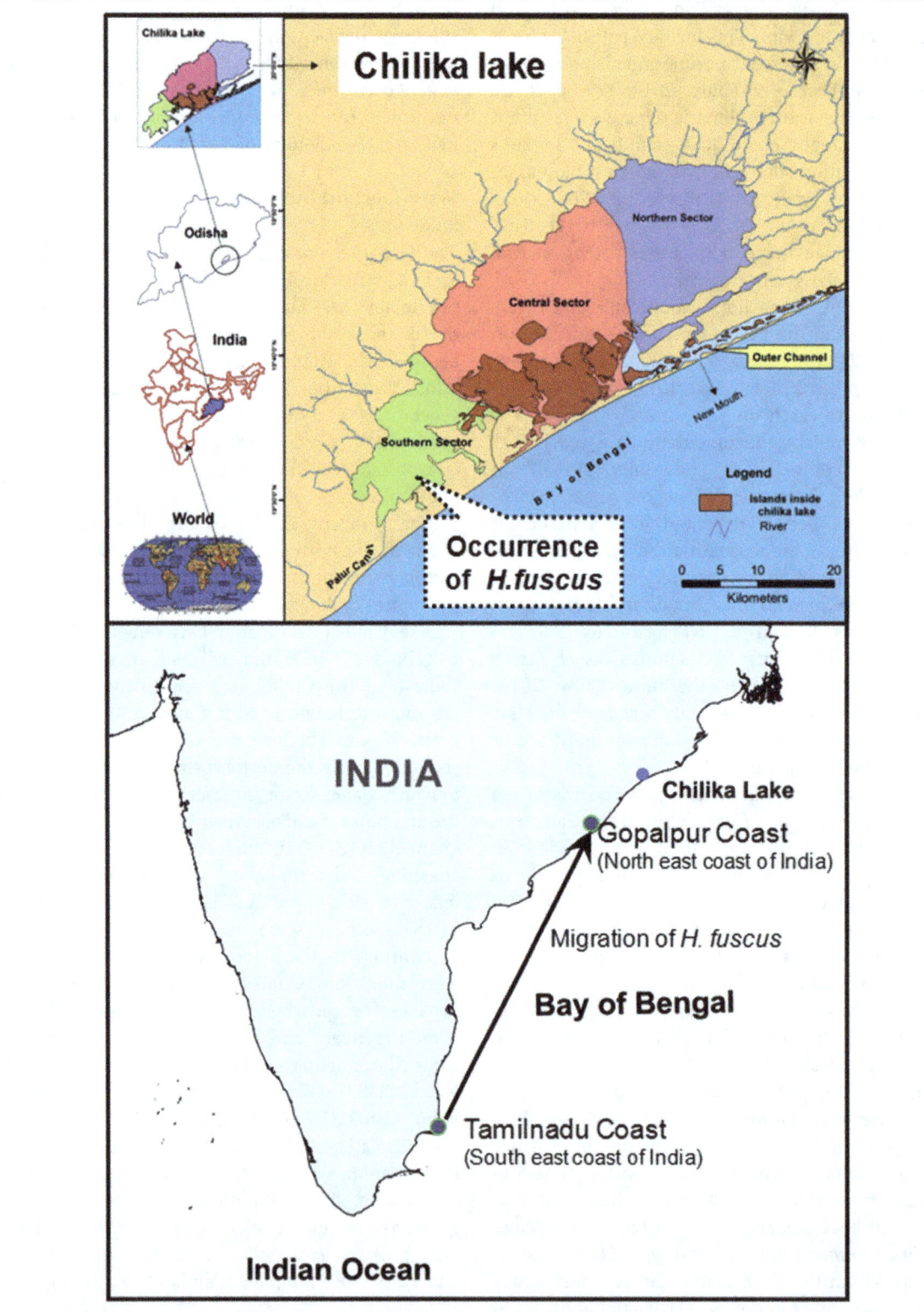

Fig. 1 Map showing the occurrence of H. fuscus in the Chilika lake (*upper*) and the migration route of H. fuscus from southeast Bay of Bengal to northeast Bay of Bengal, Gopalpur coast (*lower*)

carefully and photographs were taken immediately. A second sample was collected during a macrobenthic sampling program, using a sediment grab sampler (surface area 0.04 m^2) from the sea grass meadow region of the southern sector of Chilika Lake on 22nd March 2009. After initial observation, the sample was photographed and preserved. The specimen was then identified up to the species level by following the standard method of Lourie et al., (2004).

Results

Identification

The *Hippocampus fuscus* obtained from Gopalpur and Chilika has the following morphological features: body length equals14 cm, with 34 tail rings, 11 trunk rings, 16 dorsal fin rays and 15 pectoral fin rays. The coronet was lowly raised and slightly curved, the head was large compared to body, and the colour of the body was pale yellow to light green. The spines are slightly developed with smooth texture. After systematic analysis, both species were identified as *Hippocampus fuscus* Rüppell 1838. This specimen undergoes the following taxonomic classification (Lourie et al., 2004).

SYSTEMATICS
Class ACTINOPTERYGII
Order SYNGNATHIFORMES
Family SYNGNATHIDAE
Genus HIPPOCAMPUS (Cuvier, 1816)
Hippocampus fuscus (Rüppell 1838)

Fig. 2 Figure showing the specimen of *H. fuscus* from (**a**) Gopalpur Coast & (**b**) from Chilika Lake

(Figure 2 a from Gopalpur Coast & b from Chilika Lake) (Lourie et al., 2004).

Remarks

A coronet crown like bony crest is considered as one of the key identifying features. The body characters of this identified species were found relatively similar to three other species, namely: *H. borboniensis*, *H. hippocampus* and *H. kuda*. *H. borboniensis* has more tail rings, enlarged, knob-like spines and a better-developed coronet. Also, *H. borboniensis* has a larger head with two prominent eye spines. In contrast, the eye spine of the collected *H. fuscus* was entirely absent. The body of *H. borboniensis* was much shallower compared to the *H. fuscus*, formulating the key difference between these two species. The second species, *H. hippocampus*, has more tail rings, more dorsal fin rays, and fewer pectoral fins than *H.fuscus*. However, *H. hippocampus* has some similar characteristics with *H. fuscus*, as its distribution is restricted to the European waters (Lourie et al., 2004). However, the third species *H. kuda* has a larger body, deeper head, more tail rings, and a well-developed coronet with rounded shape as compared to the *H.fuscus*. Furthermore, the snout depth of *H.fucus* was higher than *H. kuda*.

Discussion

The present specimen *Hippocampus fuscus* was previously known as *Hippocampus brachyrhynchus* Duncker, 1914, reported from Chilika Lake by Choudhury (1916). Choudhury (1916) collected 7 sea horse species belonging to *Hippocampus brachyrhynchus* from the Rambha bay (southern sector) of Chilika Lake. Out of seven specimens, 3 were young, 3 were female and 1 was the male representative. Choudhury further confirmed that *Hippocampus fuscus* was a regular breeder inside the lake mostly in the sea grass bed region of the Lake's southern sector, also reported by Jones and Sujansingani (1954), ZSI, (1995) and Mahapatro (2016). However, comparatively limited information of its history is found in the publication. Pictorial information is entirely lacking, as well as the morphometric characters.

Thus, the study on spatial distribution of this vulnerable species in Chilika Lake and the adjoining coastal waters has been missing for last 99 years. Presently, Mahapatro et al., (2015a) found a pale yellow coloured sea horse from the sea grass bed of the Rambharatia region of Chilika Lake. Hence, the migration of sea horse from the BoB to Lake could be due to the potential feeding activities or through the tidal current. Interestingly, the specimen of *H. fuscus* was also observed in the coastal waters of Gopalpur, Bay Bengal, from the depth of 4 m as seen earlier (Mohapatro et al., 2015a). This suggests that the species *H. fuscus* was present in the

BoB and gradually shifted/drifted into the southern sector of Chilika Lake during the EICC through Palur channel. In addition, the proliferation of the sea grass bed in southern sector of Chilika Lake could support a preferable habitat for *H. fuscus*.

Consideration of the literature showed that *H. fuscus* is commonly observed in the coastal waters of Kerala and Tamilnadu (Thangaraj & Lipton, 2011; Lipton & Thangaraj, 2013). This might be a reason for the migration of *H. fuscus* from south to the northeast coast under the influence of EICC (Durand et al., 2009). A similar incident has been observed by Harasti (2015) in the Australian coastal waters, as the sea horse *Hippocampus histrix* migrated southward around 1800 km in the Great Barrier Reef due to the influence of East Australian Current (EAC). Large scale changes in the geographical distributional pattern of marine organisms which are not uncommon, as current circulation plays the key role. As seen here, EICC in the Indian east coast might have a significant role for the migration of *H. fuscus*, covering approximately 1100 km from the south to the northeast coast of India to appear within Lake Chilika.

Conclusion

The occurrence of *H. fuscus* in coastal waters of the BoB, and its extended range to the southern sector of Chilika Lake may reflect the migration from the south to the northeast coast of India under the influence of East India Coastal Current (EICC). Over the last 99 years, there has not been any information on the diversity, food, feeding habits or breeding biology of *H. fuscus* in the Chilika Lake and adjoining BoB. However, it is acknowledged that further and recurrent study of *H. fuscus* would confirm migration trends along the north eastern coast of BoB, if any, due to physical forces and/or oceanic currents in the region. In addition to the collection of in-situ datasets, satellite measured environmental parameters such as temperature, ocean colour, wind and current patterns is essential to study the effects of biophysical processes on population, community, and ecosystem dynamics (Jena et al., 2010; Mishra et al., 2003; Naik et al., 2014; Mahapatro et al., 2015b). This would help to delineate the ecological sensitive areas (ESA) for the appropriate management and conservation of *H. fuscus*.

Acknowledgments

The authors are thankful to the Head of the Department, Marine Sciences, Berhampur University, and the Chief Executive, Chilika Development Authority, Bhubaneswar, India for providing the field and laboratory facility during the study period. Authors are acknowledged with thanks to Greg Cooper, University of Southampton for his critical review and improvement throughout the manuscript. Due credit is given to Dr. P. Rajan for the preparation of the map of the study area. One of the authors RKM also thanks to the Director NCAOR for giving his encouragement towards the scientific publication.

Funding

The integrated Coastal zone Management (ICZM) project under the Chilika Development Authority (CDA) supported the fellowship during the study, collection, analysis and interpretation.

Authors contributions

The author DM collected the sample and followed the procedure to analyze the method and identified the species. RKM is analyzed and interpreted the result towards the discussion. SP was a major contributor for comments the manuscript. All authors read and approved the final manuscript.

Competing interests

The authors declare that they have no competing interests.

Author details

[1]Department of Marine Sciences, Berhampur University, Berhampur, Odisha, India. [2]National Centre for Antarctic and Ocean Research, Ministry of Earth Sciences, Government of India, Goa, India. [3]Regional CCF, Angul, Odisha, India.

References

Baillie J, Groombridge B. IUCN red list of threatened animals. Gland: IUCN (World Conservation Union); 1996.

Bailly N. *Hippocampus fuscus*, in WoRMS 2015 (World Register of Marine Species). 2015.

Balasubramanian R. Studies on seahorses with special references to *Hippocampus kelloggi* (Jordan and Synder, 1902), southeast coast of India, Ph.D Thesis. India: Annamalai University; 2002. p. 124.

Choudhury BL. Fauna of Chilika Lake: Fish part II. Memories Indian Mueseum. 1916;5:441–58.

Durand F, Shankar D, Birol F, Shenoi SSC. Spatio-temporal structure of the East India Coastal Current from satellite altimetry. J Geophys Res (C: Oceans). 2009;114(2). doi:10.1029/2008JC004807.

Fabry VJ, Seibel BA, Feely RA, Orr JC. Impacts of ocean acidification on marine fauna and ecosystem processes. ICES J Mar Sci. 2008;65:414–32.

Froese R, Pauly D. Fish Base- Hippocampus fuscus. 2014. World Wide Web electronic publication. www.Fishbase.org, (08/2014).

Harasti D. Range extension and first occurrence of the thorny sea horse *Hippocampus histrix* in New South Wales, Australia. Marine Biodiversity records, JMBUK. 2015;8:1–3. doi:10.1017/S1755267215000263.

IUCN. The IUCN Red List of Threatened Species. Version 2014.3. www.iucnredlist.org. Downloaded on 18 November 2014. 2014

Jena B, Swain D, Tyagi A. Application of Artificial Neural Networks for Sea-Surface Wind-Speed Retrieval From IRS-P4 (MSMR) Brightness Temperature. IEEE Geosci Remote Sens Lett. 2010;7(3–7):567–71. doi:10.1109/LGRS.2010.2041632.

Jena B, Sahu S, Kumar A, Swain D. Observation of oligotrophic gyre variability in the south Indian Ocean: Environmental forcing and biological response. Deep Sea Res Part I. 2013;80:1–10.

Jones S, Sujansingani KH. Fish and fisheries of the Chilka lake with statistics of fish catches for the years 1948–1950. Indian Journal of Fisheries. 1954;1(1–2):256–347.

Kendrick AJ, Hyndes GA. Patterns in the abundance and size-distribution of Syngnathid fishes among habitats in a seagrass-dominated marine environment. Estuar Coast Shelf Sci. 2003;57:631–40.

Lipton AP, Thangaraj M. Present status of seahorse fishing along the Palk Bay coast of Tamilnadu. Marine Fisheries Inf Serv. 2002;174:5–8.

Lipton AP, Thangaraj M. Distribution Pattern of Seahorse Species (Genus: *Hippocampus*) in Tamilnadu and Kerala Coasts of India. Notulae Scientia Biologicae. 2013;5:20–4.

Lourie SA, Foster SJ, Cooper EWT, Vincent ACJ. A Guide to the Identification of Seahorses. Project Seahorse and TRAFFIC North America. Washington D.C: University of British Columbia and World Wildlife Fund; 2004.

Lourie SA, Vincent ACJ, Hall HJ. Seahorses: an identification guide to the world's species and their conservation. London: Project Seahorse; 1999. 214.

Mahapatro D, Panigrahy RC, Panda S, Mishra RK. Checklist of intertidal benthic macrofauna of a brackish water coastal lagoon on east coast of India: The Chilika lake. Int J Marine Sci. 2015a;5(33):1–13.

Mahapatro D, Naik S, Behera DP, Mishra RK, Panda S. First distributional record of an Indo-Pacific porcupine puffer fish *Diodon holocanthus* (Diodontidae) from the Gopalpur coast, Bay of Bengal. Marine Biodiversity Records Marine Biol Assoc United Kingdom. 2015b;8:1–6. doi:10.1017/S1755267214001250.

Mahapatro D. Studies on the macrobenthos of chilika lake- a coastal lagoon on the east coast of India, Bay of Bengal. Berhmapur: Thesis submitted to the Berhampur University; 2016. p. 760007.

Marichamy R, Lipton AP, Ganapathy A, Ramalingam JR. Large-scale exploitation of seahorse (*Hippocapus kuda*) along the Palk Bay coast of Tamilnadu. Marine Fisheries Inf Serv Tech Extension Series. 1993;119:17–20.

Mishra RK, Shaw BP, Das SK, Rao K, Choudhary KH. Spatio –temporal variation of optically active substances in the coastal water off Orissa from Rushikulya to Dhamra (east coast of India). In J Marine Sci. 2003;32(2):133–40.

Mishra RK, Naik RK, Anilkumar N. Adaptations of phytoplankton in the Indian Ocean sector of the Southern Ocean during austral summer of 1998–2014. Frontier Earth Sci. 2015;9(4):742–52.

Murugan A, Dhanya S, Rajagopal S, Balasubramanian T. Seahorses and pipe fishes of the Tamil Nadu coast. Curr Sci. 2008;95:253–60.

Naik S, Mishra RK, Mahapatro D, Panigrahy RC. Impact of water quality on phytoplankton community and biomass in Dhamara estuary east coast of India. J Environ Biol. 2014;35:229–35.

Panda S, Bhatta KS, Rath KC, Misra CR, Samal RN. The Atlas of Chilika. Bhubaneswar: Chilika Development Authority; 2008. p. 133.

Project sea horse. 2014, http://seahorse.fisheries.ubc.ca/why-seahorses/essential-facts).

Sreepada RA, Desai UM, Naik S. The plight of Indian seahorses: Need for conservation and management. Curr Sci. 2002;82:377–8.

Thangaraj M, Lipton AP. Morphological characterization of four selected sea horse species (Genus : Hippocampus) from India. Ann Biol Res. 2011;2(4):159–67.

Thangaraj M, Lipton AP. Occurrence of the Japanese seahorse *Hippocampus mohnikei* Bleeker 1854 from the Palk Bay coast of South-eastern India. J Fish Biol. 2007;70:310–2.

Vincent ACJ. The international trade in seahorses. Cambridge: Traffic International; 1996. p. 163.

ZSI. In: Zoological Survey of India (ZSI), editor. Fauna of Chilika Lake. 1995.

25

First record of the megamouth shark, *Megachasma pelagios*, (family Megachasmidae) in the tropical western North Atlantic Ocean

Grisel Rodriguez-Ferrer[1][*] (iD), Bradley M. Wetherbee[2,5], Michelle Schärer[3], Craig Lilyestrom[1], Jan P. Zegarra[4] and Mahmood Shivji[5]

Abstract

Background: A new record of *Megachasma pelagios* is here reported for the tropical western North Atlantic Ocean from Puerto Rico.

Results: On December 10, 2016, a tourist reported an unusual stranded shark on Mojacasabe Beach, Cabo Rojo, on the southwestern coast of Puerto Rico. Visual examination of the carcass and mitochondrial DNA analysis from a dorsal fin sample revealed it to be a 457 cm female megamouth shark.

Conclusion: This record represents the first record of *M. pelagios* for the tropical western North Atlantic Ocean within the Caribbean Sea of southwest Puerto Rico and only the second record of *M. pelagios* from the North Atlantic.

Keywords: Megamouth shark, First record, Puerto Rico, Caribbean, Range extension, DNA barcoding

Background

The megamouth shark (*Megachasma pelagios*; Lamniformes: Megachasmidae) was first described based on an individual captured off Hawaii in 1976 (Taylor et al. 1983). The large filter feeding species had a number of unique characteristics and was placed in the new family Megachasmidae and genus. Since description of the holotype, the occurrence of at least 65 confirmed specimens with locations has been reported (FLMNH 2017). The International Union for Conservation of Nature (IUCN) Red List reports 102 specimens, but all of the specific locations are not provided (Simpfendorfer and Compagno 2015). The vast majority of confirmed reports are from the Indo-Pacific (FLMNH 2017). Only three have been reported from the Atlantic Ocean, two from Brazil and one from Senegal (Seret 1995, Amorim et al. 2000). Since many megamouth shark records are from fisheries interactions, further research on its ecology and habitat use is needed to better understand this species, currently listed as Least Concern by the IUCN (Simpfendorfer and Compagno 2015). On December 2016, a large shark carcass was reported on the southwestern coast of Puerto Rico. Details concerning identification of a megamouth shark, *Megachasma pelagios*, are presented.

Methods

On December 10, 2016, a tourist reported a stranded shark on Mojacasabe Beach, Cabo Rojo (17.980570 N, −67.210663 W), on the southwestern coast of Puerto Rico. Rodríguez-Ferrer verified the finding and confirmed the shark was dead and in an advanced stage of decomposition. Images of the specimen and total length were recorded and a dorsal fin sample was collected and frozen for DNA analysis. Sample was sent to and analyzed at the Nova Southeastern University, Halmos College of Natural Sciences and Oceanography.

Genomic DNA was extracted from ~25 mg of dorsal fin tissue using the QIAGEN DNeasy kit (QIAGEN Inc., Valencia, CA, USA). An approximate 655-base pair (bp)

* Correspondence: grodriguezf@drna.gov.pr
[1]Department of Natural and Environmental Resources, Recreational and Sport Fisheries Division, PO Box 366147, San Juan 00936, Puerto Rico
Full list of author information is available at the end of the article

fragment from the 5' region of the mitochondrial cytochrome c oxidase 1 gene (COI) was polymerase chain reaction (PCR) amplified using a cocktail of the primer sets FishF1 (5'-TCAACCAACCACAAAGACATTGGCAC-3'), FishF2 (5'-TCGACTAATCATAAAGATATCGGCAC-3'), FishR1 (5'-TAGACTTCTGGGTGGCCAAAGAATCA-3'), and FishR2 (5'-ACTTCAGGGTGACCGAAGAATCAGA A- 3') (Ward et al. 2005), and following procedures in Wong et al. (2009). The entire (~1300-bp) of the mitochondrial control region was PCR amplified using primers and procedures in Clarke et al. (2015). Amplicon purification and sequencing for both mitochondrial regions followed Clarke et al. (2015), with the exception that the amplicons were sequenced in one direction only using the FishR1 primer for COI and forward primer for control region. Species identity was checked by querying the National Center for Biotechnology Information (https://www.ncbi.nlm.nih.gov/) and Barcode of Life (http://www.boldsystems.org/) databases. Sequence divergence between our megamouth specimen and one sampled from the western Pacific (Chang et al. 2014) at the COI and control region was estimated in MEGA 7 (Kumar et al. 2017) as uncorrected p-distance (expressed as percent difference between the two sequences).

Results

Systematic account

Family: Megachasmidae Taylor et al. 1983
  Genus: *Megachasma* Taylor et al. 1983
  Megachasma pelagios Taylor et al. 1983
  Common name: Megamouth shark

Description

The shark carcass was in an advanced stage of decomposition. Water depth where the shark was recovered was 0.9 m. The shark carcass presented a "tadpole" body shape, large head, prominent mouth with large fleshy lips, many small triangular shaped teeth, five gill slits, brown coloration, flabby body and long upper caudal lobe that were consistent with descriptions in previous reports of *M. pelagios* specimens (Taylor et al. 1983, Nakaya et al. 1997). The specimen was a female measuring 457 cm in total length (TL). Fishing gear, scars or injuries were not observed on the body of the shark and no obvious cause of death was evident by visual inspection. Several photographs were taken for evidence (Fig. 1). After the examination and given the shark's advanced state of decomposition, the carcass was towed out to sea and discarded with the help of local commercial anglers.

Genetics

A tissue sample was taken from the dorsal fin for DNA analysis. Both COI (622 bp) and control region (613 bp) sequences confirmed the specimen as *M. pelagios* (GenBank accession numbers KY392958 and KY379851, respectively). Sequence divergence (p-distance) between the Puerto Rico specimen and the Pacific specimen was 0.0% for COI and 0.32% for the control region.

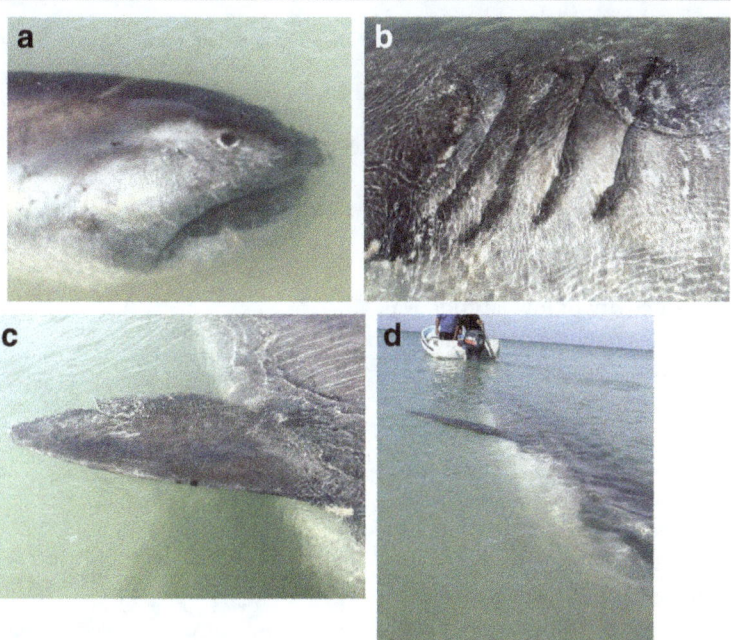

Fig. 1 Megamouth shark carcass photos taken in Cabo Rojo, Puerto Rico (December 2016), female, 457 cm TL (**a** lateral view of the head, **b** gill slits, **c** right pectoral fin, **d** carcass being towed)

Discussion

The finding of a megamouth shark in Puerto Rico expands the distribution for this species. This is only the second report of the species from the North Atlantic Ocean and the first report from the tropical western North Atlantic Ocean or Caribbean Sea. It is possible that megamouth sharks are more common in the Pacific Ocean, but the species has a wider distribution that now includes all sides of the Atlantic Ocean as well as an additional low-latitude record.

Of the three instances where megamouth sharks were observed in the Atlantic Ocean, two were males and one was of undetermined sex; therefore, this is the first confirmation of a female megamouth in the western hemisphere. Given the estimated size at maturity for females at 5 m (Nakaya et al. 1997, Smale et al. 2002, Nakaya 2008) this specimen was likely immature or a sub-adult, although the reproductive tract was not examined. No mating scars were observed on the body as reported in a larger, sexually mature female captured in Japan (Yano et al. 1997). It has been suggested that juvenile and adult megamouth sharks segregate geographically, with juveniles more common in lower latitudes, expanding their range to higher latitudes as they age (Nakaya 2008). Some records suggest year-round presence at higher latitudes and possible migration to lower latitudes during part of the year (Nakaya 2008). The three previous records from the Atlantic Ocean were in May in Senegal (Seret 1995) and September and July in Brazil (Amorim et al. 2000, Lima et al. 2009). These records, plus the present specimen stranded in December in Puerto Rico are consistent with geographical segregation by size and suggest that the species is present in the Atlantic year round, including at lower latitudes.

Based on the morphology and stomach contents of other megamouth sharks (Taylor et al. 1983, Nakaya et al. 2008, Sawamoto and Matsumoto 2012) as well as distinct daily vertical movements demonstrated by a megamouth shark tracked using acoustic telemetry (Nelson et al. 1997), these sharks are filter feeders that prey on plankton. Lower productivity of low latitude waters compared to higher latitudes may partly explain the paucity of reports of megamouth sharks at low latitudes such as the Caribbean Sea, and why the majority of records are from more temperate waters.

The p-distances provide the first estimate of mitochondrial DNA sequence divergence between Atlantic and Pacific megamouth sharks. The absence of nucleotide polymorphisms (p-distance = 0.0%) in the COI barcode and only two variable sites in the control region (p-distance = 0.32%) between widely separated individuals from different ocean basins may portend low global matrilineal genetic diversity in this enigmatic species, as seen in another pelagic filter-feeding lamniform, the basking shark (Hoelzel et al. 2006). However, confirmation of this low diversity will require further investigation with larger sample sizes.

Conclusion

The importance of the present record resides in the fact that it represents the first record of *M. pelagios* for Puerto Rico and a significant range extension into the tropical western North Atlantic Ocean.

Abbreviations
bp: Base pairs; cm: Centimeter; COI: Cytochrome c oxidase 1 gene; DNER: Department of Natural and Environmental Resources; FLMNH: Florida Museum of Natural History; PCR: Polymerase chain reaction; TL: Total length

Acknowledgments
We thank K. Finnegan for conducting the laboratory genetics analysis, the local commercial anglers who towed the shark carcass offshore, and Sgt. H. Ronda and DNER Rangers Mr. Vargas and Mr. Lugo for their support during the finding. We are grateful to the anonymous reviewer whose suggestions improved the manuscript.

Funding
The DNA analysis laboratory work was supported by funding from the Guy Harvey Ocean Foundation and Save Our Seas Foundation.

Authors' contributions
GR visually examined the carcass, took fin sample and photographs, revised literature and completed the first draft of this manuscript. BMW and MSh completed all the genetic analysis and discussion. MSc and CL provided additional references and information on the species. MSh provided the funding source. JPZ provided initial shark identification and prepared manuscript for submittal. All authors contributed edits and comments to the draft manuscript and read and approved the final manuscript.

Competing interests
The authors declare that they have no competing interests. The findings and conclusions in this article are those of the author(s) and do not necessarily represent the views of the U.S. Fish and Wildlife Service.

Author details
[1]Department of Natural and Environmental Resources, Recreational and Sport Fisheries Division, PO Box 366147, San Juan 00936, Puerto Rico. [2]Department of Biological Sciences, University of Rhode Island Kingston, 120 Flagg Road, Kingston, RI 02881, USA. [3]H.J.R Reefscaping, P. O. Box 1442, Boquerón 00622, Puerto Rico. [4]U.S. Fish & Wildlife Service, Caribbean Ecological Services Field Office, P. O. Box 491, Boquerón 00622, Puerto Rico. [5]Guy Harvey Research Institute and Save Our Seas Shark Research Center, Nova Southeastern University, 800 N Ocean Drive, Dania Beach, FL 33004, USA.

References

Amorim AF, Arfeli CA, Castro JI. Description of a juvenile megamouth shark, *Megachasma pelagios*, caught off Brazil. Environ Biol Fishes. 2000;59:117–23.

Chang CH, Shao KT, Lin YS, Chiang WC, Jang-Liaw NH. Complete mitochondrial genome of the megamouth shark *Megachasma pelagios* (Chondrichthyes, Megachasmidae). Mitochondrial DNA. 2014;25:185–7.

Clarke CR, Karl SA, Horn RL, Bernard AM, Lea JS, Hazin FH, Prodöhl PA, Shivji MS. Global mitochondrial DNA phylogeography and population structure of the silky shark, *Carcharhinus falciformis*. Mar Biol. 2015;162:945–55.

Florida Museum of Natural History. Confirmed Megamouth Shark Sightings. 2017. https://www.flmnh.ufl.edu/fish/sharks/megamouth/MegaMap.htm. Accessed 15 Jan 2017.

Hoelzel RA, Shivji MS, Magnussen JE, Francis MP. Low worldwide genetic diversity in the basking shark (*Cetorhinus maximus*). Biol Lett. 2006;2:639–42.

Kumar S, Stecher G, Tamura K. MEGA7: Molecular Evolutionary Genetics Analysis version 7.0. 2017. http://www.megasoftware.net/docs

Lima LM, Rennó B., Siciliano S. Gigante dos mares em areias fluminenses. Um dos mais raros tubarões do mundo é encontrado na costa brasileira. Ciência Hoje 2009; Ed 263.

Nakaya K. Biology of the megamouth shark. Megachasma pelagios (Lamniformes: Megachasmidae) Zool. 2008;52:603–7.

Nakaya K, Yano K, Takada K, Hiruda H. Morphology of the first female megamouth shark, *Megachasma pelagios* (Elasmobranchii: Megachasmidae), landed at Fukuoka, Japan. In: Yano K, Morrissey J, Yabumoto Y, Nakaya K, editors. Biology of the megamouth shark. Tokyo: Tokai Univ. Press; 1997. p. 51–62.

Nakaya K, Matsumoto R, Suda K. Feeding strategy of the megamouth shark *Megachasma pelagios* (Lamniformes: Megachasmidae). J Fish Biol. 2008;73: 17–34.

Nelson DR, McKibben JN, Strong Jr WR, Lowe CG, Sisneros JA, Schroeder DM, Lavenberg RJ. An acoustic tracking of a megamouth shark, *Megachasma pelagios*: a crepuscular vertical migrator. Environ Biol of Fish. 1997;49:389–99.

Sawamoto S, Matsumoto R. Stomach contents of a megamouth shark *Megachasma pelagios* from the Kuroshio Extension: evidence for feeding on a euphausiid swarm. Plankton Benthos Res. 2012;7:203–6.

Seret B. Premiere capture d'un requin grande gueule (Chondrichthyes, Megachasmidae) dans l'Atlantique, au large du Senegal. Cybium. 1995;19:425–7.

Simpfendorfer C, Compagno LJV. *Megachasma pelagios*. The IUCN Red List of Threatened Species 2015: e.T39338A2900476. http://dx.doi.org/10.2305/IUCN. UK.2015-4.RLTS.T39338A2900476.en. Accessed 15 Jan 2017.

Smale MJ, Compagno LJV, Human BA. First megamouth shark from the western Indian Ocean and South Africa. S Afr J Sci. 2002;98(7–8):349–50.

Taylor LR, Compagno LJV, Struhsaker PJ. Megamouth- a new species, genus, and family of lamnoid shark (Megachasma pelagios, family Megachasmidae) from the Hawaiian Islands. Proc Calif Acad Sci. 1983;43(8):87–110.

Ward RD, Zemlak TS, Innes BH, Last PR, Herbert PDN. DNA barcoding Australia's fish species. Phil Trans R Soc B. 2005;360:1847–57.

Wong EHK, Shivji MS, Hanner RH. Identifying sharks with DNA barcodes: assessing the utility of a nucleotide diagnostic approach. Mol Ecol Resour. 2009;9:243–56.

Yano K, Yoshitaka Y, Sho T, Osamu T, Masami F. Capture of a Mature Female Megamouth Shark *Megachasma pelagios*, from Mie, Japan. Proceedings of the 5th Indo-Pacific Fish Conference. 1997;335–49.

26

DNA barcoding of flat oyster species reveals the presence of *Ostrea stentina* Payraudeau, 1826 (Bivalvia: Ostreidae) in Japan

Masami Hamaguchi[1*], Miyuki Manabe[2], Naoto Kajihara[1], Hiromori Shimabukuro[1], Yuji Yamada[3] and Eijiro Nishi[4]

Abstract

Background: DNA barcoding is an effective method of accurately identifying morphologically similar oyster species. However, for some of Japan's *Ostrea* species there are no molecular data in the international DNA databases.

Methods: We sequenced the mitochondrial large subunit ribosomal DNA (LSrDNA) and cytochrome c oxidase subunit I (COI) gene of five known and two unidentified *Ostrea* species. Phylogenetic comparison with known *Ostrea* species permitted accurate species identification by DNA barcoding.

Results: The molecular data, which were deposited in an international DNA database, allowed for a clear distinction among native *Ostrea* species in Japan. Moreover, the nucleotide sequence data confirmed that *O. stentina* (Atsuhime-gaki) inhabits Kemi and Ibusuki, Japan.

Conclusions: This is the first record of *O. stentina* in Japan. These results provided for accurate species identification by DNA barcoding of the taxonomically problematic species *O. futamiensis*, *O. fluctigera*, *O. setoensis* and *O. stentina* in Japan.

Keywords: *Ostrea stentina*, *O. futamiensis*, *O. fluctigera*, *O. setoensis*, DNA barcoding

Background

Over the last half century, Japan's coastal ecosystems have been severely damaged by human activity. The Seto Inland Sea, which is surrounded by the Japanese main islands of Honshu, Shikoku and Kyushu, is located in the western part of Japan and is an area of human-induced ecological deterioration. The coastal areas were reclaimed for urban and industrial use during a period of rapid economic growth in the 1970s, leading to the loss of 63.7% of the natural coast (tidal flats, seagrass beds and estuary systems). Habitat loss, pollution, overfishing, invasive species and now global climate change are rapidly damaging the Seto Inland Sea. These factors have gradually decreased the biodiversity of the area, and many marine organisms have become endangered. Although many native and relict species of the last glacial epoch from the ancient East China Sea are found in the Seto Inland Sea (Botton et al. 1996; Futahashi 2011; Hamaguchi et al. 2013), other invasive alien and indigenous species have been discovered where human activity has led to the development of industrial areas along the coast (Iwasaki et al. 2004). Therefore, since 2008 we have been conducting a long-term study to monitor benthic species diversity at various tidal flats to promote the conservation of native marine fauna in the Seto Inland Sea and its adjacent marine areas supported by the Ministry of the Environment Monitoring Sites 1000 Project and the Japan Long Term Ecological Research Network.

We observed two morphologically different putative *Ostrea* species (*Ostrea* sp. A and *Ostrea* sp. B) during the field surveys. The external features of *Ostrea* sp. A were very similar to those of *O. futamiensis* Seki 1929

* Correspondence: masami@fra.affrc.go.jp
[1]National Research Institute of Fisheries and Environment of Inland Sea, Fisheries Research Agency, 2-17-5 Maruishi, Hatsukaichi, Hiroshima 739-0452, Japan
Full list of author information is available at the end of the article

while some morphological features were not identical. We considered that *Ostrea* sp. A might be a juvenile form of another *Ostrea* species present in Japan. The external features of Ostrea sp. B were very similar to those of *Crassostrea gigas* Thunberg, 1793 and we misidentified the oyster as *C. gigas* at first. Some morphological features of *Ostrea* sp. B were similar to those of *O. stentina* Payraudeau, 1826 but this species has not previously been reported from Japan. Therefore, we performed accurate species identification of the oyster specimens.

In general, Ostreidae species are economically important marine organisms, but their morphological plasticity can cause taxonomic confusion. For example, shell morphology has been used as a primary feature to distinguish different species of oyster; however, the shell is affected by habitat and environment (Tack et al. 1992; Yamaguchi 1994; Lam and Morton 2004, 2006; Liu et al. 2011). In recent years, molecular analyses have been used to accurately identify Ostreidae species. Methods such as DNA barcoding (Hebert et al. 2003; Schindel and Miller 2005) have been used to detect hidden and cryptic species, determine their distributions, monitor the biodiversity of marine fauna and reconstruct the phylogeny of taxonomically confusing *Ostrea* species (Jozefowicz and O'Foighil 1998; Hurwood et al. 2005; Lapègue et al. 2006; Polson et al. 2009; Salvi et al. 2014). Moreover, DNA barcoding can be applied to all life stages of the oyster, e.g. planktonic larvae, spat and juvenile forms.

Five flat oyster species have been reported from Japan. *O. deselamellosa* Lischke, 1869 and *O. circumpicta* Pilsbry, 1904 are fishery and aquaculture species utilized in Japan. Molecular data of these two species have been deposited in the international DNA databases (DDBJ; DNA database of Japan/EMBL; European Molecular Biology Laboratory/GenBank DNA database).

The other three species recorded are small flat oysters, viz. *O. futamiensis*, *O. fluctigera* Jousseume in Lamy, 1925 and *O. setoensis* Habe 1957 about which there is little taxonomical or ecological information. The molecular data of these three species have not as yet been deposited in the international DNA databases.

Ostrea futamiensis Seki 1929 was first discovered in Futamigaura, Hyogo Prefecture, in the eastern part of the Seto Inland Sea (Seki 1929). The oyster has a small (20–35 mm in length), moderately thick and irregularly circular- or oval-shaped shell. The World Register of Marine Species (WoRMS; http://www.marinespecies.org/) lists *O. futamiensis* as a valid species. However, this species is not commercially important in Japan, and thus ecological and chorological research on this oyster is incomplete (Okutani 2000; Iijima 2007). Wada et al. (1996) recommended that *O. futamiensis* be designated a near-threatened species. However, Henmi et al. (2014)

summarized claims made by other marine benthic researchers who maintain that *O. futamiensis* should not be designated as near-threatened because *O. futamiensis* is possibly a junior synonym of *O. denselamellosa*.

Ostrea fluctigera Jousseume in Lamy, 1925 is a hard-to-find and taxonomically problematic species. The species is small and settles on hermit crab shells. Inaba and Torigoe (2004) re-classified the species and concluded that *O. deformis* Lamarck, 1819 and *Nanostrea exigua* Harry 1985 were the synonyms of *O. fluctigera*. There has only been one paper (Kuramochi 2007) published on this topic since the reclassification by Inaba and Torigoe (2004).

Ostrea setoensis Habe 1957 is a small oyster and is also a hard-to-find species in Japan. Habe (1957) described the oyster as *O. sedea setoensis*, which is a subspecies of *O. sedea* Iredale, 1939 from Australia. However, he later transferred the oyster to the genus *Neopycnodonte* (Habe 1977). Torigoe (1983) claimed that it was an *Ostrea* species based on its anatomy and shell morphology and considered it *O. setoensis*.

As described above, species identification of *O. futamiensis*, *O. fluctigera*, *O. setoensis*, *Ostrea* sp. A and *Ostrea* sp. B by DNA barcoding has not been possible until now because no nucleotide sequence data from these oyster species has been deposited in international DNA databases.

In this study, we collected *O. futamiensis*, *O. fluctigera* and *Ostrea* sp. A from the Seto Inland Sea, and other *Ostrea* oysters including *Ostrea* sp. B and *O. setoensis* from Japanese waters elsewhere. We analyzed the nucleotide sequences of the mitochondrial large subunit ribosomal RNA (LSrRNA) and the cytochrome *c* oxidase subunit I (COI) gene to facilitate DNA barcoding of the members of the genus *Ostrea*.

Methods

Sample collection and morphological identification of Ostrea species in Japan

Ostrea sp. A and *O. fluctigera* specimens were sampled from the Kemi tidal flat in the Wakayama Prefecture. *Ostrea* sp. B were collected from Ibusuki in Kagoshima Bay. *Ostrea* sp. B was settled onto polyvinyl chloride plates used to culture *Crassostrea nippona* Seki, 1934 oysters at the Kagoshima Prefectural Fisheries Technology and Development Center. *O. futamiensis* specimens were sampled from five tidal flats (Nakatsu, Oiso, Hishiwo, Hinase and Kemi) in the Seto Inland Sea. *O. setoensis* specimens were sampled from the Tamanoura tidal flat in the Wakayama Prefecture. *O. circumpicta* and *O. denselamellosa* were collected from the Yamagata and Kumamoto Prefectures, Japan, respectively. All the oyster collection sites are shown in Fig. 1 and Table 1. *O. lurida* Carpenter, 1864 was collected from Willapa Bay, Washington State,

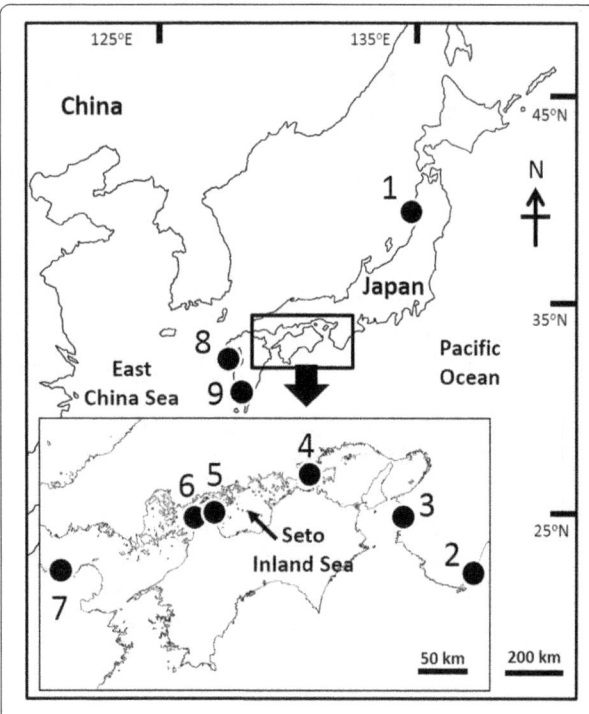

Fig. 1 Sampling sites of the *Ostrea* specimens used in this study

USA by Dr. Hori and Prof. Ruesink and compared with Japanese *Ostrea* species. We observed shell characteristics (i.e. shell shape and external features, growth lines, lamellae and ribs, umbo position and shape) and inner surface features (pallial sinus, adductor muscle scar shape and position and chomata), shell colour and hinge type. The oyster specimens were identified using these morphological features according to Seki (1929, 1930), Torigoe (1981), Inaba and Torigoe (2004) and Harry (1985). The specimens examined in this study were deposited in the Osaka Museum of Natural History (OMNH).

DNA preparation

All *O. futamiensis*, *O. fluctigera* and *O. setoensis* specimens were transported live to our laboratory in Hiroshima Prefecture, Japan. The adductor muscle of each individual organism was excised and preserved in 80% ethanol. The adductor muscle samples from *O. circumpicta*, *O. denselamellosa*, *Ostrea sp.* A, *Ostrea sp.* B and *O. lurida* obtained from each sampling site were preserved in 80% ethanol until DNA extraction. The total genomic DNA was extracted from all specimens using a DNeasy Blood & Tissue Kit (Qiagen, CA, USA) according to the manufacturer's instructions.

DNA barcoding on the basis of mitochondrial LSrRNA and COI

The mitochondrial LSrRNA and COI genes were subjected to polymerase chain reaction (PCR) amplification using our original primers (16SUF 5′-GAACTCGG CAAAATTAAACCTCGCCT-3′, 16SUR 5′-ARRGKWT TAARGGTCGAACAGA-3′) and universal primers (LCO1490 5′-GGTCAACAAATCATAAAGATATTGG-3′ and HCO2198 5′-TAAACTTCAGGGTGACCAAAA AATCA-3′) as reported by Hamaguchi et al. (2014) and Folmer et al. (1994), respectively. A MyCycler™ Thermal Cycler (Bio-Rad, CA, USA) was used to amplify PCR products in a total volume of 15 µL containing 5 U of Hot Taq™ (5 U/µL; Takara, Otsu, Japan), 10× Hot Taq™ buffer, 2.5 mM of each dNTP, 0.5–1.0 µM of each primer and 0.5 µL of template DNA. The PCR amplification cycles included denaturation at 94 °C for 1 min; 35 cycles of denaturation at 94 °C for 30 s followed by annealing at either 55 °C (LSrRNA) or 40 °C (COI) for 30 s and an extension at 72 °C for 45 s; and a final extension for 5 min at 72 °C. The PCR amplicons were checked by loading 3 µL of each sample with 3 µL of loading dye on a 2% agarose gel (Agarose S; Nippon Gene, Tokyo, Japan) containing

Table 1 Sampling sites in this study

Species	Year	Sampling site		Prefecture or State	Latitude	Longitude	N
Ostrea sp. A	2015	Kemi	Fig. 1-3	Wakayama, Japan	34.159493	135.183504	3
Ostrea sp. B	2015	Ibusuki	Fig. 1-9	Kagoshima, Japan	31.294740	130.604903	7
Ostrea futamiensis	2013	Nakatsu	Fig. 1-7	Oita, Japan	33.604920	131.237633	8
Ostrea futamiensis	2014	Hishiwo	Fig. 1-6	Hiroshima, Japan	34.380379	133.219520	12
Ostrea futamiensis	2014	Ooiso	Fig. 1-5	Hiroshima, Japan	34.398751	133.239540	12
Ostrea futamiensis	2014	Hinase	Fig. 1-4	Okayama, Japan	34.731732	134.276166	7
Ostrea futamiensis	2015	Kemi	Fig. 1-3	Wakayama, Japan	34.159493	135.183504	6
Ostrea fluctigera	2015	Kemi	Fig. 1-3	Wakayama, Japan	34.159493	135.183504	4
Ostrea setoensis	2015	Tamanoura	Fig. 1-2	Wakayama, Japan	33.568484	135.918252	3
Ostrea circumpicta	1999	Yura	Fig. 1-1	Yamagata, Japan	38.720467	139.675662	8
Ostrea denselamellosa	2008	Midori-River	Fig. 1-8	Kumamoto, Japan	32.720389	130.593348	16
Ostrea lurida	2013	Willapa Bay		Washington State, USA			12

0.5 µg/mL ethidium bromide. The remaining 12 µL of PCR product was subsequently purified using a QIAquick PCR Purification Kit (Qiagen, CA, USA).

The purified PCR amplicons were sequenced using the LSrRNA or COI primers as described above and the ABI PRISM BigDye Terminator v3.1 Cycle Sequencing Ready Reaction Kit (Applied Biosystems, CA, USA) in a Genetic Analyzer 3130 *xl* automated DNA Sequencer (Applied Biosystems, CA, USA). The final LSrRNA and COI sequences were obtained from both strands for verification, and all newly obtained sequences were deposited in the DDBJ/EMBL/GenBank databases. The accession numbers were as follows: *Ostrea* sp. A, LC051572–LC051574; *Ostrea* sp. B, LC051575–LC051581; *O. futamiensis*, AB898267–AB898274, LC051592-051609; *O. fluctigera*, LC149503–LC149510; *O. setoensis*, LC149511–LC149516; *O. denselamellosa*, AB898275–AB898279; *O. circumpicta*, AB898279–AB898282; and *O. lurida*, AB898263–AB898266.

Comparison of the molecular data of native Japanese Ostrea species with those of known Ostrea species reported worldwide

The LSrRNA sequences of our samples were compared with those of the other known *Ostrea* species using the BLAST search in GenBank. The taxonomic separation among native and other *Ostrea* species was analysed by constructing a maximum parsimony tree for the LSrRNA sequences (424 bp). The 19 nominal *Ostrea* species of which the LSrRNA sequences were compared were (accession numbers in brackets) *Ostrea* sp. A (LC051572), *Ostrea* sp. B (LC051575), *O. futamiensis* (AB898267), *O. fluctigera* (LC149507), *O. setoensis* (LC149514), *O. denselamellosa* (AB898275), *O. circumpicta* (AB898279) and *O. lurida* (AB898263), as well as the LSrRNA sequences available in the international DNA databases for *O. angasi* Sowerby, 1871 (AF052063), *O. algoensis* Sowerby II, 1871 (AF052062), *O. aupouria* Dinamani, 1981 (AF052064), *O. chilensis* Philippi in Küster, 1844 (JF808186), *O. conchaphila* Carpenter, 1857 (FJ768527), *O. edulis* Linnaeus, 1758 (DQ093488), *O. equestris* Say, 1834 (AY376603), *O. puelchana* d'Orbigny, 1842 (AF052073), *O. stentina* (JF808189 and DQ180744), *O. spreta* d'Orbigny, 1846 (DQ640402) and *Ostrea* sp. JL-2011 (HQ661001). The LSrRNA sequence for *Saccostrea glomerata* Gould, 1850 (AF353101) was used as an outgroup. The sequences obtained for each region were aligned using ClustalW (Thompson et al. 1994; gap opening penalty, 15; gap extension penalty, 6.6; transition weight, 0.5), and the MP tree based on the Tamura 3-parameter model (Tamura 1992) was reconstructed in MEGA version 6 (Tamura et al. 2013).

Estimates of evolutionary divergence between sequences within the COI of each *Ostrea* species were calculated using the Kimura 2-parameter model (K2P; Kimura 1980) in MEGA version 6. The 19 nominal *Ostrea* species of which the COI sequences were compared were *Ostrea* sp. A (LC051584), *Ostrea* sp. B (LC051590), *O. futamiensis* (AB898290), *O. circumpicta* (AB898294), *O. fluctigera* (LC149507) and *O. setoensis* (LC149514), as well as *O. angasi* (AF540598), *O. aupouria* (AF112288), *O. chilensis* (AF112286), *O. edulis* (AF120651), *O. equestris* (AY376607), *O. conchaphila* (DQ464125), *O. denselamellosa* (NC015231), *O. lurida* (NC022688), *O. puelchana* (DQ226518), *O. stentina* (DQ226522), *Ostrea* sp. MS-2011 (JF915514) and *Ostrea* sp. STH-2012 (JQ027292) whose COI sequences were available from the international DNA databases (DDBJ/EMBL/GenBank). The COI sequence for *Saccostrea glomerata* (EU007483) was used as an outgroup.

Results

Morphological features of unknown Ostrea species in Japan

We compared *Ostrea* sp. A with native *Ostrea* species and their juvenile forms. Most of the important external features of *Ostrea* sp. A were very similar to those of *O. futamiensis*; for example, the samples OMNH-Mo38148 (*Ostrea* sp. A; Fig. 2-1) and OMNH-Mo38141 (*O. futamiensis*; Fig. 2-2) both had partially embedded stones on a sandy tidal flat attached to their undersides (Kemi tidal flat; Fig. 1-3). The right valves of *Ostrea* sp. A (OMNH-Mo38148; Fig. 3-1) and those of coexisting *O. futamiensis* (OMNH-Mo38141; Fig. 3-2) were also similar and are shown in Fig. 3. Shell shapes of *Ostrea* sp. A were elliptical and flat. The left valves were very thin, and shell height and length were less than 15 and 10 mm, respectively. Chomata, of which there were approximately 15–30, were inconspicuous and restricted to both ligament sides. The umbonal cavities were shallow. The adductor muscle scars were reniform, and the dorso-anterior borders were concave. External color of

Fig. 2 *Ostrea* sp. A (1: OMNH-Mo38148) and *Ostrea futamiensis* (2: OMNH-Mo38141) in the Kemi tidal flat (Fig. 1. 3). Scale bar: 10 mm

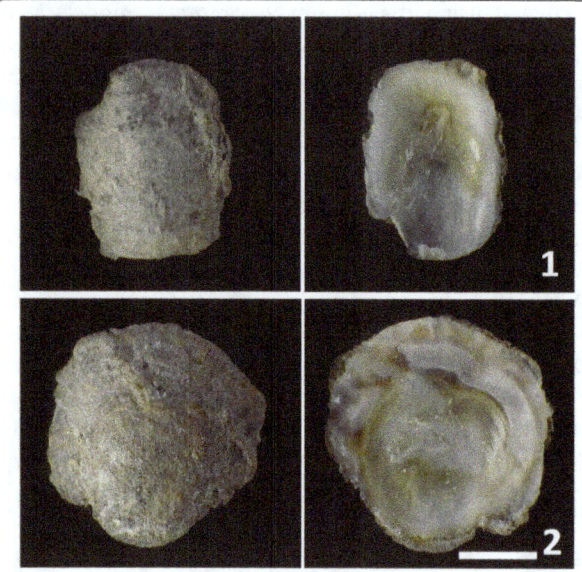

Fig. 3 The right valve of *Ostrea* sp. A (1: OMNH-Mo38148) and *Ostrea futamiensis* (2: OMNH-Mo38141) collected from the Kemi tidal flat. Scale bar: 10 mm

the right valves was opaque white to light brown with many dark brown streaks radiating from the ambo (Fig. 4). The interior shells were composed of olive to yellowish green conchiolin and a white calcareous layer, which was sometimes narrow (Fig. 4).

Fig. 4 Morphological features of *Ostrea* sp. B (OMNH-Mo38134) collected from the Kagoshima Prefectural Fisheries Technology and Development Center in Ibusuki. Scale bar: 10 mm

However, there were also several differences between *Ostrea* sp. A and *O. futamiensis* features. The shell shape of *Ostrea* sp. A was elliptical, and that of *O. futamiensis* was circular. Shape of the adductor muscle scar was very similar, but that of *Ostrea* sp. A was narrow compared with that of *O. futamiensis*. Position of the adductor muscle scar was below the center of the interior shell for *Ostrea* sp. A, whereas that of *O. futamiensis* was in the center of the interior shell. However, almost all morphological features of *Ostrea* species have been recorded from adult specimens. Therefore, we could not confirm by morphological features alone if *Ostrea* sp. A was a juvenile form of another *Ostrea* species.

The shell of *Ostrea* sp. B (OMNH-Mo38134) was orbicular and spatulate with many wrinkles, and with a height and length of less than 50 and 40 mm, respectively (Fig. 4). The external colour of the right valve was yellowish white to light brown with dark brown or black streaks radiating from the umbo. Initially, we misidentified the oyster as *C. gigas*, because the external colour and shape was very similar to that of *C. gigas*. However, the adductor muscle scar was colourless and reniform, and the dorso-anterior border was concave. The adductor muscle scar of *C. gigas* is light-coloured, purple, or brown. The chomata of the oysters were inconspicuous and restricted to each ligament side. These chomata features showed that the oyster belonged to genus *Ostrea*; *C. gigas* have no chomata. The colour of the interior shell was partly olive to yellowish green conchiolin in a white calcareous layer, whereas the interior shell of *C. gigas* is white. These morphological features clearly differed between *C. gigas* and *Ostrea* sp. B. Moreover, the external shell features of *Ostrea* sp. B were different from those of other known Japanese *Ostrea* oysters, but external and internal shell features, chomata, adductor muscle scar were similar to those of *O. stentina*.

Both *Ostrea* sp. A and *Ostrea* sp. B had inconspicuous chomata restricted to each ligament side, but other morphological features were different (Figs. 3 and 4). Although external shell features of *Ostrea* sp. B were similar to those of *O. stentina*, *O. stentina* has not been recorded from Japan. Therefore, we considered *Ostrea* sp. A and *Ostrea* sp. B to be different putative species based on morphological features, but molecular analysis by DNA barcoding is needed for accurate identification of these oysters.

DNA barcoding

The phylogenetic analysis of the LSrRNAs is shown in Fig. 5. The nucleotide sequences of both *Ostrea* sp. A and B clustered together in the *O. stentina* complex, which consisted of *O. stentina*, *O. aupouria*, *O. equestris* and *O. spreta*. *Ostrea* sp. A and B were clearly distinct

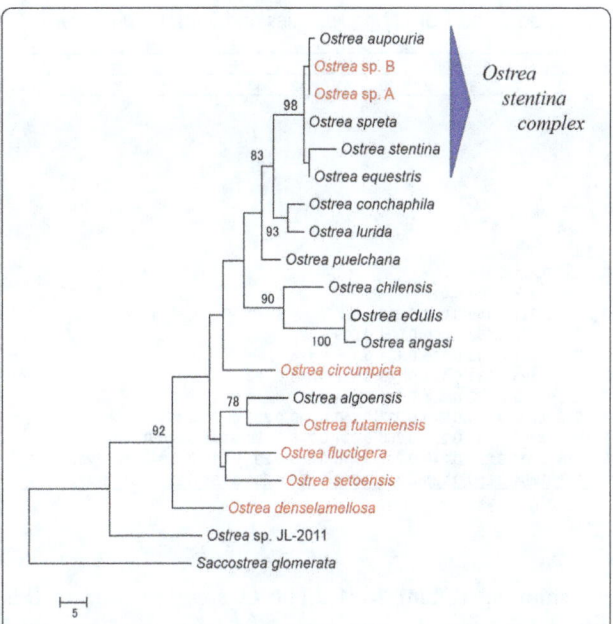

Fig. 5 Maximum parsimony tree of LSrRNA for *Ostrea* sp. rooted with *Saccostrea glomerata*. Bootstrap values above 70% are shown. *Ostrea* species native to Japan are indicated in red. Scale bar indicates five substitutions per 100 site

from the native *Ostrea* species cluster (*O. denselamellosa, O. circumpicta, O. futamiensis, O. fluctigera and O. setoensis*). *Ostrea* sp. A, *Ostrea* sp. B, *O. stentina, O. aupouria, O. equestris* and *O. spreta* were closely related. In contrast, the LSrRNA nucleotide sequences of *O. futamiensis, O. fluctigera* and *O. setoensis* revealed that these species were clearly distinct from all of the other *Ostrea* species.

Estimates of evolutionary divergence between the COI sequences obtained from the *Ostrea* species and *Saccostrea glomerata* as outgroup are shown in Table 2. The overall average evolutionary divergence was 0.205. The evolutionary divergences among *O. futamiensis, O. fluctigera, O. setoensis* and all of the other *Ostrea* species in the COI sequences ranged from 0.176 to 0.263, with clear differences between native Japanese species and known *Ostrea* species from elsewhere in the world. The *Ostrea* sp. A COI sequence was identical to that of *Ostrea* sp. B. The evolutionary divergences among *Ostrea sp.* A, *Ostrea sp.* B, *Ostrea* sp. STH-2012 (accession number JQ027292) and *O. aupouria* in the COI sequences ranged from 0 to 0.004. This indicated that these are the most closely related of the known *Ostrea* species. The WoRMS database gives *O. aupouria* Dinamani, 1981 as a synonym of *O. stentina* Payraudeau, 1826 at present. Consequently, we conclude that *Ostrea* sp. A and B are *O. stentina* Payraudeau, 1826. This is the first record of *O. stentina* from Japanese waters. Although *Ostrea* sp. A and B differed morphologically, the

molecular data clearly showed that these are the same species, viz. *O. stentina*, according to both the LSrRNA and COI genes. We concluded that *Ostrea* sp. A was a juvenile form of *Ostrea stentina*.

The genetic analysis of the native Japanese *Ostrea* species *O. denselamellosa, O. circumpicta, O. futamiensis, O. fluctigera* and *O. setoensis* clearly distinguished them from known *Ostrea* species from around the world. Furthermore, molecular data indicated that *O. futamiensis* is distributed throughout the Seto Inland Sea.

Discussion

We identified two putative unidentified *Ostrea* species, *Ostrea* sp. A and B, during our long-term study of benthic species diversity in the Seto Inland Sea and its adjacent marine areas. Although *Ostrea* sp. A and B differed morphologically, the molecular data identified them as *Ostrea stentina*. This confirmed the difficulty in identifying *Ostrea* species by morphological features alone. The Olympia oyster, *O. lurida*, is a commercially important species on the Northwest Pacific Coast of the United States of America and Canada (Bulseco 2009) and is morphologically very similar to *O. conchaphila*. Harry (1985) proposed that these two species were synonymous because of common species-specific morphological features caused by high phenotypic plasticity. Polson et al. (2009) compared the species using molecular markers and post-hoc morphological characteristics and concluded that *O. lurida* and *O. conchaphila* were separate species. In this manner, DNA markers and molecular biological methods have been used to resolve taxonomic problems caused by species identification of flat oysters based on morphological features alone (O'Foighil et al. 1999; Jozefowicz and O'Foighil 1998; Hurwood et al. 2005; Lapègue et al. 2006; Lazoski et al. 2011; Pejovic et al. 2016).

In recent years, DNA barcoding, a term coined by Hebert et al. (2003), has been used effectively to identify many animal and plant species. Furthermore, this method allows accurate species identification of morphologically similar species. DNA barcoding has previously been used to identify a various oyster species as well as newly invasive alien and cryptic species (Banks and Hedgecock 1993; O'Foighil et al. 1998; Hedgecock et al. 1999; Boundry et al. 2003; Lam and Morton 2003; Lapègue et al. 2004; Chen et al. 2011; Liu et al. 2011; Melo et al. 2010; Hong et al. 2012; Crocetta et al. 2013a,b; Gal-Vao et al. 2013; Hamaguchi et al. 2013; Sekino and Yamashita 2013; Wu et al. 2013; Hamaguchi et al. 2014; Sekino et al. 2014; Xia et al. 2014).

We identified *O. stentina* in Japanese waters via DNA barcoding and propose "Atsuhime-gaki" as its Japanese name. While this is an important discovery, the question of whether or not *O. stentina* is a native or an invasive

Table 2 Estimates of evolutionary divergence between the COI Sequences obtained from *Ostrea* species and *Saccostrea glomerata*

Species	Accession No.	Abbriviation	1	2	3	4	5	6	7	8	9	10	11	12	13	14	15	16	17	18
Ostrea sp. A	LC051584	1																		
Ostrea sp. B	LC051590	2	0.002																	
Ostrea futamiensis	AB898290	3	0.249	0.246																
Ostrea fluctigera	LC149507	4	0.237	0.240	0.218															
Ostrea setoensis	LC149514	5	0.219	0.222	0.213	0.204														
Ostrea circumpicta	AB898294	6	0.252	0.256	0.237	0.262	0.231													
Ostrea denselamellosa	NC015231	7	0.227	0.224	0.212	0.240	0.250	0.246												
Ostrea conchaphila	DQ464125	8	0.162	0.164	0.222	0.219	0.176	0.234	0.222											
Ostrea stentina	DQ226522	9	0.009	0.011	0.234	0.234	0.216	0.237	0.221	0.159										
Ostrea aupouria	AF112288	10	0.004	0.005	0.249	0.234	0.216	0.249	0.227	0.164	0.013									
Ostreola equestris	AY376607	11	0.009	0.011	0.234	0.234	0.216	0.237	0.221	0.159	0.000	0.013								
Ostrea sp. STH-2012	JQ027292	12	0.000	0.002	0.249	0.237	0.219	0.252	0.227	0.162	0.009	0.004	0.009							
Ostrea puelchana	DQ226518	13	0.126	0.128	0.263	0.247	0.228	0.224	0.221	0.143	0.118	0.131	0.118	0.126						
Ostrea lurida	NC022688	14	0.146	0.149	0.228	0.210	0.187	0.234	0.195	0.034	0.143	0.149	0.143	0.146	0.136					
Ostrea edulis	AF120651	15	0.237	0.240	0.219	0.213	0.213	0.218	0.210	0.222	0.234	0.234	0.234	0.237	0.216	0.213				
Ostrea chilensis	AF112286	16	0.240	0.237	0.237	0.256	0.247	0.240	0.200	0.260	0.240	0.237	0.240	0.240	0.231	0.231	0.154			
Ostrea sp. MS-2011	JF915514	17	0.246	0.249	0.240	0.246	0.213	0.227	0.195	0.237	0.234	0.250	0.234	0.246	0.198	0.244	0.204	0.259		
Ostrea angasi	AF540598	18	0.250	0.253	0.219	0.225	0.225	0.218	0.207	0.222	0.240	0.247	0.240	0.250	0.207	0.213	0.020	0.159	0.207	
Saccostrea glomerata	EU007483	19	0.296	0.300	0.279	0.315	0.323	0.337	0.304	0.280	0.286	0.304	0.286	0.296	0.283	0.256	0.327	0.341	0.290	0.330

Red character show the native Ostrea oyster species in Japan

alien species remains unanswered. This species is widely distributed along Atlantic, Mediterranean, North African, New Zealand and South American coasts (Lapègue et al. 2006; Gofas et al. 2011; Crocetta et al. 2013a, b; Pejovic et al. 2016). In several cases, supposedly distinct *Ostrea* species in separate geographical areas have been revised to a single species by molecular analysis. Kenchington et al. (2002) reported that the European flat oyster *O. edulis* and *O. angasi* are conspecific based on their molecular analysis. Using mitochondrial COI sequences, O'Foighil et al. (1999) proved that *O. chilensis* is widely distributed from New Zealand to Chile, and they discussed genetic exchanges within transoceanic ranges that occur as a result of rafting. In DNA databases, the COI nucleotide sequences of an oyster collected from Taiwan (*Ostrea* sp. STH-2012, accession number JQ027292) were identical to those of *O. stentina* in Japan. The Kuroshio Current flows past Taiwan Island to the southern part of Japan. In recent years, as a result of global warming, a northward shift in the distribution patterns of tropical marine benthic species has been observed in Japan. Ibusuki in the Kagoshima Prefecture and Kemi in the Wakayama Prefecture, where the *O. stentina* were collected for this study, are located in the southern part of Japan, where subtropical and tropical oyster species have been observed (Hamaguchi et al. 2014). Thus, *O. stentina* from Taiwan could ride the warm Kuroshio Current to Ibusuki and Kemi either by dispersion of planktonic larvae or rafting (O'Foighil et al. 1999). If this is the case, it is likely that *O. stentina* is a native oyster in Japan. Our preliminary survey, in which *O. stentina* was identified along coasts exposed to the Kuroshio Current, supports this hypothesis.

However, many invasive alien species of marine organism have been introduced in Japan by various human activities (Iwasaki et al. 2004); many of these have been introduced by ballast water, hull fouling and sea chests via shipping (Otani 2004). The *O. stentina* used in this study, for example, were collected from the Kemi tidal flat and Ibusuki. An oil storage facility and a private steel plant are located near these sites, and either oil tankers or iron ore ships may have introduced *O. stentina* to the area from the Arabian Sea or from countries bordering the Indo-Pacific Ocean such as Asia, South America and Oceania. In the near future, we will survey the distribution of *O. stentina* in Japan to determine if the oyster is a native or an invasive alien species. If *O. stentina* is a newly invasive alien species, it will undoubtedly impact Japan's native ecosystems (Ruesink et al. 2005).

Another aim of this study was to develop DNA barcoding for the taxonomically confusing species *O. futamiensis*, *O. fluctigera* and *O. setoensis*. Habe and Itoh (1965), Habe and Kosuge (1967) claimed, based on morphological similarities, that *O. futamiensis* was an ecological variant of the sympatric *O. denselamellosa*. Torigoe and Inaba (1975) compared the electrophoretic patterns of muscle proteins and some morphological features of larvae and of adult shells of three native *Ostrea* species (*O. denselamellosa*, *O. circumpicta* and *O. futamiensis*) and concluded that these were separate species. *O. fluctigera* and *O. setoensis* are small oysters and were re-classified by Torigoe (1983), Inaba (1995) and Inaba and Torigoe (2004). The taxonomic status of both these oyster species is currently unknown.

In this study, we determined the nucleotide sequences of *Ostrea* LSrRNA and COI regions, which are widely used for DNA barcoding. The data confirmed that *O. futamiensis*, *O. fluctigera*, *O. setoensis*, other native *Ostrea* species and the newly found *O. stentina* were distinct from each other. These results strongly support the findings of Torigoe and Inaba (1975) and Inaba and Torigoe (2004). Moreover, the LSrRNA and COI nucleotide sequences both proved that *O. futamiensis*,

O. fluctigera and *O. setoensis* were distinct from the known foreign *Ostrea* species deposited in the DNA database. Habe (1957) reported that *O. setoensis* was a subspecies of *O. sedea*. Iredale, 1939, for which sequence information was not available to us. The molecular data in this study indicated that *O. setoensis* was a separate species. If molecular data for *O. sedea* become available, the taxonomic status of Japan's *O. setoensis* can be confirmed.

We suggest that for the accurate identification of *Ostrea* species with high phenotypic plasticity, both traditional morphological methods and current molecular methods should be used. At present, information on the distribution patterns and ecology of four oysters, *O. stentina*, *O. futamiensis*, *O. fluctigera* and *O. setoensis*, is incomplete. Additionally, planktonic *O. futamiensis* larvae have distinctive morphological features and coloration and are easy to distinguish from other *Ostrea* species (Torigoe and Inaba 1975). We found *O. futamiensis*-like larvae in planktonic samples collected from Matsushima Bay, in the northern part of Japan. Bussarawit and Cedhagen (2010, 2012) reported that they detected *O. futamiensis*-like larvae in samples collected from Phuket, Thailand; however, surprisingly, the adult species could not be found in any of the samples. In fact, these oysters may be widely distributed from Japan to Southeast Asia. Therefore, DNA barcoding by using our new molecular data of small *Ostrea* oyster species could be useful in surveys of these *Ostrea* species inhabiting Korea, China and other Asian countries.

In the near future, we intend to revise the taxonomic status of the Japanese *Ostrea* species using more molecular data than was included in this study, e.g. the nucleotide sequences of complete mitochondrial DNA, multilocus analysis of mitochondrial DNA and nuclear DNA, and rRNA sequence-structure models (Milbury and Gaffney 2003; Wu et al. 2010; Ren et al. 2009, 2010; Danic-Tchaleu et al. 2011; Wu et al. 2012; Salvi et al. 2014).

Conclusions

In addition to clearly establishing that *O. futamiensis*, *O. fluctigera*, *O. setoensis* and *O. stentina* are species distinct from the other native oyster species, we also reported the occurrence of *O. stentina*, a new oyster species to Japanese waters. Furthermore, the nucleotide sequence data obtained in this study, which provides significant information on *O. stentina*, *O. futamiensis*, *O. fluctigera and O. setoensis*, may prove useful for monitoring species diversity in marine fauna. Finally, we offer our results as proof of the need to more fully incorporate the use of DNA barcoding in field studies and monitoring efforts conducted on oyster species.

Acknowledgements
The authors would like to thank Prof. Jennifer Ruesink (Washington State University), Dr. Masakazu Hori (National Research Institute of Fisheries and Environment of Inland Sea) and Mr. Shin-Ichiro Toi (Hiroshima University) for their assistance in collecting and providing samples. The authors also thank Dr. Torigoe for his comments and suggestions about the taxonomy of *Ostrea* oyster species. Our molecular analysis study was supported by "A feasibility study on biodiversity assessment methods in fishing ground environment" from the Fisheries Agency, Japan.

Funding
This study was supported by "A feasibility study on biodiversity assessment methods in fishing ground environment" from the Fisheries Agency, Japan.

Authors' contributions
MH carried out the molecular analysis on all of the specimens and drafted the manuscript. MM discovered and collected the *Ostrea* sp. B specimens in Kagoshima Prefecture. NK, HS and EN carried out the morphological identification of specimens. All authors collected specimens at various collection sites in Japan. All authors have read and approved the final manuscript.

Competing interests
The authors declare that they have no competing interests.

Author details
[1]National Research Institute of Fisheries and Environment of Inland Sea, Fisheries Research Agency, 2-17-5 Maruishi, Hatsukaichi, Hiroshima 739-0452, Japan. [2]Kagoshima Prefectural Fisheries Technology and Development Center, 160-10 Takada-ue, Ibusuki, Kagoshima 891-0315, Japan. [3]Kurashikiminami High School, 330 Yoshioka, Kurashiki, Okayama 710-0842, Japan. [4]Yokohama National University, 79-1 Tokiwadai, Yokohama, Kanagawa 240-8501, Japan.

References
Banks MA, Hedgecock D. Discrimination between closely related Pacific oyster species (*Crassostrea*) via mitochondrial DNA sequences coding for large subunit rRNA. Mol Mar Biol Biotech. 1993;2:129–36.
Botton ML, Shuster Jr CN, Sekiguchi K, Sugita H. Amplexus and mating behavior in the Japanese horseshoe crab, *Tachypleus tridentatus*. Zool Sci. 1996;13:151–9.
Boundry P, Heurtebise S, Lapègue S. Mitochondrial and nuclear DNA sequence variation of presumed *Crassostrea gigas* and *Crassostrea angulata* specimens; a new oyster species in Hong Kong? Aquaculture. 2003;228:15–25.
Bulseco A. A synopsis of the Olympia oyster (*Ostrea lurida*). *Hohonu-A Journal of Academic Writing University of Hawai'i at Hilo*. Aquaculture. 2009;262:63–72.
Bussarawit S, Cedhagen T. The oyster fauna of Thailand. Kyoto: Kyoto University Press; 2010. p. 43.
Bussarawit S, Cedhagen T. Larvae of commercial and other oyster species in Thailand (Andaman Sea and Gulf of Thailand). Steenstrupia. 2012;32:95–162.

Chen J, Li Q, Kong L, Yu H. How DNA barcodes complement taxonomy and explore species diversity: the case study of a poorly understood marine fauna. PLos ONE. 2011;6:e21326. 10.137/journal.pone.0021326.

Crocetta F, Mariottini P, Salvi D, Oliverio M. Does GenBank provide a reliable DNA barcode reference to identify small alien oysters invading the Mediterranean Sea? J Mar Biol Assoc UK. 2013a;2015(95):111–22.

Crocetta F, Bitar G, Zibrowius H, Oliverio M. Biogeographical homogeneity in the eastern Mediterranean Sea. II. Temporal variation in Lebanese bivalve biota. Aquat Biol. 2013b;19:75–84. +1-26 (supplementary files).

Danic-Tchaleu G, Heurtebise S, Morga B, Lape'gue S. Complete mitochondrial DNA sequences of the European flat oyster Ostrea edulis confirms Ostreidae classification. BMC Res Notes. 2011;4:400 (http://www.biomedcentral.com/1756-0500/4/400).

Folmer O, Black M, Hoeh W, Lutz R, Vrijenhoek R. DNA primers for amplification of mitochondrial cytochrome c oxidase subunit I from diverse metazoan invertebrates. Mol Mar Biol Biotech. 1994;3:294–9.

Futahashi R. A revisional study of Japanese dragonflies based on DNA analysis. Tombo. 2011;53:67–74.

Gal-Vao MS, Pereira OM, Hilsdorf AWS. Molecular identification and distribution of mangrove oysters (Crassostrea) in an estuarine ecosystem in Southeast Brazil; implications for aquaculture and fisheries management. Aquac Res. 2013;44:1589–601.

Gofas S, Moreno D, Salas C. Moluscos marinos de Andalucía. Volumen II. Servicio de Publicaciones e Intercambio Científico. Malaga: Universidad de Málaga; 2011 [In Spanish].

Habe T. Description of four new Bivalves from Japan. Venus. 1957;19:117–82.

Habe T, Itoh K. Shells of the world, I. The northern Pacific. Japan: Osaka, Hoikusha; 1965. p. 176 pp. 56 pls [In Japanese].

Habe T, Kosuge S. Common shells of Japan. Japan: Osaka, Hoikusha; 1967. p. 223 pp., 64 pls [In Japanese].

Habe T. Systematic of Mollusca in Japan, bivalvia and Scaphopoda. Tokyo: Hokuryuukan; 1977 [In Japanese].

Hamaguchi M, Shimabukuro H, Kawane M, Hamaguchi T. New record of Kumamoto oyster, Crassostrea sikamea, in Seto Inland Sea. Mar Biodivers Rec. 2013;6:e16. doi:10.1017/S1755267212001297.

Hamaguchi M, Shimabukuro H, Usuki H, Hori M. Occurrences of the Indo–West Pacific rock oyster Saccostrea cucullata in mainland Japan. Mar Biodivers Rec. 2014;7:e84. doi:10.1017/S1755267214000864.

Harry H. Synopsis of the supraspecific classification of living oyster (Bivalvia: Gryphaeidae and Ostreidae). Veliger. 1985;28:121–58.

Hebert PDN, Cywinska A, Ball SI, de-Waard JR. Biological identifications through DNA barcodes. Proc Natl Acad Sci U S A. 2003;270:313–21.

Hedgecock D, Li G, Banks MA, Kain Z. Occurrence of the Kumamoto oyster Crassostrea sikamea in the Ariake Sea, Japan. Mar Biol. 1999;133:65–8.

Henmi Y, Itani G, Iwasaki K, Nishikawa T, Sato M, Sato S, Taru M, Fujita Y, Fukuda H, Kubo H, Kimura T, Kimura S, Maenosono T, Matsubara F, Nagai T, Naruse T, Nishi E, Osawa M, Suzuki T, Wada K, Watanabe T, Yamanishi R, Yamashita H, Yanagi K. The present status and problems of threatened benthic animals in the tidal flats of Japan. Jap J Benthol. 2014;69:1–17 [In Japanese with English abstract].

Hong J-S, Sekino M, Sato S. Molecular species diagnosis confirmed the occurrence of Kumamoto oyster Crassostrea sikamea in Korean waters. Fisheries Sci. 2012;78:259–67.

Hurwood DA, Heasman MP, Mather PB. Geneflow, colonization and demographic history of the flat oyster Ostrea angasi. Mar Freshwater Res. 2005;56:1099–106.

Iijima A. The 7th National Survey on the Natural Environment: Shallow Sea Survey (Tidal Flats), Biodiversity Center of Japan, Nature Conservation Bureau, Ministry of the Environment. 2007 [In Japanese].

Inaba A. On a white small oyster. What is the scientific name? Chiribotan. 1995;26:37–43 [In Japanese].

Inaba A, Torigoe K. Oysters in the world, Part 2: systematic description of the recent oysters. Bull Nishinomiya Shell Mus. 2004;3:1–63 [In Japanese].

Iwasaki K, Kimura T, Kinoshita K, Yamaguchi H, Nishikawa T, Nishi E, Yamanishi R, Hayashi I, Okoshi K, Kosuge J, Suzuki T, Henmi Y, Furota T, Mukai H. Human-mediated introduction and dispersal of marine organisms in Japan: results of a questionnaire survey by the committee for the preservation of the natural environment, the Japanese Association of benthology. Jap J Benthol. 2004;59:22–44 [In Japanese with English abstract].

Jozefowicz CJ, O'Foighil DO. Phylogenetic analysis of southern hemisphere flat oysters based on partial mitochondrial 16S rDNA gene sequences. Mol Phylogenet Evol. 1998;10:426–35.

Kenchington E, Bird CJ, Osborne J, Reith M. Novel repeat elements in the nuclear ribosomal RNA operon of the flat oyster Ostrea edulis C. Linneaus, 1758 and O. angasi Sowerby 1871. J Shellfish Res. 2002;21:697–705.

Kimura M. A simple method for estimating evolutionary rate of base substitutions through comparative studies of nucleotide sequences. J Mol Evol. 1980;16:111–20.

Kuramochi T. Ecology of an oyster, Ostrea fluctigera (Mollusca; Bivalvia), collected from Sagami Bay, central Japan. Sessile Organ. 2007;24:155–7 [In Japanese].

Lam K, Morton B. Mitochondrial DNA and morphological identification of a new species of Crassostrea (Bivalvia: Ostreidae) cultured for centuries in the Pearl River delta, Hong Kong, China. Aquaculture. 2003;228:1–13.

Lam K, Morton B. The oysters of Hong Kong (Bivalvia: Ostreidae and Gryphaeidae). Raffles B Zool. 2004;52:11–28.

Lam K, Morton B. Morphological and mitochondrial-DNA analysis of the indo-west Pacific rock oysters (Ostreidae: Saccostrea species). J Mollus Stud. 2006;72:235–45.

Lapègue S, Batista FM, Heurtebise S, Yu Z, Boudry P. Evidence for the presence of the Portuguese oyster, Crassostrea angulata, in northern China. J Shellfish Res. 2004;23:759–63.

Lapègue S, Salah IB, Batista FM, Heurtebise S, Neifar L, Boudry P. Phylogeographic study of the dwarf oyster, Ostreola stentina, from Morocco, Portugal and Tunisia: evidence of a geographic disjunction with the closely related taxa, Ostrea aupouria and Ostreola equestris. Mar Biol. 2006;150:103–10.

Lazoski C, Gusmao J, Boudry P, SolèCava AM. Phylogeny and phylogeography of Atlantic oyster species: evolutionary history, limited genetic connectivity and isolation by distance. Mar Ecol Prog Ser. 2011;426:197–212.

Liu J, Li Q, Kong L, Yu H, Zheng X. Identifying the true oysters (Bivalvia: Ostreidae) with mitochondrial phylogeny and distance-based DNA barcoding. Mol Ecol Resour. 2011;11:820–30.

Melo AGC, Varela ES, Beasley CR, Schneider H, Sampaio I, Gaffney PM, Reece KS, Tagliaro CH. Molecular identification, phylogeny and geographic distribution of Brazilian mangrove oysters (Crassostrea). Genet Mol Biol. 2010;33:564–72.

Milbury CA, Gaffney PM. Complete mitochondrial DNA sequence of the eastern oyster Crassostrea virginica. Mar Biotech. 2003;7:679–712.

O'Foighil D, Gaffney PM, Wilbur AE, Hilbish TJ. Mitochondrial cytochrome oxidase I gene sequences support and Asian origin for the Portuguese oyster Crassostrea angulata. Mar Biol. 1998;131:497–503.

O'Foighil D, Marshall BA, Hilbish TJ, Pino MA. Trans-Pacific range extension by rafting is inferred for the flat oyster Ostrea chilensis. Biol Bull. 1999;196:122–6.

Okutani T. Marine mollusks in Japan. Tokyo: Tokai University Press; 2000 [In Japanese].

Otani M. Introduced marine organisms in Japanese Coastal waters, and the processes involved in their entry. Jap J Benthol. 2004;59:45–57 [In Japanese with English abstract].

Pejovic I, Ardura A, Miralles L, Arias A, Borrell YJ, Garcia-Vazquez E. DNA barcoding for assessment of exotic molluscs associated with maritime ports in northern Iberia. Mar Biol Res. 2016;12:168–76.

Polson MP, Hewson WE, Eernisse DJ, Baker PK, Zacherl DC. You say Conchaphila, I say Lurida: molecular evidence for restricting the Olympia oyster (Ostrea lurida Carpenter 1864) to temperate Western North America. J Shellfish Res. 2009;28:11–21.

Ren J, Liu X, Zhang G, Liu B, Guo X. 'Tandem duplication-random loss' is not a real feature of oyster mitochondrial genomes. BMC Genomics. 2009;10:84.

Ren J, Liu X, Jiang F, Guo X, Liu B. Unusual conservation of mitochondrial gene order in Crassostrea oysters: evidence for recent speciation in Asia. BMC Evol Biol. 2010;10:394. http://www.biomedcentral.com/1471-2148/10/394.

Ruesink JL, Lenihan HS, Trimble AC, Heiman KW, Fiorenza-Micheli F, Byers JE, Kay MC. Introduction of non-native oysters: Ecosystem effects and restoration implications. Annu Rev Ecol Evol S. 2005;36:643–89.

Salvi D, Macali A, Mariottini P. Molecular phylogenetics and systematics of the bivalve family Ostreidae based on rRNA sequence-structure models and multilocus species tree. PLoS One. 2014;9:e108696.

Schindel DE, Miller SE. DNA barcoding a useful tool for taxonomists. Nature. 2005;435:17.

Seki H. Description of a new species of oyster from Japan. Proc Imperial Acad. 1929;5:477–9.

Seki H. On Ostrea futamiensis Seki. Venus. 1930;2:10–3 [In Japanese].

Sekino M, Yamashita H. Mitochondrial DNA barcoding for Okinawan oysters: a cryptic population of the Portuguese oyster Crassostrea angulata in Japanese waters. Fisheries Sci. 2013;79:61–76.

Sekino M, Ishikawa H, Fujiwara A, Doyola-Solis EFC, Lebata-Ramos MJH, Yamashita H. The first record of a cupped oyster species *Crassostrea dianbaiensis* in the waters of Japan. Fisheries Sci. 2014;81:267–81.

Tack JF, Berghe E, Polk PH. Ecomorphology of *Crassostrea cucullata* (Born, 1778) (*Ostreidae*) in a mangrove creek (Gazi, Kenya). Hydrobiologia. 1992;247:109–17.

Tamura K. Estimation of the number of nucleotide substitutions when there are strong transition-transversion and G + C-content biases. Mol Biol Evol. 1992;9:678–87.

Tamura K, Stecher G, Peterson D, Filipski A, Kumar S. MEGA6: molecular evolutionary genetics analysis version 6.0. Mol Biol Evol. 2013;30:2725–9.

Thompson JD, Higgins DG, Gibson TJ. CLUSTAL W: improving the sensitivity of progressive multiple sequence alignment through sequence weighting, position-specific gap penalties and weight matrix choice. Nucleic Acids Res. 1994;22:4673–80.

Torigoe K. Oysters in Japan. J Sci Hiroshima Univ Ser B Div I. 1981;29:291–481.

Torigoe K. Systematic position of *Ostrea sedea setoenis* Habe, 1957. Venus. 1983;41:29–295.

Torigoe K, Inaba A. A comparison between *Ostrea denselamellosa* and *Ostrea futamiensis* using electrophoresis on muscle proteins. Jap J Malacol. 1975;34:93–8.

Wada K, Nishihira M, Furota T, Nojima T, Yamanishi R, Nishikawa Y, Goshima S, Suzuki T, Kato M, Shimamura T, Fukuda H. Annual report of marine benthos in tidal flat in Japan. WWF Jpn Sci Rep. 1996;3:1–182 [In Japanese].

Wu X, Xu X, Yu Z, Wei Z, Xia J. Comparison of seven *Crassostrea* mitogenomes and phylogenetic analyses. Mol Phylogenet Evol. 2010;57:448–54.

Wu X, Li X, Li L, Xu X, Xia J, Yu Z. New features of Asian *Crassostrea* oyster mitochondrial genome: A novel alloacceptor rRNA gene recruitment and two novel ORFs. Gene. 2012;507:112–8.

Wu X, Xiao S, Yu Z. Mitochondrial DNA and morphological identification of *Crassostrea zhanjiangensis* sp. nov. (Bivalvia: Ostreidae): a new species in Zhanjiang, China. Aquat Living Resour. 2013;26:273–80.

Xia J, Wu X, Xian S, Yu Z. Mitochondrial DNA and morphological identification of a new cupped oyster species *Crassostrea dianbaiensis* (Bivalvia; Ostreidae) in the South China Sea. Aquat Living Resour. 2014;27:41–8.

Yamaguchi K. Shell structure and behaviour related to cementation in oysters. Mar Biol. 1994;118:89–100.

27

A Chinese mitten crab *Eriocheir sinensis* from the River Clyde – the first record of this invasive, non-native species in Scotland

William E. Yeomans[1]* and John Clark[2]

Abstract

The remains (probably a moult) of a female Chinese mitten crab were found in the freshwater River Clyde. This is the first report of a Chinese mitten crab from the wild in Scotland and is well outside the previously recorded range of the species in Great Britain. The origin of this specimen remains enigmatic but this finding has potentially serious implications for the Clyde catchment, the biota of which is recovering from centuries of poor water quality and structural modification. Work is now required to determine whether the Clyde has been colonised by the Chinese mitten crab and, if so, to establish the size of the resident population, its associated threat to local biodiversity and the implications for river management under the European Union Water Framework Directive.

Keywords: Invasive, Non-native, Alien, Chinese mitten crab, Freshwater, Scotland, Great Britain, United Kingdom

Introduction

The Chinese mitten crab, *Eriocheir sinensis* H. Milne Edwards, 1853 is native to east Asia but has spread to such an extent that it is listed among the 100 most invasive and undesirable species in the world (Lowe et al. 2004; de Lafontaine 2005). The United Kingdom Technical Advisory Group for Alien Species, which provides guidance relating to the use of non-native species in Water Framework Directive waterbody classification, has placed *E. sinensis* on the "High Impact List" of aquatic, non-native species. It identified this species as having the potential to impact freshwater, transitional and coastal environments and species. The adult crab breeds in estuaries and the newly-settled juveniles migrate into freshwater where they spend the majority of their life (3–5 years) before migrating downstream again when sexually mature to spawn and subsequently die. The history of the Chinese mitten crab in the British Isles and beyond has been reviewed by Clark et al. (1998); Herborg et al. (2003, 2005, 2007); Gilbey et al. (2008); and Dittel and Epifanio (2009). *Eriocheir sinensis* is the only crab species currently likely to be found in freshwaters in the British Isles; known associated negative economic and ecological impacts elsewhere are largely due to its habit of burrowing into soft banks, its catholic diet, and the disruption caused to fishing and irrigation activities during the spawning migration (Dittel and Epifanio 2009). Pre-invasion data from native ecosystems is often absent, making the ecological impacts of mitten crab invasion poorly understood (Gilbey et al. 2008) although comprehensive risk assessments do exist (e.g. Buoma and Soes 2010).

Materials and methods

Data from a number of recording initiatives have been pooled by others to determine the national picture of Chinese mitten crab distribution in Great Britain. Until now, there have been no records of *E. sinensis* from Great Britain north of the River Tyne on the east coast, and the Duddon Estuary in Cumbria on the west coast (http://mittencrabs.org.uk/distribution) (Fig. 1).

Results

The remains (probably a moult) of a female Chinese mitten crab were recovered by Mr John Clark from the "Skurlie Pool" on the River Clyde in western Scotland at United Kingdom National Grid Reference NS 63042 62514 [Lat 55° 50′ 10.122″; Long −4° 11′ 18.118″] on

* Correspondence: william.yeomans@glasgow.ac.uk
[1]Clyde River Foundation, Graham Kerr Building, University of Glasgow, Glasgow G12 8QQ, Scotland, UK
Full list of author information is available at the end of the article

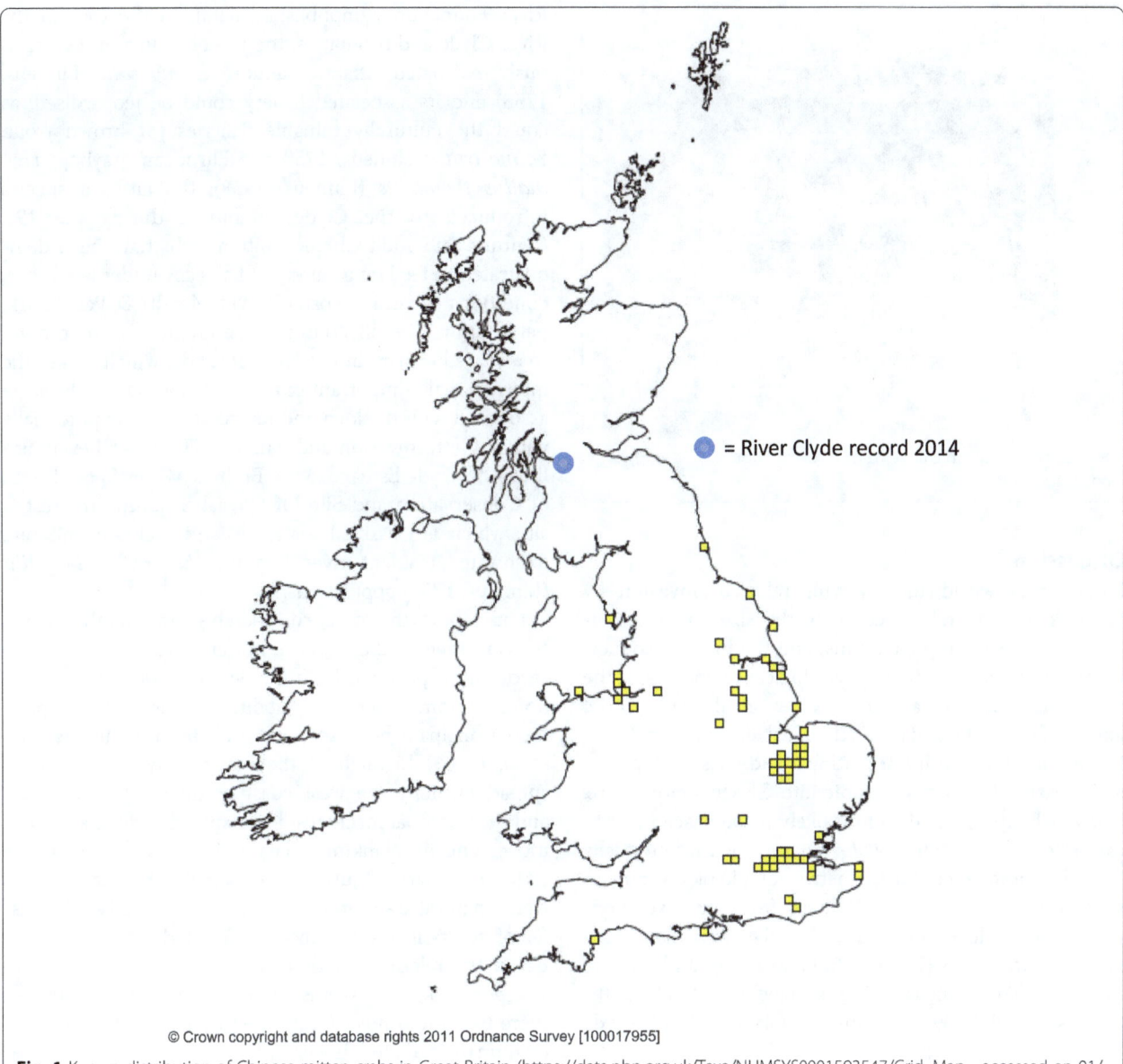

© Crown copyright and database rights 2011 Ordnance Survey [100017955]

Fig. 1 Known distribution of Chinese mitten crabs in Great Britain (https://data.nbn.org.uk/Taxa/NHMSYS0001593547/Grid_Map - accessed on 01/07/2014)

23 June 2014 (Fig. 1). The river forms the boundary between the City of Glasgow and the County of South Lanarkshire at this point, approximately 4 km above the limit of salt water penetration (Dalmarnock Roadbridge). The affected reach is still influenced by freshwater backing up during high Spring tides. The remains were discovered in a complete state but disaggregated when removed from the water. It was identified as a female because the setal mats on the chela did not completely circumvent the propodus, as is the case with male crabs (Paul Clark, pers. comm.). Many of the hard parts were recovered, although the carapace and underside fractured into several pieces, making it

impossible to measure the carapace width to give an accurate indication of the size of the crab. The remains (largely claws, legs and fragments of carapace; Fig. 2) were passed to the Clyde River Foundation for confirmation of identification on 25 June 2014 and a second opinion was received on photographic evidence from Paul Clark (Natural History Museum, London) on 26 June 2014 (Fig. 3). The remains were initially preserved in 80% Industrial Methylated Spirit solution by the Clyde River Foundation, and subsequently 80% ethanol when lodged with The Hunterian (Zoology Museum) at the University of Glasgow (Entry Number 1424) on 3 July 2014.

Fig. 2 Chinese mitten crab remains recovered from the River Clyde on 22 June 2014

Discussion

Large areas worldwide are vulnerable to invasion by Chinese mitten crabs because of the size and distribution of the existing populations, and the high reproductive rate and wide range of physiological tolerances of the species (Herborg et al. 2007; Gilbey et al. 2008; Dittel and Epifanio 2009). The Firth of Clyde and the Clyde Estuary are busy with international and coastal shipping, and pleasure boating is possible into Glasgow city centre. The Clyde system is therefore likely to be susceptible to extensive colonisation by *E. sinensis*, with potentially serious consequences for the native aquatic biodiversity in watercourses already under threat from invasive, non-native species like North American signal crayfish, *Pacifastacus leniusculus* (Dana, 1852) and European bullhead, *Cottus gobio* Linneus, 1758 (Crawford et al. 2006; Gladman et al. 2009; Yeomans and Gladman 2011; Clyde

Fig. 3 Claw of female Chinese mitten crab recovered from the River Clyde on 22 June 2014

River Foundation, unpublished data). In the case of the River Clyde and tributaries, the recent return of the previously extirpated Atlantic salmon (*Salmo salar* Linneus, 1758) and its associated fishery could be jeopardised, as could the culturally valuable fisheries for brown trout, *Salmo trutta* Linneus, 1758 and European grayling, *Thymallus thymallus* (Linneus, 1758); the latter a species introduced to the Clyde catchment during the 19th Century. Sub-adult Chinese mitten crabs have been demonstrated to feed on a variety of fish eggs under laboratory conditions (Jessica Webster, David Morritt & Paul Clark, pers. comm.). Locally, the River Leven links Loch Lomond to a shared estuary in the Firth of Clyde, which makes the internationally-important biodiversity and natural heritage associated with the loch and its tributaries also potentially vulnerable to invasion and damage. The River Leven also links the Clyde Estuary to the Endrick Water Special Area of Conservation and Site Of Special Scientific Interest, a site which is classified for its Atlantic salmon and rare freshwater resident river lamprey, *Lampetra fluviatilis* (Linneus, 1758) populations.

This report, therefore, considerably expands the known British range of the Chinese mitten crab and is the first record of a specimen from the wild in Scotland. There is, however, some uncertainty about the origin of the specimen, principally because it was probably a moult. Herborg et al. (2005) highlighted the inevitability of *E. sinensis* spreading along the west coast of the United Kingdom and the potential mechanisms for spread include advective movement of planktonic larvae in coastal waters; the transfer of larvae, juveniles or adults in ships' ballast water; natural transfer and/or the human-mediated transfer of adults in the live food trade (Herborg et al. 2007). Given the relative isolation of the Clyde from the other UK populations, it seems unlikely that the crab arrived there by natural means, rather either from hull fouling or by human-mediated movement.

The precise nature of the introduction is likely to remain enigmatic but inward transfer in ballast water cannot be ruled out. It is an offence under Section 14 of the Wildlife and Countryside Act 1981 (as amended) in Scotland "*for any person to release, allow to escape from captivity or otherwise cause any animal outwith the control of any person to be at a place outwith its native range*"; which makes it illegal to release Chinese mitten crabs to the wild. It is not, however, currently illegal to sell live Chinese mitten crabs (http://www.scotland.gov.uk/Topics/Environment/Wildlife-Habitats/InvasiveSpecies/legislation). Live mitten crabs have been sold in Glasgow in recent times (Colin Bean, pers. comm.).

The chance that the crab was deliberately released to the river cannot be discounted, and there is evidence from elsewhere that such introductions have taken place in order to initiate commercial fisheries or even as part of religious rituals (Shiu and Stokes 2008). The Clyde mitten crab was also found in the vicinity of an illegal eel trap, which leads to the consideration that the crab could either have been used to bait the trap or that it was caught in the trap and discarded by the trapper back into the river. In the former case, it is likely that alternative baits (e.g. fish) would be more commonly used; if the latter is true, then such traps may provide a mechanism for future monitoring of the extent and density of the population. Furthermore, the possibility that the remains were planted as a hoax has been considered and rejected because the area where they were found is relatively inaccessible and the prospect of detection before complete decomposition was considered small. The River Clyde also has one of the most urban Atlantic salmon fisheries in the world, and such a hoax is more likely to have been carried out in a more accessible, busier location.

It is important however, in the absence of further live specimens, not to place too much emphasis on this single occurrence. Rudnick et al. (2003) document cases from areas in North America where *"a few"* Chinese mitten crabs have been collected but where self-sustaining populations were not established. The River Clyde is relatively large at the sampling point and the probability of a solitary individual being found is low, which suggests an established population but one which has escaped detection thus far. Low densities of crabs at the upstream edges of populations make detection difficult (Rudnick et al. 2003) and an assessment of the claw size (Fig. 3) suggests that this specimen was an immature animal. Given that mature crabs migrate downstream in the autumn and winter and their immature conspecifics migrate upstream in spring (Herborg et al. 2005), it appears likely that this individual was either resident or migrating upstream.

The value of a monitoring scheme for *E. sinensis* has been well demonstrated on the River Thames in England (Clark et al. 1998). Herborg et al. (2005) also emphasised the importance of monitoring around rare records such as this because of potentially rapid future population increase under more favourable conditions and/or the exceedance of a critical population density.

This may be a rare opportunity to document the dynamics of a Chinese mitten crab invasion in Great Britain from a relatively small population and this is not purely of academic interest. The site from where the carcass was recovered is within the "River Clyde (North Calder to Tidal Weir)" waterbody as described for the European Union Water Framework Directive legislation (http://ec.europa.eu/environment/water/water-framework/index_en.html). For water management purposes, this waterbody is currently classified under the Water Framework Directive as being of "Poor Ecological Potential with overall ecological status of Poor". The overall status comprises many tiers of classification data and the component for "Alien Species" in the waterbody is "High" currently (i.e. free of the non-native species which are considered to negatively impact this parameter). The presence of a self-sustaining population of Chinese mitten crabs would also reduce that component of the overall classification. For that reason, this record will be followed by an intensive site-specific search for live mitten crabs. If that proves positive, a long-term monitoring programme for *E. sinensis* in the River Clyde system should be implemented, aimed specifically to describe the size and distribution of the population and to quantify its interaction with native biological communities. When the extent of the Clyde population is known, options for its containment and control can be considered. For example, the issues and practicalities surrounding commercial exploitation as a means to deplete an English *E. sinensis* population have already been examined (Clark et al. 2008; Clark 2011).

Acknowledgements
David Bailey, Colin Bean, Paul Clark and Phil Rainbow provided comments on an early manuscript. Katriona Lundberg clarified some aspects of the Water Framework Directive status of the Chinese mitten crab in the Clyde catchment.

Funding
The work was funded privately from Clyde River Foundation resources.

Authors' contributions
JC discovered the remains of the specimen and was involved in the drafting of the manuscript. WY wrote the manuscript. Both authors read and approved the final manuscript.

Competing interests
The authors declare that they have no competing interests.

Author details
[1]Clyde River Foundation, Graham Kerr Building, University of Glasgow, Glasgow G12 8QQ, Scotland, UK. [2]11 Kildary Avenue, Glasgow G44 3AX, Scotland, UK.

References
Buoma S, Soes DM. A risk assessment of the Chinese mitten crab in The Netherlands. Bureau Wartenberg bv for Ministry of Agriculture, Nature and Food Quality, Wageningen, Project 09–772. 2010. p. 51.

Clark PF. The commercial exploitation of the Chinese mitten crab *Eriocheir sinensis* in the River Thames, London: damned if we don't and damned if we do. In: Galil BS, Clark PF, Carlton JT, editors. In the wrong place – alien marine crustaceans: distribution, biology and impacts. Invading nature – Springer series in invasion ecology, vol. 6. London: Springer Science and Business Media; 2011. p. 537–80.

Clark PF, Rainbow PS, Robbins RS, Smith B, Yeomans WE, Thomas M, Dobson G. The alien Chinese mitten crab, *Eriocheir sinensis* (Crustacea: Decapoda: Brachyura), in the Thames Catchment. J Mar Biol Assoc UK. 1998;78:1215–21.

Clark PF, Campbell P, Smith B, Rainbow PS, Pearce D, Miguez RP. The commercial exploitation of Thames mitten crabs: a feasibility study. A report for The Department for Environment, Food and Rural Affairs by the Department of Zoology, The Natural History Museum, London. DEFRA Reference FGE 274. Pp1-81 + Appendices 1–6. 2008.

Crawford L, Yeomans WE, Adams CE. The impact of introduced signal crayfish *Pacifastacus leniusculus* on stream invertebrate communities. Aquat Conserv. 2006;16:611–21.

De Lafontaine Y. First record of the Chinese Mitten Crab (Eriocheir sinensis) in the St. Lawrence River, Canada. J Great Lakes Res. 2005;31:367-70.

Dittel AI, Epifanio CE. Invasion biology of the Chinese mitten crab *Eriocheir sinensis*: A brief review. J Exp Mar Biol Ecol. 2009;374:79–92.

Gilbey V, Attrill MJ, Coleman RA. Juvenile Chinese mitten crabs (*Eriocheir sinensis*) in the Thames estuary: distribution, movement and possible interactions with the native crab *Carcinus maenas*. Biol Invasions. 2008;10:67–77.

Gladman Z, Adams C, Bean C, Sinclair C, Yeomans W. Signal crayfish in Scotland. In: Brickland J, Holdich DM, Inhoff E, editors. Crayfish Conservation in the British Isles. Proceedings of a Conference held on 25th March 2009 in Leeds, UK. 2009. p. 43–8.

Herborg L-M, Rushton SP, Clare AS, Bentley MG. Spread of the Chinese mitten crab (*Eriocheir sinensis* H. Milne Edwards) in Continental Europe: analysis of a historical data set. Hydrobiologia. 2003;503:21–8.

Herborg L-M, Rushton SP, Clare AS, Bentley MG. The invasion of the Chinese mitten crab (*Eriocheir sinensis*) in the United Kingdom and its comparison to continental Europe. Biol Invasions. 2005;7:959–68.

Herborg L-M, Rudnick DA, Siliang Y, Lodge DM, MacIsaac HJ. Predicting the range of Chinese mitten crabs in Europe. Conserv Biol. 2007;21:1316–23.

Lowe S, Browne M, Boudjelass S, De Poorter M. 100 of the World's Worst Invasive Species Database. Published by The Invasive Species Specialist Group (ISSG), a specialist group of the Species Survival Commission (SSC) of the World Conservation Union (IUCN); 2004. p. 12.

Rudnick DA, Hieb K, Grimmer KF, Resh VH. Patterns and processes of biological invasion: the Chinese mitten crab in San Francisco Bay. Auckland: Basic Appl Ecol. 2003;4:249–62.

Shiu H, Stokes L. Buddhist animal release practices: historic, environmental, public health and economic concerns. Contemporary Buddhism. 2008;9:181–96.

Yeomans WE, Gladman Z. North American signal crayfish, *Pacifastacus leniusculus* (Dana) in the River Kelvin, Glasgow. The Glasgow Naturalist. 2011;25:104–5.

28

First record of *Heptranchias perlo* (Bonnaterre 1788) in Guatemala's Caribbean Sea

Ana Hacohen-Domené[1]* (iD), Francisco Polanco-Vásquez[1] and Rachel T. Graham[2]

Abstract

Background: This report represents the first record of the sharpnose sevengill shark *Heptranchias perlo* in Guatemala's Caribbean Sea.

Methods: Two *H. perlo* specimens were captured by artisanal fishermen of the coastal community, El Quetzalito. All specimens were captured with a trammel net, in waters of 200 m depth

Results: Both specimens were female with total lengths of 280 and 370 mm. Details regarding the identification and measurement of both specimens are presented.

Conclusion: These specimens represent the first record of both species in Guatemalan waters. Also, this report further increases the species' range of distribution in the Caribbean and Central America.

Keywords: First record, Deep-water sharks, Hexanchidae, Caribbean

Background

The family Hexanchidae includes three genera and four described species: *Hexanchus* Rafinesque 1810, *Heptranchias* Rafinesque 1810, and *Notorynchus* Ayres 1855 (Ebert and Stehmann 2013). Hexanchidae sharks have a worldwide distribution in cold temperate to tropical seas. Most species in the family are deepwater inhabitants of the outer continental shelves, upper continental slopes, insular shelves and slopes, and submarine canyons down to at least 2500 m depth, occurring in both benthic and neritic (Carpenter 2002; Ebert and Stehmann 2013).

The sharpnose sevengill shark, *Heptranchias perlo* (Bonnattere1788), is uncommon through its range and many aspects of its biology are poorly known. *H. perlo* has been known to occur primarily in deep waters down to 1000 m (Compagno et al. 2005; Ebert et al. 2013).

The sharpnose sevengill shark occurrence in the Western Atlantic has been reported in Mexico, Jamaica, Bahamas (USA), Cuba and Panama (Bonfil 1997; Claro and Parenti 2001; Paul and Fowler 2003; McLaughlin and Morrissey

2004; Kyne et al. 2012; Benavides et al. 2014). The importance of this study lies in the fact that it represents the first confirmed record of occurrence of *H. perlo* in Guatemala's Caribbean Sea. Currently, *H. perlo* is listed as "Near Threatened" by the International Union for the Conservation of Nature's Red List (Paul and Fowler 2003).

Methods

On the 20th March 2016, two female sevengill sharks were captured by artisanal fishermen in Guatemala's Caribbean Sea and landed in the coastal village of El Quetzalito, Izabal (15° 49.776 N, -88° 12.340 W) (Fig. 1). All specimens were captured with a 1000 m long bottom trammel net of 3.5 inches mesh size and one panel. Specimens were captured at approximately 200 m depth, based on known length of net deployed. All specimens were examined and identified to species level using identification guides (Compagno 1984; Compagno et al. 2001).

Both specimens were preserved in formaldehyde (10%) and subsequently transferred to ethyl alcohol (70%) for final preservation. Both sevengill shark specimens were deposited in the Laboratory of Biological Science and Oceanography, Centro de Estudios del Mar y Acuicultura (CEMA) of the Universidad San Carlos de Guatemala

* Correspondence: ahacohen@fundacionmundoazul.com
[1] Fundación Mundo Azul, Blvd. Rafael Landívar 10-05 Paseo Cayalá Zona 16, Edificio D1 Oficina 212, Guatemala City, Guatemala
Full list of author information is available at the end of the article

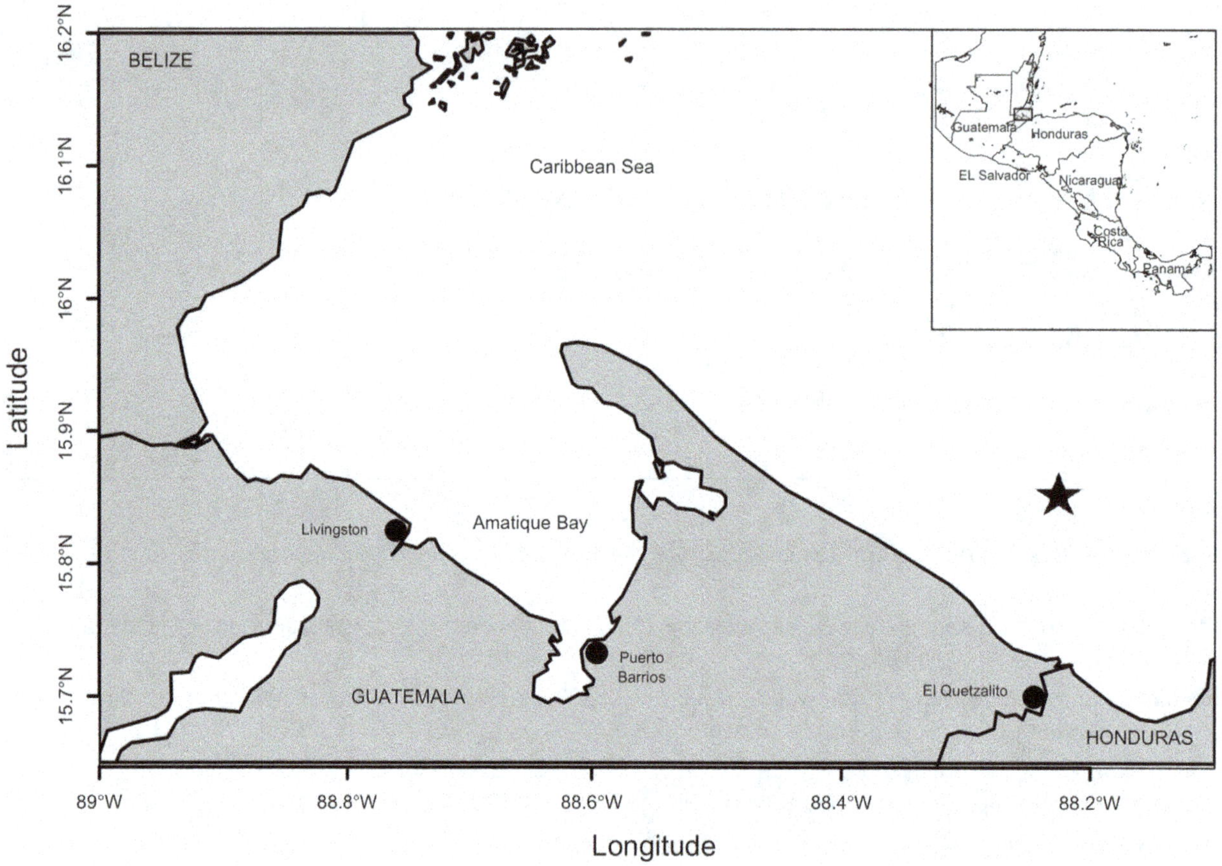

Fig. 1 Study area with location (★) of capture in relation to the fishing village of El Quetzalito

(USAC). The specimens are part of the collection registered to the Consejo Nacional de Áreas Protegidas (CONAP) under the reference number (Rf) 250 and 251. Measurements were taken on the sevengill shark specimens using a vernier calipers or measuring tape, as proposed by Compagno (2001). A total of 80 morphometric measurements were taken (Compagno 2001).

Results
Systematic account
   Family: Hexanchidae
   Genus: Heptranchias Rafinesque, 1810
   Heptranchias perlo (Bonnaterre, 1788)
   Common name: Sharpnose sevengill shark, cañabota (local name)

Fig. 2 *Heptranchias perlo*: **a**. Rf250, female, 370 mm TL and (**b**). Rf251, female, 280 mm TL

Table 1 Morphometric measurements (mm) of individuals of two female specimens of *Heptranchias perlo*

Measurements	Rf250	Rf251
Total length	370	280
Fork length	280	210
Precaudal fin length	260	190
Pre-first dorsal-fin length	179.6	136.2
Head length	80.5	62.9
Prebranchial length	62.6	51.6
Prespiracular length	55.4	44.1
Preorbital length	21.9	18.5
Prepectoral-fin length	73.3	59.1
Prepelvic-fin length	157.1	166.7
Snout-vent length	171.1	123
Preanal-fin length	199.7	143.1
Dorsal caudal-fin space	45.9	37
Pectoral-fin pelvic-fin space	60.3	35
Pelvic-fin anal-fin space	25.2	16
Anal-fin caudal-fin space	34.4	25.3
Pelvic-fin caudal-fin space	72.9	56.2
Vent caudal-fin length	149.5	127.5
Prenarial length	10.8	6.5
Preoral length	19.4	16.5
Eye length	20.2	14.7
Eye height	7	4.1
Intergill length	22.3	10.3
First gill slit height	25.5	16.6
Second gill slit height	21.6	13.1
Third gill slit height	18.8	10.8
Fourth gill slit height	16.9	9.1
Fifth gill slit height	15.4	8.9
Sixth gill slit height	14.4	7.3
Seventh gill slit height	13	6.4
First dorsal-fin length	30.5	21.1
First dorsal-fin anterior margin	29.1	22.9
First dorsal-fin base	23.5	16.3
First dorsal-fin height	16.7	10.8
First dorsal-fin inner margin	7	5.4
First dorsal-fin posterior margin	15.2	12.8
Pectoral-fin anterior margin	44.6	32.2
Pectoral-fin base	38.2	26.2
Pectoral-fin inner margin	22.2	12.6
Pectoral-fin posterior margin	32.8	22.4
Pectoral-fin height	35.9	24.3
Dorsal caudal-fin margin	112.8	92.5
Preventral caudal-fin margin	28.6	21.8

Table 1 Morphometric measurements (mm) of individuals of two female specimens of *Heptranchias perlo (Continued)*

Upper postventral caudal-fin margin	57.1	42.9
Lower postventral caudal-fin margin	16.8	8.2
Caudal-fin fork width	20.3	19.1
Caudal-fin fork length	27.5	24.1
Subterminal caudal-fin margin	16.5	13
Subterminal caudal-fin width	8.2	4.2
Terminal caudal-fin margin	12.4	11
Terminal caudal-fin lobe	17.5	12.3
Pelvic-fin length	34.9	27.1
Pelvic-fin anterior margin	15.4	10.6
Pelvic-fin base	20.6	14.5
Pelvic-fin height	9.8	9.5
Pelvic-fin inner margin [length]	12.3	10.5
Pelvic-fin posterior margin [length]	24.6	16.9
Anal-fin length	26.7	18.4
Anal-fin anterior margin	15.4	10.7
Anal-fin base	20.2	13.6
Anal-fin height	9.9	7.6
Anal-fin inner margin	6.3	4.7
Anal-fin posterior margin	15.4	12.6
Head height	28.7	20.5
Trunk height	20.7	12.9
Abdomen height	20.8	11.8
Tail height	17.8	10.8
Caudal-fin peduncle height	14.2	9.3
Mouth length	36.9	28.4
Mouth width	22.7	17.4
Nostril width	5.7	3.5
Internarial space	11	6.4
Interorbital space	20	14.9
Spiracle length	1.9	0.9
Eye spiracle space	12.5	9.4
Head width	29.7	19.7
Trunk width	24.9	18.1
Abdomen width	9.6	8.3
Tail width	14.5	10.7
Caudal-fin peduncle width	6.1	4.8

Material examined

Rf 250: female, 370 mm TL (Fig. 2a, Table 1); Rf 251: female, 280 mm TL (Fig. 2b, Table 1).

Description

Small shark with seven pairs of gill openings. Slender body. Head extremely narrow and pointed, with large eyes and narrow mouth. Dorsal coloration brownish grey, with lighter coloration below. One spineless dorsal fin with black apex.

Discussion

This report represents the first record of *H. perlo*, in Guatemala's Caribbean Sea. In the Caribbean of Central America, *H. perlo* has only been reported for Panama (Benavides et al. 2014). During a deep water fishery survey using bottom trawling along the Caribbean coast of Central America, three *H. perlo* individuals were captured in Panama (Benavides et al. 2014); one female (670 mm TL) and two male (750–820 mm TL), sizes greater than the specimens reported in this study. Additionally, the capture depth of *H. perlo*, approx. 200 m (this study), coincides with the range describe by the species (Ebert and Stehmann 2013; Benavides et al. 2014). Knowledge of *H. perlo* biology is limited. A reproductive study conducted in the southwestern waters of Kyushu, Japan report *H. perlo* maturity of female is reached between 950 mm and 1050 mm (Tanaka and Mizue 1977). Moreover, birth size of *H. perlo* is 260–270 mm (Tanaka and Mizue 1977), close to that obtained in this study for Rf 251 specimen (280 mm TL).

Globally, this species is of minor commercial importance and occurs mostly as bycatch in bottom trawl and longline fisheries which may have caused population declines where deepwater fisheries have been practiced in the last decade. According to artisanal fishers in the region, *H. perlo* is rarely captured by the area's local fishermen, and is only captured incidentally when fishing with trammel nets. When captured, fishers rarely utilize the meat due to the species' small size although they render the liver for shark oil.

Conclusion

To date, reports regarding shark diversity of Guatemala´s Caribbean Sea are scare (Thorson et al. 1966; Hacohen-Domené et al. 2016; Polanco-Vásquez et al. 2017). This study forms the first confirmed records of *H. perlo* in Guatemala's Caribbean Sea and increase the number of known shark species in Guatemala. This report further increases the species' range of distribution throughout the Caribbean and Central America. This study highlights the need for comprehensive deep-sea research surveys to obtain a more complete assessment of the region's deep-water elasmobranchs.

Abbreviations
Cm: Centimeter; M: Meter; Mm: Millimeter; TL: Total length

Acknowledgements
Fundación Mundo Azul would like to thank the community of El Quetzalito for their constant support to the research program in the area. Also, we would like to thank the Centro de Estudios del Mar y Acuicultura (CEMA) of the Universidad San Carlos de Guatemala (USAC) for providing the space to take the morphometric measurements.

Funding
Fundación Mundo Azul provided funding for the elasmobranch monitoring project in 2016. The Whitley Fund for Nature provided funding for RTG's time.

Authors' contributions
AH and FP participated in the identification of the species, recorded the morphometric data of all specimens, and contributed to draft the manuscript. RTG contributed to draft the manuscript. All authors read and approved the final manuscript.

Competing interest
The authors declare that they have no competing interests.

Author details
[1]Fundación Mundo Azul, Blvd. Rafael Landívar 10-05 Paseo Cayalá Zona 16, Edificio D1 Oficina 212, Guatemala City, Guatemala. [2]MarAlliance, 32 Coconut Drive, PO Box 283, San Pedro, Belize.

References
Benavides R, Brenes CL, Márquez A. Análisis de la población de condrictios (Vertebrata: Chondrichthyes) de aguas demersales y profundas del Caribe centroamericano, a partir de faenas de prospección pesquera con redes de arrastre. Rev Cienc Mar Cost. 2014;6:9–27.

Bonfil R. Status of shark resources in the Southern Gulf of Mexico and Caribbean: implications for management. Fish Res. 1997;29(2):101–17.

Carpenter KE. The living marine resources of the Western Central Atlantic (Vol. 3). Rome: FAO; 2002.

Claro R, Parenti LR. The marine ichthyofauna of Cuba. Ecology of the marine fishes of Cuba, 21-57. In: Claro R, Lindeman KC, Parenti LR, editors. Ecology of marine fishes of Cuba. Washington and London: Smithsonian Institution Press; 2001.

Compagno LJV. FAO species catalogue. Vol. 4. Sharks of the world. An annotated and illustrated catalogue of shark species known to date. Part 1. Hexanchiforms to Lamniformes. Rome, Italy: FAO Fisheries Synopsis; 1984;4:1–249.

Compagno LJV. Sharks of the world. An annotated and illustrated catalogue of shark species known to date. Vol. 2. Bullhead, mackerel and carpet sharks (Heterodontiformes, Lamniformes and Orectolobiformes). FAO Species Catalogue for Fishery Purposes no. 1. FAO: Rome, Italy: 2001;2:1–269.

Compagno LJV, Dando M, Fowler S. Sharks of the world. New Jersey: Princeton University Press; 2005.

Ebert DA, Stehmann MFW. FAO Species Catalogue for Fishery Purposes No. 7. Sharks, Batoids and Chimaeras of the North Atlantic. Rome: Food and Agriculture Organization of the United Nations; 2013.

Ebert DA, Fowler S, Compagno L. Sharks of the world: a fully illustrated guide. Plymouth: Wild Nature Press; 2013.

Hacohen-Domené A, Polanco-Vásquez F, Graham RT. First report of the whitesaddled catshark *Scyliorhinus hesperius* (Springer 1966) in Guatemala's Caribbean Sea. Mar Biodivers Rec. 2016;9(1):101.

Kyne PM, Carlson JK, Ebert DA, Fordham SV, Bizzarro JJ, Graham RT, Kulka DW, Tewes EE, Harrison LR, Dulvy NK. The conservation status of North American, Central American, and Caribbean Chondrichthyans, Technical report. Vancouver: IUCN Species Survival Commission Shark Specialist Group; 2012.

McLaughlin DM, Morrissey JF. New records of elasmobranchs from the Cayman Trench. Jamaica Bull Mar Sci. 2004;75:481–5.

Paul L, Fowler S. (SSG Australia & Oceania Regional Workshop, March 2003). *Heptranchias perlo*. The IUCN Red List of Threatened Species 2003: e. T41823A10572878. http://dx.doi.org/10.2305/IUCN.UK.2003.RLTS. T41823A10572878.en. Accessed 5 Jan 2017.

Polanco-Vásquez F, Hacohen-Domené A, López T, Pacay A, Graham RT. First record of the chimaera *Neoharriota carri* (Bullis and Carpenter 1966) in the Caribbean of Guatemala. Mar Biodivers Rec. 2017;10:1.

Tanaka S, Mizue K. Studies on sharks. 11. Reproduction in female *Heptranchias perlo*. Bull Fac Fish Nagasaki Univ. 1977;42:1–9.

Thorson TB, Cowan CM, Watson DE. Sharks and sawfish in the lake Izabal-Rio Dulce system. Guatemala Copeia. 1966;3:620–2.

29

First recorded occurrence of *Cheirocratus robustus* Sars, 1894 in the British Isles

Alan A. Myers[1*], David McGrath[2] and Will Musk[3]

Abstract

Background: Collections of the amphipod genus *Cheirocratus* from the North Sea and Ireland proved to include *C. robustus* Sars a species previously known only from Norway and Sweden.

Results: Material of *C.robustus* is described and figured from the Humber and Ireland together with the closely related species *C. sundevalli* (Rathke). A key to males of the *Cheirocratus* species of the North East Atlantic and Mediterranean is provided.

Conclusions: *C. robustus* is shown to be widespread in the eastern North Atlantic where it was previously overlooked.

Keywords: Amphipoda, *Cheirocratus robustus*, British Isles, New record

Background

Collections of *Cheirocratus* from the Humber region of the North Sea and from several localities on the West Coast of Ireland, proved to include specimens of *C. robustus* Sars, a species previously recorded only from Norway and Sweden and probably overlooked elsewhere.

Methods

Specimens were preserved in 70% ethanol. Dissection was made under a Wild stereomicroscope and body parts were mounted on microscope slides in glycerine for drawing with a drawing tube on a Nikon compound microscope. In the diagnoses, character states that distinguish *C. robustus* from *C. sundevallii* are listed in bold.

Material is deposited in the National Museum of Ireland, Natural History. (NMINH) and Goteborgs Naturhistorika Museum (GNM) Sweden.

Results

Systematics

Order Amphipoda Latreille, (Latreille 1816)
Suborder Senticaudata Lowry & Myers, (Lowry & Myers 2013)

Infraorder Hadziida S. Karaman, (Karaman 1932)
Superfamily Calliopioidea Sars, (Sars 1893)
Family Cheirocratidae d´Udekem d'Acoz, (D'Udekem d'Acoz 2010)
Cheirocratus robustus Sars.
(Figs. 1, 2 and 3)
Cheirocratus robustus Sars, (Sars 1894): 526, pl. 185, fig. 2.—Oldevig, (Oldevig 1932): 186, pl.2, fig. 2.

Material examined

Three males, six females (NMINH 2016.16.1), RSMP H 0205 Baseline (53.431843°N, 0.38073°E), Humber region of North Sea, 10 m depth, gravel, 23.09.2014, IECS (collected by MESL) one male, one female (NMINH 2016.16.2) RSMP H 0293 Baseline (53.414395°N, 0.52727°E), Humber region of North Sea, 12 m depth, gravel, 23.09.2014, IECS (collected by MESL); one female (NMINH 2016.16.3), RSMP H 0211 Baseline (53.437086°N, 0.398443°E), Humber region of North Sea, 11 m depth, gravel, 23.09.2014, IECS (collected by MESL); one male (NMINH 2016.16.4), Marine Harvest salmon farm, Inishdoonver, Clew Bay, Co Galway, Ireland, 21.5 m depth, current 17 cm/.sec, 50 m from edge of salmon cage, 07.08.2013; one male (NMINH 2016.16.5), JN1067, Rutland Island, 01.09.2010, RUG38; three males, eight females, two immature (NMINH 2016.16.6), JN1006, Kilkieran, 14.10.2010, KKG 17; two males, three females (NMINH 2016.16.7), JN1066, Valentia 16.9.2010, VAG 14; three males, (NMINH 2016.16.8) Hum Agg, 2014, sample

* Correspondence: bavayia@gmail.com
[1]School of Biological, Earth and Environmental Sciences, University College Cork, Cork Enterprise Centre, Distillery Fields, North Mall, Cork, Ireland
Full list of author information is available at the end of the article

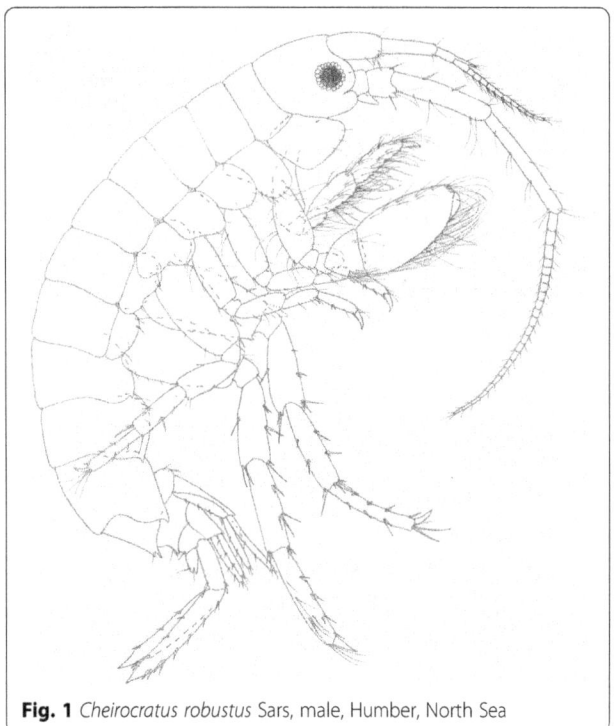

Fig. 1 *Cheirocratus robustus* Sars, male, Humber, North Sea

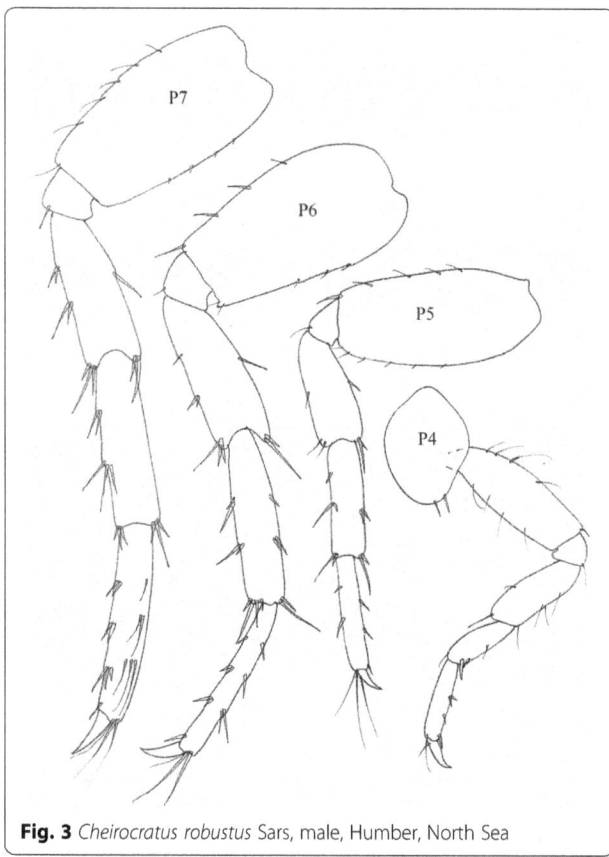

Fig. 3 *Cheirocratus robustus* Sars, male, Humber, North Sea

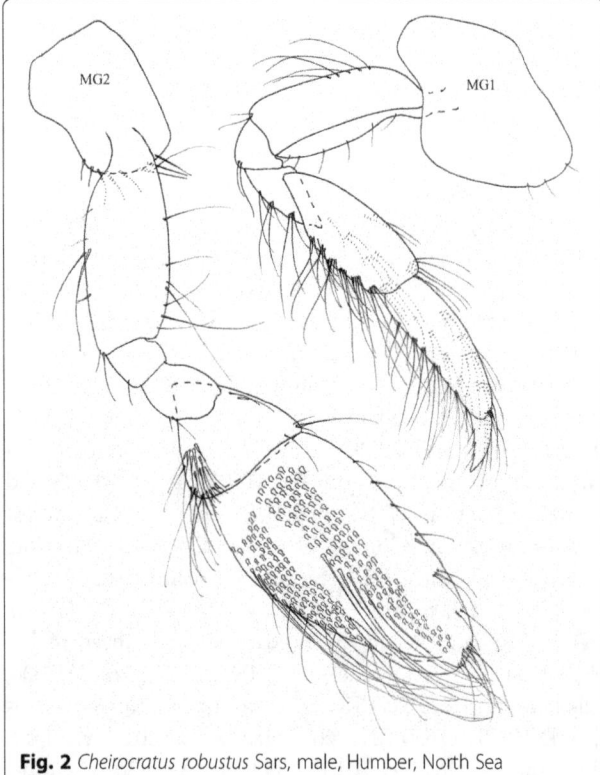

Fig. 2 *Cheirocratus robustus* Sars, male, Humber, North Sea

523; 11 males, seven females, (GNM 9907), Gullmarfjord, Gullmar strömmar, Sweden, 58°15′10″N 11°30′00″E, 15 m, stone, gravel and sand, living and dead algae, 31.07.1921, Hugo Oldevig; two males, three females, (GNM 9908) Gullmarfjord, Gullmar strömmar, Sweden, 58°15′10″N 11°30′00″E, 6–10 m, dead algae, gravel. 31.07.1921, Hugo Oldevig.

Diagnosis

Head with cheek notch; antenna 2 much longer than antenna 1; male gnathopod 2 much larger than gnathopod 1; **male gnathopod 1 robust**, basis subovoid, **without anterodistal spine, carpus and propodus subequal in length, dactylus stout; male gnathopod 2 propodus inner face heavily clothed in long setae, inner face without medial ridge, spine or robust setae, but with small bifid protrubence distally**; pereopods 5–7 robust, pleon segment 1 with three strong dorsal spines; uropod 3 biramous, rami long, subequal in length, distally acute.

Discussion

This is the first record of *C. robustus* Sars from the British Isles. It was previously known only from Norway, Sars, (Sars 1894) (59°91′23″N, 10°74′92″E to 63°43′05″N, 10°39′51″E) and Sweden (see material examined). It can be distinguished in general from its close congener *C. sundevallii* by its much more robust appendages. The male gnathopod 1 lacks an anterodistal spine on the basis, has the carpus and propodus subequal in length, and has a stout dactylus (*C. sundevallii* male gnathopod 1 has anterodistal spine, carpus much longer than propodus and a slender dactylus). In *C. robustus*, the gnathopod 2 has dense setae over much of the inner face of the propodus that lacks ridges, spine or robust setae medially on the inner face. It does have a small protrubence on the distal end of the inner face but this cannot be viewed without removal of some of the dense setation (*C. sundevallii* has dense setae restricted to the outer margin of the inner face of the propodus and has a ridge on the inner face bearing medially a spine and two robust setae and distally a blunt irregular spine bearing a robust seta). The absence of *C. robustus* from the diagnostic key to Irish and British marine Amphipoda in Lincoln (Lincoln 1979) and the superficial similarity of *C. robustus* to *C. sundevallii* probably explains why *C. robustus* was overlooked in the past and confused with *C. sundevalili*. All previous records of *C. sundevallii* in British and Irish waters must be regarded with caution.

Cheirocratus sundevallii (Rathke)
(Fig. 4)
Gammarus sundevallii Rathke, (Rathke 1843): 65.

Cheirocratus sundevallii: Stebbing, (Stebbing 1888): 204.– Stebbing, (Stebbing 1906): 418.– Chevreux & Fage, (Chevreux & Fage 1925): 223.– Lincoln, (Lincoln 1979): 308, fig. 144.– Karaman, (Karaman 1982): 267, fig. 182.
Cheirocratus sundewallii: Sars, (Sars 1894): 524, pl. 184, 185.
Liljeborgia shetlandica Bate & Westwood, (Bate & Westwood 1863): 206.
Protomedeia whitei Bate, (Bate 1862): 169.

Material examined
Three males, five females (NMINH 2016.16.9), JN1066, Valentia, 16.9.2010, VAG 13; one male (NMINH 2016.16.10), JN1066, Valentia, 16.9.2010, VAG 16; two males, one female (NMINH 2016.16.11), JN1066, Valentia,16.9.2010, VAG 14; one male, one female (NMINH 2016.16.12), Carnsore point, C72 52.267 N 6.213 W in 29 m, 1977, Gravel, D. McGrath.

Diagnosis
Head with cheek notch; antenna 2 much longer than antenna 1; male gnathopod 2 much larger than gnathopod 1; **male gnathopod 1 very slender**, basis subovoid, **with strong anterodistal spine, carpus much longer than propodus, dactylus slender; male gnathopod 2 basis with small anterodistal spine, propodus heavily clothed in long setae on the posterior margin of the**

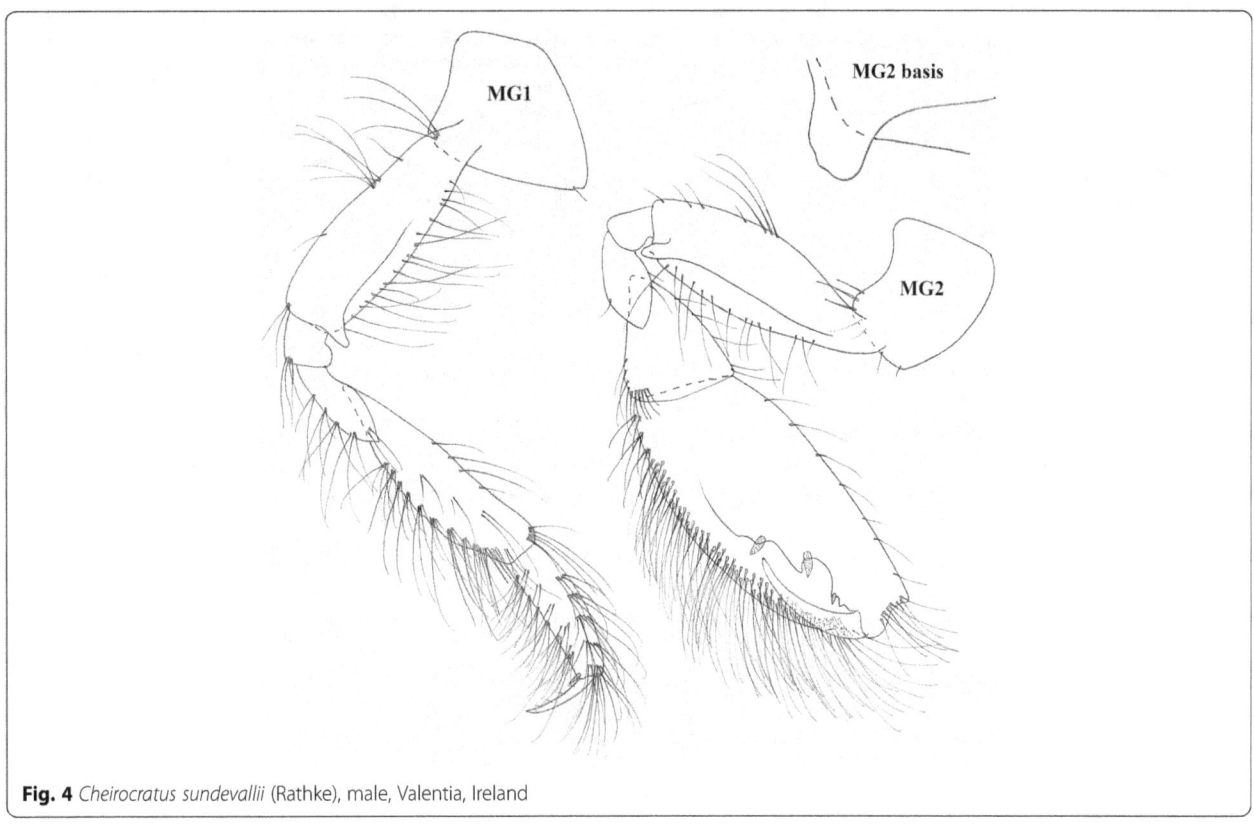

Fig. 4 *Cheirocratus sundevallii* (Rathke), male, Valentia, Ireland

inner face, lacking long setae on the centre of the inner face, but with scalloped ridge bearing a spine and two (Lincoln (1979) figures three) robust setae medially and a small, blunt, irregular spine distally that bears a robust seta; pereopods 5–7 relatively feeble, pleon segment 1 with three strong dorsal spines; uropod 3 biramous, rami long, subequal in length, distally acute.

Discussion

C. sundevalli is widespread in the North East Atlantic and Mediterranean. For differences between *C. sundevalli* and *C. robustus,* see the remarks for that species.

Key to the male *Cheirocratus* of the N.E. Atlantic and Mediterranean

1. Urosome segment 1 with median dorsal spine...*C. monodontus*
   Urosome segment 1 with three dorsal spines...2
2. Gnathopod 2 propodus palm with multiple spines……......................................*C. assimilis*
   Gnathopod 2 propodus palm without spines......................3
3. Gnathopod 2 propodous broadest proximally, dactylus not folding across face of propodus......*C. intermedius*
   Gnathopod 2 propodous sub-ovoid, dactylus folding across face of propodus …….........................4
4. Gnathopod 1 basis without anterodistal spine, propodus equal to carpus; gnathopod 2 basis without anterodistal spine, propodus inner face clothed in dense and very long setae and lacking medial spine or robust setae……………….….…......................................…......*C. robustus*
   Gnathopod 1 basis with anterodistal spine, propodus half the length of carpus; gnathopod 2 basis with anterodistal spine, propodus with dense very long setae on posterior margin of inner face only and with a medial ridge bearing a spine and 2–3 robust setae……………...*C. sundevallii*

Conclusions

Cheirocratus robustus Sars previously known only from Norway and Sweden is now shown to be widespread in the British Isles, occurring in the North Sea and along the west coast of Ireland.

Abbreviations

G1–2: Gnathopod 1–2; M: Male; P3–7: Pereopods 3–7

Acknowledgements

The authors would like to thank Mark Russell from British Marine Aggregate Producers Association (BMAPA), Tarmac Marine Ltd and Hanson Aggregates Marine Ltd for permission to use their PSA and survey data, Keith Cooper from the Centre for Environment, Fisheries and Aquaculture Science (Cefas), for providing PSA and survey data, and Marine Ecological Surveys Limited (MESL) for collecting the North Sea specimens. We wish to thank Dr Kennet Lundin for the loan of *C. robustus* from the Goteborg Naturhistorika Museum and for locality data and Dr. Eddie McCormack of Aquafact International Services Ltd. for making available Irish *C. robustus* material.

Funding

No funding was available.

Authors' contributions

Taxonomic expertise, descriptions and illustrations of taxa AM, Taxonomic expertise DMcG, collection and ecological input WM. All authors read and approved the final manuscript.

Competing interests

The authors declare that they have no competing interests.

Author details

[1]School of Biological, Earth and Environmental Sciences, University College Cork, Cork Enterprise Centre, Distillery Fields, North Mall, Cork, Ireland. [2]Department of Natural Sciences, Galway and Mayo Institute of Technology, Dublin Road, Galway, Ireland. [3]Institute of Estuarine & Coastal Studies (IECS), University of Hull, Cottingham Road, Hull HU6 7RX, UK.

References

Bate CS. Catalogue of the Specimens of Amphipodous Crustacea in the Collection of the British Museum. London: Trustees, British Museum; 1862. p. 1–399.

Bate CS, Westwood JO. A History of the British Sessile-eyed Crustacea. London: John van Voorst; 1863. p. 1–507.

Chevreux E, Fage L. Amphipodes. Faune de France. 1925;9:1–488.

D'Udekem d'Acoz C. Contribution to the knowledge of European Liljeborgiidae (Crustacea, Amphipoda), with considerations on the family and its affinities. Bull Inst Roy Sci Nat Belg. 2010;80:127–259.

Karaman S. Beitrage zur Kenntnis der Süsswasser-Amphipoden. Prirod Razp. 1932;2:179–232.

Karaman GS. Family Gammaridae. In: Ruffo S, editor. The Amphipoda of the Mediterranean, Part III, vol. 13. Monaco: Mém Inst Océanogr; 1982. p. 245–364.

Latreille PA. Amphipoda. In: Nouveau Dictionaire d'histoire naturelle, appliquée aux Arts, à l'Agriculture à l'Économie rurale et domestique à la Médecine etc. Par une societé de Naturalistes et d'Agriculteurs. 2nd edition, volume 1. Paris: Deterville; 1816. p. 467–9.

Lincoln RJ. British Marine Amphipoda: Gammaridea. London, British Museum (Natural History). 1979. p. 1–658.

Lowry JK, Myers AA. A Phylogeny and Classification of the Senticaudata subord. nov. (Crustacea: Amphipoda). Zootaxa. 2013;3610(1):1–80.

Oldevig H. Sveriges Amphipoder. Got Kungl Vetenskaps Vitterh-Samhalles Handl. 1932;ser. B,3(4):1–282.

Rathke H. Beitrage zur Fauna Norwegens. Verh Kaiser Leop-Carol Akad Natur, Breslau. 1843;20(1):1–264,264b,264c.

Sars GO. An account of the Crustacea of Norway, with short descriptions and figures of all the species. Part 16 Paramphithoidae, Epimeridae (part); Part 17 Epimeridae (concluded), Syrrhoidae (part); Part 18 Syrrhoidae (concluded), Pardaliscidae (part); Part 19 Pardaliscidae (concluded), Eusiridae; Part 20 Calliopiidae (part); Part 21 Calliopiidae (concluded), Atylidae. Christiania and Copenhagen (Cammermeyers).1893;p.341–472, pls.121–168.

Sars GO. An account of the Crustacea of Norway, with short descriptions and figures of all the species. Part 22 Gammaridae (part); Part 23 Gammaridae (continued); Part 24 Gammaridae (concluded), Photiidae (part); Part 25/26 Photiidae (concluded), Podoceridae (part); Part 27/28 Podoceridae (concluded), Corophiidae, Cheluridae; Part 29/30 Dulichiidae, Caprellidae, Cyamidae. Christiania and Copenhagen (Cammermeyers). 1894;p. 473–671. pls169~240.

Stebbing TRR. Report on the Amphipoda collected by H.M.S. Challenger during the years 1873–1876. Rep Sci Res H.M.S. Challenger 1873–76. Zoology. 1888; 29:1–1737. pls 1–210.

Stebbing TRR. Amphipoda. I. Gammaridea. Das Tierreich. 1906;2:1–806.

30

First live records of the ruby seadragon (*Phyllopteryx dewysea*, Syngnathidae)

Greg W. Rouse[1]* [iD], Josefin Stiller[1] and Nerida G. Wilson[1,2,3]

Abstract

Until recently, only two species of seadragon were known, *Phycodurus eques* (the leafy seadragon) and *Phyllopteryx taeniolatus* (the common seadragon), both from Australia. In 2015, we described a new species of seadragon, *Phyllopteryx dewysea* (the ruby seadragon). Although the leafy and common seadragons are well known and commonly seen in aquarium exhibits world-wide, the ruby seadragon was known only from four preserved specimens, leaving many aspects of its biology unknown. Based on specimen records, it was speculated that the ruby seadragon normally lives at depths beyond recreational SCUBA diving limits, which may also explain why it went undiscovered for so long. The ruby seadragon also bears a superficial resemblance to the common seadragon, with a number of specimens misidentified in museum collections. The only recent live-collected specimen was trawled from the Recherche Archipelago, a cluster of over 100 islands in Western Australia. We took a small remotely operated vehicle (miniROV) to this locality to obtain the first images of live ruby seadragons. We made observations on the seadragon habitat and behavior, including feeding. We also provide new key observations on their morphology, notably that they lack dermal appendages and have a prehensile tail. We recommend that the ruby seadragon be protected as soon as practicable.

Keywords: Australia, Biodiversity, Marine, Syngnathid, Seadragon

Background

The original description of the ruby seadragon (*Phyllopteryx dewysea*, Stiller, Wilson and Rouse 2015) was based on four specimens, three from near Perth in Western Australia. One of these had been collected in 1919 as a beach-washed specimen and the other two were trawled from 72 m in 1956. The fourth specimen (holotype) was a male carrying a brood of eggs, which was trawled in 2007 from ~50 m depth in the Recherche Archipelago, Western Australia. This locality is than more than 1000 km to the east of where the older specimens had been collected. The holotype and brood were preserved in ethanol, allowing for DNA sequencing. As well as marked DNA evidence for the ruby seadragon as a new species, there were clear diagnostic morphological features, including its red color, 18 trunk rings, forward-pointing dorsal spines on the 11th trunk ring, and an enlarged pectoral area. Dermal appendages are seen in the common and leafy seadragons and function as camouflage in the temperate algal reef habitats

(Kuiter 2000). Although the museum specimens of ruby seadragons possessed the enlarged bony spines to which these appendages attach in the other species, the appendages themselves were absent. It was not clear if the appendages had been lost post-mortem. Other outstanding questions included aspects of their behavior and what kind of habitat they occupied. To obtain the first live observations of the ruby seadragon, we went to the Recherche Archipelago in Western Australia in early April 2016.

Materials and methods

The holotype of *P. dewysea* had been collected on a biodiversity survey and trawl coordinates were available. Collection details for the other known specimens were less detailed, and these were collected many years ago without any subsequent records despite neighbouring an urban environment (Stiller et al. 2015). We therefore targeted the Recherche Archipelago as the best locality to find living ruby seadragons. This idea was reinforced by the fact that after the ruby seadragon was described, another specimen washed ashore in February, 2015 at Culver Cliffs, east of the Recherche

* Correspondence: grouse@ucsd.edu
[1]Scripps Institution of Oceanography, UCSD, La Jolla, CA 92037, USA
Full list of author information is available at the end of the article

Archipelago (Additional file 1) (Della Vedova 2015) and another washed up near Esperance, and was donated to the Western Australian Museum (WAMP34456.001X). As the 50+ meters depth made SCUBA diving activities impractical, we deployed a Teledyne SeaBotix vLBV300 (Anonymous 2015) remotely operated vehicle (miniROV) with a low-light video camera to make observations. As the boat charter was for a restricted period and inclement weather prevented most operations, we only had a single day (April 7) at the site close to the type locality. We undertook four dives of up to 1-h duration on this day. Plans to attempt to catch ruby seadragons using the mini-ROV were abandoned owing to the sea conditions.

Results

The ocean swell had a marked influence at the 50+ meters depth at which the miniROV was operating (Additional file 2: Movie S1). The habitat was mostly deeply-rippled sand over a hard reef substrate at a depth of more than 50 m and a bottom temperature of 18° Celsius. Numerous large demosponges, gorgonians and hydroid cnidarians, erect fenestrate and foliose bryozoan colonies, and algae (including *Caulerpa* cf. *longifolia*) were present. On the second dive, at 54 m depth, we observed a ruby seadragon near a sponge and proceeded to follow the fish. This specimen (Fig. 1a), sex indeterminate, had no dermal appendages and had a curled, likely prehensile tail with a yellow tip (as seen in the beach-washed specimen, Additional file 1). The fish tended to move slowly away from the miniROV, turning backwards and forwards to hold position in the surge, but when it encountered a large object such as an erect sponge, it would often linger before moving on (Additional file 2: Movie S1). After 8 min, we encountered a second ruby seadragon (Fig. 1b), possibly female because of the deeper and more circular body shape, with a notably darker red color and more obvious vertical bars. The two fish stayed in the same vicinity for about 1 min (Additional file 2: Movie S1). We then followed the second fish for 20 min. This fish also lacked dermal appendages and also had a curled tail (Fig. 1c). We did not directly observe either fish hold onto an object with their tail, but they regularly trailed the tail over an object, in a manner which resulted in a mild recoil from the object as the fish departed (Additional file 2: Movie S1). During nearly 30 min of observations on the two ruby seadragons we saw 10 feeding strikes, mostly near the bottom, but in some cases the fish rose slightly above the benthos to strike (Additional file 2: Movie S1). The other three dives that day in the same locality (a further 100 min of observation) revealed no other ruby seadragons.

Fig. 1 The first live observations of the ruby seadragon, (*Phyllopteryx dewysea*), in the Recherche Archipelago, Western Australia. **a** The first specimen set against a large sponge, showing the absence of dermal appendages. *Arrow points* to yellow-tipped tail that is uncoiled. **b** Both individuals in the typical habitat. **c** The second specimen showing the coiling of the tail. Note the lack of dermal appendages. The *inset* shows a detail of the apparently prehensile tail

Discussion

The documentation of living ruby seadragons at more than 50 m depth confirms that these fish live at deeper depths than leafy and common seadragons and in a very different habitat. Leafy seadragons occur along the coast of Western and South Australia from south of Perth to east of Adelaide and are generally found in depths from 3 to 25 m near brown algae, seagrass, and sand (Kuiter 2000; Stiller et al. 2017; Connolly et al. 2002a). Leafy seadragons are known from as deep as 30 m and with a reddish coloration (Kuiter 2000). Ultrasonic tracking of leafy seadragons showed that they stay over *Posidonia* seagrass, macroalgae-covered reefs and sand, but seemed to avoid *Amphibolis* seagrass and boulders with brown algae (Connolly et al. 2002b). Common seadragons have wider geographic range, extending east to the central New South Wales coast and also around Tasmania (Wilson et al. 2016). They show a similar depth preference to leafy seadragons and are also known for living associated with rocky reefs, sand patches, kelp and seagrass (Kuiter 2000; Sanchez-Camara and Booth 2004). The video results for the ruby seadragon clearly show that these fish live in a very different habitat compared to their relatives. They also lack dermal appendages as opposed to the prominent appendages of leafy and common seadragons. In the sparse habitat they occupy, appendages would serve little purpose as camouflaging agents and could add significant costs in drag or fluid resistance, particularly in strong surge. It appears that at these low-light depths, an efficient camouflage strategy for ruby seadragons is to rely on cryptic red coloration.

Surprisingly, we also saw that the ruby seadragon has what appears to be a prehensile tail. Although neither fish was observed to use it in a direct hold, the conditions on the day were relatively calm. The fish would often be exposed to much stronger surge, and then may well use their tails to stop from being swept off the very limited reefal habitats. A prehensile tail is found across a range of Syngnathidae and appears to have evolved at least five times from an ancestral state where the tail possessed a tail fin (Neutens et al. 2014). With regards to seadragons, the phylogeny and transformation for tail shown by Neutens et al. (2014) shows the leafy and common seadragon as a clade tail lacking a tail fin and without grasping capacities. This condition was reasonably interpreted as a single loss of prehensile ability by Neutens et al. (2014) since pipehorses (*Solegnathus* and *Syngnathoides*) have prehensile tails and are the closest relatives of the seadragon clade. In these pipehorses, the ability to curl the tail appears to be facilitated by a reduction of the plates, either ventrally (*Solegnathus*) or entirely (*Syngnathoides*) (Neutens et al. 2014). The ruby seadragon holotype (Stiller et al. 2015) lacks the very tip of the tail (ca. 1 cm) and so the available microCT scan cannot detect this. Nonetheless, the plates do not appear to be reduced as in pipehorses and all rows seem to be present along the tail in the ruby seadragon.

The discovery that the ruby seadragon has a prehensile tail complicates the scenario of the evolution of prehensile tails in this group, as it is the closest relative to the common seadragon (Stiller et al. 2015), which cannot bend its tail. One parsimonious explanation is that the absence of a prehensile tail in the leafy and common seadragons has independently evolved in each species and that the prehensile tail of the ruby seadragon is a retained plesiomorphic condition. Alternatively, the prehensile tail may have been lost in an ancestor of all seadragons, and the ruby seadragon has re-acquired a prehensile tail. To help choose between these scenarios, detailed observations of leafy and common seadragon tails could be compared to the available microCT scan of the ruby seadragon (Stiller et al. 2015).

Conclusions

Our observations on living ruby seadragons, as well the specimens either trawled (Stiller et al. 2015) or washed ashore (Della Vedova 2015), including a male with a brood of young, suggest there is a viable population in the vicinity of the Recherche Archipelago. The earlier records from further west (Stiller et al. 2015) suggest the ruby seadragon may have a widespread distribution in Western Australia, though there is little contemporary evidence to support this. As with the leafy and common seadragons we encourage the protection of the ruby seadragon as soon as practicable. We are proceeding with actions to nominate the ruby seadragon for listing at state and federal levels, to afford it the protection already available to the other two seadragon species.

Additional files

Additional file 1: A ruby seadragon, *Phyllopteryx dewysea* that washed up on the Point Culver cliffs in Western Australia. Photos taken by Zoe Della Vedova. A. The specimen appeared to be fresh and still showed the bright red coloration and light snout markings. B. Although the curling of the tail seems to disappear after death, a yellow tip of the tail was visible (arrow).

Additional file 2: Movie S1. Underwater footage of living ruby seadragons, *Phyllopteryx dewysea* illustrating their typical habitat and behavior. The video was taken using a miniROV at ca. 54 m depth in Recherche Archipelago, Western Australia.

Acknowledgements
We thank Dewy White (Lowe Family Foundation) for funding this expedition. We thank Total Marine Technology (TMT), especially Chris Porter, for their technical support, supplying the miniROV and pilot Daffyd Philips for expert technique. We also thank Colin Hampson and Kinglsey Whitta (crew of the *Southern Conquest*) and Peter and Jaimen Hudson of Esperance Diving and Fishing. Zoe Della Vedova is thanked for the photos of the beach-cast specimen. We thank two anonymous reviewers for their valuable comments on the manuscript. We have no competing interests.

Funding
Lowe Family Foundation.

Authors' contributions
GWR, JS, NGW conceived the project, collected field data and drafted the manuscript. All authors gave final approval for publication.

Competing interests
The authors declare that they have no competing interests.

Author details
[1]Scripps Institution of Oceanography, UCSD, La Jolla, CA 92037, USA. [2]Western Australian Museum, Locked Bag 49, Welshpool DC, WA 6986, Australia. [3]University of Western Australia, Crawley, WA 6009, Australia.

References

Anonymous: vLBV300. http://www.teledynemarine.com/lbv300-5/?BrandID=19. Accessed 5 Dec 2016.

Connolly RM, Melville AJ, Preston KM. Patterns of movement and habitat use by leafy seadragons tracked ultrasonically. J Fish Biol. 2002a;61:684–95.

Connolly RM, Melville AJ, Keesing JK. Abundance, movement and individual identification of leafy seadragons *Phycodurus eques* (Pisces: Syngnathidae). Mar Freshwater Res. 2002b;53:777–80.

Della Vedova Z. Rare ruby seadragon washes up on WA cliffs. http://www.australiangeographic.com.au/blogs/ag-blog/2015/03/rare-ruby-seadragon-washes-up-on-wa-cliffs. Accessed 30 May 2016.

Kuiter RH. Seahorses, pipefishes and their relatives - a comprehensive guide to syngnathiformes. Chorleywood: TMC Publishing; 2000.

Neutens C, Adriaens D, Christiaens J, De Kegel B, Dierick M, Boistel R, Van Hoorebeke L. Grasping convergent evolution in syngnathids: a unique tale of tails. J Anat. 2014;224:710–23.

Sanchez-Camara J, Booth DJ. Movement, home range and site fidelity of the weedy seadragon *Phyllopteryx taeniolatus* (Teleostei: Syngnathidae). Environ Biol Fishes. 2004;70:31–41.

Stiller J, Wilson NG, Rouse GW. A spectacular new species of seadragon (Syngnathidae). Royal Society Open Science. 2015;2:140458.

Stiller J, Wilson NG, Donnellan SC, Rouse GW. The leafy seadragon, *Phycodurus eques*, a flagship species with low but structured genetic variability. J Hered. 2017. doi:10.1093/jhered/esw1075.

Wilson NG, Stiller J, Rouse GW. Barriers to gene flow in common seadragons (Syngnathidae: *Phyllopteryx taeniolatus*). Conserv Genet. 2016. doi:10.1007/s10592-10016-10881-y.

31

First documentation of encrusting specimen of *Cliona delitrix* on Curaçao: a cause for concern?

Benjamin Mueller🆔

Abstract

The coral excavating sponge *Cliona delitrix* is one of the most aggressive and conspicuous excavating sponges on Caribbean reefs. While *C. delitrix* is very prominent displaying its typical encrusting growth form (β-stage) on the Caribbean island of Bonaire, it is rather elusive and only exhibits a papillated habitus (α-stage) on the neighboring island of Curaçao. Here I document the first two encrusting specimen of *C. delitrix* on Curaçao and discuss potential explanations for island-specific differences in its habitus and occurrence. An increase of encrusting specimen could have profound consequences for Curaçaoan reefs and should thus be monitored closely.

Keywords: *Cliona delitrix*, Sponge, Growth stage, Coral reef, Curaçao

Background

The coral excavating sponge *Cliona delitrix* (Pang 1973) (Hadromerida, Demospongiae) is one of the most destructive bioeroders on Caribbean coral reefs (e.g. Chaves-Fonnegra et al. 2007). It can excavate 10–12 cm into coral skeletons and/or the limestone framework (Pang 1973; Zilberberg et al. 2006), and spreads laterally at mean rates of ~1.5 cm y^{-1} (Chaves-Fonnegra and Zea 2011; Rützler 2002). While doing so it can kill adjacent corals by undermining tissue fronts and filaments (Chaves-Fonnegra and Zea 2007). Initially, *C. delitrix* forms discrete ostial and oscular papillae (α-stage) (Fig. 1a). Quickly these papillae start to fuse (Fig. 1b), until eventually all papillae are fused and connected by a tough layer of thin tissue, which starts to overgrow the substrate in an encrusting manner (β-stage) (Fig. 1c-d).

Due to its bright red-orange color and increasing abundance, *C. delitrix* is one of the most conspicuous excavating sponges on many coral reefs throughout the Caribbean and Eastern and North Eastern coast of Brazil (Chaves-Fonnegra et al. 2007; Rose and Risk 1985; Ward-Paige et al. 2005; see map in Van Soest 2010 for distribution range of *C. delitrix*). Interestingly, *C. delitrix* is rather elusive on the fringing reefs of the Southern Caribbean island of Curaçao, yet it is very prominent on

the neighboring island of Bonaire, merely 41 km off Curaçao (Mueller et al. 2014) (Fig. 2). Fringing reefs on both islands are situated in the same physicochemical province (Chollett et al. 2012) and are characterized by a similar reef geomorphology with a narrow reef terrace and steep reef slope (20–50°) (Van Duyl 1985), as well as comparable reef community structures and reef health conditions (Jackson et al. 2014; Sandin et al. 2008). Despite the close proximity and similarities in environmental conditions of these two islands, *C. delitrix* on Curaçao appears to solely display the α-stage, with mostly unfused papillae, whereas large encrusting individuals occur commonly around Bonaire. It should be noted that the exclusively-papillated excavating sponge *Cliona laticavicola* (Pang 1973) (Hadromerida, Demospongiae) closely resembles the α-stage of *C. delitrix*. However, genetic and spicule analysis confirm that papillated specimen on Curaçao are indeed *C. delitrix* (Chaves-Fonnegra et al. 2015). Moreover, due to the fact that both species were described from the same reef on Jamaica (*C. laticavicola* from shallow-rocky habitats and *C. delitrix* from the reef slope), *C. laticavicola* has been proposed to be an early life history stage or ecophenotype of *C. delitrix* (Zea et al. 2014).

This study reports the first documentation of encrusting specimen of *C. delitrix* from Curaçao and discusses

---

Correspondence: muellerb@ymail.com
Carmabi Foundation, Willemstad, Curaçao

Fig. 1 Typical succession of growth phases in the excavating sponge *Cliona delitrix*. **a** Discrete ostial and oscular papillae (*a*-stage). **b** Papillae start quickly to fuse (*a*-stage transition to *β*-stage). **c** All papillae are fused and connected by a tough layer of thin tissue (*β*-stage). **d** Encrusting sponge continues to overgrow the substrate (*β*-stage). Tissue often harbors a dense population of whitish zoanthids at this stage. Pictures were taken on the fore reef slope of Bonaire, Southern Caribbean, between 10 and 15 m depth

potential explanations for island-specific differences in its habitus and occurrence.

Material and methods

Two encrusting specimen of *C. delitrix* were photographed during a dive on November 3, 2016 in front of the Sea Aquarium Park on Curaçao (12° 05' N, 68° 53' W). The site is characterized by a diverse coral community with >30% live coral cover, high structural complexity, and a steep fore reef slope (>50°) (Van Duyl 1985). The coral community at the drop-off (approx. 8 m depth) is dominated by *Orbicella* spp.

Results

Two encrusting specimen (*β*-stage) of *C. delitrix* were recorded on a single colony of *Orbicella faveolata* (formerly *Montastraea faveolata*) at 8 m depth (Fig. 3). Both specimen were located in two separate areas of partial coral-mortality and were surrounded by a band of turf algae. No additional encrusting specimen of *C. delitrix* were encountered in the vicinity during the 60 min dive.

Discussion

Cliona delitrix has been reported to occur on Curaçao (Bruckner and Bruckner 2006; Chaves-Fonnegra et al. 2015; Van Soest 1981), yet large encrusting specimen have not been recorded so far and were not encountered in the coral reef monitoring program and/or other research activities of the Carmabi Research Station (pers. comm. M. Vermeij). While it cannot be excluded that more *C. delitrix* displaying the *β*-stage exist or existed on Curaçaoan reefs, it is safe to say that this habitus is rare. This raises the question why encrusting *C. delitrix* are not as abundant as on Bonaire, where such specimen are commonly encountered and densities of 0.03 individuals m^{-2} have been documented (Mueller et al. 2014)? Possible explanations could include (1) differences in environmental conditions between the two islands and/or (2) genetic differences between the local populations of *C. delitrix*. However, environmental conditions including geomorphology and community structure are very similar on both islands (Chollett et al. 2012; Sandin et al. 2008; Van Duyl 1985). Given the suggested positive effect of anthropogenic disturbances (e.g. organic

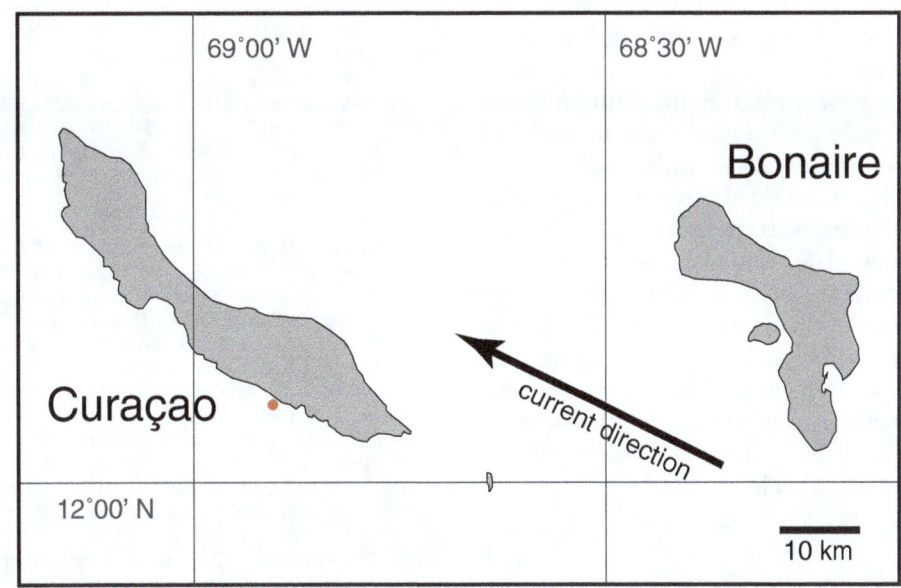

Fig. 2 Map of the Southern Caribbean islands of Curaçao and Bonaire. Red circle indicates the position of the observed encrusting specimen of *Cliona delitrix* in front of the Sea Aquarium Park on Curaçao. Predominant current direction of the Caribbean Current is indicated with an arrow

Fig. 3 Photographs of two encrusting specimen of *Cliona delitrix* growing on one single colony of *Orbicella faveolata* on Curaçao, Southern Caribbean. **a** Overview of the two specimen in two separate areas of partial coral-mortality and were surrounded by a band of turf algae. **b** Close-up of the larger specimen

pollution) on the abundance of *C. delitrix* (Chaves-Fonnegra et al. 2007; Rose and Risk 1985; Ward-Paige et al. 2005), a more than five times higher human population density (Centraal Bureau voor de Statistiek 2016; Central Bureau of Statistics Curaçao 2016), more industrial development, and a less restrictive marine resource management policy would rather suggest more favorable conditions on Curaçao than on Bonaire. Slightly higher dissolved organic carbon (DOC) and bacterial concentrations (Mueller et al. 2014) further suggest that food limitation is not a likely cause for the lower prevalence of *C. delitrix* on Curaçaoan reefs. Moreover, a planktonic larval period between 1 and 10 days (Mariani et al. 2006; Warburton 1958) in combination with the strong Caribbean Currents have been proposed to enable gene flow between populations of *Cliona delitrix* of up to 500 km across the Southern Caribbean (Chaves-Fonnegra et al. 2015). It is therefore highly likely that the Caribbean Currents flowing from Bonaire to Curaçao with up to 70 cm s^{-2} (Fratantoni 2001) (Fig. 2), should allow for a good connectivity between *C. delitrix* populations of both islands (Chaves-Fonnegra et al. 2015), as reported for local coral populations (e.g. Baums et al. 2005; Baums et al. 2006). Thus, as larvae of encrusting specimen of *C. delitrix* from Bonaire can be expected to seed Curaçaoan reefs, genetic differences are unlikely to be the reason for the lack of specimen displaying the β-stage on Curaçao.

In addition, *C. delitrix* is reported to spread particularly in the aftermath of catastrophic episodical disturbances, such as hurricanes and bleaching events, where recently deceased coral are rapidly colonized

(Chaves-Fonnegra et al. 2015; Chaves-Fonnegra and Zea 2011). In general, Curaçao has not been as strongly affected by bleaching events as other places in the Caribbean. However, during the 2010 bleaching event 12–30% of all coral colonies were affected and on average 10% of those subsequently died (Vermeij 2012). Despite this substantial opening of suitable substrate, no encrusting specimen of *C. delitrix* were recorded until now. This raises the question if the here reported occurrence constitutes an isolated event or marks the onset of an ongoing trend? Given the fierce competitiveness of *Cliona delitrix*, its capability to kill live coral, as well as its high excavation rate, this could potentially have profound consequences for Curaçaoan benthic communities and their calcium carbonate budgets and should therefore be monitored closely.

Acknowledgements

I thank the staff of Carmabi for their logistic support. Fieldwork was performed under the research permit (#2012/48584) issued by the Curaçaoan Ministry of Health, Environment and Nature (GMN) to the CARMABI Foundation.

Funding

The author declares that there was no funding received for this study.

Author's contributions

BM conceived and performed the experiment, analyzed the data, and wrote the manuscript.

Competing interests

The author declares that he has no competing interests.

References

Baums IB, Miller MW, Hellberg ME. Regionally isolated populations of an imperiled Caribbean coral *Acropora palmata*. Mol Ecol. 2005;14:1377–90.

Baums IB, Paris CB, Cherubin L. A bio-oceanographic filter to larval dispersal in a reef-building coral. Limnol Oceanogr. 2006;51:1969–81.

Bruckner A, Bruckner R. The recent decline of *Montastraea annularis* (complex) coral populations in western Curaçao: a cause for concern? Rev Biol Trop. 2006;54:45–58.

Chaves-Fonnegra A, Feldheim K, Secord J, Lopez JV. Population structure and dispersal of the coral excavating sponge. *Cliona delitrix* Mol Ecol. 2015. doi:10.1111/mec.13134.

Chaves-Fonnegra A, Zea S. Observations on reef coral undermining by the Caribbean excavating sponge *Cliona delitrix* (Demospongiae, Hadromerida) Porifera research: biodiversity, innovation and sustainability. Série Livros. 2007; 28:247–64.

Chaves-Fonnegra A, Zea S. Coral colonization by the encrusting excavating Caribbean sponge *Cliona delitrix*. Mar Ecol. 2011;32:162–73.

Chaves-Fonnegra A, Zea S, Gómez M. Abundance of the excavating sponge *Cliona delitrix* in relation to sewage discharge at San Andrés Island, SW Caribbean Colombia. Bol Investig Mar Costeras. 2007;36:63–78.

Centraal Bureau voor de Statistiek. Bevolkingsontwikkeling Caribisch Nederland. 2016. http://statline.cbs.nl. Retrieved 17 December 2016.

Central Bureau of Statistics Curaçao. Population tables. 2016. http://www.cbs.cw. Retrieved 17 December 2016.

Chollett I, Mumby PJ, Muller-Karger FE, Hu C. Physical environments of the Caribbean Sea. Limnol Oceanogr. 2012;57:1233–44.

Fratantoni DM. North Atlantic surface circulation during the 1990's observed with satellite-tracked drifters. J Geophys Res Oceans. 2001;106:22067–93.

Jackson J, Donovan M, Cramer K, Lam V. Status and trends of Caribbean coral reefs: 1970–2012. Switzerland: Global Coral Reef Monitoring Network, IUCN; 2014.

Mariani S, Uriz M-J, Turon X, Alcoverro T. Dispersal strategies in sponge larvae: integrating the life history of larvae and the hydrologic component. Oecologia. 2006;149:174–84.

Mueller B, de Goeij JM, Vermeij MJ, Mulders Y, van der Ent E, Ribes M, van Duyl FC. Natural diet of coral-excavating sponges consists mainly of dissolved organic carbon (DOC). PLoS One. 2014;9:e90152.

Pang RK. The ecology of some jamaican excavating sponges. Bull Mar Sci. 1973; 23:227–43.

Rose CS, Risk MJ. Increase in *Cliona delitrix* infestation of *Montastraea cavernosa* heads on an organically polluted portion of the Grand Cayman fringing reef. Mar Ecol. 1985;6:345–63.

Rützler K. Impact of crustose clionid sponges on Caribbean reef corals. Acta Geol Hispanica. 2002;37:61–72.

Sandin SA, Sampayo EM, Vermeij MJ. Coral reef fish and benthic community structure of Bonaire and Curaçao, Netherlands Antilles. Caribb J Sci. 2008;44:137–44.

Van Duyl FC. Atlas of the living reefs of Curaçao and Bonaire (Netherlands Antilles). vol 117. Studies of the flora and fauna of Surinam and the Netherlands Antilles. Utrecht: Uitgaven van de Stichting "Natuurwetenschappelijke Studiekring voor Suriname en Curaçao"; 1985. http://trove.nla.gov.au/work/11232553?selectedversion=NBD891739.

Van Soest R. A checklist of the Curaçao sponges (Porifera, Demospongiae) including a pictorial key to the more common reef forms. Verslagen entechnische gegevens/Instituut voor Taxonomische Zoölogie (Zoölogisch Museum), Universiteit van Amsterdam. 1981;31:1–44. http://www.repository. naturalis.nl/document/550110

Van Soest RWM. *Cliona delitrix* Pang, 1973. In: Van Soest RWM, Boury-Esnault N, Hooper JNA, Rützler K, de Voogd NJ, Alvarez de Glasby B, Hajdu E, Pisera AB, Manconi R, Schoenberg C, Klautau M, Picton B, Kelly M, Vacelet J, Dohrmann M, Díaz M-C, Cárdenas P, Carballo JL. World Porifera database; 2010. http:// www.marinespecies.org/porifera/porifera.php?p=taxdetails&id=170437. Accessed 12 Mar 2017.

Vermeij MJA. The Current State of Curaçao's Coral Reefs. 2012

Warburton FE. Reproduction of fused larvae in the boring sponge, *Cliona celata*. Grant Nature. 1958;181:493–4.

Ward-Paige CA, Risk MJ, Sherwood OA, Jaap WC. Clionid sponge surveys on the Florida reef tract suggest land-based nutrient inputs. Mar Pollut Bull. 2005;51: 570–9. doi:10.1016/j.marpolbul.2005.04.006.

Zea S, Henkel TP, Pawlik JR. The sponge guide: a picture guide to Caribbean sponges. 3rd ed. 2014. Available online at, http://www.spongeguide.org.

Zilberberg C, Maldonado M, Solé-Cava AM. Assessment of the relative contribution of asexual propagation in a population of the coral-excavating sponge *Cliona delitrix* from the Bahamas. Coral Reefs. 2006;25:297–301.

32

First report of the whitesaddled catshark *Scyliorhinus hesperius* (Springer 1966) in Guatemala's Caribbean Sea

Ana Hacohen-Domené[1][*] [iD], Francisco Polanco-Vásquez[1] and Rachel T. Graham[2]

Abstract

Background: The present study represents the first record of *Scyliorhinus hesperius* in Guatemala's Caribbean Sea.

Methods: Five male whitesaddled catsharks, *S. hesperius*, were captured in 200 m deep waters of Guatemala's Caribbean coast.

Results and Conclusion: All specimens were male with total lengths ranging from 420 mm to 510 mm. These fish represent the first record of mature male *S. hesperius*, the first record for this species in Guatemalan territorial waters, and a range extension in the Western Central Atlantic.

Keywords: Deep-water chondrichthyans, First record, Range extension, Caribbean,

Background

Scyliorhinidae (catsharks) constitute the largest shark family with at least 160 species distributed across 17 genera (Ebert et al. 1996). These species are broadly distributed throughout temperate and tropical waters, inhabiting the bottom of shallow and deep waters over 100 m (Nakaya 1975). Catsharks are small, demersal species, and relatively poor swimmers (Compagno et al. 2005).

The genus *Scyliorhinus* Blainville 1816, is comprised of 16 species distributed in cold, subtemperate to tropical waters (Ebert et al., 2015; Soares et al., 2016) including the eastern and western Atlantic and the Mediterranean (Rodríguez-Cabello et al. 2007; Ebert et al. 2015). In the western Atlantic, *Scyliorhinus* is most diverse and at least six species of *Scyliorhinus* are distributed in the Caribbean (Compagno, 1984) with three species occurring throughout the Central America Caribbean: *Scyliorhinus boa* Goode and Bean 1896, *S. retifer* Garman 1881, and *S. hesperius* Springer 1966 (Compagno, 1984).

S. hesperius was described by Springer (1966) based on an immature female holotype of 415 mm total length

(TL). Additionally Springer (1966) examined 12 specimens, sex not specified, ranging in total length from 177–460 mm. All specimens were captured between 274 m and 530 m depth in the Western Caribbean near Jamaica and Honduras and southward towards Panama and Columbia (Springer 1966). Later, Ross and Quattrini (2009) reported sightings of three individual sharks *S. hesperius* resting on thick coral rubble between 580–604 m depth, off of Jacksonville, Florida, while conducting deep water dives on deep reefs along the southeastern US continental shelf slope.

Maximum size for this species is based on a singular female *S. hesperius* (470 mm), with no information on its biology (Leandro 2004). Compagno (1988) reports two immature specimens from Nicaragua, one female and one male, 159 mm and 356 mm respectively. Size range of specimens reported in this study was 420–510 mm TL, all adult male sharks. This study represents the first report of adult male *S. hesperius* and also the largest *S. hesperius* specimen collected to date, based on morphometric data of specimen Rf. 252.

The whitesaddled catshark *S. hesperius* is currently listed by the International Union for the Conservation of Nature's Red List as Data Deficient due to insufficient information available to assess the species population status (Leandro 2004). No current information exists for

* Correspondence: ahacohen@fundacionmundoazul.com
[1] Fundación Mundo Azul, Blvd. Rafael Landivar 10-05 Paseo Cayala Zona 16, Edificio D1 Oficina 212, Guatemala City, Guatemala
Full list of author information is available at the end of the article

this species' biology, and distributional limits are poorly known. This paper reports the first record of *S. hesperius* in Guatemalan waters, representing a range extension in Central America and the Caribbean.

Methods

On March 20th 2016, five whitesaddled catsharks were captured by artisanal fishermen of El Quetzalito, Izabal Department, in Guatemala (Fig. 1). These specimens were captured with a 1000 m wide bottom trammel net, consisting of one panel with 3.5 in. mesh size, set at 200 m depth.

All specimens were initially kept on ice prior to preservation in formaldehyde (10%) for 3 weeks and transferred to ethyl alcohol (70%) for final preservation. The specimens were donated to the Laboratory of Biological Science and Oceanography, Centro de Estudios del Mar y Acuicultura (CEMA) of the Universidad San Carlos de Guatemala (USAC). The specimens are part of the collection registered to the Consejo Nacional de Áreas Protegidas (CONAP) under the reference numbers (Rf) 252, 253, 254, 255 and 256.

The five specimens were sexed and measured after being preserved. A total of 91 morphometric measurements were taken (Table 1) as proposed by Compagno (2001). Measurements are expressed as percentages of total length (%TL). All specimens were examined and identified using identification guides developed by Compagno (1984, 2001). Maturity for males was determined by the full calcification of claspers.

Results

Systematic account
Family Scyliorhinidae Gill, 1862
Scyliorhinus Blainville, 1816
Scyliorhinus hesperius Springer, 1966
Common name: Whitesaddled catshark.

Material examined

Rf 252 specimen: male, mature, 510 mm TL (Fig. 2a);
Rf 253 specimen: male, mature, 455 mm TL (Fig. 2b);
Rf 254 specimen: male, mature, 429 mm TL (Fig. 2c);
Rf 255 specimen: male, mature, 435 mm TL (Fig. 2d);
Rf 256 specimen: male, mature, 420 mm TL (Fig. 2e).

All specimens were caught approximately 15 Km north of El Quetzalito, Izabal, Guatemala (15° 49.776 N,−88° 12.340 W), at approximately 200 m, based on known length of net deployed.

Description

Color pattern variable, of seven to eight dark saddles with large light spots concentrated in the saddle marks. Background coloration is light brown on the dorsal surface and paler on the ventral surface. First dorsal fin originates behind pelvic fins, and larger than second dorsal fin (Fig. 2). Lower labial furrows present (Fig. 3).

Discussion

This study provides multiple first records for *S. hesperius* with the largest of the species described based on Reference 252 and also the first mature males. This study further supports the extension of the known range

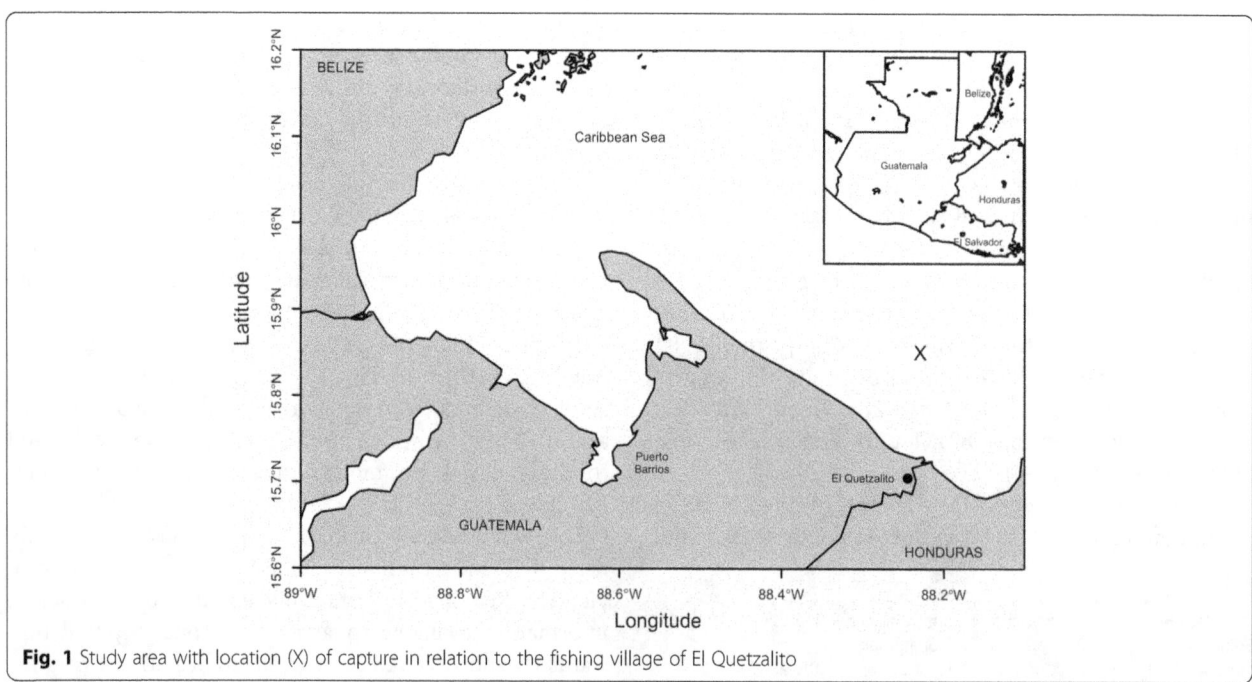

Fig. 1 Study area with location (X) of capture in relation to the fishing village of El Quetzalito

Table 1 Morphometric measurements (mm) of individuals of five male specimens of *Scyliorhinus hesperius*

Measurements	Rf. 252	Rf. 253	Rf. 254	Rf. 255	Rf. 256
Total length (mm)	510.0	455.0	429.0	435.0	420.0
Fork length	440.0	401.0	381.0	390.4	370.0
Precaudal fin length	400.0	358.0	335.0	345.5	330.0
Pre-second dorsal-fin length	350.0	315.0	293.3	303.1	272.2
Pre-first dorsal-fin length	247.6	235.2	220.3	234.9	204.7
Head length	98.1	93.8	91.8	97.2	90.0
Prebranchial length	69.4	67.2	66.3	68.8	58.2
Prespiracular length	52.1	50.4	52.2	53.5	49.4
Preorbital length	28.8	26.9	26.4	27.3	26.4
Prepectoral-fin length	91.3	86.8	83.8	82.1	75.3
Prepelvic-fin length	205.2	186.1	178.8	189.7	175.2
Snout-vent length	221.2	200.3	190.9	204.8	197.9
Preanal-fin length	304.3	277.9	268.2	278.2	253.2
Interdorsal space	55.1	53.8	45.0	47.9	40.1
Dorsal caudal-fin space	30.4	23.6	23.0	22.0	20.0
Pectoral-fin pelvic-fin space	96.4	86.2	79.4	82.0	76.6
Pelvic-fin anal-fin space	70.6	64.2	58.0	62.1	52.7
Anal-fin caudal-fin space	48.3	39.5	45.5	39.3	43.0
Pelvic-fin caudal-fin space	155.5	139.1	130.4	132.0	127.0
Vent caudal-fin length	184.9	167.8	168.6	159.8	164.0
Prenarial length	14.5	12.4	13.4	12.5	12.3
Preoral length	22.2	20.9	18.9	20.7	18.2
Eye length	20.3	18.2	18.4	20.0	19.9
Eye height	6.7	5.1	5.1	5.2	4.2
Intergill length	30.3	27.7	23.4	28.4	23.8
First gill slit height	7.5	6.5	5.2	5.8	4.6
Second gill slit height	6.5	5.3	4.5	4.8	3.8
Third gill slit height	8.7	7.9	5.1	5.9	5.2
Fourth gill slit height	6.3	5.7	3.8	5.3	4.5
Fifth gill slit height	4.9	4.5	3.7	5.1	4.5
First dorsal-fin length	45.7	39.8	39.9	38.4	34.3
First dorsal-fin anterior margin	46.8	42.5	44.1	41.8	36.4
First dorsal-fin base	31.3	27.4	29.2	28.2	25.4
First dorsal-fin height	34.4	24.2	21.2	23.7	23.5
First dorsal-fin inner margin	14.7	14.2	14.2	12.3	9.8
First dorsal-fin posterior margin	29.2	22.7	22.1	22.2	19.8
Second dorsal-fin length	41.2	31.8	35.4	32.2	30.6
Second dorsal-fin anterior margin	31.5	29.6	32.3	31.2	28.4
Second dorsal-fin base	25.5	20.7	23.1	21.0	21.3
Second dorsal-fin height	21.6	15.5	13.0	14.9	14.8
Second dorsal-fin inner margin	14.1	12.2	10.6	10.6	10.1
Second dorsal-fin posterior margin	16.3	16.0	14.5	14.3	11.0
Pectoral-fin anterior margin	67.3	61.2	60.1	59.1	54.0
Pectoral-fin base	57.0	52.0	51.3	49.7	46.7
Pectoral-fin inner margin	29.0	26.6	24.8	26.1	23.5
Pectoral-fin posterior margin	48.2	45.7	41.5	42.9	38.5

Table 1 Morphometric measurements (mm) of individuals of five male specimens of *Scyliorhinus hesperius (Continued)*

Pectoral-fin height	50.9	50.8	47.7	51.0	40.3
Dorsal caudal-fin margin	107.5	97.4	96.9	93.7	89.0
Preventral caudal-fin margin	14.8	11.5	13.8	11.4	8.3
Upper postventral caudal-fin margin	24.2	28.1	25.6	22.4	27.9
Lower postventral caudal-fin margin	46.9	44.5	45.3	40.0	36.4
Caudal-fin fork width	26.5	26.2	26.1	25.6	20.5
Caudal-fin fork length	49.6	46.7	45.1	45.5	42.3
Subterminal caudal-fin margin	22.9	18.9	19.0	20.2	18.8
Subterminal caudal-fin width	11.4	9.2	10.3	9.2	8.2
Terminal caudal-fin margin	29.0	26.4	23.2	24.2	20.4
Terminal caudal-fin lobe	36.7	30.0	28.9	30.8	27.3
Pelvic-fin length	58.8	55.7	48.4	51.4	49.0
Pelvic-fin anterior margin	36.9	35.8	32.2	33.7	31.1
Pelvic-fin base	31.1	30.7	28.8	26.8	25.8
Pelvic-fin height	29.0	19.8	28.2	18.7	18.1
Pelvic-fin inner margin (length)	27.9	23.4	22.7	20.8	20.8
Pelvic-fin posterior margin (length)	39.4	35.9	34.5	35.2	35.5
Anal-fin length	53.5	49.9	45.8	47.5	42.0
Anal-fin anterior margin	34.4	34.7	36.0	36.0	28.8
Anal-fin base	41.7	37.5	35.6	34.9	32.9
Anal-fin height	19.6	16.4	16.8	14.5	15.3
Anal-fin inner margin	12.5	14.3	12.5	12.0	11.2
Anal-fin posterior margin	28.0	25.7	23.1	24.5	17.4
Head height	33.5	33.7	34.0	29.8	30.3
Trunk height	47.1	39.0	41.3	43.0	27.3
Abdomen height	31.4	25.8	28.9	26.4	26.2
Tail height	35.7	30.0	31.5	27.3	26.8
Caudal-fin peduncle height	15.5	16.8	14.3	13.4	12.8
Mouth length	16.9	15.6	15.2	14.7	18.4
Mouth width	32.0	25.4	28.2	26.8	23.3
Upper labial-furrow length	2.2	3.0	3.0	2.3	1.5
Lower labial-furrow length	6.0	5.0	5.1	6.1	5.2
Nostril width	9.9	9.8	9.0	10.3	10.3
Internarial space	8.0	7.9	7.8	7.9	5.7
Interorbital space	29.9	26.6	28.8	23.9	26.6
Spiracle length	3.5	2.8	2.9	2.2	1.9
Eye spiracle space	4.0	4.1	4.2	4.1	3.5
Head width	49.6	44.0	40.3	44.4	39.9
Trunk width	40.5	43.4	41.4	40.9	35.9
Abdomen width	19.6	15.2	14.4	13.8	12.9
Tail width	26.5	21.0	22.6	21.9	21.4
Caudal-fin peduncle width	8.0	7.7	9.0	6.1	6.8
Clasper outer length	22.7	20.4	19.4	20.3	20.1
Clasper inner length	37.5	33.7	35.0	31.4	34.1
Clasper base width	4.9	4.8	4.0	4.3	3.7

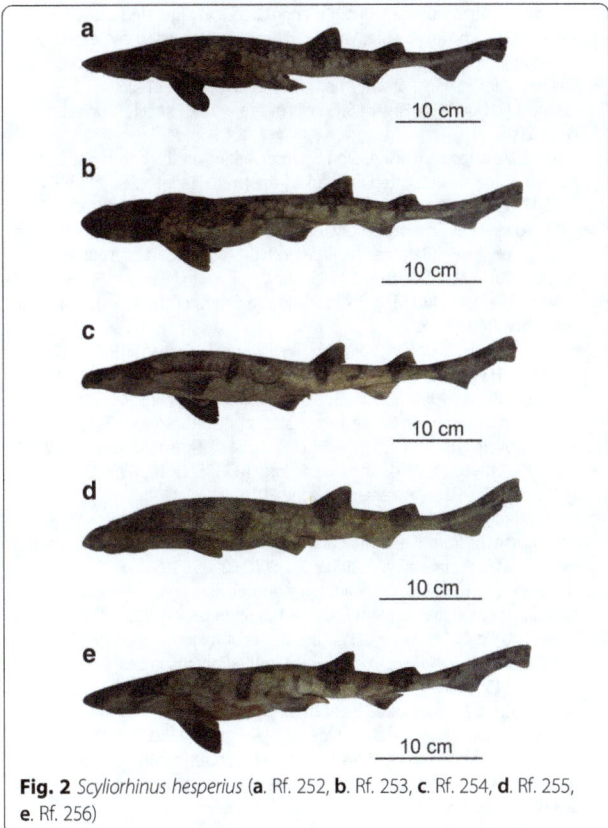

Fig. 2 *Scyliorhinus hesperius* (**a**. Rf. 252, **b**. Rf. 253, **c**. Rf. 254, **d**. Rf. 255, **e**. Rf. 256)

of *S. hesperius* in the Caribbean while including a first record for the species in Guatemala. Although the whitesaddled catshark has been recorded in the Western Central Atlantic in Honduras, Panamá and Colombia (Kyne et al. 2012), Ross and Quattrini's regional study

(2009) and this study suggest that the species' range may be more extensive than originally thought.

Knowledge of *S. hesperius* basic biology is limited. By comparison, considerable literature exists on *S. canicula*, a relatively abundant catshark species distributed throughout the Eastern North Atlantic and Mediterranean (Sims et al. 2001; Rodríguez-Cabello et al. 2007; Ebert et al. 2015). In south-west Ireland, an acoustic tagging study of four *S. canicula* revealed that two tagged females exhibited alternative behavioural strategies compared to the tagged two males, a difference resulting in spatial segregation of the sexes by habitat (Sims et al. 2001). Sims (2003) further reports that sexual segregation in this species occurs primarily as a consequence of male avoidance by females. In the western Mediterranean, segregation between juveniles and adults occurs for *S. canicula* where juveniles are found in depths greater than 100 m while adults almost exclusively occupy shallower depths Massutí and Moranta (2003). By comparison, in the Northern Aegean Sea, the pattern of vertical distribution of *S. canicula* showed that individuals did not exhibit any sexual segregation and juveniles and adults were found together in the bathyal zone, often located swimming near the benthos D'Onghia et al. (1995). Considering the range of behavioral strategies demonstrated by members of the genus *Scyliorhinus*, it is currently unclear if *S. hesperius* exhibits sexual and size segregation in the Caribbean. This study's record of male-only specimens raises the question if *S. hesperius* segregate by sex in waters 200 m deep in Guatemala's Caribbean waters. According to artisanal fishers interviewed, this species is rarely captured in

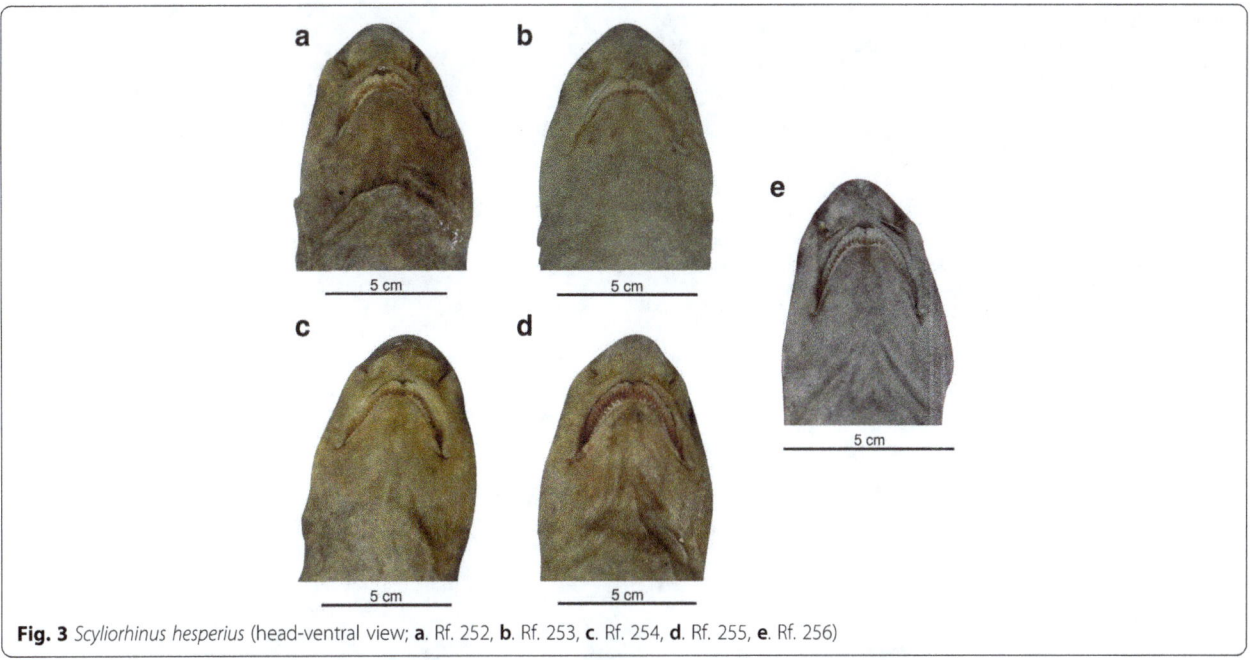

Fig. 3 *Scyliorhinus hesperius* (head-ventral view; **a**. Rf. 252, **b**. Rf. 253, **c**. Rf. 254, **d**. Rf. 255, **e**. Rf. 256)

the fishery and never utilized due to the species' small size. These results suggests a further need for fisheries-independent studies to elucidate habitat preferences and distribution by sex and size of *S. hesperius*.

Conclusions

This paper provides noteworthy multiple firsts records of *S. hesperius* in Guatemalan waters, that represents a range extension in Central America and the Caribbean, the largest *S. hesperius* and the first mature males collected to date. Future studies are needed to identify the behavior and ecology of *S. hesperius* in the Caribbean in light of increasing fisheries effort.

Abbreviations
Km: Kilometer; M: Meter; Mm: Millimeter; TL: Total length

Acknowledgements
Fundación Mundo Azul would like to thank the community of El Quetzalito for their constant support to the research program in the area. Also, we would like to thank the Centro de Estudios del Mar y Acuicultura (CEMA) of the Universidad San Carlos de Guatemala (USAC) for providing the space to take the morphometric measurements.

Funding
Fundación Mundo Azul provided funding for the elasmobranch monitoring project in 2016. MarAlliance provided funding for RTG's time.

Authors' contributions
AH and FP participated in the identification of the species, recorded the morphometric data of all specimens, and contributed to draft the manuscript. RTG contributed to draft the manuscript. All authors read and approved the final manuscript.

Competing interest
The authors declare that they have no competing interests.

Author details
[1]Fundación Mundo Azul, Blvd. Rafael Landivar 10-05 Paseo Cayala Zona 16, Edificio D1 Oficina 212, Guatemala City, Guatemala. [2]MarAlliance, 32 Coconut Drive, PO Box 283, San Pedro, Belize.

References
Compagno LJV. FAO species catalogue. Vol. 4. Sharks of the world. An annotated and illustrated catalogue of shark species known to date. Part 1. Hexanchiforms to Lamniformes. FAO Fish Synop. 1984;4:1–249.

Compagno LJV. *Scyliorhinus comoroensis* sp. n., a new catshark from the Comoro Islands, western Indian Ocean (Carcharhiniformes, Scyliorhinidae). Bull Mus Natl Hist Nat. 1988;10(3):603–25.

Compagno LJV. Sharks of the world. An annotated and illustrated catalogue of shark species known to date. Vol. 2. Bullhead, mackerel and carpet sharks (Heterodontiformes, Lamniformes and Orectolobiformes). FAO Species Catalogue for Fishery Purposes no. 1. Rome: FAO. 2001;2:1–269.

Compagno LJV, Dando M, Fowler S. Sharks of the world: Princeton University Press. 2005.

D'Onghia G, Matarrese A, Tursi A, Sion L. Observations on the depth distribution pattern of the small-spotted catshark in the North Aegean Sea. J Fish Biol. 1995;47:421–6.

Ebert DA, Cowley PD, Compagno LJV. A preliminary investigation of the feeding ecology of skates (Batoidea: Rajidae) off the west coast of southern Africa. S Afr J Mar Sci. 1996;17:233–40.

Ebert DA, Fowler S, Dando M. A Pocket Guide to Sharks of the World. Princeton University Press; 2015.

Kyne PM, Carlson JK, Ebert DA, Fordham SV, Bizzarro JJ, Graham RT, Kulka DW, Tewes EE, Harrison LR. Dulvy NK (eds). Central American, and Caribbean Chondrichthyans. IUCN Species Survival Commission Shark Specialist Group, Vancouver, Canada: The Conservation Status of North American; 2012.

Leandro L. *Scyliorhinus hesperius*. The IUCN Red List of Threatened Species. 2004: e. T44590A10910276.2004. http://dx.doi.org/10.2305/IUCN.UK.2004.RLTS. T44590A10910276.en. Accessed 21 June 2016.

Massutí E, Moranta J. Demersal assemblages and depth distribution of elasmobranchs from the continental shelf and slope off the Balearic Islands (western Mediterranean). ICES J Mar Sci. 2003;60(4):753–66.

Nakaya K. Taxonomy, comparative anatomy and phylogeny of Japanese catsharks, Scyliorhinidae. Mem Fac Fish Hokkaido Univ. 1975;23(1):1–94.

Rodríguez-Cabello C, Sánchez F, Olaso I. Distribution patterns and sexual segregations of *Scyliorhinus canicula* (L.) in the Cantabrian Sea. J Fish Biol. 2007;70(5):1568–86.

Ross SW, Quattrini AM. Deep-sea reef fish assemblage patterns on the Blake Plateau (Western North Atlantic Ocean). Mar Ecol. 2009;30:74–92.

Sims DW. Tractable models for testing theories about natural strategies: foraging behaviour and habitat selection of free-ranging sharks. J Fish Biol. 2003;63(s1):53–73.

Sims D, Nash J, Morritt D. Movements and activity of male and female dogfish in a tidal sea lough: alternative behavioural strategies and apparent sexual segregation. Mar Biol. 2001;139(6):1165–75.

Soares KD, Gomes UL, De Carvalho MR. Taxonomic review of catsharks of the *Scyliorhinus haeckelii* group, with the description of a new species (Chondrichthyes: Carcharhiniformes: Scyliorhinidae). Zootaxa. 2016;4066(5):501–34.

Springer S. A review of western Atlantic cat sharks, Scyliorhinidae, with descriptions of a new genus and five new species. Fish Bull. 1966;65(3):581–624.

The first authenticated record of Pygmy Killer Whale (*Feresa attenuata* Gray 1874) in Mozambique; has it been previously overlooked?

Gary A. Allport[1*], Christopher Curtis[2], Tiago Pampulim Simões[3] and Maria J. Rodrigues[4]

Abstract

Background: The cetacean fauna of the poorly-studied waters off eastern Africa is still being described. Information on the cetacean species occurring in specific range states is important for understanding their geographical distribution ranges and for implementing national and international conservation and management measures. This report presents the first authenticated record of the Pygmy killer whale in Mozambican waters and the first record on the eastern coast of southern Africa since 1970.

Methods: As a part of regular informal surveys for birds and other marine life from Maputo, Mozambique, three Pygmy killer whales were seen, approached and photographed north of Inhaca Island (25°52'54.22"S 33° 8'33.62"E), on 23 April 2017.

Results: The animals were seen interacting on the surface for 35 min, travelling at ca. 1 km/h along the shelf edge in water 235 m deep. All three animals had been overlooked by the authors earlier in the day, misidentified as Spinner dolphins (*Stenella longirostris*).

Conclusion: This is the first authenticated record of Pygmy killer whale in Mozambican waters and the first recent record on the eastern coast of southern Africa since 1970, emphasising the lack of knowledge of offshore marine biodiversity in Mozambique. Previous reported records of the species in Mozambique lie outside Mozambique Exclusive Economic Zones or lack evidence. The species should be included in relevant conservation planning. Available identification material focusses on the separation of this species from Melon-headed whale (*Peponocephala electra*) and fails to note the great similarities of this species with smaller dolphins. At-sea observers are encouraged to consider this species when identifying 'dolphins'. Pygmy killer whales are easily overlooked.

Keywords: Pygmy killer whale, Feresa attenuata, Mozambique, First record, Identification

Background

The biodiversity of the national marine Exclusive Economic Zones (EEZs) of eastern southern Africa is generally poorly known and documented, especially in offshore areas, mostly due to the logistical and financial challenges of working in the region as well as the lack of local scientists. The cetacean fauna of Tanzania has recently been reviewed confirming 20 cetacean species (Brualik, Kasuga, Wittich, Said Shahib Said, Macaulay, Gordon and Gillespie: A Nationwide survey of Cetaceans in Tanzania. WCS Report, unpublished) although several of these species have only been verified on few occasions. The cetacean fauna of Mozambique has barely been described with no formal listing. The best known marine fauna in the region is South Africa where the cetaceans are relatively well-documented (Stuart and Stuart 2007; Best 2007).

National conservation legislations apply to EEZs and rely on an understanding of the species occurring therein for policy implementation. Many cetaceans are highly-mobile and may travel across multiple range

* Correspondence: Gary.allport@birdlife.org

[1]BirdLife International, David Attenborough Building, Pembroke Street, Cambridge CB2 3QZ, UK

Full list of author information is available at the end of the article

states and international conservation agreements have been developed, such as The Convention on Migratory Species and CITES, which rely on the cooperation, participation, and coordination of each of the range states, and so the confirmed list of species included under such agreements is fundamental to their implementation too.

Observations of cetaceans within these relatively poorly known areas also offer opportunities to better document the biology of little-known species, with potential to advance knowledge of identification, habitat preferences and behaviour.

Here we present details of an observation of three Pygmy Killer Whales (*Feresa attenuata* Gray 1874) in Mozambique in April 2017. The animals were recorded on an informal pelagic trip from Maputo, and constitute the first authenticated record for Mozambique.

The Pygmy killer whale is a rare and poorly known cetacean (Donahue and Perryman 2002; Ross and Leatherwood 1994; McSweeney et al. 2009). The species was only known from two skulls collected in the 1800s until the 1950s when a single animal was harpooned by a blackfish-boat from Taiji, Japan (Yamada 1954). Subsequently there have been a scattering of records in tropical and subtropical waters, but particularly in the eastern tropical Pacific, the Hawaiian Archipelago, off Sri Lanka, Taiwan and Japan. Records have mostly been gathered opportunistically from strandings, bycatch and deliberate trapping, with sight records more recently as the species has become better known.

There have been periodic reviews collating records of the species in the eastern Pacific (Van Waerebeek and Reyes 1988), Atlantic (Caldwell and Caldwell 1971) and Indian Oceans (Leatherwood et al. 1991) as well globally (Caldwell and Caldwell 1971; Brownell, Yao, Lee and Wang: Worldwide review of Pygmy Killer Whales, Feresa attenuata, mass strandings reveals Taiwan hotspot, unpublished; Jeyabaskaran et al. 2011). A small number of animals have been individually photo identified (McSweeney et al. 2009) and satellite tagged (Baird et al. 2011) off Hawai'i. The species remains, however, extremely little known and is classified as 'Data Deficient' by the International Union for Conservation of Nature (IUCN) (Taylor et al. 2008) and whilst it can occur locally at a density of 4.49 animals 1000 Km^{-2} (within the Hawai'i Main Island strata; Barlow 2006) it is considered thinly distributed and rare globally (Donahue and Perryman 2002; McSweeney et al. 2009). The species is listed in Appendix II of CITES.

Record gathering has not only been limited due to infrequent encounters but the Pygmy killer whale is also difficult to identify. The species is very similar to two other 'Blackfish' the Melon-headed Whale (*Peponocephala electra*) and, to a lesser extent, the False killer whale (*Pseudorca crassidens*) and there has been re-identification

of animals in the photographic record (Siciliano et al. 2007; Baird 2010; Siciliano and Brownell 2015). The mainstream identification guides have focussed on providing informative, helpful and accurate information on the separation of these three similar Blackfish species (Carwardine 1995; Shirihai et al. 2006) aiding correct identification at-sea significantly.

Pygmy killer whales are mostly recorded in warm, deep waters >250 m and only close to shore around oceanic islands (Donahue and Perryman 2002; McSweeney et al. 2009). Group size is mostly less than 50 animals and small groups of 3–10 animals are regularly encountered. Evidence of group cohesion and faithfulness has been found with individually identified animals cohorting for more than 21 years off Hawai'i Main Island (McSweeney et al. 2009).

The species is known from the Indian Ocean with a relatively well known population off Sri Lanka and a scattering of records across the region from the Gulf of Aden in the northwest, Indonesia in the east, and as far south as southern Africa (Leatherwood et al. 1991). On the Indian Ocean coast of Africa the species is known from a short series of records in South Africa (Best 1970; Findlay et al. 1992) and one record from the French Southern Ocean Territories (GBIF Backbone Taxonomy 2016). The species is listed as occurring in Madagascar (Taylor et al. 2008) but there is no published supporting evidence for this.

The existing knowledge of the distribution, ecology and behaviour of this species has come mostly from opportunistic encounters so it is important that all observations are documented to broaden our knowledge.

Methods

Fourteen pelagic trips have been undertaken from Maputo since May 2011 looking for seabirds, sea mammals and other marine life. The trips last a full day and the route varies with local conditions but always range across Maputo Bay and then out over the shelf to the east into water of 250–750 m depth. A global positioning system is used to record the track and photos are taken of all wildlife of interest.

On 23 April 2017 the route plan was to depart at 05.30 am Central Africa Time (CAT) steam quickly to the north of Inhaca Island, then head more slowly eastwards out into deeper waters looking for birds and other wildlife, chum for seabirds and then head southward along the 400 m depth contour before returning back to Maputo later in the day. The weather and sea conditions were unusually favourable with <1 m swell, light winds <5 knots and sea state 2 all day.

Results

At 06.40 am CAT whilst steaming outbound northwards GA spotted two or more small cetaceans logging on the surface at least c1.5 km distant eastwards from the planned route. They appeared to be Spinner dolphins (*Stenella longirostris*), a species seen on most trips (9 out of 14), and often common. The animals were not approached as they were not noteworthy enough to warrant a significant deviation from the planned route. GA noted, however, that the animals were quite dark in colour – attributed to the early morning light at the time – and were resting stationary on the surface, unusual for Spinner dolphins.

At 12.56 CAT the return route took the vessel 4 km further eastward of the location of the earlier dolphin sighting when three small cetaceans were spotted (25° 52′54.22″S 33°08′33.62″E; Fig. 1) and called as Spinner dolphins by GA and others. A gentle approach was made as the animals were resting on the surface. GA was sure they were the same animals seen six hours earlier. They were moving slowly, socialising and GA quickly became unsure of the identification, noting that the characteristic bottle of a dolphin had not been visible when the animals turned on the surface. Once within 100 m one animal spy hopped showing a rounded head shape and no bottle (Fig. 2).

The three animals (Fig. 3) were identified as Pygmy killer whales based on small size, rounded head shape, white lips and rounded front flipper tips (Fig. 4) (Carwardine 1995). In addition the white to pinkish white belly (Fig. 5) and small group size supported the identification. Efforts were made to photograph and video [see Additional file 1] the animals, and a careful approach yielded a friendly encounter.

The sighting continued for 35 min during which time the three animals moved 0.57 km from the initial location, slowly (0.98 km/h) southwards staying on the surface interacting with each other intensively, mostly ignoring the boat, but on two occasions approaching the bow inquisitively within 2–3 m. Water temperature was 25.8 °C and depth at the location was 235 m on the shelf edge.

There were three animals, two of which were similar in size, estimated at c.2 m length. One animal was more lightly built, darker in colour with a smaller dorsal fin (Fig. 6). It was estimated to be 5–10% smaller than the other two. All three animals showed characteristic scarring (Fig. 4). The plate in Shirihai et al. (2006) suggests that such scarring is characteristic of male Pygmy killer whales, although no supporting evidence is given.

The three animals exhibited an interesting behaviour, staying within very close proximity of each other, often surfacing synchronously and bunching up tightly. The group would come to a dead stop and at which time one animal would clamber over the top of the other two, starting from a parallel position and crossing the other two animals on or near the surface, which made no particular attempt to move. The animal would finish the manoeuvre crossing to the other

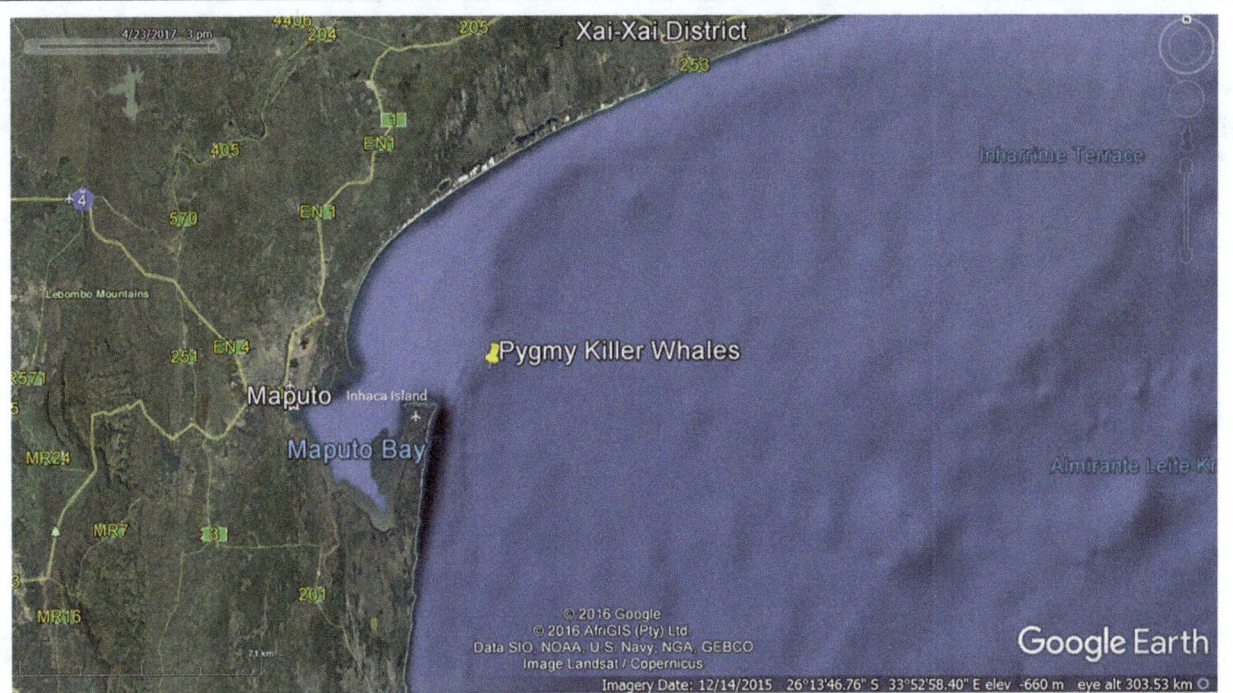

Fig. 1 Map of location of Pygmy killer whales *Feresa attenuata* off Maputo, Mozambique 23 April 2017. Google Earth copyright attribution as per image

Fig. 2 Pygmy killer whale *Feresa attenuata* off Maputo, Mozambique 23 April 2017. Half-breaching, note pale snout. Photo C. Curtis

side of the other two animals at or near to the rear of their dorsal fins. An additional movie file shows this behaviour [see Additional file 1]. Close examination of the video showed that it was the smaller darker animal that was undertaking this cross-over behaviour.

The decision was taken to leave the group of whales after 35 min. Afterwards a group of 5–6 Spinner Dolphins were seen 7 km further to the south west, in typical mode, moving quickly and not seen well, and a group of c15 Bottle-nosed Dolphins *Tursiops truncatus* were seen 11 km to the south west which were boat friendly and rode the bow for several minutes.

The photos and videos of the Pygmy killer whales were shared with experts who confirmed the identification (R. Baird, A. Martin in litt. 2017.).

Discussion

Records in Mozambique

In a collation of records worldwide Jeyabaskaran et al. (2011) cites two records of Pygmy killer whale in Mozambique. However, both these records are cited as from Leatherwood et al. (1991) the first of which, of nine individuals, appears to be a misinterpretation of the data table in Leatherwood et al. (1991) – there being no record of nine individuals described therein – and the second record of 25–30 animals seen on 6th July 1985 (reported by J. Beadon to S. Leatherwood and not assigned to a country) is from a location that lies within French Southern Ocean Territories off Europa Island and not in the Mozambique EEZ. There is a second record on 13th June 1985 from the same locality reported as

Fig. 3 Three Pygmy killer whales *Feresa attenuata* off Maputo, Mozambique 23 April 2017. Note synchronous surfacing [see video in Additional file 1]. Photo (video grab) G. Allport

Fig. 4 Pygmy killer whale *Feresa attenuata* off Maputo, Mozambique 23 April 2017. Note white lips and snout, scratches on body as well as rounded tips to pectoral fins. Photo C. Curtis

a gillnet catch in Leatherwood and Reeves (1989) not reported in Jeyabaskaran et al. (2011). Both these latter records are within the Mozambique Channel but are not in Mozambique.

The only other possible record in Mozambique is of an animal photographed at Pemba, Mozambique which is undated and not formally published (Reichelt and Baines 2012) although looks very likely to be a Pygmy killer whale.

The sighting reported herein is therefore the first authenticated record of this rare cetacean for Mozambique and is of significance in flagging both the limited knowledge of Mozambique's marine EEZ and the likely significant marine biodiversity therein.

Tagging studies of two animals in Hawai'i showed Pygmy killer whales travelled at an average speed of 3.1 km/h and ranging up to 79 and 106 km straight line distance in 10 and 21 days of tagging respectively. Both animals stayed within close proximity of the island along the shelf edge (Baird et al. 2011). Recent sighting reports

from Australia also found four groups of this species on the shelf edge within a relatively finite area (Owen and Donnelly 2014). It is also worth stressing that within the known home range of the resident Hawai'i group, sightings were only reported once per 35 observation days (McSweeney et al. 2009). It is therefore quite possible that a small group of these rare animals are resident in this area of southern Mozambique and have passed unnoticed until now.

Shirihai et al. (2006) report the species as showing boat shyness and if this were to be the case then it would compound the problems of confirming the presence of this species. However, the evidence for this is lacking and the animals in our encounter were not boat shy, indeed were mildly inquisitive.

It is worth clarifying the other records from the Indian Ocean coast off South Africa here. The four records reported by Ross (1984) are summarised in Table 1. Leatherwood et al. (1991) may not have had access to Ross (1984) as their summary table only includes two records

Fig. 5 Pygmy killer whales *Feresa attenuata* off Maputo, Mozambique 23 April 2017. Note pinkish white rear belly and rounded tips to pectoral fins. Photo C. Curtis

Fig. 6 Pygmy killer whales *Feresa attenuata* off Maputo, Mozambique 23 April 2017. Showing smaller darker invidual (rear, right). Photo C. Curtis

from 1968/9 with question marks on an undated record from Port Elizabeth. Jeyabaskaran et al. (2011) follows Leatherwood et al. (1991) but includes a record from 1974 attributed to Ross (1979) – the latter is a study of the Pygmy and Dwarf sperm whales (*Kogia* spp.) and makes no reference to Pygmy killer whale so this record should be discounted. Whilst not in the Indian Ocean, it should also be noted that Jeyabaskaran et al. (2011) incorrectly categorise a record of this species from Namibia (Oosthuizen 2004) in their tabulation as from South Africa.

It is notable that there have been no records from the region since 1970 even with greater coverage by knowledgeable observers. Three of the four historical records from the region are in April/May, as is the record herein. The Pygmy killer whale is considered a warm water species (Donahue and Perryman 2002) and it may be that the species is a wanderer to the region's seas in small groups when water temperature and prey availability are suitable. A warm water incursion was attributed to the first record of Pygmy killer whale in the Gulf of California in

2014 (Elorriaga-Verplancken et al. 2016). However, water temperatures off Maputo were not notably high at the time of the observation (GA pers. obs., http://www.fishtrack.com/fishing-charts/south-africa-east_62420) but it may be that cooler water temperatures to the south mean that the species only wanders further in to the subtropics when warmer water incursions take place.

Identification

The identification of Pygmy killer whale, as with all cetaceans, has been greatly advanced with the production of advanced field guides. Carwardine (1995) and Shirihai et al. (2006) are the two most useful guides for the Mozambique Channel region, and both carry high quality illustrations of the Pygmy killer whale and clear indication of the key features for separating it from the very similar Melon-headed whale at sea. However, both guides fail to emphasise the similarity of Pygmy killer whale to small dolphins; Shirihai et al. (2006) says "could be overlooked among dolphins". With limited space in their publications it is clear that focus of the

Table 1 Records of Pygmy killer whale *Feresa attenuata* in the Indian Ocean coast of southern Africa

Date	Location	Type	Number	Reference	Specimen no
06.03.1957	Schoemakerskop, Port Elizabeth, South Africa	Skull specimen	1	Ross 1984, Findlay et al. 1992	PEM 1514/09
16.05.1968	Richard's Bay, South Africa	Stranding, specimen (rostrum mandible and teeth)	1	Bass 1969; Best 1970; Ross 1984; Leatherwood et al. 1991; Findlay et al. 1992; Jeyabaskaran et al. 2011	SAM 35601
16.08.1969	At sea off Port St Johns, 31°38′S, 30° 07′E, South Africa	At sea sighting	11	Best 1970; Ross 1984; Findlay et al. 1992; Leatherwood et al. 1991; Jeyabaskaran et al. 2011	
09.04.1970	2 km east of Woody Cape, nr. Port Elizabeth, South Africa	Stranding, specimens	2	Ross 1984; Findlay et al. 1992	PEM 1515/41 & 1515/42

identification material should be on the sibling species of Blackfish most similar to Pygmy killer whale. However, even experienced marine mammal observers are advised to consider this species whenever dolphins are found at-sea and especially where the frontal head shape of the animals encountered remains unseen. It is worth stressing that unlike Melon-headed whale, Pygmy killer whales do not normally lift the full face above the water as they surface to breathe (Shirihai et al. 2006 and this observation) so it is not easy to confirm the lack of a bottle. Furthermore, in calmer waters the small bow wave pushed in front of the face looks like a bottle from a distance [see Additional file 1]. In our encounter, the first confirmation of the head shape was when an animal breached, and high quality digital photos were examined closely. The frontal area of the head is more bulbous and melon-shaped than in the confusingly named Melon-headed whale but the bulbous head shape is still not very evident when the Pygmy killer whales are moving and especially when heading away from the observer, as is very often the case in at-sea conditions. Small dark dolphins should always be checked carefully with Pygmy killer whale in mind.

Conservation issues

IUCN lists the Pygmy killer whale as Data Deficient (Taylor et al. 2008). In Mozambique, there are a number of pieces of legislation directly and indirectly applicable. All whales are protected by the Recreational and Sports Fishing Regulation (Decree 51/99 of 31 August 1999) but only for recreational and sports fishers. The international trade of this species is controlled through Appendix II of CITES, of which Mozambique is signatory. The Indian Ocean Tuna Commission, of which Mozambique is also signatory, only requests that in the case that cetaceans are caught, then the incident should be duly reported. The species is not listed by Convention of Migratory Species.

The record of Pygmy killer whale reported here lies close (7 nautical Miles [nM]) to the northern boundary of one of Mozambique's premier marine protected areas, the Ponta do Ouro Partial Marine Reserve, which includes Inhaca Island and the offshore waters up to 3 nM. Whilst it is possible that the species occurs within the reserve, the shelf edge marine habitat that this species frequents is on the very seaward boundary of the reserve.

There have been reports in Hawai'i of Pygmy killer whales showing signs of fishing line cuts to the head and of hooks from long lining gear embedded in the mouth (Cascadia Research Collective 2017). This species is therefore likely to be vulnerable to long lining activities targeting billfish and tuna in Mozambican waters.

Evidence from stranded individuals of several similar species indicates that they have swallowed discarded plastic items, which may eventually lead to death (e.g. Scott et al. 2001); this species may also be at risk.

Pygmy killer whale, like beaked whales, is likely to be vulnerable to loud anthropogenic sounds, such as those generated by navy sonar and seismic exploration (Cox et al. 2006) and have been a part of multi-species unusual stranding events in Taiwan (Wang and Yang 2006). There is some evidence that this is a potential problem in Mozambique (Raba 2006).

This species does not appear to be particularly abundant anywhere that it has been sighted. In Hawai'i, sub-populations appear to be small (McSweeney et al. 2009), and this, along with their limited movements (Baird et al. 2011), suggests the species may be particularly vulnerable to human impacts regionally (Taylor et al. 2008).

Conclusions

This is the first authenticated record of Pygmy killer whale in Mozambican waters emphasising the lack of knowledge of offshore marine biodiversity in Mozambique and the potential for the country to hold as yet undiscovered rare marine biodiversity. The species should be included in relevant conservation planning.

Available identification material focusses on the separation of Pygmy killer whale from Melon-headed Whale and fails to note the similarities of this species with other smaller dolphins. At-sea observers are encouraged to consider this species when identifying any 'dolphins' they encounter.

Abbreviations
CAT: Central Africa Time; CITES: Convention on International Trade in Endangered Species; E: East; EEZ: Exclusive Economic Zone; GPS: Global positioning system; IUCN: International Union for the Conservation of Nature; km: kilometres; km/h: kilometres per hour; m: metres; nM: nautical Miles; S: South

Acknowledgements
Thanks go to all members of the trip that made it viable, including Tim Petersen and Manuel Costeira da Rocha. Anouk Ilangakoon, Robin Baird, Gill Braulik, Tim Collins, Tony Martin, Hugo Rainey, Melinda Rekdahl and Dylan Walker helped with identification, information sources and gave other advice.

Funding
World Wildlife Fund – Mozambique covered the costs of publication of this paper.

Authors' contributions
GA led the research trip, gathered basic trip data, identified the animals in the field, took photographs and video, undertook literature review and wrote the paper. CC took photographs and video. TPS spotted the animals and skippered the boat in a cetacean friendly manner. MJM wrote the section on conservation in Mozambique. All authors read and approved the final manuscript.

Competing interests

The authors declare that they have no competing interests.

Author details

[1]BirdLife International, David Attenborough Building, Pembroke Street, Cambridge CB2 3QZ, UK. [2]Casa 31, Condominio Sommerschield II, Rua do Palmar No. 817, Maputo, Mozambique. [3]Number One Serviços Maritimos, Av. Karl Marx nr. 478, 6 andar, Maputo, Mozambique. [4]WWF Mozambique, Av. Kenneth Kaunda 1174, Maputo, Mozambique.

References

Baird RW. Pygmy Killer Whales (Feresa attenuata) or False Killer Whales (Pseudorca crassidens)? Identification of a group of small cetaceans seen off Ecuador in 2003. Aquat Mamm. 2010;36(3):326–7. http://dx.doi.org/10.1578/AM.36.3.2010.326.

Baird RW, Schorr GS, Webster DL, McSweeney DJ, Hanson MB, Andrews RD. Movements of two satellite-tagged pygmy killer whales (Feresa attenuata) off the island of Hawai'i. Mar Mamm Sci. 2011;E332–7. http://dx.doi.org/10.1111/j.1748-7692.2010.00458.x

Barlow J. Cetacean abundance in Hawaiian waters estimated from a summer/fall survey in 2002. Mar Mamm Sci. 2006;22:446–64.

Bass A. A rare whale stranded in Zululand. Bull S Afr Assoc Mar Bio Res. 1969;7:36.

Best PB. Records of the Pygmy killer whale, Feresa attenuata, from Southern Africa, with notes on behaviour in captivity. Ann S Afr Mus. 1970;57:1–14.

Best PB. Whales and Dolphins of the Southern African Subregion. Cambridge, United Kingdom & Cape Town, South Africa: University Press; 2007.

Caldwell DK, Caldwell MC. The Pygmy killer whale, Feresa attenuata, in the western Atlantic, with a summary of world records. J Mammal. 1971;52:206–9.

Carwardine M. Whales, dolphins and porpoises. London: Dorling Kindersley; 1995.

Cascadia Research Collective. Pygmy killer whales in Hawai'i. http://www.cascadiaresearch.org/hawaiian-cetacean-studies/pygmy-killer-whales-hawaii. Accessed 16 May 2017.

Cox TM, Ragen TJ, Read AJ, Vos E, Baird RW, Balcomb K, Barlow J, Caldwell J, Cranford T, Crum L, D'Amico A, D'Spain A, Fernández J, Finneran J, Gentry R, Gerth W, Gulland F, Hildebrand J, Houser D, Hullar T, Jepson PD, Ketten D, Macleod CD, Miller P, Moore S, Mountain D, Palka D, Ponganis P, Rommel S, Rowles T, Taylor B, Tyack P, Wartzok D, Gisiner R, Mead J, Benner L. Understanding the impacts of anthropogenic sound on beaked whales. J Cetacean Res Manag. 2006;7(3):177–87.

Donahue MA, Perryman WL. Pygmy Killer Whale (Feresa attenuata). In: Perrin WF, Würsig B, Thewissen JGM, editors. Encyclopedia of marine mammals. San Diego: Academic; 2002. p. 1009–10.

Elorriaga-Verplancken FR, Rosales-Nanduca H, Paniagua-Mendoza A, Martínez-Aguilar S, Nader-Valencia AK, Robles-Hernández R, Gómez-Díaz F, Urbán RJ. First Record of Pygmy Killer Whales (Feresa attenuata) in the Gulf of California, Mexico: Diet Inferences and Probable Relation with Warm Conditions During 2014. Aquat Mamm. 2016;42:20–6.

Findlay KP, Best PB, Ross GJB, Cockcroft VG. The distribution of small Odontocete Cetaceans off the coasts of South Africa and Namibia. S Afr J Mar Sci. 1992;12:237–70. doi:10.2989/02577619209504706.

GBIF Backbone Taxonomy. GBIF Secretariat. Checklist Dataset. 2016. https://doi.org/10.15468/39omei Accessed via GBIF.org on 17 May 2017.

Gray JE. Description of the skull of a new species of dolphin (Feresa attenuata). Annu Mag Nat Hist. 1874;4:238–9.

Jeyabaskaran R, Vivekanandan PS, Yousuf KSSM. First record of Pygmy killer whale Feresa attenuata Gray, 1874 from India with a review of their occurrence in the World Oceans. J Mar Biol Assoc India. 2011;53:208–17.

Leatherwood S, Reeves RR. Marine mammal research and conservation in Sri Lanka 1985–1986. UNEP Mar Mamm Tech Rep. 1989;1:1–138.

Leatherwood S, Mcdonald D, Prematunga WP, Girton P, Ilangaakoon A, Mcbrearty D. Records of the "Blackfish" (Killer, False Killer, Pygmy Killer and Melon-headed Whales) in the Indian Ocean. UNEP Mar Mamm Tech Rep. 1991;3:33–65.

McSweeney DJ, Baird RW, Mahaffy SD, Webster DL, Schorr GS. Site fidelity and association patterns of a rare species: Pygmy killer whales (Feresa attenuata) in the main Hawaiian Islands. Mar Mamm Sci. 2009;25:557–72.

Oosthuizen WH. Progress report on cetacean research, January 2004 to December 2004, with statistical data for the calendar year 2004. 2004. http://www.iwcoffice.org/_documents/sci_com/2005progreports/SC-57-ProgRepSouthAfrica.pdf. 17 May 2017.

Owen K, Donnelly D. The most southerly worldwide sightings of Pygmy killer whales (Feresa attenuate). Mar Biodivers Rec. 2014;7:e46. doi:10.1017/S1755267214000463.

Raba N. Mass stranding of bottlenose dolphins in Mozambique. MARMAM. 2006. https://lists.uvic.ca/pipermail/marmam/2006-October/000704.html. Accessed 17 May 2017.

Reichelt M, Baines M. Pygmy killer whale breaching. 2012. http://www.wildscope.com/ocean-life/Mozambique.html. 17 May 2017.

Ross GJB. Records of Pygmy and Dwarf sperm whales, genus Kogia, from Southern Africa, with biological notes and some comparisons. Ann Cape Prov Mus (Nat Hist). 1979;11:259–337.

Ross GJB. The smaller cetaceans of the south east coast of southern Africa. Ann Cape Prov Mus. 1984;15:173–410.

Ross GJB, Leatherwood S. Pygmy killer whale Feresa attenuata Gray, 1874. In: Ridgway SH, Harrison RJ, editors. Handbook of marine mammals. 5th ed. London: Academic; 1994. p. 387–404.

Scott MD, Hohn AA, Westgate AJ, Nicolas JR, Whitaker BR, Campbell WB. A note on the release and tracking of a rehabilitated Pygmy sperm whale (Kogia breviceps). J Cetacean Res Manag. 2001;3:87–94.

Shirihai H, Jarrett B, Kirwan GM. Whales, dolphins, and other marine mammals of the world. Princeton: Princeton University Press; 2006.

Siciliano S, Brownell RL. Getting to know you: Identification of Pygmy Killer Whales (Feresa attenuata) and Melon-headed Whales (Peponocephala electra) under challenging conditions. Braz J Oceanogr. 2015;63(4):511–4.

Siciliano S, Moreno IB, Silva E. Early sightings of the Pygmy killer whale (Feresa attenuata) off the Brazilian coast: a correction to Rossi-Santos et al. (2006). Mar Biodivers Rec. 2007;1:1–3.

Stuart C, Stuart T. Field Guide to the Mammals of Southern Africa. 4th ed. South Africa: Struik Nature; 2007.

Taylor BL, Baird R, Barlow J, Dawson SM, Ford J, Mead JG, Notarbartolo di Sciara G, Wade P, Pitman RL. Feresa attenuata. The IUCN Red List of Threatened Species. 2008;e.T8551A12921135. http://dx.doi.org/10.2305/IUCN.UK.2008.RLTS.T8551A12921135.en. Downloaded on 14 May 2017.

Van Waerebeek K, Reyes JC. First record of the Pygmy killer whale, Feresa attenuata Gray, 1875 from Peru, with a summary of distribution in the eastern Pacific. Z Säugetierkunde. 1988;53:253–5.

Wang JY, Yang SC. Unusual cetacean stranding events of Taiwan in 2004 and 2005. J Cetacean Res Manag. 2006;8:283–92.

Yamada M. An account of a rare porpoise Feresa (Gray) from Japan. Sci Rep Whales Res Inst. 1954;9:59–88.

34

Seeing the invisible: *Chriolepis lepidota* (Gobiidae), literally as never seen before

J. Tavera*[iD] and S. Rojas-Vélez

Abstract

Background: For the first time, almost half a century after its discovery and description, the poorly known endemic gobiid fish *Chriolepis lepidota* was seen alive at Malpelo Island.

Methods: During a 12-day expedition on March 2017, 18 specimens of this species were observed and photographed at different depths by means of SCUBA diving.

Results: Species maximum size and habitat preference are herein documented.

Conclusions: This sighting represents the first record of the species in the wild. Also, this report increases our knowledge on the ecology and biology of an unknown species.

Keywords: Malpelo Island, Endemic, Pretty goby, Fish

Background

The family Gobiidae includes about 210 genera and at least 1950 species, and is considered one of the most diverse groups among bony fishes (Nelson, 2006). Small size coupled with ecological and physiological flexibility has allowed members of this family to live in many different and sometimes harsh habitats (Thomson et al., 2000). Despite tropical habitats are rich in gobiids, they are often inconspicuous because of their tiny size and ecology; generally, gobies are cryptic species occupying crevices or interstices in the sand, reef or rocky substrates (Thacker, 2011). Even though, gobiids are an important component of the biodiversity in almost every environment, they are often poorly known and frequently misidentified. Amongst gobies the New World genus *Chriolepis* Gilbert 1892 are small, secretive, fishes with cryptobenthic lifestyle. They are sedentary species found in primarily insular and spatially restricted areas of reef-rock and rubble habitats in moderately-deep to deep shelf waters, typically known from only a few specimens (Findley, Unpub. PhD Diss), (Hastings & Findley, 2013; Hastings & Findley, 2015). Recent molecular studies (Tornabene et al., 2016) recognized the non-monophyly of *Chriolepis*, recovering the

Atlantic species *Pycnomma rooseveltii*, and the Pacific *Pycnomma semisquamatum* (now *Chriolepis rooseveltii* and *C. semisquamata*, respectively) nested within this genus. *Chriolepis* has divided pelvic fins, although the inner bases of the fins are closely approximated; typically, species have seven spines in the first dorsal fin (Findley, Unpub.PhD Diss). As currently defined, *Chriolepis* differs from other genera of seven-spined gobies in lacking head pores in all species but *C. rooseveltii* (Ginsburg, 1939) and *C. semisquamata* (Rutter, 1904) and by having at least some pelvic-fin rays branched (Hastings & Findley, 2015; Tornabene et al., 2016). Most species can be distinguished by extent of squamation or a combination of this character and color pattern. The inactive behavior of these secretive fishes, combined with small body size related to tight-crevice and rock-interspace inhabitation, has favored morphological adaptation and geographical isolation in these fishes (Findley, Unpub. PhD Diss). Eight species of *Chriolepis* occurs in the tropical eastern Pacific, three of which are endemics of oceanic islands with two of them only known from one or two specimens (e.g. pretty goby *Chriolepis lepidota* Findley 1975, Malpelo Island; and mystery goby *Chriolepis tagus* Ginsburg 1953, Galapagos Islands). This work presents for the first time, after its discovery, habitat data and photos in situ of the poorly known Malpelo endemic *C. lepidota*.

* Correspondence: jose.tavera@correounivalle.edu.co
Laboratorio de Ictiología, Grupo de Investigación SEyBA, Departamento de Biología, Universidad del Valle, Cali, Colombia

Methods

Located on the Malpelo Ridge, a volcanic submarine crest that extends northeast-southwest, Malpelo (4°0′07″N; 81° 36′27″W) is a small Colombian oceanic island in the East Pacific Ocean, separated from mainland by approximately 500 km and depths greater than 3300 m (Graham, 1975). The island is part of a Colombian National Natural Parks system and actually is considered as a wildlife sanctuary, declared by UNESCO as a natural World Heritage Site. The present island is the remnant of a much larger structure, once eight to ten times bigger than its present size (Stead, 1975). Cocos, Malpelo and Carnegie Ridges are interpreted to be traces that began to form when the Galápagos hotspot initiated at ~20–22 Ma (Hey, 1977; Lonsdale et al., 1978). The Malpelo Ridge was separated from the Carnegie Ridge in the Miocene by now-extinct seafloor spreading (Lonsdale et al., 1978). Volcanic rocks from Malpelo Island yielded ages around 17 ~ 15 Ma (Hoernle et al., 2002).

Malpelo is one of several oceanic volcanic islands in the tropical eastern Pacific that have never been connected, even by shallow water, with any other islands or the mainland (Graham, 1975). Weather has eroded the island forming steep cliffs and sea caves along its sides (Stead, 1975). The site in which *C. lepidota* was observed is known as "El Arrecife" (4°0′15.81″N; 81°36′15.80″W, Fig. 1). This location has Malpelo's largest coral formation, located between 4 m and 30 m depth (Chasqui et al., 2007), and a flat area with cobbles and rubble interspersed with large-grained sand consisting of eroded coral and shells. The flat area (5–10 m depth) extends to the east about 150 m and sinks in a steep slope down to 30 m.

A total of 40 dives and 60 belt transects (20 × 2 m) were made along different sites of the island, assessing Malpelo's endemic fishes. Transects were performed at different depths ranks, along which fish were counted. According to depth of detection, two arbitrarily categories were designated: shallow for individuals seen above 10 m and deep for those observed below 10 m. Additionally, the diameter of individual grains of sediment, where the fish were hidden, was considered and its classification follows Wentworth scale (Wentworth, 1922). Finally, fish size was estimated by means of a PVC tube labeled each cm.

A combination of remoteness plus enough evolutionary time has driven speciation biogeographic isolation at Malpelo Island, as a consequence five endemic fish species, one of them *C. lepidota*, exist on the island (Chasqui et al., 2011; Robertson & Allen, 2015). This species had only been seen dead, after two specimens were collected using rotenone-based ichthyocide and SCUBA diving during the 1972 Smithsonian Institution-U.S. Navy Expedition to Malpelo Island. Findley (1975) (Findley, 1975) described the species, and the holotype and paratype (the only known specimens, up to this day) were deposited in the National Museum of Natural History (USNM), Smithsonian Institution, Washington, D. C., under catalog numbers, USNM 211456 and 211,457. Until this record, a very good sketch of the species illustrated by Jeanean Thomson was the only image available of this fish (Fig. 2) (Findley, 1975). *Chriolepis lepidota* can be easily distinguished from the additional three species of gobies found at Malpelo Island (*Bollmania spA, Coryphopterus urospilus* and *Lythrypnus dalli*) by having 7 spines on dorsal fin; pelvic fins completely separated; and by its marbled coloration, containing small black and white spots scattered over head and body; 2 black spots on pectoral base; 5 brown bands on body made-up by 2 dark areas separated by a whitish spotted line; yellow-whitish interspaces between bands with a brownish narrow vertical mid-line; a dorsal white thin line over head just behind the eyes; and a dark bar at the base of the caudal fin across the caudal peduncle. Dorsal (VII,11), anal (10 total elements) and pectoral (20) fin counts (made on photographs), as well as color pattern corresponds in many ways to original drawing and description of the species (see Fig. 2 and Fig. 3) (Findley, 1975).

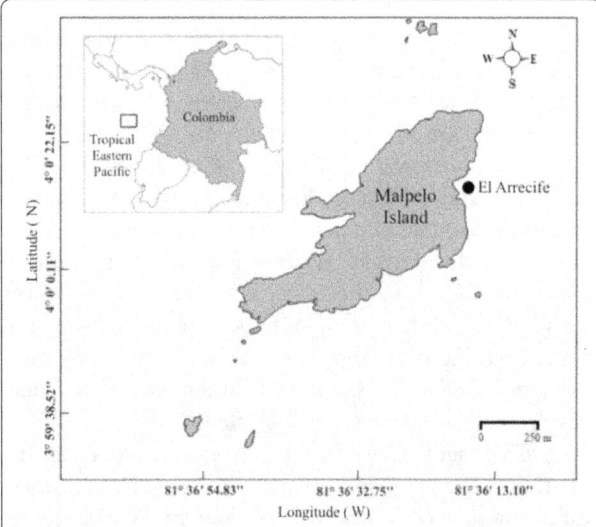

Fig. 1 El Arrecife, site where the endemic gobiid fish *Chriolepis lepidota* was observed. Malpelo Island, Colombia

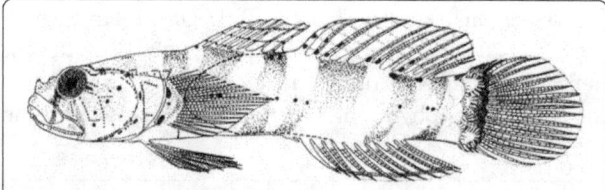

Fig. 2 *Chriolepis lepidotus*, holotype, USNM 211456, male, 30.0 mm SL. Drawing by Jeanean Thomson. Reproduced with permission of Lloyd Findley

Fig. 3 Photographs of *Chriolepis lepidota* at Malpelo Island, Colombia. Pictures credit belongs to Stephania Rojas

Results

Almost half a century passed before *C. lepidota*, a fish that had never been seen alive, could be detected. During a 12-day expedition made from the 5th to the 16th of March 2017, several specimens of the species were photographed (Fig. 3). Coincidentally the only two specimens known of this species were also collected in March (2nd and 3rd) back in 1972. The banded color pattern shown by *C. lepidota*, certainly related to its cryptic mode of life, kept this species out of sight for 45 years. *C. lepidota* was discovered on the east side of the island and its identity was confirmed using (Robertson & Allen, 2015; Findley, 1975) and (Findley, Unpub. PhD Diss). The small marbled looking fish was observed, at depths between 8 m to 18 m, over rocky bottom with some calcareous sand and very sparse algal growth. Ten small (2–4 cm) specimens were detected, on the flat portion of El Arrecife, posed over cobbles where they seek refuge underneath few seconds after being recognized. Eight bigger specimens (4–7 cm), were found deeper over the slope hidden in crevice-like interspaces under or at the bases of boulder size rocks.

The habitat in which *C. lepidota* was found corresponds accurately with previous descriptions of preferred habitats in which most eastern Pacific *Chriolepis* have been collected (Findley, Unpub. PhD Diss). Our observations extend species maximum size to approx. 6 cm.

Discussion

C. lepidota distribution suggests that habitat segregation might be related to size, with smaller individuals (*n* = 10) living on flat shallower bottoms underneath cobbles, and bigger ones (*n* = 8) found over the slope hidden in the interspaces under boulders and rocks crevices sheltered at the interface between sand and rock. Possibly, three major environmental gradients, acting together or independently, influence species segregation: substrate inclination, depth and grain size (Fig. 4). This preliminary result prompts a testable hypothesis for future ecological studies.

According to the IUCN red list of threatened species *C. lepidota* is considered as Vulnerable (Findley & Van Tassell, 2010), being the increased duration and frequency of ENSO events the mayor threats identified, given the restricted range of this species. Despite being an endemic to Malpelo Island, regional assessment includes it under the category Data Deficient given the

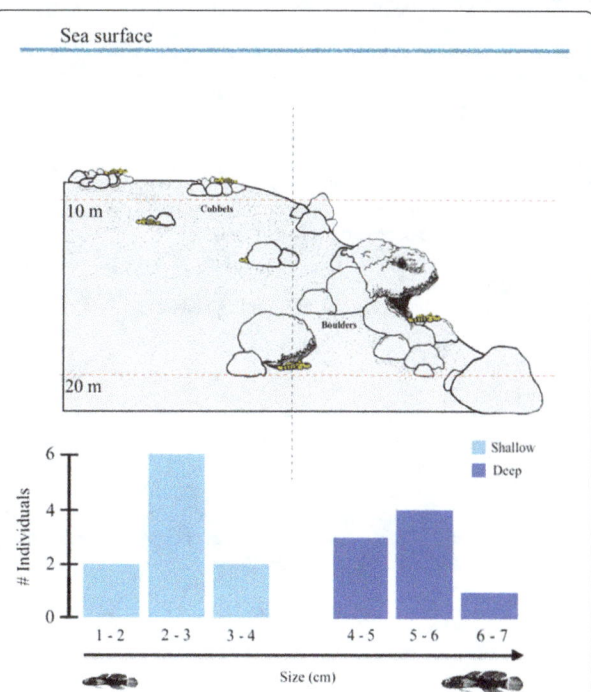

Fig. 4 Schematic diagram of environmental factors inferred to affect *Chriolepis lepidota* habitat segregation (substrate inclination, depth and grain size). The black dashed line indicates an abrupt change on substrate inclination; red dashed lines show 10 m depth increments; cobbles: grain size <250 mm, boulder: grain size >250 mm. Lower bar plot shows total of fish detected by size intervals colored according to depth: shallow correspond to observations above 10 m and deep below 10 m

lack of information available for the species (Zapata & Chasqui, 2017). This paper constitutes the first data published for this species since its discovery.

Finally, there is much more to investigate about this endemic species. Quoting J. L. B. Smith (1958) in (Findley, Unpub. PhD Diss): *"The gobioid fishes are one of the major trials of ichthyologists... Being of little or no economic significance, although normally abundant, especially in tropical areas, these fishes are virtually unknown to any but the expert seeking them".*

Conclusions

Up to this date, reports regarding *Chriolepis lepidota* were lacking and the existence of the species was even questioned. This report represents the first record of *C. lepidota* after its description (1975) and increases our biological and ecological knowledge on this cryptic species. It also highlights the need for a comprehensive assessment of Pacific Colombian fish diversity, which have been overseen for decades.

Acknowledgments
We would like to thank Fernando Zapata and his research group "Ecología de Arrecifes Coralinos", all personnel from SFF Malpelo, people from Fundación Malpelo y otros Ecosistemas Marinos and to Maria Patricia vessel crew. Also, to Arturo Acero and two anonymous reviewers for reading and commenting on an early version of this manuscript. Finally, Lloyd Findley which kindly allow us to use *C. lepidota* original drawing.

Funding
Not applicable.

Authors' contributions
JT sight and identified the fish, collected data and drafted the manuscript; SRV collected data photographed *C. lepidota* and reviewed the manuscript. Both authors read and approved the final manuscript.

Competing interests
The authors declare that they have no competing interests.

References
Chasqui L, Gil-Agudelo DL, Nieto R. Endemic shallow reef fishes from Malpelo Island: abundance and distribution. Bol de Investig Marinas y Costeras – INVEMAR. 2011;40:107–16.

Chasqui L, Zapata FA. Tamaño y composición de dos formaciones coralinas del Santuario de Fauna y Flora Malpelo, Pacífico colombiano. In: INVEMAR, editor. Informe del estado de los ambientes marinos y costeros en Colombia: Año 2006. Santa Marta: INVEMAR; 2007. p. 96–101.

Findley L, Van Tassell J. *Chriolepis lepidota*. In: *The IUCN Red List of Threatened Species 2010*. 2010. https://doi.org/10.2305/IUCN.UK.2010-3.RLTS. T183325A8093991.en. Accessed 21 Apr 2017.

Findley LT. A new species of goby from Malpelo Island (Teleostei: Gobiidae: *Chriolepis*). Smithson Contrib Zool. 1975;176:94–8.

Graham JB. The biological investigation of Malpelo Island, Colombia. Washington: Smithsonian Institution Press; 1975.

Hastings PA, Findley LT. *Chriolepis bilix*, a new species of goby (Teleostei: Gobiidae) from deep waters of the western Atlantic. Zootaxa. 2013; doi:10. 11646/zootaxa.3745.5.8.

Hastings PA, Findley LT. *Chriolepis prolata*, a new species of Atlantic goby (Teleostei: Gobiidae) from the north American continental shelf. Zootaxa. 2015; doi:10.11646/zootaxa.3904.4.8.

Hey R. Tectonic evolution of the Cocos-Nazca spreading center. Geol Soc Am Bull. 1977; doi:https://doi.org/10.1130/0016-7606(1977)88<i:TEOTCS>2.0.CO;2.

Hoernle K, van den Bogaard P, Werner R, Lissinna B, Hauff F, Alvarado G, Garbe-Schönberg D. Missing history (16-71 ma) of the Galápagos hotspot: implications for the tectonic and biological evolution of the Americas. Geology. 2002; doi:https://doi.org/10.1130/0091-7613(2002)030<0795: MHMOTG>2.0.CO;2.

Lonsdale P, Klitgord KD, et al. Geol Soc Am Bull. 1978; doi:https://doi.org/10.1130/ 0016-7606(1978)89<981:SATHOT>2.0.CO;2.

Nelson J. Fishes of the world. 4th ed. Hoboken: John Wiley & Sons; 2006.

Robertson DR, Allen GR. Shorefishes of the Tropical Eastern Pacific: online information system. Version 2.0 Smithsonian Tropical Research Institute, Balboa, Panamá. 2015. http://biogeodb.stri.si.edu/sftep/es/pages. Accessed 21 Apr 2017.

Stead JA. Field observations on the geology of Malpelo Island. Smithson Contrib Zoolo. 1975;176:17–20.

Thacker CE. Systematics of Gobiidae. In: Patzner RA, Van Tassell JL, Kovacic M, Kapoor BG, editors. Biology of gobies. Boca Raton: Science Publishers; 2011. p. 129–36.

Thomson DA, Findley LT, Kerstitch AN. Reef fishes of the sea of Cortez. The rocky-shore fishes of the Gulf of California. Revised ed. Austin: The University of Texas Press; 2000.

Tornabene L, Van Tassell JL, Gilmore RG, Robertson DR, Young F, Baldwin CC. Molecular phylogeny, analysis of character evolution, and submersible collections enable a new classification for a diverse group of gobies (Teleostei: Gobiidae: Nes subgroup), including nine new species and four new genera. Zool J Linnean Soc. 2016; doi:https://doi.org/10.1111/zoj.12394.

Wentworth CK. A scale of grade and class terms for Clastic sediments. J Geology. 1922; doi:https://doi.org/10.1086/622910.

Zapata FA, Chasqui L. Chriolepis lepidota. In: Chasqui L, Polanco A, Acero A, Mejía-Falla PA, Navia A, Zapata LA, Caldas JP, editors. Libro rojo de peces marinos de Colombia. Santa Marta: Instituto de Investigaciones Marinas y Costeras Invemar; 2017. p. 413–5.

35

First record of Omura's whale, *Balaenoptera omurai*, in Sri Lankan waters

Asha de Vos

Abstract

An unusually coloured, small baleen whale was documented off the southern coast of Sri Lanka in February 2017 during routine field surveys. Based on five distinct morphological characteristics including jaw asymmetry, presence of a prominent central rostral ridge, blaze on right side, asymmetrical chevron on left and right sides and a strongly falcate dorsal fin the individual was positively identified as an Omura's whale (*Balaenoptera omurai*). This discovery represents the first confirmed sighting of Omura's whale in Sri Lankan and therefore central Northern Indian Ocean waters.

Keywords: Northern Indian Ocean, Sri Lanka, Omura's whale, Sri Lanka, Distribution, *Balaenoptera omurai*

Background

Originally misclassified as a pygmy form of Bryde's whale (*Balaenoptera edeni*) in the 1970s, the Omura's whale has since been described as a distinct baleen whale species of the family Balaenopteridae (Wada et al. 2003). Sasaki et al. (2006) showed that in fact the Omura's whale represents an ancient independent lineage that diverged around 17 million years ago within the Balaenopteridae.

This species is currently confirmed from the northeastern and South Atlantic, western Pacific and Indian Ocean. The records from the Indian Ocean are largely from the eastern Indian Ocean and more recently, with the discovery of a resident population of Omura's whales in Madagascar, the southwestern Indian Ocean and one from Iran in the northwest Indian Ocean.

This discovery represents the first confirmed documentation of Omura's whale within Sri Lankan waters and therefore the first from the central Northern Indian Ocean.

Results

An Omura's whale was photographed on 5 February 2017 during routine blue whale photo-identification surveys. The solitary individual was documented approximately 7 km from shore in water between 55–65 m deep (Fig. 1). As the research vessel was switched off, the

animal approached the boat enabling a series of photographs highlighting a number of key morphological characteristics to be taken. These characteristics include, jaw asymmetry (Fig. 2a and b), prominent single central rostral ridge and not three as found in *B. edeni* (Fig. 2b), blaze on right side (Fig. 2c), asymmetrical chevron on both right and left sides (Fig. 2c and d) and falcate dorsal fin (Fig. 2e). These characteristics allow us to morphologically distinguish this individual from Bryde's whales that are commonly recorded in Sri Lankan waters.

The following morphologically diagnostic features enabled the identification of this individual as an Omura's whale (Fig. 2).

1. Jaw asymmetry
   As described by Cerchio et al. (2015), this individual showed evidence of asymmetrical colouration of the lower jaws, with a darkly pigmented left jaw (Fig. 2 a) and lightly pigmented right jaw (Fig. 2 b).
2. Presence of a prominent single medial ridge
   The prominent rostral ridge and absence of pronounced lateral ridges (only lightly visible) enabled differentiation from the more commonly sighted Bryde's whale with central ridge and lateral ridges (Wada et al. 2003).
3. Presence of right side blaze
   As described by Cerchio et al. (2015) white pigmentation is more extensive on the right side of the body compared to the left. This individual

Correspondence: ashadevos@gmail.com
Oceanswell and The Sri Lankan Blue Whale Project, 131 W.A.D. Ramanayake Mawatha, Colombo 2, Sri Lanka

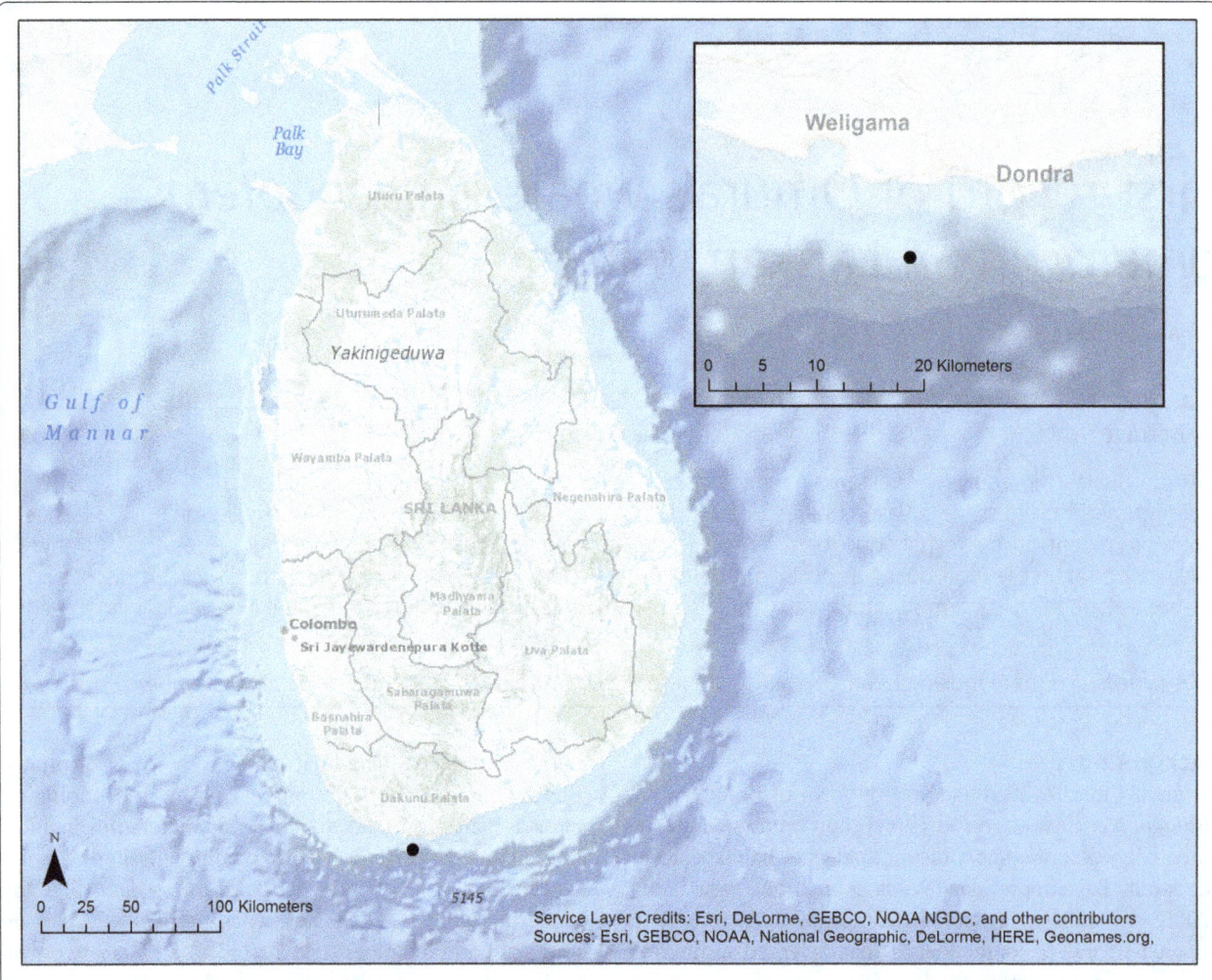

Fig 1 Map showing sighting location of Omura's whale off southern Sri Lanka encountered on 5 February 2017. Black dot on southern coast represents the location of the sighting, which was approximately 6.9 km offshore in water that was between 55–65 m deep

also showed evidence of a lightly pigmented blaze anterior to the eye only present on the right side and multiple dark stripes bisecting the blaze.
4. Asymmetrical chevron on both right and left sides
Lightly pigmented chevron anterior to the dorsal fin evident on both sides but more pronounced on right side of the body, displaying a double-banded pattern also described by Cerchio et al. (2015).
5. Strongly falcate dorsal fin
The dorsal fin was pointed, strongly falcate and backswept (Wada et al. 2003; Cerchio et al. 2015; Jefferson et al. 2008; Ranjbar et al. 2016)

Apart from the distinguishing characteristics, this individual also possessed a mark that resembled a tyre mark on its left dorsal flank and an entanglement scar on the tip of its left rostrum. Both these markings appear to be made by external interactions and are not characteristics of this species.

Discussion

The morphological characteristics described through the photographs taken on 5 February 2017 provide evidence for the presence of Omura's whales in Sri Lankan waters and thereby the central Northern Indian Ocean. A previous record from Iran already indicated presence of this species within the western Northern Indian Ocean (Ranjbar et al. 2016). This is the only confirmed record of an Omura's whale in Sri Lankan waters to date. This is a new record within an expected range within which few sightings/records have previously been available.

The detailed description of the external features of this species provided by Cerchio et al. (2015) were useful for the identification of this individual within Sri Lankan waters. The key features of interest that enable us to discern between Omura's and Bryde's that are often recorded in these waters include the prominent dorsal ridge on the rostrum (Bryde's whales often have three head ridges; the central ridge is flanked by two lateral

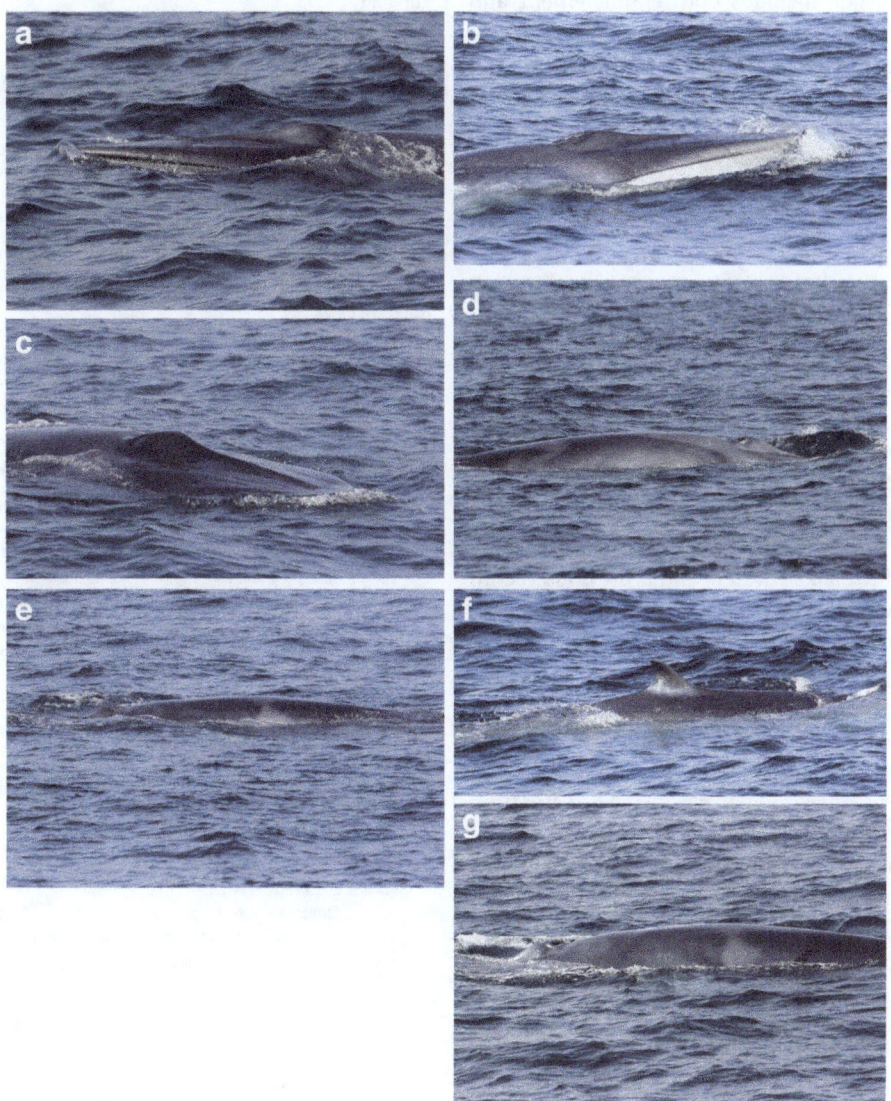

Fig 2 Omura's whale documented off southern Sri Lanka on 5 February 2017. The morphological characteristics captured in these images distinguish this individual from Bryde's whales that are commonly seen in Sri Lankan waters. These characteristics include; Jaw asymmetry with **a** *left jaw* being dark in colour compared to the **b** *right jaw* which is light in colouration; **c** prominent single ridge on rostrum and weak lateral ridges on each side; Chevron on **d** *right* (more prominent) and **e** *left sides*; and **f** strongly falcate dorsal fin. Other markings of note include **a** entanglement scar on *left upper jaw* and **g** 'tyre mark' on *left dorsal flank*

rostral ridges), body colouration (Bryde's whales are dark throughout their bodies), and the shape of the dorsal fin (which is small and falcate in relation to that of the Bryde's whale) (Yamada 2009).

The 'tyre' like markings observed on the left dorsal flank of this individual may represent attachment sites of a remora (Echeneidae) as speculated by Cerchio et al. (2015). Remoras are commonly seen attached on blue whales in Sri Lankan waters, but do not leave any visible markings of this nature likely due to a difference in the physical characteristic of the dermis (Cerchio, pers. Comm..).

This individual also showed evidence of an entanglement scar on the left side of its upper jaw indicating that this is a potential threat for this species in these waters. Because the range of this species is still unknown the threats they face are yet unclear making the documentation of this entanglement scar particularly important. Bycatch in local fisheries has been reported from Songkhla, Thailand (Adulyanukosol et al. 2012), and given its penchant for shallow water habitats, bycatch is likely a threat throughout its range (Cerchio et al. 2015). In Sri Lankan waters, ship-strike is the leading population-level threat to blue whales, followed by incidental catch which includes both entanglement and bycatch (de Vos et al. 2016). Given the smaller size of Omura's whales compared to blue whales it can be considered a particularly pertinent threat to this species in Sri Lankan waters.

The population of Omura's whales off northwest Madagascar was preferentially seen in water that was 4–202 m deep with SST between 27.4 and 30.2 °C (Cerchio et al. 2015). The sighting reported here was made in waters 55–65 m deep within 7 km of the coast providing further evidence that these whales prefer shallow shelf waters.

Cerchio et al. (2015) suggested that because extensive genetic sampling of Bryde's whale populations in the North Indian Ocean did not reveal evidence of *B. omurai* (Kershaw et al. 2013), the distribution of this species is discontinuous with the Madagascar population being potentially isolated from the eastern populations. However, this record from Sri Lankan waters and a previous record from Iran (Ranjbar et al. 2016) may provide some evidence of connectivity across their range.

Given the rarity of this sighting it is important to continue to monitor and record sightings of this species, document resightings of individuals across years and within seasons to estimate population abundance and define movements and ranges, clarify the distinction between Omura's and Bryde's whales to ensure accurate records and identify the threats faced by this species.

The images illustrate the characteristic features of this species and highlight the importance of field surveys and photo-identification work that enable the discovery and description of new species and provide opportunity to expand our knowledge of the marine mammals inhabiting our oceans. As such, please submit any images of Bryde's whales or Omura's whales from Sri Lankan waters, to the respective catalogues by contacting the corresponding author.

Conclusions

This is the only confirmed record of Omura's whales from Sri Lankan waters and the central Northern Indian Ocean. It is a new record within an expected range within which few sightings/records have previously been available and may provide some evidence of connectivity with the populations in the eastern Indian Ocean.

Acknowledgements
All research reported in this manuscript was conducted under a Department of Wildlife Conservation, Sri Lanka permit (number WL/3/2/1/18). I would like to acknowledge Robert Brownell Jr. and Salvatore Cerchio for help in confirming the sighting. Further, I wish to thank my field team Ariesha Wikramanayake, Ben Yexley and Rosalind Bown for their assistance through the field period. Finally, I wish to thank Andrew Lewin for help in producing the map indicating the location of sighting.

Funding
Fieldwork was conducted using funds from a National Geographic Emerging Explorer grant.

Authors' contributions
AdV conducted the fieldwork, analysed the photos and wrote the paper.

Competing interests
The author declares that she has no competing interests.

References
Adulyanukosol K, Thongsukdee S, Passada S, Prempree T, Wannarangsee T. Bryde's whales in Thailand. Bangkok, Thailand: Aksornthai Printing Co.; 2012.

de Vos A, Brownell Jr RL, Tershy BR, Croll DA. Anthropogenic threats and conservation needs of blue whales, *Balaenoptera musculus indica*, around Sri Lanka. J Mar Biol. 2016;2016:12.

Cerchio S, Andrianantenaina B, Lindsay A, Rekdahl M, Andrianarivelo N, Rasoloarijao T. Omura's whales (*Balaenoptera omurai*) off northwest Madagascar: ecology, behaviour and conservation needs. R Soc Open Sci. 2015;2:150301.

Jefferson TA, Webber MA, Pitman RL. Marine mammals of the world: A comprehensive guide to their identification. Canada: Academic; 2008.

Kershaw F, Leslie MS, Collins T, Mansur RM, Smith BD, Minton G, Baldwin R, LeDuc RG, Anderson RC, Brownell Jr RL, Rosenbaum HC. Population Differentiation of 2 Forms of Bryde's Whales in the Indian and Pacific Oceans. J Hered. 2013;104:755–64.

Ranjbar S, Dakhteh MS, Waerebeek KV. Omura's whale (*Balaenoptera omurai*) stranding on Qeshm Island, Iran: further evidence for a wide (sub)tropical distribution, including the Persian Gulf. Journal of Marine Biology and Oceanography. 2016;5:1–9.

Sasaki T, Nikaido M, Wada S, Yamada TK, Cao Y, Hasegawa M, Okada N. *Balaenoptera omurai* is a newly discovered baleen whale that represents an ancient evolutionary lineage. Mol Phylogenet Evol. 2006;41:40–52.

Wada S, Oishi M, Yamada TK. A newly discovered species of living baleen whale. Nature. 2003;426:278–81.

Yamada TK. Omura's whale, *Balaenoptera omurai*. In: Ohdachi S, Ishibashi Y, Iwasa M, Saitoh T, editors. The Wild mammals of Japan. Kyoto, Japan: Shoukahoh Book Sellers and Mammalogical Society of Japan; 2009. p. 330–1.

36

Range extension of the sesarmid crab *Clistocoeloma villosum* along the eastern Pacific coast of the Izu Peninsula, Japan

Takeshi Yuhara[1*], Hiroyuki Yokooka[2] and Masanori Taru[3]

Abstract

Background: The Pacific coastline along the southern Izu Peninsula, Japan, is strongly influenced by warm tropical waters of the Kuroshio Current. A new easternmost record of the near-threatened sesarmid crab *Clistocoeloma villosum* is reported from the southern part of Izu Peninsula.

Methods: The present study was conducted in August 2014 and February 2015, on tidal flats in the mouth of the Aono River, draining the southern part of Izu Peninsula. Crabs were collected by hand on the tidal flat substrate, under cobble stones and on the periphery of associated mangrove forests.

Results and conclusion: Body sizes and morphological characteristics closely matched existing descriptions of *C. villosum*, the distribution range having been extended ca. 350 km eastward from the Kii Peninsula (traditional eastern boundary of the species), suggesting broad northeastwardly directed planktonic larval transport by the warm Kuroshio Current along the Pacific coast of Japan. The survival and settlement of larvae of this southern species along the southern coast of the Izu Peninsula, is evidence of the suitability of the small gravel dominated tidal flats in the region as habitat for the species.

Keywords: *Clistocoeloma villosum*, Brachyura, Near-threatened species, New record, Izu Peninsula, Kuroshio current, Larval dispersion

Background

The coastline of southern Izu Peninsula, southeastern Honshu Island, Japan is strongly influenced by the Kuroshio Current, which carries warm tropical waters in a northeastwardly direction (Yamano et al., 2011). Although the short steeply graded rivers characteristic of the peninsula largely restrict the formation of tidal flats at the river mouths, the Aono River mouth is characterized by very small tidal flats, including salt marshes and semi-mangrove forests. General investigations of benthic macro-invertebrates on the tidal flats have included accounts of a number of crab species (Tanaka et al., 2004; Ito, 2014; Yokooka et al., 2015), as well as snails (Nishiwaki et al., 1991; Murase & Yuhara, 2011). However, detailed investigations of the tidal flats and associated semi-mangrove areas have not yet been undertaken.

The present report details the first records of the sesarmid crab *Clistocoeloma villosum* (A. Milne Edwards, 1869) on the tidal flats and in the semi-mangrove forest area of the Aono River mouth, southern Izu Peninsula. Previously, the distribution range of this crab in Japan had been recognized as between the Ryukyu Islands and Kii Peninsula, southern Honshu Island (see Karasawa et al., 2006), approximately 350 km distant from Izu Peninsula.

Methods

The present study was conducted in August 2014 and February 2015. *Clistocoeloma villosum* was monitored along the Aono River (34°38′06″ N, 138°53′11″ E) at Minami-Izu Town, Izu Peninsula, Shizuoka, Japan (Fig. 1), the intertidal zone habitat of the river mouth being characterized by mud flats, gravels, cobble stones, oyster beds and salt marshes. Plants included the common reed (*Phragmites australis*), semi-mangrove plants, such as *Hibiscus hamabo* (Nakanishi, 2001), and the

* Correspondence: yugo88@nifty.com
[1]Tohoku University Graduate School of Life Science, 6-3 Aoba, Aramaki, Aoba-ku, Sendai, Miyagi 980-8578, Japan
Full list of author information is available at the end of the article

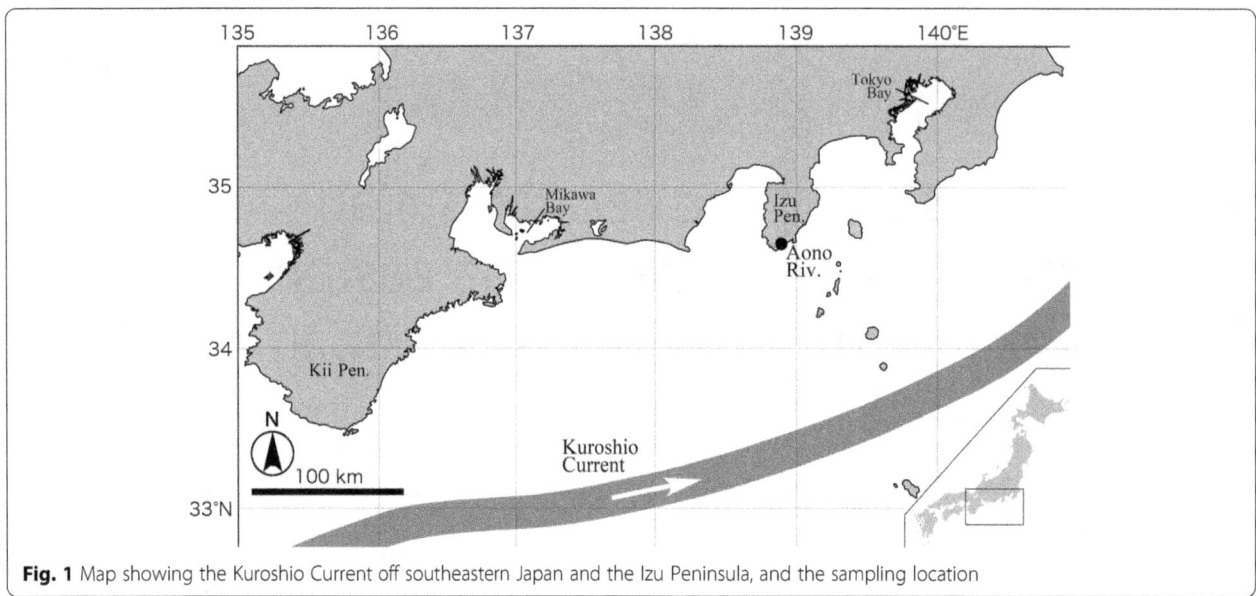

Fig. 1 Map showing the Kuroshio Current off southeastern Japan and the Izu Peninsula, and the sampling location

mangrove *Kandelia obovata*, introduced from Iriomote Island, Japan in 1959 (Masuda, 1999).

Brachyuran crabs were collected by hand from the mud flat surface, under gravels and cobble stones around the mangrove forest edges. All crabs collected were stored in 70% ethanol and representative specimens deposited in the Osaka Museum of Natural History. Familial and generic classification followed the guidelines proposed by de Grave et al. (2009). Descriptions and specific keys used for identifying *C. villosum* included Tesch (1917), Crosnier (1965), Nomoto et al. (1999), Komai et al. (2004) and Lee et al. (2010).

Results
Systematics
Order DECAPODA (Latreille, 1802)

  Infraorder BRACHYURA (Linnaeus, 1758)

  Family SESARMIDAE (Dana, 1851)

  Genus *Clistocoeloma* (A. Milne Edwards, 1873)

  Clistocoeloma villosum (A. Milne Edwards, 1869) Fig. 2

  Sesarma villosum A. Milne-Edwards, 1869:31.

  Sesarma (*Holometopus*) *villosa*: Tesch, 1917:208, pl. 17, Fig. 2.

  Sesarma (*Holometopus*) *villosum*: Crosnier, 1965:55, figs 75, 76, 77a, 78.

  Chiromantes villosum: Nomoto et al., 1999:9, pls. 1-6; Kishino et al., 2001:17, pl. 2,2; Shokita et al., 2002:78, photo 4A-1.

  Clisotocoeloma villosum: Davie, 2002:221; Komai et al., 2004:39, fig. 3; Lee et al., 2010:180, Fig. 1; Maenosono & Saeki, 2016:5, Fig. 2d.

  Material examined (all from Aono River mouth, southern Izu Peninsula, Shizuoka, Japan): OMNH-Ar10119, 1 male, CW: 11.6 mm, CL: 9.6 mm, 1 female,

CW: 10.3 mm, CL: 8.4 mm, upper intertidal, mud bottom under cobble stone, coll, T. Yuhara, 31 August 2014; OMNH-Ar10120, 1 male, CW: 15.1 mm, CL: 12.7 mm, 1 female, CW: 9.1 mm, CL: 7.8 mm, upper intertidal, mud bottom under cobble stone, coll, T. Yuhara, 7 February 2015.

Diagnosis
Carapace rectangular (Fig. 2a), greatest width across middle, about 1.2 times length; dorsal surface with numerous very short stiff setae, often in small groups; lateral margins slightly sinuous, anterior part slightly convex, lined with short stiff setae similar to those on dorsal surface; frontal margin moderately deflexed with faint median notch. No trace of epibranchial tooth. Antennule and antenna contiguous. Basal antennular segment subrectagular. Antenna set oblique; flagellum relatively long, extending into orbit. Third maxilliped with well development flagellum on expond.

Chelipeds of male (Fig. 2b) and female (Fig. 2c) subequal; palm of male with 1 lined, partially pectinated ridge along entire length of dorsal surface, comprising a few small granules and 20-38 small to large corneous teeth; inner surface comprising a row of small tubercles adjacent to dorsal margin (Fig. 2d); palms of females with a few small granules only on dorsal surface; upper border of dactylus of male bearing ca. 30 rectangular tubercles in a line toward lower tip (Fig. 2e), that of female bearing 15 small tubercles in a line becoming base toward middle of dactylus.

Ambulatory legs (Fig. 2f) (second to fifth pereopods) moderately short, third and fourth legs of similar length, longer than second and fifth, covered with short stiff setae; merus broad, terminal of anterior upper margin

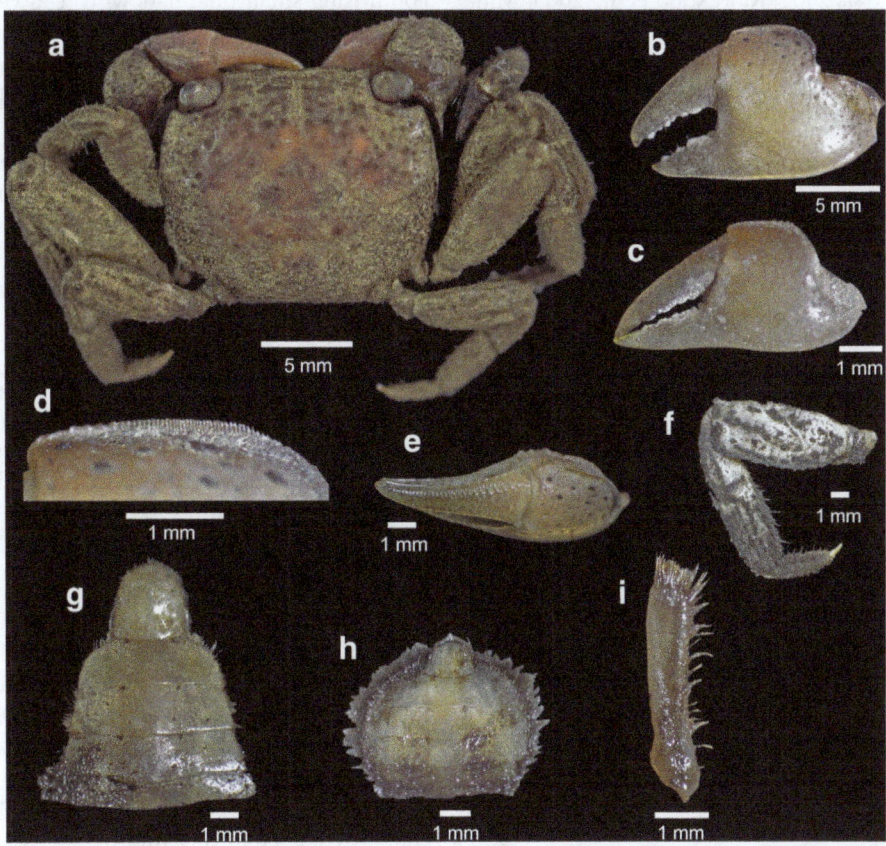

Fig. 2 *Clistocoeloma villosum* (A. Milne Edwards, 1869), male, CW 15.1 mm, CL 12.7 mm, female, CW 9.1 mm, CL 7.8 mm (OMNH-Ar10120), Aono River. **a** morphogy of dorsal view of carapace of male; **b** left cheliped of male, outer view; **c** left cheliped of female, outer view; **d** inner dorsal margin of *left* palm of male, showing crest, outer view; **e** left cheliped of male, dorsal view; **f** left fifth perepod of male, dorsal view; **g** abdomen of male; **h** abdomen of female; **i** left first pleopod, outer view

without corneous spinules; propodus and dactylus covered with numerous short stiff setae, upper and lower margins lined with rows of longish setae, dactylus terminating in corneous claw, tip without hairs.

Abdomen of male (Fig. 2g) wide, narrowest at base of telson, covered with short setae; fifth and sixth abdominal somites of similar length; telson slightly longer than basal width. Abdomen of female (Fig. 2h) with telson evenly rounded, as long as basal width and longer than mid-line length of sixth abdominal somite. The anterior of sixth abdominal somite gently sunk and put the telson in it.

First gonopod (Fig. 2i) stout, nearly straight; terminal process short, with shallow notch distally.

Crosnier (1965) and Lee et al. (2010) reported that the male chelipeds in this species have 15-16 corneous teeth on the dorsal surface of the palm, whereas Tesch (1917) and Komai et al. (2004) reported males with 20-30 teeth. The present male specimens had 20-38 such teeth, the number varying with body size.

The collected specimens corresponded closely to the descriptions of *Sesarma* (*Holometopus*) *villosa* provided by Tesch (1917), *Sesarma* (*Holometomus*) *villosum*, provided by Crosnier (1965), *Chiromantes villosum* provided by Nomoto et al. (1999), and *Clistocoeloma villosa*, provided by Komai et al. (2004) and Lee et al. (2010).

Coloration
Generally in life, carapace muddy colour; palm of cheliped purplish.

Distribution
Clistocoeloma villosum is widely distributed in the Indo-Pacific Ocean, including Madagascar, Aceh, Sumatra, New Guinea, Queensland, Australia, Caroline Islands, Samoa Islands and Korea (Jejudo Island) (Tesch, 1917; Crosnier, 1965; Komai et al., 2004; Lee et al., 2010). Previous records along Japanese coastal regions include the central and southern Ryukyu Islands (Miyako, Iriomote, Ishigaki, Okinawa and Amami-Ohshima Islands), Kyushu Island (Miyazaki Prefecture, Nagasaki Prefecture), Shikoku Island (Ehime Prefecture) and the Kii Peninsula, extending from the Honshu mainland (Wakayama Prefecture) (Kishino et al., 2001; Shokita et

al., 2002; Komai et al., 2004; Kawakubo et al., 2005; Ministry of the Environment, 2005; Karasawa et al., 2006; Miura & Jitsumasa, 2010; Wada, 2013; Maenosono & Saeki, 2016).

Ecological note

Clistocoeloma villosum dwells under stones on the upper intertidal zone of tidalflats, in salt marshes and on the landward edges of semi-mangrove forests. Uncommon, except in the southern Ryukyu Islands, the species is recognized as near threatened in Japan (Japanese Association of Benthology, 2012; Ministry of the Environment, 2017). The new locality is approximately 350 km northeast of the Kii Peninsula, being the easternmost reported to date (see Karasawa et al., 2006).

Discussion

Although the Kii Peninsula has been the traditionally-recognized eastern boundary of *Clistocoeloma villosum*, that species, along with several other crabs, i.e., *Ptychognathus capillidigitatus* (Yokooka et al., 2015), *Macrophthalmus banzai* (Yokooka & Nomoto, 2013), *Tubuca arcuata* (Yuhara & Aizawa, 2016) and *Austruca lactea* (Tamura & Narita, 2013), is now established as having extended northeastward to the Izu Peninsula and Tokyo Bay, although still rare at that site. In addition, other southern-based marine invertebrates have been reported from corals (Yamano et al., 2011), nemerteans (Yamamori et al., 2013) and snails (Hayase et al., 2013).

The Kuroshio Current, which flows southwest to northeast along the Pacific coast of Japan, has enriched the communities of southern-based species on the southern Izu Peninsula area by providing aquatic environments suitable for their larvae. Planktonic crab larvae are generally dispersed over relatively long periods in the water column (Cuesta et al., 2006). Although larval development of *C. villosum* have not yet been reported, that of a closely related species [*Clistocoeloma sinense* (three zoeal stages; Saba, 1972, Cuesta et al., 2006)] has been observed. Because the entire brachyuran larval stage (including three zoeal stages) has been estimated as ca. 16 days (Fukuda, 1980), it is likely that a portion of juveniles originating from around the Kii Peninsula can survive transportation via coastal waters to the Izu Peninsula because of the rapid speed of the Kuroshio Current, which sometimes exceeds 2 m/s (Teramoto, 1987). This suggests that the warm Kuroshio Current can transport planktonic larvae from the southwestern subtropical zone to the northeastern temperate zone along the Pacific coast of Japan. The larvae of southwestern-based crabs can settle and survive along the coast of the southern Izu Peninsula near the Kuroshio Current due to the favorable habitat,

comprising small tidal flats with reed vegetation, gravels, cobble stones and semi-mangrove communities.

Clistocoeloma villosum dwells under stones on the upper intertidal region of tidal flats and salt marshes, as well as under leaf litter over moist soil on the landward of edges of semi-mangrove forests (Komai et al., 2004, Lee et al., 2010, Japanese Association of Benthology, 2012). However, tidal flats encompassing such environments are generally rare along the Pacific coast in approximately 250 km between Mikawa Bay and Tokyo Bay. Furthermore, an earlier study that revealed a salient genetic differentiation in the closely-related saltmarsh crab *C. sinense*, suggested the existence of a barrier to larval transport between these two bays (Yuhara et al., 2014). Therefore, the upper intertidal area of Aono River mouth in southern Izu Peninsula provides a vital habitat for this northeasternmost population of *C. villosum.*

Conclusion

The present study has clarified that the distribution range of the near-threatened sesarmid crab *Clistoceloma villosum* had been extended eastward from the Kii Peninsula to the Izu Peninsula. It suggests that the larvae of this crab can be transported northeastwardly along the Pacific coast of Japan by the warm Kuroshio Current in planktonic period and can be settled in the southern Izu Peninsula due to the existence of suitable habitat, comprising small tidal flats with reed vegetation, gravels, cobble stones and semi-mangrove communities.

Acknowledgments
We wish to thank the Biology Club of Shimoda High School for field assistance. We are also grateful to Ms. Mieko Takezawa for valuable advice about the Aono River. We also thank Dr. Graham S. Hardy (Whangarei, New Zealand) revised an early draft of the manuscript. We especially thank the Shizuoka Prefectural Government for cooperation to conduct this study along the semi-mangrove forest area of Aono River.

Funding
Not applicable.

Authors' contributions
TY, HY and MT collected the specimens. TY drafted the manuscript. TY and HY took the photographs of the specimens, checked the identification of the specimens and helped to improve the manuscript. MT prepared the map and helped to improve the manuscript. All authors read and approved the final manuscript.

Competing interests
The authors declare that they have no competing interests.

Author details
[1]Tohoku University Graduate School of Life Science, 6-3 Aoba, Aramaki, Aoba-ku, Sendai, Miyagi 980-8578, Japan. [2]IDEA Consultants, Inc. Institute of Environmental Ecology, 1334-5 Riemon, Yaizu, Shizuoka 421-0212, Japan. [3]Faculty of Sciences, Tokyo Bay Ecosystem Research Center, Toho University, Miyama 2-2-1, Funabashi, Chiba 274-8510, Japan.

References

Crosnier A. Crustaces decapodes. *Grapsidae et Ocypodidae*. Faune Madagascar. 1965;18:1–143.

Cuesta JA, Guerao G, Liu HC, Schubart CD. Morphology of the first zoeal stages of eleven sesarmidae (Crustacea, Brachyura, Thoracotremata) from the Indo-West pacific, with a summary of familial larval characters. Invertebr Reprod Dev. 2006;49:151–173.

De Grave S, Pentcheff ND, Ahyong ST, Chan T, Crandall KA, Dworschak PC, Felder DL, Feldmann RM, Fransen CHJM, Goulding LYD, Lemaitre R, Low MEY, Martin JW, Ng PKL, Schweitzer CE, Tan SH, Tshudy D, Wetzer R. A classification of living and fossil genera of decapod crustaceans. Raffles B Zool. 2009;21:1–109.

Fukuda Y. An attempt to estimate the length of platktonic larval period in Brachyura (1). Calanus. 1980;7:9–12.

Hayase Y, Kageyama Y, Kimura S. Molluscan fauna in the mouth of Ihara river, Shizuoka city: the northward range extension of Neritimorpha. Rep Nagoya Shell Club. 2013;38:23–32.

Ito M. Distribution of the fiddler crabs *Uca arcuata* and *Uca lactea* in Shizuoka prefecture, Japan. Jpn J Bentho. 2014;69:76–84.

Japanese Association of Benthology. Threatened animals of Japanese tidal flats: red data book of seashore benthos. Kanagawa: Tokai University Press; 2012.

Karasawa T, Kimura S, Kuroda M, Nomoto A. Record of *Clistocoeloma villosum* (Crustacea, Brachyura, Sesarmidae) from the Waka river estuary in Wakayama prefecture Japan. Nanki-Seibutu. 2006;48:60–62.

Kawakubo A, Otani T, Nakahara Y, Yone H, Shimojo K, Hashiguchi K, Ito T, Kurusaki Y, Yamaguchi Y. Record of the crustacean and the mollusks from south Kujukushima isles area. Trans Nagasaki Biol Soc. 2005;60:17–27.

Kishino T, Yonezawa T, Nomoto A, Kimura S, Wada K. Twelve rare species of brachyuran crabs recorded in the brackish waters of Amami-oshima island, Kagoshima prefecture, Japan. Nanki-seibutu. 2001;43:15–22.

Komai T, Nagai T, Yogi A, Naruse T, Fujita Y, Shokita S. New records of four grapsoid crabs (Crustacea: Decapoda: Brachyura) from Japan, with notes on four rare species. Nat Hist Res. 2004;8:33–63.

Lee SY, Jung J, Kim W. A new report on sesarmid crab *Clistocoeloma villosum* (Crustacea: Decapoda: Brachyura) from Korea. Kor J Syst Zool. 2010;26:179–181.

Maenosono T, Saeki T. The sesarmid (Crustacea: Decapoda: Brachyura) fauna of Ishigaki-jima island, Ryukyu archipelago, Japan, with new distribution records. Fauna Ryukyuana. 2016;33:1–13.

Masuda S. Plantation of mangrove at the Aono river in Minami-Izu town, Shizuoka prefecture, Japan. Macro Rev. 1999;11:63–70.

Ministry of the Environment. The 6th national survey on the natural environment (Ehime). Biodiversity center, Nature conservation bureau. Yamanashi prefecture: Ministry of the Environment; 2005.

Ministry of the Environment. The red list of the threatened marine species. http://www.env.go.jp/press/103813.html (2017). Accessed 26 Mar 2017.

Miura T, Jitsumasa T. Benthic mollusks and crustaceans recorded from the Hitotsuse river estuary, Miyazaki, Japan. B Fac Agr, Miyazaki Univ. 2010;56:29–44.

Murase A, Yuhara T. Record of an endangered gastropod, *Cerithidea rhizophorarum*, from estuarine habitat of Ogamo river, Southern Izu Peninsula, Central Japan. Biogeogr Soc Jpn. 2011;67:261–264.

Nakanishi H. Distribution and population of *Hibiscus hamabo* Siebold et Zucc. In: Studies on the vegetation of alluvial plains. (ed, The association to commemorate the retirement of Prof. Dr. Shigetoshi Okuda). Yokohama: The association of commemorate the retirement of Prof. Dr. Sigetoshi Okuda; 2001. p. 37–46.

Nishiwaki S, Hirata T, Ueda H, Tsuchiya Y, Sato T. Distribution and its factor of *Clithon retropictus* (Prosobranchia: Nertidae) in the rivers of Izu peninsula. B Col Med Tech Nurs Univ of Tsukuba. 1991;12:51–57.

Nomoto A, Yodo S, Kimura S, Kishino T, Sakano M, Wada K. Six rare brachyuran species of the family Grapsidae, recorded from the Kinokawa river estuary, Wakayama prefecture. Nanki-seibutu. 1999;41:5–9.

Saba M. *Umore-benkeigani no koukihassei* [Studies on the post-embryonic development of *Clisotocoeloma merguiense* de Man]. Mie-seibutsu. 1972;22:25–29.

Shokita S, Nagai T, Fujita Y, Naruse T, Ito A, Nagamatsu T, Yamazaki T, Shinjo M, Nagata Y. Distribution and abundance of crustaceans in the Ohura river mangrove swamp of Okinawa island, Japan. In: A comprehensive study for mangrove ecosystem in Okinawa. Naha: Research institute for subtropics; 2002. p. 73–86.

Tamura M, Narita A. New record of the fiddler crab *Uca lactea*, including the adult male, female and the juvenile crabs, at the mouth of Obitsu river, Kisaradu, Chiba. B Biol Soc Chiba. 2013;62:79.

Tanaka H, Shibagaki K, Ikezawa H, Kanezawa A, Wada K. Occurrence of two species of fiddler crab (Decapoda, Ocypodidae) on the tidal flat of the Aono river, Izu peninsula, Shizuoka prefecture, Japan. Jpn J Benthol. 2004;59:8–12.

Teramoto T. Kuroshio. In: Wadachi K, editor. Encyclopedia of oceanography. Tokyo: Tokyodo; 1987. p. 185.

Tesch JJ. Synopsis of the genera *Sesarma, Metasesarma, Sarmatium* and *Clistocoeloma*, with a key to the determination of the Indo-Pacific species. Zool Med. 1917;3:127–260. pls.15–17.

Wada T. Record of sesarmid crab *Clistocoloma villosum* from the Osaka Nankou bird sanctuary, Japan. Nat Stud. 2013;59:126–127.

Yamamori L, Hirose M, Kajihara H. Northward range extension of *Baseodiscus hemprichii* (Nemertea: Heteronemertea)—due to rising sea-surface temperatures? Mar Biodivers Rec. 2013;6:e100.

Yamano H, Sugihara K, Nomura K. Rapid poleward range expansion of tropical reef corals in response to rising sea surface temperatures. Geophys Res Lett. 2011;38(4):L04601. doi:10.1029/2010GL046474.

Yokooka H, Nomoto A. First record of *Macrophthalmus banzai* (Crustacea, Brachyura, Macrophthalmidae) in Shizuoka prefecture, Japan. Nanki-seibutu. 2013;55:137–140.

Yokooka H, Yuhara T, Tagashira R. First record of *Ptychognathus capillidigitatus* (Crustacea: Decapoda: Varunidae) in Shizuoka prefecture, Japan. Cancer. 2015;24:39–45.

Yuhara T, Aizawa K. Confirming of the fiddler crab *Uca arcuata* at the mouth of Obitsu river in Tokyo bay. B Biol Soc Chiba. 2016;65:52–54.

Yuhara T, Kawane M, Furota T. Genetic population structure of local populations of the endangered saltmarsh sesarmid crab *Clistocoeloma sinense* in Japan. PLoS One. 2014;9(1):e84720. http://doi.org/10.1371/journal.pone.0084720.

37

First record of *Thysanozoon brocchii* (Platyhelminthes: Polycladida) from Indian waters

Reshma Pitale[*] and Deepak Apte

Abstract

This work reports the occurrence of *Thysanozoon brocchii* from the rocky intertidal coast of Ratnagiri and Dwarka, the West Coast of India, for the first time. Two morphotypes were found, the first morph has a buff brownish papillate dorsal surface, with few specimens having white spots. The second morph has light coloured papillae which form a distinct cross marking along the dorsal surface. The species possesses double male copulatory apparatus with seminal vesicle, prostatic vesicle and sclerotized stylet. This cosmopolitan species has previously been recorded from Italy and Mediterranean, Japan, South and West Africa, Florida, New Zealand, Brazil, and United Kingdom. Synonymized species and older descriptions have been compared to examine similarities and dissimilarities. Considering the existence of varied colour morphs of this species, a detailed comparative analysis of morphological characters, reproductive histology and molecular framework is recommended.

Keywords: Pseudocerotidae, Colour patterns, Intraspecific variations, Reproductive anatomy, Cosmopolitan, Maharashtra, Gujarat

Introduction

Polyclad flatworms are free living members of the phylum Platyhelminthes. These coral reef and rocky shore inhabitants are more diverse in the tropical environment (Prudhoe, 1985). Cryptic behavior, apparent specificity with food preference, aposematic colouration and mimicry with opisthobranch molluscs and fishes (Ang & Newman 1998; Newman & Canon, 2005) make them potentially significant in intertidal or reef ecology.

Members of the family Pseudocerotidae are peculiar for their brilliant colour patterns. However, colour variation is commonly seen forming species complex in their systematics. Perhaps such complexities arising in the taxonomy of these worms can be resolved by careful observations of external as well as reproductive anatomical features. Newman and Canon (1995) described three species of the genus *Pseudoceros* from the Indo-Pacific region showing remarkable similarity within external appearance. They, too, emphasized the significance of precise documentation of colour and patterns. Litvaitis, et al. (2010) tested the coloration pattern of *Pseudoceros bicolor* complex against molecular evidence and emerging complexities in the taxonomy of the same.

Grube (1840) described the genus *Thysanozoon* considering the papillate dorsal surface, unlike the rest of the Cotylean genera. Type species of the genus was *Thysanozoon diesingii*, which has later been synonymized for *T. brocchii*. This genus includes about 23 species worldwide (Tyler, 2013). Laidlaw (1902) reported *Thysanozoon plehni* from the Laccadive Island of India. However, this species has been reviewed and assigned to *Acanthozoon plehni*. After a gap of about a century, Apte and Pitale (2011) mentioned a member of this genus from Kavaratti, Lakshadweep Island, India. Later, Sreeraj and Raghunathan (2013) reported *Thysanozoon nigropapilosum* from South Andaman.

The present study reports *Thysanozoon brocchii* for the first time from the Indian shoreline. Apart from external morphology and reproductive anatomy, the study encompassed the review of characteristic features of some of the synonyms and a few previous descriptions, and their comparisons with the two different morphs presented herein.

* Correspondence: rd.pitale@bnhs.org
Bombay Natural History Society, Hornbill House, S.B. Singh Road, Mumbai, Maharashtra 400 001, India

Materials and methods

Collection was carried out during the daytime low tide by handpicking the worm using a paint brush, within the littoral area of Ratnagiri (Mandvi 16.98758° N 73.27486° E), Maharashtra state and Dwarka (22.240323° N 68.957424° E), Gujarat state, West Coast of India (Fig. 1). Presence of heterogeneous habitat is a common characteristic of both these shores, and pebbles, cobbles and rock pools of various sizes and depths are found at various zones. Luxuriant growth of algae and patches of coral-rubble were observed during surveys.

Photographs were taken in the wild as well as ex situ to record true colour and pattern. Animals were then fixed in 10% frozen buffered formalin and later stored in 70% ethanol for long term preservation (Quiroga et al., 2004). Studies on key anatomical features were carried out using a Stereo Microscope (Leica EZ4 D). The systematic classification system established by Faubel (1984) was followed. Longitudinal serial sections of the reproductive system (6 μm) were obtained by specimen (Pclad-0044) embedding in paraplast and staining with hematoxilin and eosine. Remaining specimens are deposited in collections of the Bombay Natural History Society.

SYSTEMATICS.
Order POLYCLADIDA Lang, 1884
Suborder COTYLEA Lang, 1884
Family Pseudocerotidae Lang, 1884
Genus *Thysanozoon* Grube, 1840
Thysanozoon brocchii Risso, 1818

Material examined

Two specimens (11.49 mm × 8.32 mm and 11.30 mm × 8 mm, preserved) found under rock pebble at Mandavi, Ratnagiri, 9 May 2012 (BNHS_Pclad-0044) as 18 histological slides.

One specimen (10.2 mm × 8.5 mm preserved) found under rock pebble within algae at Dwarka, Gujarat, 12 December 2012 (BNHS_Pclad-0080).

Diagnosis

Brown-buff to cream dorsal surface with yellowish brown to dark brown papillae cover, margin with pinkish tint; ends with white dotted line. Papillae from median longitudinal line are lighter thus distinct, transverse line of light coloured papillae about 1/3rd posterior to longitudinal length also present in single

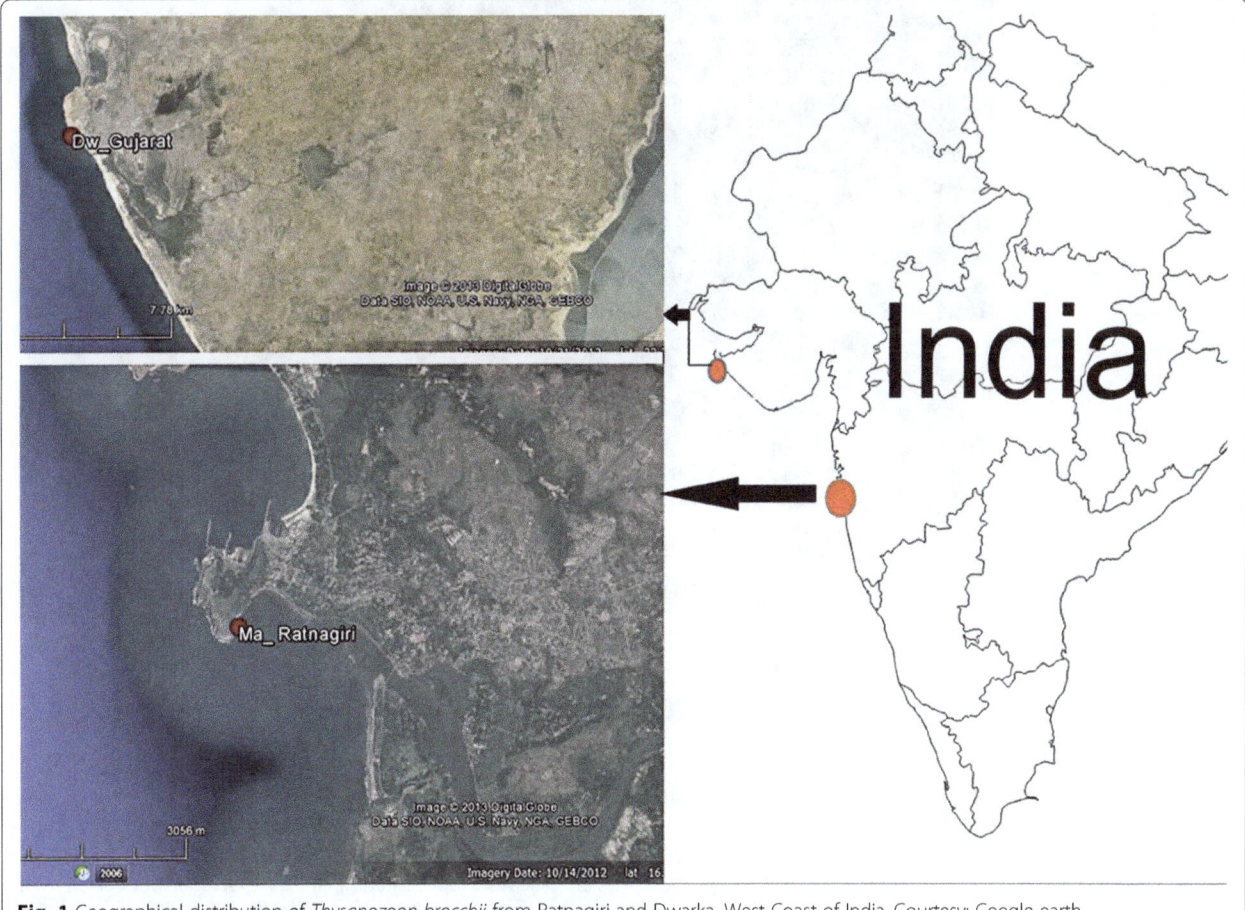

Fig. 1 Geographical distribution of *Thysanozoon brocchii* from Ratnagiri and Dwarka, West Coast of India. Courtesy: Google earth

specimen (Pclad-0080); double male reproductive structures; elongated vagina.

Description

The first morph (BNHS-Pclad-0044) is broadly oval, oblong body and found slightly raised medially. Dorsal surface covered with papillae which are aggregated in median region, diminish in size, and become scarce towards margin and found absent a little above the margin. Papillae are approximately cylindrical (0.3 mm–0.4 mm), knob-like or even tapering in the same specimen.

Dorsal surface, ground colour is buff-brown-creamish with black median longitudinal stripe. Yellowish reticulation of intestinal branches observed prominently towards margin. Pinkish tint found towards periphery, and a dotted white line on the rim. Papillae are buff-brown to dark brown having grayish outline and some possess white spots around the tip. Papillae present over median longitudinal stripe are mostly distinct, tapering, and lack brown pigments. These papillae are creamish or whitish, possess white dots and run antero-posteriorly. Pseudotentacles are held erect; they are ear-like, creamish brown and with white tip. On the inner margin of each pseudotentacle, a fine black line that runs towards the median stripe can be observed. Cerebral eyespots (30–34) arranged in horseshoe-shaped cluster (size 0.27 mm), present within the colourless area just posterior to the pseudotentacles (Fig. 2c). Pseudotentacular eyes are found distributed as four clusters dorsally and two clusters ventrally. Dorsal cluster bears about 40–45 eyespots whereas ventral cluster bears about 50–60 eyespots.

Another morph (BNHS-Pclad-0080) has creamish dorsal surface and yellowish reticulation noticed especially towards margin (Fig. 2b). Dotted white line delineates the margin. Brown coloured papillae are cylindrical, pointed, with white spots around the tip. Papillae present over

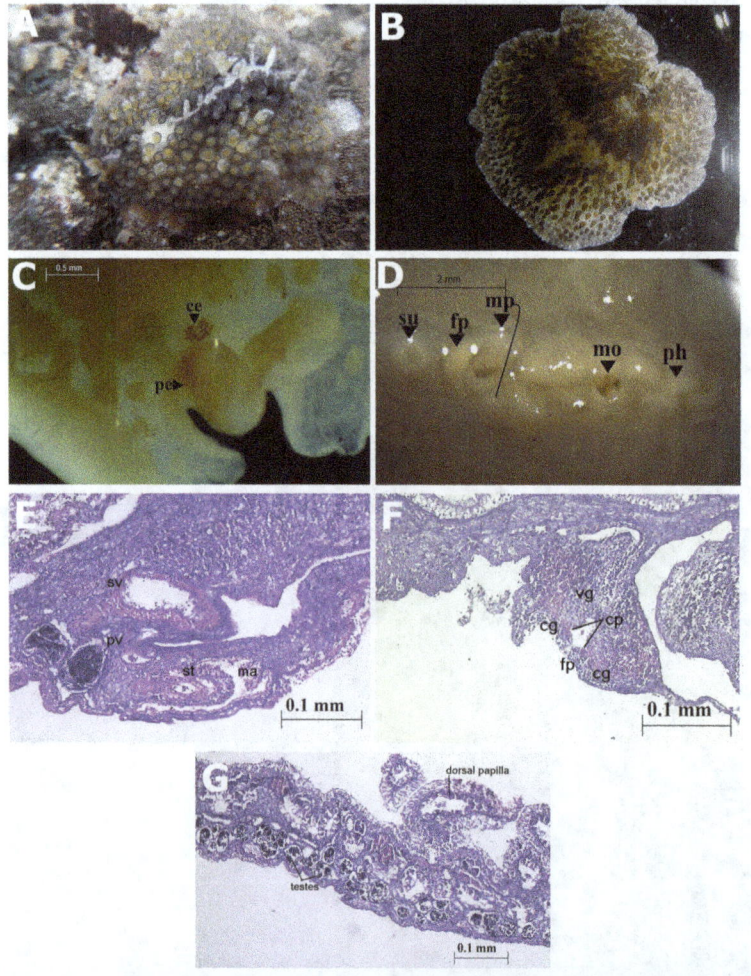

Fig. 2 *Thysanozoon brocchii* Risso, 1818: **a** (1st morph) under pebble within the rock pool (**b**) 2nd morph (**c**) anterior end with cerebral eyespots (ce), pseudotentacular eyespots (pe); (**d**) ventral surface showing pharynx (ph), Mouth (mo) male gonopore (mp), female gonopore (fp), sucker (su); (**e**) Sagittal section of male reproductive system with male antrum (ma), prostatic vesicle (pv), stylet (st); (**f**) female reproductive structure with vagina (vg), cement pouch (cp), cement gland (cg), Intestine (it); (**g**) ventral testes and dorsal papillae

median longitudinal and transverse line are creamish, possess white spots and form a cross-like marking. Black-brown pigments of the median stripe can be seen over anterior region and on the pseudotentacles.

The two morphs described above are similar in terms papillae arrangement, median lighter papillae and margin colouration. The papillae shape in second morph is cylindrical and white dots present over each papillae are more numerous than the first morph. Ground colour is creamish and papillae colour is darker and colour tone is more even in second morph as in the first one. Considering the distinctive colour pattern of second morph the single specimen has been kept intact and not been sectioned for histology.

Semi-transparent whitish ventral surface, with a mouth, opening medially within pharynx and about 3.01 mm distant from anterior margin (Fig. 2d). Pharynx is in the form of 4–5 simple and shallow folds and about 4.68 mm long. Two male gonopores are present immediately behind the pharynx, on either side of the median line and about 1.24 mm distant from each other. Medially placed, female gonopore is present 0.71 mm posterior to the male gonopore followed by sucker (0.46 mm from fp).

Double male copulatory system with numerous testes (0.02 mm to 0.04 mm in length) arranged in 2–3 rows located ventrally. Seminal vesicle (0.14 mm × 0.064 mm) is muscular, slightly oval, bent and pointed at its anterior end (Fig. 2e). Vas deference arranged laterally, visible through ventral side and runs posteriorly. Small and oval prostatic vesicle (0.0435 mm × 0.0342 mm) found ventral to the seminal vesicle. Penis with stylet found further down to the seminal vesicle, probably contracted or bent during fixation, as it is visible only in the form of a rounded structure (Fig. 2e). Male atrium is shallow (0.045 mm).

Female reproductive system consists of ovaries which are found scattered dorsally. Female antrum is narrow, with minute lateral invagination for cement pouches. Vagina elongated backward (0.2 mm) and cement glands are seen spread around vagina (Fig. 2f).

Taxonomic remarks

In taxonomic study of pseudocerotidae, the genus *Thysanozoon* is difficult, probably due to the unavailability of enough details in the older literature. Body colour pattern, papillae shape, their colour and distribution are important features in the taxonomic study of this genus (Brusa et al., 2009). The specimens (BNHS_Pclad-0044 and 0080) somewhat fit with the original description of Risso, 1818 and specimens described later (Palombi, 1928; Pearse, 1938; Marcus and Marcus, 1968; Brusa et al., 2009 and Bahia et al., 2012, 2014) from different parts of the world.

About twenty species described earlier have later been synonymized for *T. brocchii* (Faubel, 1984). Table 1 gives the review of the characters noted from these synonyms.

Tergipes dicquemare, Planaria dicquemari and *Thysanozoon dicquemaris* although very poorly described, seems to be similar in terms of dorsal and ventral colour pattern. *Planaria tuberculata, Planaria verrucosa* are similar in terms of colour and pattern. *Eolidiceros panormis* and *Thysanozoon diesingii* are nearly identical due to possession of yellow and purple pigment on dorsal surface as well as the purple dash lines on the margin. The colour pattern of *Eolidiceros brochii* is slightly different from the other species due to possession of reddish-brown papillae and purple dots all over the dorsal surface. Four of the synonyms viz. *E. panormis, T. diesingii, T. lagidium* and *T. fockei* are found possessing purple pigmentation on dorsal surface. Transverse bands of lighter papillae forming a cross were observed within *E. panormis, T. brochii* var. *cruciatum* and *T. lagidium*. The remarks section of Table 1 elaborates the comments from the authors about the resemblance of their respective specimens, either with *T. brocchii* or previously described allied species. Diesing (1850) reports about *Thysanozoon tuberculatum* and synonymizes three previously known species viz. *Planaria tuberculata, Thysanozoon dicquemaris* and *Planaria dicquemari. Planaria brocchi* and *Eolidiceros brocchi* are synonymised for the species *Thysanozoon brocchii* and *Eolidiceros panormus* for *Thysanozoon panormus* respectively. Further, Diesing (1862) while separately mentioning the genus *Thysanozoon*, synonymizes species which are described earlier under the varied genera viz. Planariae spec. by Delle Chiaje, Stylochi spec. by Diesing Eolidiceri spec. by Quatrefages.

Table 2 gives an account of characters mentioned from some of the older descriptions of *Thysanozoon brocchii*. The descriptions by Quatrefage (1845) and Lang (1884) are unvarying and collected from type locality of the concerned species. Material obtained from Japan (Yeri and Kaburaki, 1916; Kato, 1944) was found resembling the above two descriptions, especially in terms of possessing purple pigment and papillae colour pattern. Palombi (1928) differs with Lang (1884), Quatrefage (1845), Yeri & Kaburaki (1916), and Marcus and Marcus (1968) due to having exclusive brown dorsal pigment and ventral eyespot arrangements. Meanwhile, Pearse (1938) recorded the three morphs (pigmentation- brown, purple and intermediate) of the species from Crooked Island. Marcus and Marcus (1968) elaborated the reason for presence of broader female antrum in Marcus (1949) material of *T. lagidium*. They also discussed the external and internal anatomical, intraspecific variations within this species and finally merged *T. lagidium* with *T. brocchii*. Faubel (1984) while reviewing the polyclad systematics followed the same criteria. However, Prudhoe (1985)

Table 1 Review of the characters from synonymized taxa of *Thysanozoon brocchii*

Synonymized taxa	Body form	Colour-pattern	Eye spots/arrangement	Habitat/Locality	Remarks
Tergipes dicquemare Risso, 1818	oblong, flattened, Dorsal: covered with many sessile papillae serving as respiratory organ, tentacles earlike, Ventral: greyish	Dorsal White- yellowish Intestinal canal - white –reddish	-	Under stone Nice sea, France	-
Planaria tuberculata Delle chiaje, 1828	compressed, ovate, wide Dorsal: papillate; absent towards margin, longitudinal median line present	Dorsal Papillae: few with white dots, median line white colour Margin: white band even on pseudotentacles	-	Algae dweller Naples, Italy	Not identical to *Planaria brocchii* described by Risso 1818, although has certain similar characters in body colouration
Thysanozoon tuberculatum Delle Chiaje, 1828	Elliptical Dorsal: long tentacles, median papillae with conical tip	Dorsal: Blackish Papillae: median with white tip and bluish at the base	-	Found among algae	-
Planaria verrucosa Delle chiaje, 1829	Dorsal: papillate; more on back and smaller towards margin	Dorsal Margin: milky white and black	-	Naples, Italy	differs from *P. tuberculata* being small, colourful margin without papillae and papillae bears white dots at the tip
Stylochus papillosus Diesing, 1836	½ - 1-1/2 in. narrows ovate flattened Tentacles: filament like, medially raised (keel on the middle) Papillae: numerous papillae present on the keel area	Dorsal: light yellow or reddish-brown keel: white	Indefinite number on the disc shape bright spot	Adriatic Sea	-
Thysanozoon diesingii Grube, 1840	flat, leaf like Dorsal: covered with papillae (pedicels)	Dorsal: longitudinal bright yellowish line Margin: white with purple dash lines	Eye spots 31–28	Palermo-Italy east coast of Ceylon (Sri-Lanka) Indian Ocean	compared with Mediterranean *P. brochii* and *P. tuberculata* and mentioned the similarity with *P. brochii* except for the absence of papillae opening
Planaria dicquemari Delle Chiaje, 1841	Papillae: larger in the middle and smaller towards margin	Dorsal: Yellow, blackish speckled Margin: whitish	In two groups at the base of the tentacles		differs from *P. tuberculata* (Delle Chiaje, 1828) in terms of size, papillae and colour
Thysanozoon diequemaris (Delle Chiaje, 1841) Diesing, 1850	Elliptical, Papillae: long, conical, tentacles somewhat sickle shape	Dorsal: Body greyish Papillae: yellowish white Ventral: Grey	-	In stones- Mediterranean Sea	-
Eolidiceros brochii Quatrafages, 1845	regularly elliptical, slightly curved in the middle Tentacles: triangular, Papillae	Dorsal: light yellowish brown, edges are transparent and bluish Upper side of the head region is white; tinted	Numerous CE: 20–25 TE: group of 5–6 eyespots between the tentacles	In *Fucus* Naples	May be similar to *Planaria tuberculate* (Delle Chiaje, 1828)

Table 1 Review of the characters from synonymized taxa of *Thysanozoon brocchii* (*Continued*)

Taxon	Size / shape	Colour / pattern	Eyespots / ocelli	Locality	Remarks
	(appendages): fusiform, longer and smaller bigger and smaller towards edges Size: 18 mm	with greenish, sides are light brown Papillae: reddish-brown but tip is yellowish white; blackish purple dots all over the surface			
Eolidiceros panormis Quatrefage, 1845	elliptical, triangular head and straight large tentacles Dorsal: papillate (cylindrical appendages) Size: 6 mm × 3 mm	Dorsal: mid region yellow-greenish speckled with brown, head region white, anterior end with purple and brown pigments, transverse pale yellow band present 1/3 posterior of the body Papillae: similar colour like body but less numerous than *E. brocchii* Margin: brown speckled and purple dashes	Dorsal: 3 large eyespots surrounded by smaller ones; one is posterior and two are at the base of the pseudotentacles Ventral: arranged between and at the base of the pseudotentacles	on rock Palermo, Italy	Compared with *E. brocchii* for eyespots, body shape, colour pattern and arrangement of dorsal papillae
Thysanozoon fockei Diesing, 1850	Sub elliptical Dorsal: papillated	Papillae: yellow-purple papillae	-	Tergesti, Italy	-
Planeolis panormus, Stimpson 1857	Papillae: scattered over the body; head larger and with large tentacles	-	ocelli on the tentacles at the base of the tentacles and between the tentacles	-	information about genus is given but refers to *Eolidiceros panormus* Quatrefage, 1845 for the species description
Thysanozoon sp. (*Eolidiceros quatrefages*) Moseley, 1877	Size: when expanded 10 cm × 6 cm	Dorsal: dark purple, tubercles with white tip	-	Zamboangan, Philippines	-
Thysanozoon brochii var. *cruciatum* Laidlaw, 1906	Size: 8–16 mm	Dorsal: yellow-dark gray, narrow longitudinal white stripe and 1/3 posterior transverse band present	-	under rock Porto Praya, Cape Verde Island	-
Thysanozoon lagidium Marcus, 1949	Dorsal: papillate Size: 13x7mm	Dorsal: Grayish brown-purplish Papillae: darker, few with white tip light coloured papillae forming longitudinal stripe and medial transverse stripe present	-	Brazil	-
Thysazonoon cf. *lagidium* Quiroga et al. 2004 Based on image presented in the paper	-	Dorsal Papillae: black- brown midline papillae whitish; whitish papillae forming cross are present around 1/3rd of body length	-	Coloumbia; Brazil	-

Table 2 Review of the characters noted from older literature of *Thysanozoon brocchii*

Author	Body form	Colour -Pattern	Eye spots/arrangement	Habitat/Locality	Remark
Tergipes brocchii Risso, 1818	Oval, oblong, tubercles on the dorsal surface; tentacles ear shaped	Dorsal: violet brown; Papillae: white dots at the end of the tubercles Ventral: transparent	small, black	Under stones Nice sea, France and Naples, Italy	Original description
Quatrefages, 1845	Elliptical, raised medially Papillae: dense in the middle surface and scarce towards margin longitudinal mid-line papillae are fusiform and longer; small tubers towards margin Size: 16 mm × 8-9 mm	Dorsal: yellowish-brown reddish; purple pigments Papillae: brown reddish with yellowish white tip Margin: dark brown ends with white line Cerebral area is colourless triangular portion between pseudotentacles	CE: two groups 20-25; 5-6 PE: 5-6 smaller eyes near the edges of the pseudotentacles Ventral-5-6 eyes placed on each tentacle; line of 7-8 large eyespots	Naples, Italy	-
Diesing, 1850	Elongated plane Papillae: fusiform Tentacles- thick and tuberculate	Dorsal: yellowish Margin: white-blue Papillae: red-brown to black -blue	Beneath, above and at the base of the tentacles	In rocks Nice Sea; Toulouse	-
Lang, 1884	Compressed, broadly oval, oblong Dorsal: papillate (tubers or tubes) Papillae: bulgy at the base and tapers further; small and sparse in the longitudinal and transverse midline Pseudotentacles: Bead shaped, sharp	Dorsal: dark brown-violet with white margin; lighter than the papillae Papillae: whitish spots on the darker papillae, lighter in transverse and longitudinal bands, pore at the tip might be the intestinal diverticulum leading outside Pseudotentacles: Whitish grey or dark brown-black with light colour spots Cerebral area is lighter and clearly demarcated within pseudotentacular area Ventral: dirty gray-blue-brownish or yellowish	CE: Horseshoe shape PE: two roundish groups each on either side of mid-line	shallow water in association with ascidian and sabellids, Naples and coast of Posilipo, Italy	wide variations observed within same individuals for various characters (papillae shape, number, colour and size vary greatly), *Eolidiceros panormus* is young specimen of *T. brochii*; sexual maturation occurs at different size
Yeri and Kaburaki, 1916	Broadly Oval, frilled margin Size: 35 mm × 21 mm Papillae: slender-conical, all over the dorsal surface, longer in the mid region; smaller and sparse towards the margin	Dorsal: purplish gray or yellowish purple, yellowish-whitish; longitudinal median stripe Papillae: dark gray with purple tint, base colourless, few with whitish tip Some specimens with whitish papillae forming transverse line Ventral: Lighter and comparatively darker at the margin	CE: two clusters above the brain in colourless area PE: numerous	Misaki and Matsuwa, Japan	Internal anatomy same as Mediterranean specimens, intestinal branches extending in to the papillae

Table 2 Review of the characters noted from older literature of *Thysanozoon brocchii* (*Continued*)

Palombi, 1928	Flatten with free wavy margin	Dorsal: surface chestnut brown except for the marginal area, space between the pseudotentacles lighter	CE: two clusters placed centrally within colourless area PE: irregular arrangement ventral: present within each pseudotentacles two groups of eye clusters, further adjoining row of eyes run along the anterior margin	Suez Canal	Specimen observed matches Lang's description of *T. brocchii*. *T. dissingii* Grube, 1840 can be considered as *T. brochii*
Pearse, 1938	Dorsal: Papillate Size: 28-33 × 8-12 mm	Dorsal: cream with light yellow reticulum, black pigmentation, median longitudinal dark stripe with light line through it, anterior tentacular region blackish with unpigmented cerebral patch containing eyespots Papillae: light brown Papillae: purplish-brown become light brown towards margin, some with white spots and dark tips and forms 'T' Margin: brown-purple redial bands	-	Eel grass, Crooked Island, near Florida	Three colour morphs were collected, immature specimens
Palombi, 1939	-	-	-	Shelley beach (East London) South Africa	compare with Palombi, 1938 and Kato's Japanese specimen and mentioned cosmopolitan distribution
Kato, 1944	Size: 50-60 mm	Dorsal: Brown colour Papillae: brown with whitish yellow especially at median line and spread otherwise Margin: bluish	-	Misaki, Susaki near Simoda, Sugasima, Sima, Seto- Japan	gives cross section of male reproductive system and papillae showing intestinal branch
Hyman, 1952	-	Dorsal: purplish maroon Papillae: brown some with white spots Margin: white spots	-	Under stones county causeway/ Biscayne Bay; Florida	immature specimen thus not sectioned for reproductive histology however, intestinal branches going to papillae is observed uncertainty about the identification
Marcus & Marcus, 1968	Dorsal: Papillate, scares towards margin Size: 15 mm × 5 mm and 24 mm × 16 mm	Dorsal: papillae are darker and become lighter towards the margin, white spots on the papillae	-	Algae and mangroves from Piscadera bay, Curacao and Florida	Broader cement pouches resulting in complex female antrum can be considered as usual intraspecific variation thus *T. lagidium* can be merged with *T. brocchii*. Review of Lang,1884; Marcus, 1949 and Pearse, 1938 and Hyman, 1952 descriptions were made

Table 2 Review of the characters noted from older literature of *Thysanozoon brocchii* (Continued)

Reference	Morphology	Dorsal	Eyespots	Locality	Remarks
Vera et al. 2008	elongated	Dorsal: darker to light brown ground colour, mid-dorsal longitudinal line and perpendicular white line forming cross	-	Canary Islands (Eastern Atlantic Ocean)	active swimmer compared with Quiroga et al. 2004 *Thysanozoon* cf. *lagidum* from Colombia
Brusa et al. 2009	Oval, papillate, slightly undulated margins. Papillae: Mid-line papillae are longer and decreases towards the periphery. Pseudotentacles: pointed ear like. Size: 25 mm × 15 mm	Dorsal: yellowish brown, margin with discontinuous black line. Pseudotentacles: dark brown with white tip	CE: Horseshoe shape. PE: four dorsal and two ventral clusters	Puerto Pirámides and Puerto Madryn, Argentina	Followed the Marcus and Marcus, 1968 and considered the *T. lagidium* as synonym of *T. brocchii*
Bulnes et al. 2011	Oval. Pseudotentacles: enlarged distally and no pigmentation. Papillae: big bulky conical, longer and dense around dorsal bulge; declines towards margin (length and number); no papillae on the margin. Size: 26 mm × 15 mm	Dorsal: overall light brown but yellowish brown in the central became transparent towards margin; round black spots over the surface especially towards margin	CE: two separated triangular clusters. \TE: numerous single row dorsally. Ventral: scattered	In mussel and ascidians community Mar del Plata harbour, Argentina	Hypothesised papillae on surface for gas exchange and digestion as also mentioned by Prodhoe.
Bahia et al., 2012	Oval, elongated. Size: 12-17 mm × 10-14 mm. Papillae: dorsal surface with smaller towards the margin. Pseudotentacles: ear like	Dorsal: Whitish with dark brown to yellow brown pigmentation median longitudinal whitish line, some specimens with transverse line also forming cross. Some papillae with white spot. Ventral: white	CE- Horseshoe shape, 30–50 eyespots. PE- four dorsal(20–25) and two ventral clusters (65–70)	rocky and reef-flat areas Santa Rita and Búzios, Brazil	Differ in eye spots arrangement with Palombi, 1928; Considered the presence of complex female antrum as intraspecific variation; Emphasis on the revision of species considering the varied colour pattern and its cosmopolitan occurrence
Noreña et al. 2014	oval, oblong. Papillae: acorn-like	Colour photo reference is also available		Atlantic coast of the Iberian Peninsula	
Bahia et al. 2014	Papillae: size decrease towards margin. Size: 25 × 19 mm to 7 × 6 mm	Dorsal: brown to yellowish brown with longitudinal median line cream coloured, cream transversal line, some papillae with white spots	-	Praia das Conchas, Cabo Frio, Brazil	-

considered *T. lagidium* as a separate species. Two recent findings (Brusa et al., 2009 & Bahia et al., 2012, 2014) from Argentina and Brazil respectively show absence of purple pigments mentioned in most of the older records. Current specimens (First morph) found intermediate of all the older descriptions, but shows approximate resemblance with Bahia et al., 2012, 2014 in terms of colour-pattern and eyespots arrangement. Second morph shows close similarity with Pearse (1938) (brown morph) Quiroga et al., 2004 and Brusa et al., 2009.

Overall, compilation and comparison certainly represents the combinations of colour patterns exist between the *T. brocchii* species complex (Table 3). Perhaps these combinations are accountable for the current taxonomic confusion. Nonetheless, several currently synonymized species have also been synonymized previously and their progress (from genera - *Tergipes*, *Eolidiceros* or *Planeolis*) towards the genus *Thysanozoon* (after 1840) is clearly

evident as one approaches from Risso (1818) to Marcus and Marcus (1968). But, several older descriptions are not even and do not cover all the criteria which can be commonly used to compare species within the complex. Faubel (1984) while revising polyclad systematics includes twenty such species as synonyms of *T. brocchii*. Prudhoe (1985) proposed the *T. lagidium* as different species based on the transfer line forming 'T' shape colour pattern of papillae. However, with reference to the information presented in the Table1 few more species exist with similar character and not found stated by Prodhoe (1985). Thus, for a time being *T. lagidium* should be consider as the synonym for *T. brocchii* as mentioned by Faubel (1984).

Currently, this species is known from Naples, Italy (type locality), other parts of the Mediterranean, Algeria, Suez Canal (Palombi, 1928), South and West Africa, Florida, Brazil, Argentina, Borneo, Japan, Vietnam, New Zealand (Prudhoe, 1989) and with this report, extend to the Indian Coast.

Discussion

Comparative examinations clearly indicate and support the fact of existence of variable colour morphs of this species. Certainly, usual intraspecific variation and two or more species sharing similar colour patterns are two contended facts that exist, particularly in the pseudocerotid polyclads. Thus, the allocation of all presently noted morphs for *T. brocchii* is only convincing when external colour patterns, characters of reproductive anatomy and molecular framework reveal the similarity. In this context, revision of this species is urgently required. Perhaps fresh collection of specimens from the similar or adjacent places of previously described region can contribute to resolve mystery of colour patterns. Cladistical analysis using external characters and molecular data techniques are beneficial. Bulnes et al., 2011 raised important query about the cosmopolitan distribution and temperature as limiting factor in polyclad species distribution. Thus, ecological data inferring the habitat and food preference should be encouraged.

Table 3 Colour variation on each part of the colour pattern observed in the literature cited

Major colour and pattern combinations observed	
Dorsal ground colour	whitish; cream; light yellow; brown; light brown; grey; dark gray; blackish; dark purple; chestnut brown; white-yellowish; yellowish brown; yellow-greenish; yellowish-brown-reddish; yellowish purple; reddish-brown; grayish-brown-purple; violet-brown; purplish gray; purplish maroon;
Ground colour-pattern	
Speckled	Black; brown
Pigments	Purple; brown; dark brown to yellow brown
Reticulation	light yellow
Median stripe	dark
Round spots	black
Papillae colour	whitish; dark gray; yellowish-white; yellow-purple, reddish-brown; black-brown; brown reddish; black-blue; purplish-brown, brown; Bluish base
Papillae colour - pattern	
Dots	White; blackish purple
Tip	White; yellowish white
Base	colourless
Tints	purple
Margin Colour	white; Milky white to black; white-purple; white-blue; bluish; dark brown; brown-purple; transparent
Margin colour-pattern	
Band	White; radial
Dash	purple
Speckled	brown
Spots	white
Line	discontinuous black

Acknowledgments

The authors are grateful to Dr. Marcela Bolaños for providing valuable guidance and required literature. Special thanks to Ms. Juliana Bahia from the Zoologische Staatssammlung, München, Germany for sharing some of the references and to Ashok Bhagat for expertly providing histological sections. We greatly appreciate Dr.Vishal Bhave for collection and Mr. Rajendra Pawar, Mr. Vishwas Shinde and Mr. Rajesh Parmar for assistance during fieldwork and collection.

Authors' contributions

RP first collected, identified and reported species. RP drafted the manuscript. DA guided and provided funds required for the study and carried out final editing of the manuscript. Both authors read and approved the final manuscript.

Competing interests

The authors declare that they have no competing interests.

References

Ang HP, Newman LJ. Warning colouration in pseudocerotid flatworms (Platyhelminthes, Polycladida), a preliminary study. Hydrobiologia. 1998;383: 29–33.

Apte D, Pitale R. New records of Polyclad flatworms (Platyhelminthes: Turbellaria) from coral reefs of Lakshadweep Island, India. J Bombay Nat Hist Soc. 2011;108(2):109–13.

Bahia J, Padula V, Delgado M. Five new records and morphological data of polyclad species (Platyhelminthes: Turbellaria) from Rio Grande do Norte, northeastern Brazil. Zootaxa. 2012;3170:31–44.

Bahia J, Padula V, Lavrado HP, Quiroga S. Taxonomy of Cotylea (Platyhelminthes: Polycladida) from Cabo Frio, southeastern Brazil, with the description of a new species. Zootaxa. 2014;3873(5):495–525.

Brusa F, Damborenea C, Quiroga S. First records of Pseudocerotidae (Platyhelminthes: Polycladida: Cotylea) from Patagonia, Argentina. Zootaxa. 2009;2283:51–9.

Bulnes VN, Albano MJ, Obenat SM, Cazzaniga NJ. Three pseudocerotid (Platyhelminthes, Polycladida, Cotylea) species from the Argentinian coast. Zootaxa. 2011;2990:30–44.

Delle CS. Descrizione e notomia degli animali invertebrati della Sicilia citeriore osservati vivi negli anni. Naples: Batteli & Co.; 1841. p. 1–8.

Delle Chiaje S. (1822–1829) Memorie sulla storia e notomia degli animali senza vertebre del regno di Napoli. Napoli 1822–1829. [the atlas with 109 pl. appeared 1822; the 4 vols of text appeared as Vol 2; 1828 Vol 3; 1830 117–214. Fratelli Fernand.

Diesing CM. Helminthologische Beiträge. Nova Acta Acad Leopoldina. 1836;18:316.

Diesing C. M. (1850) Systema Helminthum. I. *Vindobonae*, 679.

Diesing K. Revision Der Turebellarien. IN COMMISSION BEI KARI, GEROLD'S SOHN, BÜCHHÄNDLER DER KAISERL. AKADEMIE DER WISSENSCHAFTEN. 1862;2:485–578.

Faubel A. The Polycladida, Turbellaria. Proposal and establishment of a new system. Part II. The Cotylea. Mitteilungen des Hamburgischenzoologischen Museums und Instituts. 1984;81:89–259.

Grube AE. (1840) Actinien, Echinodermen und Wuermer des adriatischen und Mittelmeers, nach eigenen Sammlungen beschreiben. Koenigsberg. 92.

Hyman LH. Further notes of the turbellarian fauna of the Atlantic Coast of the United States. Biol Bull. 1952;103:195–200.

Kato K. Polycladida of Japan. *J Sigenkagaku Kenkyusyo* I. 1944:257–318.

Laidlaw F. The marine Turbellaria, with an account of the anatomy of some species. Fauna Geology Maldive Laccadive Archipelagoes. 1902;1:282–312.

Laidlaw, F. (1906) On the marine fauna of the Cape Verde Islands, from collections made in 1904 by Mr C. Crossland. The polyclad Turbellaria. Proceedings of Zoological Society London, 705–719.

Lang A. Die Polycladen des Golfes von Neapel und der angrenzenden Meeresabschnitte. Eine Monographie. *Fauna Flora des Golfesv.* Neapel, Leipzig. 1884;11:1–688.

Litvaitis M, Bolaños M, Quiroga S. When names are wrong and colours deceive: unravelling the Pseudoceros Bicolor species complex (Turbellaria: Polycladida). J Nat Hist. 2010;44(13):829–45. doi:10.1080/00222930903537074.

Marcus E. Turbellaria brasileiros (7). Boletim da Faculdade de Filosofia, Ciências e Letras da Universidade de São Paulo Zoologia. 1949;14:7–155.

Marcus E, Marcus E. Polycladida from Curaçao and faunistically related regions. Stud Fauna Curaçao Caribb Isl. 1968;101:1–133.

Moseley HN. On *Stylochus Pelagicus*, an new species of pelagic planarian, with notes on other pelagic species, on the larval forms of Thysanozoon, and of a Gymnosomatous Pteropod. Mciroscopical J. 1877;17:23–32.

Newman LJ, Cannon LRG. Color pattern variation in tropical flatworm, *Pseudoceros* (Platyhelminthes: Polycladida) with description of three new species. Raffles Bull Zool. 1995;43:435–46.

Newman LJ, Cannon LRG. Fabulous flatworms: a guide to marine polyclads. Version 1. Canberra and Melbourne, Australia: ABRS and CSIRO Publishing, CD-ROM; 2005.

Noreña C, Marquina D, Pérez J, Almon B. First records of Cotylea (Polycladida, Platyhelminthes) for the Atlantic coast of the Iberian peninsula. ZooKeys. 2014;404:1–22. doi:10.3897/zookeys.404.7122.

Palombi A. Report on the Turbellaria (Cambridge expedition to the Suez Canal, 1924). Transact Zool Soc London. 1928;22:579–632.

Palombi A. Turbellari del sud Africa. Secondo contributo. Arch Zool Ital. 1938;25:124–49.

Palombi A. Turbellaria del Sud Africa. Policladi di East London. Terzo contributo Arch zool Italiano. 1939;28:123–49.

Pearse A. Polyclads of the east coast of North America. Proc U S Nat Mus. 1938;86:67–98.

Prudhoe S. A monograph on Polyclad Turbellaria. Oxford: Oxford University Press; 1985. p. 259.

Prudhoe S. Polyclad turbellarians recorded from African waters. Bull Br MusnatHist (zool). 1989;55(1):47–96.

Quatrefages De A. Études sur les types inférieurs de l'embranchement des annelés: mémoire sur quelques planairées marines appartenant aux genres Tricelis (Ehr.), Polycelis (Ehr.), Prosthiostomum (Nob.), Proceros (Nob.), Eolidiceros (Nob.), et Stylochus (Ehr). Annales des Sciences Naturelles, (3) Zool. 1845;4:129–84.

Quiroga S, Bolaños M, Litvaitis M. A checklist of polyclad flatworms (Platyhelminthes: Polycladida) from the Caribbean coast of Colombia, South America. Zootaxa. 2004;633:1–12.

Risso A. Mémoire sur quelques Gastéropodes nouveaux Nudibranches et Tectibranches observés dans la mer de Nice. J Physique, Chimie, d'Hist natt et Arts. 1818;87:272–376.

Sreeraj C, Raghunathan C. Pseudocerotid polyclads (Platyhelminthes, Turbellaria, Polycladida) from Andaman and Nicobar Islands, India. Proc Int Acad Ecol Environ Sci. 2013;3(1):36–41.

Stimpson W. Prodromus descriptionis anumalium evertebratorum quea in Expeditione as Oceanum, Pacificum Septentrionalem a Republica Federata missa, Johanne Rodgers Duce, observavit et descripsit. Part I Turbellaria Dendrocoela. Proc Acad Natl Sci Phila. 1857;9:19–31.

Tyler S. (2013) Thysanozoon Grube, 1840. Accessed through: World Register of Marine Species at http://www.marinespecies.org/aphia.php?p=taxdetails&id= 142242 on 2013-10-01.

Vera A, Moro L, Bacallado JJ, Hernández F. Contribución al conocimiento de la biodiversidad de políclados (Platyhelminthes, Turbellaria) em las Islas Canarias. Revista de la Academia Canaria de Ciencias. 2008;20(4):45–59.

Yeri M, Kaburaki T. (1916) Description of some Japenese polyclad turbellaria. J Coll Sci Tokyo Imperial Univ. XXXIX, 1–54.

38

First record of *Syllis vittata* (Polychaeta: Syllidae) in the Dutch North Sea

Inês Maia Dias[1*] ⓘ, Martijn Spierings[1], Joop W. P. Coolen[1,2], Babeth Van Der Weide[1] and Joël Cuperus[3]

Abstract

Background: *Syllis vittata* is present from British Waters to the Mediterranean Sea, Morocco and the Canary Islands and recorded from the South African coast and Indian Ocean.

Results and conclusion: In this paper, *S. vittata* is reported for the first time in the Dutch EEZ.

Keywords: Syllidae, *Syllis vittata*, polychaeta, North Sea, Gas platform, Distribution, First record

Background

Syllidae is a highly diverse family of polychaete worms with over 700 species described in 74 different genera (Aguado and San Martín 2009; San Martín and Aguado 2014). They are easily recognizable by their proventriculus, which is situated behind the pharynx and works as a suction pump during feeding processes (Aguado and San Martín 2009; Fauchald and Jumars 1979; San Martín 2003), often visible through the body wall. However, the taxonomy of the Syllidae is often problematic due to the high variability of its morphological features and life history traits (San Martín 2003). It has been recently divided into five subfamilies (Anoplosyllinae, Autolytinae, Eusyllinae, Exogoninae and Syllinae) according to phylogenetic relationships (Aguado and San Martín 2009).

Syllidae mostly reproduce by epitoky, within which two modes are recognized: epigamy (characteristic for Anoplosyllinae, Eusyllinae and Exogoninae) and schizogamy, i.e., stolonization (characteristic for Syllinae and Autolytinae; Aguado and San Martín 2009; Franke 1999; Musco et al. 2010). The reproductive strategy of *Syllis vittata* Grube 1840 is still unknown, however it is believed to be similar to the majority of other Syllinae species (Musco et al. 2010). Furthermore, these authors suggest that *S. vittata* might be a simultaneous hermaphrodite, though not necessarily with internal or self-fertilization.

Syllids are primarily observed in shallow waters in a variety of marine habitats worldwide, although they can also be found in deep waters (San Martín 2003). *S. vittata*, in particular, is present in the Eastern Atlantic, from British Waters to Morocco and the Canary Islands (López et al. 1996; San Martín 2003), in South African coast (Berrisford 1969), in Indian Ocean (Mozambique and Natal; (López et al. 1996; Musco and Giangrande 2005; San Martín 2003), and in the Mediterranean Sea (López et al. 1996; Musco and Giangrande 2005; San Martín 2003). The type locality is Palermo (Italy). In the North Sea, early records suggest that this species was only found on the coast of Norfolk (Ostler 2005). In this study, *Syllis vittata* is recorded for the first time in the Dutch area of the North Sea.

Methods

Two specimens of *Syllis vittata* were found in the Dutch area of the North Sea, on a gas platform (platform L15-A, situated 11 km off the coast of the island of Vlieland; Fig. 1), in samples processed for the RECON (Reef effects of structures in the North Sea: Islands or connections?) project. Samples were collected with a surface supplied airlift, as described in Coolen et al. (2015a) and fixed in a borax buffered 6% formaldehyde solution. In the laboratory, before the identification process, specimens were placed in a 70% ethanol and 3% glycerol solution. For the identification, Keys provided by San Martín (2003) and San Martín and Worsfold (2015) were used. Both specimens were stored at the Wageningen Marine Research benthic reference collection.

* Correspondence: inesmaiadias@gmail.com
[1]Wageningen Marine Research, P.O. Box 571780 AB Den Helder, The Netherlands
Full list of author information is available at the end of the article

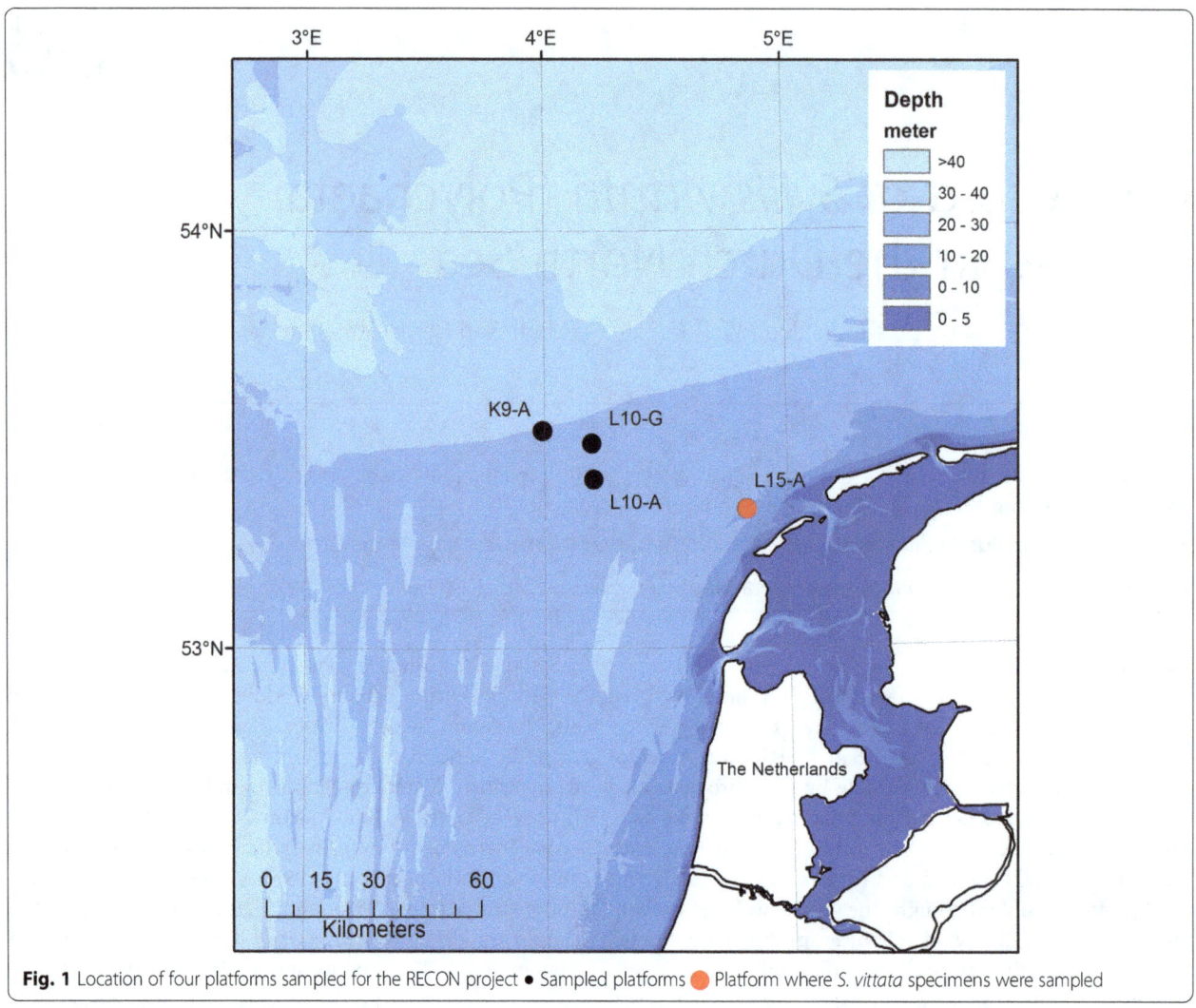

Fig. 1 Location of four platforms sampled for the RECON project ● Sampled platforms ● Platform where *S. vittata* specimens were sampled

Results

Material examined

L15-A platform, The Netherlands, 53.3295°N, 4.8301°E (WGS84), 5 and 6 June 2014, 2 specimens, 7 and 20 m depth.

SYSTEMATICS
Order PHYLLODOCIDA Dales 1962
Suborder NEREIDIFORMIA
Family SYLLIDAE Grube 1850
Subfamily SYLLINAE Rioja 1925
Genus *Syllis* Savigny in Lamarck 1818
Syllis vittata Grube 1840

Syllis (Typosyllis) vittata Fauvel 1923:263–264, fig. 98i–l. Day 1967:252, Fig. 12.4.m–o
Syllis aurita Claparède, 1864: 539-540, plate V fig. 5
Syllis buskii McIntosh, 1908
Syllis vittata San Martín 2003:430–432, figs. 236–237.

General distribution

From British Waters to the Mediterranean Sea, Morocco and the Canary Islands (Eastern Atlantic) and recorded on the South African coast, Mozambique and Natal (Indian Ocean)

Type locality

Palermo (Italy)

Description of examined material

Length up to 10.75 mm, width 0.9 mm. Body broad, robust and cylindrical, 40 chaetigers, with a dorsal dark transverse stripe per segment (Fig. 2a). Prostomium wider than long, with two pairs of eyes in a trapezoidal arrangement. Antennae moniliform; median antenna inserted in the middle of the prostomium, with 30 articles; lateral ones inserted before the anterior eyes, with 23–25 articles. Dorsal tentacular cirri with 35–38 articles and ventral cirri with 21–24 articles, very similar to dorsal ones. Dorsal

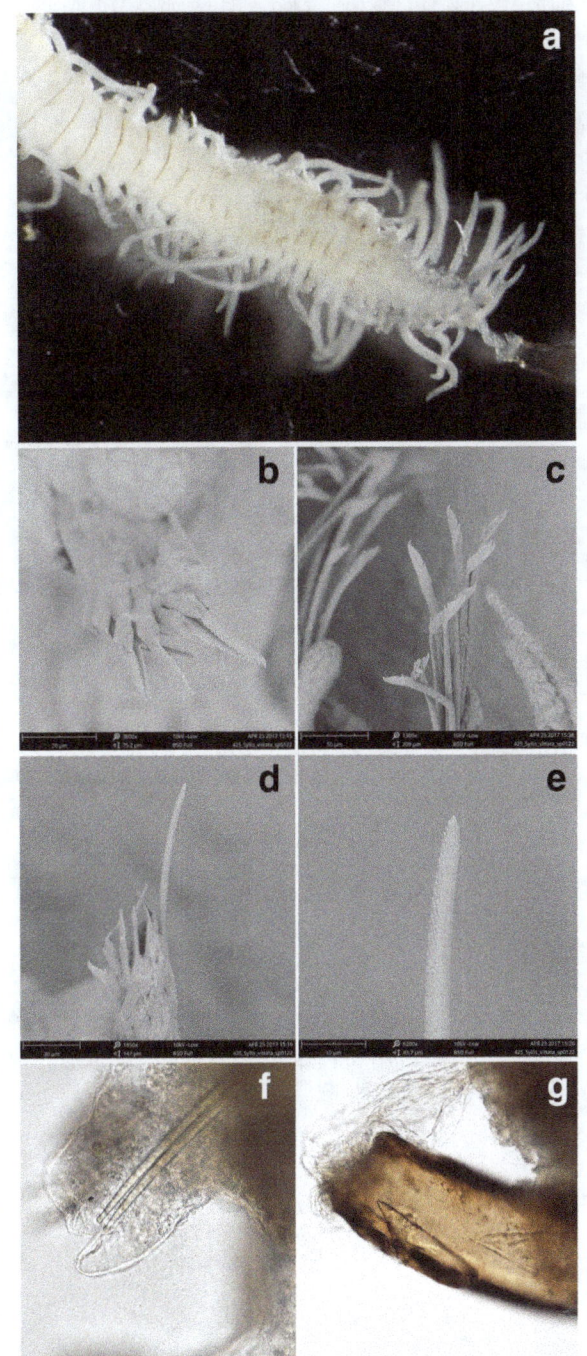

Fig. 2 *Syllis vittata* : **a**) specimen 1, **b**) anterior chaetae, **c**) mid body chaetae, **d**) posterior chaetae, **e**) simple chaeta, **f**) posterior aciculae, **g**) pharyngeal tooth

serrated dorsal simple chaeta (Fig. 2e), only present in posterior parapodia. Aciculae possess a rounded tip (Fig. 2f). and their number decreases towards the posterior end, with only one in the last chaetigers. Pharynx extending through 10 segments, with a large, acute tooth, encircled by 10 large marginal papillae, placed near the pharyngeal opening (Fig. 2g). Proventriculus extending through 9 segments, with 38 rows of muscle cells.

Discussion

The presence of the polychaete *Syllis vittata* off the coast of the island of Vlieland represents its first record in the Dutch Exclusive Economic Zone (EEZ). With this record, this species has only been found on two separate locations in the North Sea, with the other observation being old records off the coast of Norfolk (United Kingdom) by the Joint Nature Conservation Committee (JNCC), in 1993 (Ostler 2005).

The scarce observations of this species in the North Sea may be a result of its apparent preference for hard-bottom substrates, as *Syllis vittata* is usually found on rocky shores (Antoniadou et al. 2004; Cosentino 2011; López et al. 1996; Simon et al. 2014), calcareous substrates (Cardell and Gili 1988; López et al. 1996; San Martín 2003) and on *Sabellaria* and *Mytilus* colonies (López et al. 1996; San Martín 2003). Thousands of other artificial hard substrates are present in the Dutch North Sea (Coolen et al. 2015b) but natural hard substrates are scarce (Coolen et al. 2016). The biodiversity of hard substrates is understudied in the Netherlands. Its predilection for coastal habitats and a lack of research effort may justify why this species wasn't found on the other two platforms analysed.

The wide distribution of *Syllis vittata* suggests that this might be a cryptogenic species. However, more information is needed to confirm this hypothesis.

Nevertheless, it is likely that *Syllis vittata* will colonise all the available hard substrates, and spread throughout the North Sea. The distribution limits of this species is, then, yet unclear.

Conclusion

Two specimens of *Syllis vittata* were observed on a gas platform 11 km off the coast of the island of Vlieland. This is the first record of this species in the Dutch EEZ. Ongoing monitoring is needed to verify the spread or loss of this species in the North Sea.

Acknowledgments
ENGIE Exploration & Production Nederland B.V. allowed and facilitated us to sample their installations and we are especially grateful to Ed Schmidt, Nathalie Kaarls, Ulf Sjöqvist, Maico Vrijenhoeff, Ben Waardenburg and Kees van Braak for their help arranging this cooperation.
Further help was provided by Guillermo San Martín who confirmed the identification.

cirri alternating long (longer than the width of the corresponding segment), with 30–35 articles, and short (as long as the corresponding segment), with 20–24 articles. Anterior parapodia with 20 compound chaetae (Fig. 2b), mid body parapodia with 15 (Fig. 2c) and 12 in posterior parapodia (Fig. 2d), all unidentate. One bifid

Funding

The work reported in this publication was funded through the Wageningen UR TripleP@Sea Innovation program (J.C., KB-14-007) and by the Dutch Department of Economic Affairs (J.C., KB-24-16), the Nederlandse Aardolie Maatschappij BV, Wintershall Holding GmbH, Energiebeheer Nederland B.V. and the INSITE North Sea fund via the RECON project.

Authors' contributions

IMD and MS carried out the identification of all Polychaeta present in all the samples. JWPC did the sampling and designed the study. BvdW and JC assisted with identifying the samples. All authors read and approved the final manuscript.

Competing interests

The authors declare that they have no competing interests.

Author details

[1]Wageningen Marine Research, P.O. Box 571780 AB Den Helder, The Netherlands. [2]Wageningen University, Chair group Aquatic Ecology and Water Quality Management, Droevendaalsesteeg 3a, 6708 PD Wageningen, The Netherlands. [3]Rijkswaterstaat, Ministry of Infrastructure and the Environment, Zuiderwagenplein 2, 8224 AD Lelystad, The Netherlands.

References

Aguado MT, San Martín G. Phylogeny of Syllidae (Polychaeta) based on morphological data. Zool Scr. 2009;38(4):379–402.

Antoniadou C, Nicolaidou A, Chintiroglou C. Polychaetes associated with the sciaphilic alga community in the northern Aegean Sea: Spatial and temporal variability. Helgol Mar Res. 2004;58(3):168–82.

Berrisford CD. Biology and zoogeography of the vema seamount: a report on the first biological collection made on the summit. Trans R Soc South Africa. 1969;38(4):387–98.

Cardell MJ, Gili JM. Distribution of a population of annelid polychaetes in the "trottoir" of the midlittoral zone on the coast of North-East Spain, Western Mediterranean. Mar Biol. 1988;99(1):83–92.

Claparède, É. Glanures zootomiques parmi les annélides de Port-Vendres (Pyrénées Orientales). Mémoires de la Société de Physique et d'Histoire Naturelle de Genève. 1864;17(2):463–600, plates I-VIII.

Coolen JWP, Bos OG, Glorius S, Lengkeek W, Cuperus J, Van der Weide BE, et al. Reefs, sand and reef-like sand: a comparison of the benthic biodiversity of habitats in the Dutch Borkum Reef Grounds. J Sea Res. 2015a;103:84–92.

Coolen JWP, Lengkeek W, Lewis G, Bos OG, van Walraven L, van Dongen U. First record of Caryophyllia smithii in the central southern North Sea: artificial reefs affect range extensions of sessile benthic species. Mar Biodivers Rec. 2015b;8(e140):4.

Coolen JWP, Lengkeek W, Degraer S, Kerckhof F, Kirkwood RJ, Lindeboom HJ. Distribution of the invasive Caprella mutica and native Caprella linearis on artificial hard substrata in the North Sea: separation by habitat. Aquat Inv. 2016;11(4):437–49.

Cosentino A. Microhabitat selection in a local syllid assemblage with the first record of Syllis hyllebergi (Syllinae) in the central Mediterranean. Ital J Zool. 2011;78(May):267–79.

Dales RP. The polychaete Stomodeum and the inter-relationships of the families of Polychaeta. Proc Zool Soc London. 1962;139(3):389–428.

Day JH. A monograph on the Polychaeta of Southern Africa. Br Mus. 1967;656:1–878.

Fauchald K, Jumars P. The diet of worms: a studie of Polychaete feeding gilds. Oceanogr Mar Biol Annu Rev. 1979;17:193–284.

Fauvel P. Quatrième note préliminaire sur les Polychètes provenant des campagnées de l'Hirondelle et de la Princesse-Alice, ou deposées dans le Musée Océanographique de Monaco. Bull de l'Inst Océanog. 1923;270:1–80.

Franke HD. Reproduction of the Syllidae (Annelida: Polychaeta). Hydrobiologia. 1999;402(2):39–55.

Grube AE. Actinien, Echinodermen und Würmer des Adriatischen- und Mittelmeers: nach eigenen Sammlungen beschrieben. 1840. p. 1–92.

Grube AE. The families of the annelids. Arch Nat Hist. 1850;16(1):249–364.

Lamarck J-B. Histoire naturelle des animaux sans vertèbres présentant les caractères généraux et particuliers de ces animaux, leur distribution, leurs classes, leurs familles, leurs genres, et la citation des principales espèces qui s'y rapportent; précédée d'une introduction offrant la détermination des caractères essentiels de l'animal, sa distinction du végétal et des autres corps naturels, enfin, l'exposition des principes fondamentaux de la zoologie. Verdière. 1818; p. 612

López E, San Martín G, Jiménez M. Syllinae (Syllidae, Annelida, Polychaeta) from Chafarinas Islands (Alborán Sea, W. Mediterranean). Misc Zool. 1996;19(1):105–18.

McIntosh, WC. A monograph of British Annelids. Ray Society of London, II. Part I. Polychaeta. Nephthydidae to Syllidae. 1908;1–232.

Musco L, Giangrande A. Mediterranean Syllidae (Annelida: Polychaeta) revisited: Biogeography, diversity and species fidelity to environmental features. Mar Ecol Prog Ser. 2005;304:143–53.

Musco L, Lepore E, Gherardi M, Sciscioli M, Mercurio M, Giangrande A. Sperm ultrastructure of three Syllinae (Annelida, Phyllodocida) species with considerations on syllid phylogeny and Syllis vittata reproductive biology. Zoomorphology. 2010;129(2):133–9.

Ostler R. Marine Nature Conservation Review (MNCR) and associated benthic marine data held and managed by JNCC. Jt. Nat. Conserv. Committee, Cent. Ecol. Hydrol. Aberdeenshire, UK. 2005. http://www.emodnet-biology.eu/. Accessed 16 Jun 2016.

Rioja E. Anelidos poliquetos de San Vicente de la Barquera (Cantabrico). Trabajos del Museo Nacional de Ciencias Naturale. Ser Zool. 1925;53:1–62.

San Martín G. Annelida, Polychaeta II: Syllidae. In: Ramos MA et al. (Eds). Fauna Iber. vol. 21. Madrid. Museo Nacional de Ciencias Naturales. CSIC. 2003; p. 554.

San Martín G, Aguado MT. Family Syllidae. In: Schmidt-Rhaesa A, editor. En Phyllodocida Nereidiformia. Handbook of Zoology, Annelida. A Natural History of the Phyla of the Animal Kingdom. Zürich: Verlag Walter der Gruyter GmbH & Co; 2014. p. 1–52.

San Martín G, Worsfold TM. Guide and keys for the identification of Syllidae (Annelida, Phyllodocida) from the British Isles (reported and expected species). Zookeys. 2015;29(488):1–29.

Simon C, San Martín G, Robinson G. Two new species of Syllis (Polychaeta: Syllidae) from South Africa, one of them viviparous, with remarks on larval development and vivipary. J Mar Biol Assoc UK. 2014;94(04):729–46.

39

Vagrancy in paradise: documentation of the chevron butterflyfish *Chaetodon trifascialis* in Kaneohe Bay, Oahu, Hawaiian islands

Erik C. Franklin

Abstract

The chevron butterflyfish *Chaetodon trifascialis* is an obligate corallivore found on Indo-Pacific coral reefs. The typical northern extent of the distributional range of the chevron butterflyfish overlaps with the occurrence of one of its preferred corals *Acropora cytherea* on reefs of French Frigate Shoals in the Northwestern Hawaiian Islands. Here I document the occurrence of a single *C. trifascialis* in Kaneohe Bay, Oahu outside its typical range and discuss the role of biogeographic vagrants as potential colonizers that provide direct evidence of inter-island larval connectivity in the Central Pacific Ocean.

Keywords: *Chaetodon trifascialis*, Butterflyfish, Coral reef, Oahu, Vagrant species, Biogeography, Hawaiian islands

Background

The chevron butterflyfish *Chaetodon trifascialis* Quoy & Gaimard, 1825 is an obligate corallivore found on Indo-Pacific coral reefs. The species exhibits a strong ecological relationship with corals of the genus *Acropora*, in particular *A. hyacinthus*, and *A. cytherea* (Berumen & Pratchett, 2008; Berumen et al., 2012; Lawton et al., 2012; Pratchett et al., 2013; Pratchett et al., 2006). Typically, harems of *C. trifascialis* occupy a home range and feed primarily on *Acropora* corals (Yabuta & Berumen, 2013). Due to the specialist diet on corals, *C. trifascialis* has been suggested as an indicator species for the condition of coral reefs (Reese, 1981; Ohman et al., 1998). Globally, *C. trifascialis* is in "near threatened" status by the IUCN due to the susceptibility of corals to bleaching events (Carpenter & Pratchett, 2010). For example, the *C. trifascialis* was not observed from reefs that experienced massive bleaching events in follow-up surveys at sites where they had been previously documented (Pratchett et al., 2006). While *C. trifascialis* is distributed throughout the Indo-Pacific, it only occurs commonly in one location of the Hawaiian archipelago, French Frigate Shoals, where *Acropora cytherea* is well established (Asher et al., 2012; Randall, 2007; Grigg, 1981).

This study documents the occurrence of a single *C. trifascialis* in Kaneohe Bay, Oahu, Hawaiian Islands and discusses the role of biogeographic vagrants in this region as potential colonizers that demonstrate inter-island larval connectivity.

Methods

One specimen of *C. trifascialis* was observed and photographed during a dive on September 4, 2013 in a coral patch reef at 21°27′5.04″N and 157°47′25.537″W in Kaneohe Bay, Oahu, Hawaiian Islands (Fig. 1). The specimen was not collected because the author did not possess a Hawaii state collection permit for that particular species at the time of observation. The site was characterized by a stand of *Pocillopora meandrina* corals in the back reef of Kaneohe Bay near Sampan Channel.

Results

One specimen of *C. trifascialis* was observed and photographed within the branches of a *Pocillopora meandrina* coral colony at 4 m depth (Fig. 2). The fish was visually estimated as 5 cm TL and not collected. The coral colony was located in back reef habitat with isolated coral patches among surrounding extents of sand-bottom in Kaneohe Bay. No additional specimens of *C. trifascialis* were encountered during four dives (total time 2 h) in the general vicinity of the initial observation.

Correspondence: erik.franklin@hawaii.edu
Hawaii Institute of Marine Biology, School of Ocean and Earth Science and Technology, University of Hawaii at Manoa, Kaneohe, HI, USA

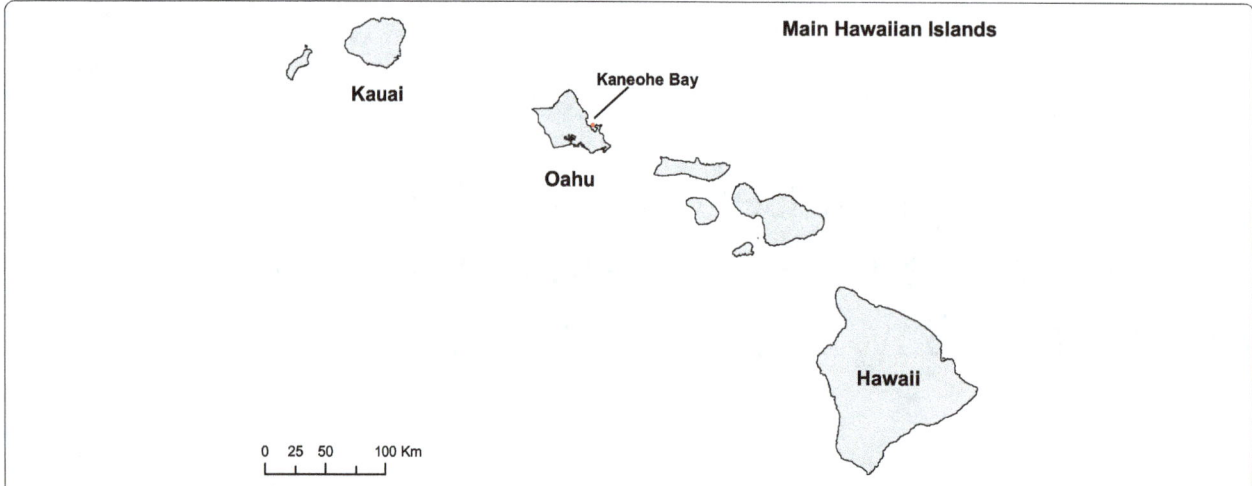

Fig. 1 Map of the island of Oahu in the Hawaiian Islands. Red circle indicates the position of the observed specimen of *Chaetodon trifascialis* in Kaneohe Bay

Discussion

The northern range of *Chaetodon trifascialis* has been observed on reefs of French Frigate Shoals in the Northwestern Hawaiian Islands, yet documented evidence of specimens has not been recorded elsewhere in the Hawaiian archipelago. For example, a regional-scale reef fish monitoring survey program of the Hawaiian Islands by NOAA from 2000 to 2016 did not observe a single specimen of *C. trifascialis* in the MHI (I. Williams pers. comm.). Prior anecdotes of *C. trifascialis* on Oahu reefs were described by Reese (Reese, 1981) but not supported with primary evidence, so this record represents the first documented observation of the species in the main

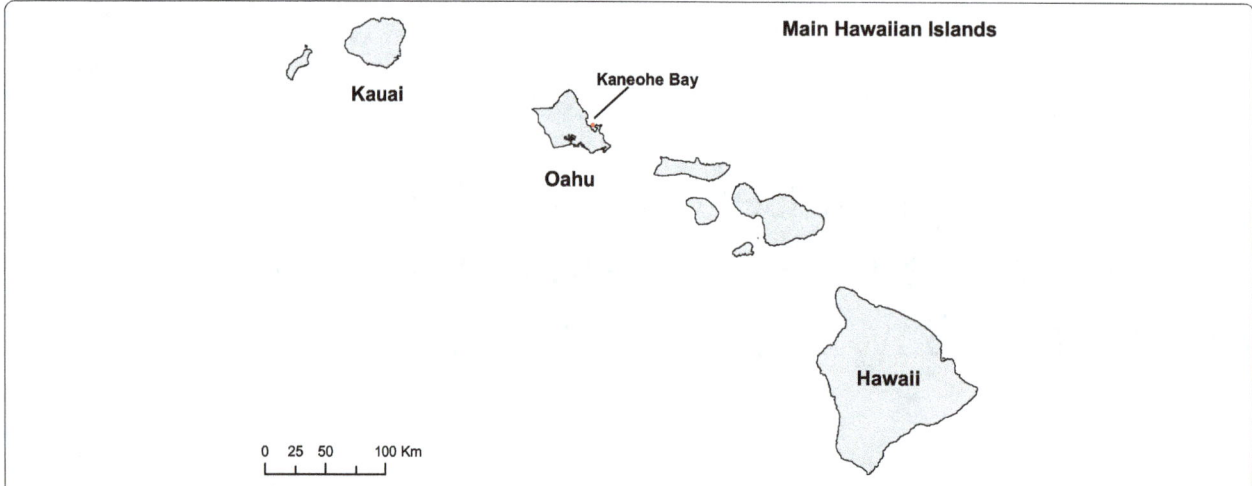

Fig. 2 Photograph of a specimen of *Chaetodon trifascialis* observed within the branches of *Pocillopora meandrina* in Kaneohe Bay, Oahu, HI

Hawaiian Islands. Other observations of *C. trifascialis* on Oahu reefs were made in Kaneohe Bay in 1983 (R. Kosaki pers. comm.) and Hanauma Bay in the late 1980's or early 1990's (B. Mundy pers. comm.). Given known larval connectivity pathways, the specimen of *C. trifascialis* was most likely a vagrant species from Johnston Atoll (Kobayashi, 2006; Wren et al., 2016).

Vagrant butterflyfish species are not uncommon. A review of geographical distribution shifts in 52 fish families found that butterflyfish (family: Chaetodontidae) had a significantly higher proportion of vagrants than expected by species richness given each family's total number of species (Feary & Pratchett, 2014). For example, in the Azores and Madeira Islands off the western African coast, only vagrant butterflyfish species have been observed and there are no known permanent populations (Kulbicki et al., 2013). The vagrant species observed in the Azores, *Chaetodon sedentarius*, is commonly distributed in the western Atlantic and Caribbean region (Allen et al., 2010). Two butterflyfishes, *C. lunula* (Pyle et al., 2010) and *C. unimaculatus* (Myers & Pratchett, 2010a), are Indo-Pacfic species that are biogeographic vagrants in the Galapagos Islands. Furthermore, a single individual of the black butterflyfish *C. flavirostris* has been observed as a vagrant species at Easter Island (Myers & Pratchett, 2010b). These examples are not comprehensive but provide a sample to demonstrate the relative commonness of butterflyfish vagrant species.

In the Hawaiian archipelago, larval connectivity from Johnston Atoll represents the most likely source for periodic colonization of reef fauna, such as butterflyfish (family: Chaetodontidae) and stony corals of family Acroporidae (Kobayashi, 2006; Wren et al., 2016). The overlap in butterflyfish species composition between the Hawaiian archipelago and Johnston Atoll is nearly complete, with the exception of *C. fremblii* which has not been observed at Johnston Atoll (Randall, 2007; Kosaki et al., 1991; Randall et al., 1985; Wagner et al., 2014). The lined butterflyfish *C. lineolatus* has been rarely observed at Johnston Atoll (Randall et al., 1985), so could be described as a biogeographic vagrant there. The *Acropora* corals that are the primary prey for *C. trifascialis* are commonly found at French Frigate Shoals in the Northwestern Hawaiian Islands but have been very rarely sighted in the main Hawaiian Islands (Asher et al., 2012; Grigg et al., 1981). Isolated *Acropora cytherea* colonies have been recorded off the islands of Kauai (Grigg et al., 1981; Maragos, 1977; Kenyon et al., 2007) and Oahu (Kosaki et al., 2013) but attempted resightings for these colonies have failed. While most singleton species arrivals to the Hawaiian Islands probably perish, several large colonies of *Acropora gemmifera* were observed in April 2013 on the Kona coast of Hawaii Island (Walsh et al., 2013). The size of the *A. gemmifera* corals suggested that they had been present for decades in the area. Grigg (Grigg, 1981) hypothesized that *Acropora* were in the process of post-Pleistocene recolonization from larval dispersal outside of the archipelago. Given the obligate feeding relationship of *C. trifascialis* on *Acropora* corals, a successful colonization of *C. trifascialis* would require the pre-establishment of *Acropora* to settlement sites in the main Hawaiian Islands. The probability of this occurrence is low but not impossible especially if range expansions of these *Acropora* species to the MHI are facilitated by the predicted increases of sea surface temperatures under changing climates (Baird et al., 2012; Yamano et al., 2011). Under future warming conditions with enough larval supply and survival, the successful colonization and establishment of both *Acropora* species and *C. trifascialis* from Johnston Atoll to sites in the Main Hawaiian Islands may be possible as a continued recolonization of the Hawaiian archipelgo (Grigg, 1981; Kobayashi, 2006; Maragos & Jokiel, 1986).

Acknowledgements

I thank J. Randall for confirmation of the species identification, R. Kosaki, B. Mundy, I. Williams, P. Ayotte, A. Gray, J. Asher, and K. Lino for sharing survey information about Chaetodontidae of the Hawaiian Islands and Johnston Atoll, and two anonymous reviewers for improving this manuscript. This is HIMB contribution #1701 and SOEST contribution #10110.

Funding

NOAA award #NA10NMF4520163.

Author's contributions

EF conceived and performed the experiment, analyzed the data, and wrote the manuscript.

Competing interests

The author declares that he has no competing interests.

References

Allen GR, Floeter S, McEachran JD: *Chaetodon sedentarius. The IUCN Red List of Threatened Species* 2010, 2010:e.T155220A4749409.

Asher J, Maragos J, Kenyon J, Vargas-Angel B, Coccagna E. Range extensions for several species of Acropora in the Hawaiian archipelago and Papahanaumokuakea marine National Monument. Bull Mar Sci. 2012;88:337–8.

Baird AH, Sommer B, Madin JS. Pole-ward range expansion of Acropora spp along the east coast of Australia. Coral Reefs. 2012;31:1063.

Berumen ML, Pratchett MS. Trade-offs associated with dietary specialization in corallivorous butterflyfishes (Chaetodontidae : Chaetodon). Behav Ecol Sociobiol. 2008;62:989–94.

Berumen ML, Trip EDL, Pratchett MS, Choat JH. Differences in demographic traits of four butterflyfish species between two reefs of the great barrier reef separated by 1,200 km. Coral Reefs. 2012;31:169–77.

Carpenter KE, Pratchett M: *Chaetodon trifascialis. The IUCN Red List of Threatened Species* 2010, 2010:e.T165712A6098323.

Feary DA, Pratchett MS. J Emslie M, fowler AM, Figueira WF, Luiz OJ, Nakamura Y, booth DJ: latitudinal shifts in coral reef fishes: why some species do and others do not shift. Fish Fish. 2014;15:593–615.

Grigg RW. Acropora in Hawaii .2. Zoogeography. Pac Sci. 1981;35:15–24.

Grigg RW, Wells JW, Wallace C. Acropora in Hawaii .1. History of the scientific record, Systematics, and ecology. Pac Sci. 1981;35:1–13.

Kenyon J, Godwin S, Montgomery A, Brainard R. Rare sighting of Acropora Cytherea in the main Hawaiian islands. Coral Reefs. 2007;26:309.

Kobayashi DR. Colonization of the Hawaiian archipelago via Johnston atoll: a characterization of oceanographic transport corridors for pelagic larvae using computer simulation. Coral Reefs. 2006;25:407–17.

Kosaki RK, Pyle RL, Randall JE, Irons DK. New records of fishes from Johnston atoll, with notes on biogeography. Pac Sci. 1991;45:186–203.

Kosaki RK, Wagner D, Leonard JC, Hauk BB, Gleason KA. First report of the table coral <I>Acropora Cytherea</I> (Scleractinia: Acroporidae) from Oahu (main Hawaiian islands). Bull Mar Sci. 2013;89:745–6.

Kulbicki M, Vigliola L, Wantiez L, Hubert N, Floeter SR, Myers RF. Biogeography of butterflyfishes: a global model for reef fishes? In: Prachett MS, Berumen ML, Kapoor BG, editors. Biology of Butterflyfishes. Boca Raton, FL: CRC Press; 2013.

Lawton RJ, Cole AJ, Berumen ML, Pratchett MS. Geographic variation in resource use by specialist versus generalist butterflyfishes. Ecography. 2012;35:566–76.

Maragos J. Order Scleractinia. In: EL DDW, editor. Reef and shore fauna of Hawaii section 1: protozoa through Ctenophora. Honolulu: Bishop Museum; 1977. p. 158–241.

Maragos JE, Jokiel PL. Reef corals of Johnston atoll: one of the world's most isolated reefs. Coral Reefs. 1986;4:141–50.

Myers R, Pratchett M: *Chaetodon unimaculatus. The IUCN Red List of Threatened Species* 2010, 2010a:e.T165714A6099340.

Myers R, Pratchett M: *Chaetodon flavirostris. The IUCN Red List of Threatened Species* 2010, 2010b:e.T165688A6091974.

Ohman MC, Rajasuriya A, Svensson S. The use of butterflyfishes (Chaetodontidae) as bio-indicators of habitat structure and human disturbance. Ambio. 1998; 27:708–16.

Pratchett MS, Graham NA, Cole AJ. Specialist corallivores dominate butterflyfish assemblages in coral-dominated reef habitats. J Fish Biol. 2013;82:1177–91.

Pratchett MS, Wilson SK, Baird AH. Declines in the abundance of Chaetodon butterflyfishes following extensive coral depletion. J Fish Biol. 2006;69: 1269–80.

Pyle R, Craig MT, Pratchett M: *Chaetodon lunula. The IUCN Red List of Threatened Species* 2010, 2010:e.T165651A6080984.en.

Randall JE. Reef and shore fishes of the Hawaiian islands. Honolulu: Sea Grant College Program, University of Hawai'i; 2007.

Randall JE, Lobel PS, Chave EH. Annoted checklist of the fishes of Johnston Island. Pac Sci. 1985;39:24–80.

Reese ES. Predation on corals by fishes of the family Chaetodontidae: implications for conservation and management of coral reef ecosystems. Bull Mar Sci. 1981;31:594–604.

Wagner D, Kosaki RK, Spalding HL, Whitton RK, Pyle RL, Sherwood AR, Tsuda RT, Calcinai B. Mesophotic surveys of the flora and fauna at Johnston atoll, Central Pacific Ocean. Marine Biodiversity Records. 2014;7:e68.

Walsh WJ, Cotton S, Jackson L, Lamson M, Martin R, Osada-D'Avella K, Preskitt L. First record of Acropora Gemmifera in the main Hawaiian islands. Coral Reefs. 2013;33:57.

Wren JL, Kobayashi DR, Jia Y, Toonen RJ. Modeled population connectivity across the Hawaiian archipelago. PLoS One. 2016;11:e0167626.

Permissions

All chapters in this book were first published in MBR, by BioMed Central; hereby published with permission under the Creative Commons Attribution License or equivalent. Every chapter published in this book has been scrutinized by our experts. Their significance has been extensively debated. The topics covered herein carry significant findings which will fuel the growth of the discipline. They may even be implemented as practical applications or may be referred to as a beginning point for another development.

The contributors of this book come from diverse backgrounds, making this book a truly international effort. This book will bring forth new frontiers with its revolutionizing research information and detailed analysis of the nascent developments around the world.

We would like to thank all the contributing authors for lending their expertise to make the book truly unique. They have played a crucial role in the development of this book. Without their invaluable contributions this book wouldn't have been possible. They have made vital efforts to compile up to date information on the varied aspects of this subject to make this book a valuable addition to the collection of many professionals and students.

This book was conceptualized with the vision of imparting up-to-date information and advanced data in this field. To ensure the same, a matchless editorial board was set up. Every individual on the board went through rigorous rounds of assessment to prove their worth. After which they invested a large part of their time researching and compiling the most relevant data for our readers.

The editorial board has been involved in producing this book since its inception. They have spent rigorous hours researching and exploring the diverse topics which have resulted in the successful publishing of this book. They have passed on their knowledge of decades through this book. To expedite this challenging task, the publisher supported the team at every step. A small team of assistant editors was also appointed to further simplify the editing procedure and attain best results for the readers.

Apart from the editorial board, the designing team has also invested a significant amount of their time in understanding the subject and creating the most relevant covers. They scrutinized every image to scout for the most suitable representation of the subject and create an appropriate cover for the book.

The publishing team has been an ardent support to the editorial, designing and production team. Their endless efforts to recruit the best for this project, has resulted in the accomplishment of this book. They are a veteran in the field of academics and their pool of knowledge is as vast as their experience in printing. Their expertise and guidance has proved useful at every step. Their uncompromising quality standards have made this book an exceptional effort. Their encouragement from time to time has been an inspiration for everyone.

The publisher and the editorial board hope that this book will prove to be a valuable piece of knowledge for researchers, students, practitioners and scholars across the globe.

List of Contributors

Vanessa Ochi Agostini
Universidade Federal do Rio Grande (FURG), Programa de Pós-Graduação em Oceanografia Biológica (PPGOB), Instituto de Oceanografia (IO), Avenida Itália, Km 8, CEP 96203-900 Rio Grande, RS, Brazil

Carla Penna Ozorio
Universidade Federal do Rio Grande do Sul (UFRGS), Instituto de Biociências, Departamento de Zoologia, Avenida Bento Gonçalves, 9500, CEP 91501-970 Porto Alegre, RS, Brazil

Leonardo Ortega
Dirección Nacional de Recursos Acuáticos, Constituyente 1497, C.P 11200 Montevideo, Uruguay

Luis Rubio
Dirección Nacional de Recursos Acuáticos, Constituyente 1497, C.P 11200 Montevideo, Uruguay
Museo Nacional de Historia Natural, C.C. 399, C.P 11000 Montevideo, Uruguay

Valentina Leoni, Wilson Serra and Silvana González
Dirección Nacional de Recursos Acuáticos, Constituyente 1497, C.P 11200 Montevideo, Uruguay
Museo Nacional de Historia Natural, C.C. 399, C.P 11000 Montevideo, Uruguay
InvBiota. Invertebrados del Uruguay, Montevideo, Uruguay

Fabrizio Scarabino
Dirección Nacional de Recursos Acuáticos, Constituyente 1497, C.P 11200 Montevideo, Uruguay
Museo Nacional de Historia Natural, C.C. 399, C.P 11000 Montevideo, Uruguay
Centro Universitario Regional Este, Sede Rocha, Universidad de la República, Ruta 9, Km 208, C.P 27000 Rocha, Uruguay

Gabriela Failla Siquier
Lab. Zoología de Invertebrados, Dpto. de Biología Animal, Facultad de Ciencias, Universidad de la República, CP 11400 Montevideo, Uruguay

Alicia Dutra and Ana Gabriella Alonzo Campi
Prof. de Ciencias Biológicas egresada del I.P.A, Montevideo, Uruguay

Martin Abreu
COENDU, Conservación de Especies Nativas del Uruguay, Montevideo, Uruguay

Sergio N. Stampar
Departamento de Ciências Biológicas, Faculdade de Ciências e Letras, Unesp - Univ Estadual Paulista, Assis, Av. Dom Antonio, 2100, Assis 19806-900, Brazil

André C. Morandini
Departamento de Zoologia

Allen W. L. To
WWF-Hong Kong, 15/F, Manhattan Centre, 8 Kwai Cheong Road, Kwai Chung, New Territories, Hong Kong

Stanley K. H. Shea
BLOOM Association, c/o, ADMCF, 9 Queen's Road Central, Hong Kong

M. Natalia Rincón-Díaz and Adriana Santos-Martínez
Universidad Nacional de Colombia, Sede Caribe. San Luis Free Town # 52-44, San Andrés Isla, Colombia

Brigitte Gavio
Universidad Nacional de Colombia, Sede Caribe. San Luis Free Town # 52-44, San Andrés Isla, Colombia
Departamento de Biología, Universidad Nacional de Colombia, sede Bogotá, Ciudad Universitaria, Bogotá, Colombia

Michael J. Wynne
University of Michigan Herbarium, Ann Arbor, MI 48108, USA

H. Arraj
Department of Marine Biology, High Institute of Marine Researches (HIMR), Tishreen University, Latakia, Syria

H. Mayhoob and A. Abbas
Department of Botany, Faculty of Science, Tishreen University, Latakia, Syria

F. Hoe Chang
Biodiversity and Biosecurity Group, National Institute of Water & Atmospheric Research Ltd., P. O. Box 14-901, Kilbirnie, Wellington 6241, New Zealand

Lisa Northcote
Ocean Sediments Group, National Institute of Water & Atmospheric Research Ltd., P. O. Box 14-901, Kilbirnie, Wellington 6241, New Zealand

Leonardo Manir Feitosa
Departamento de Biologia, Universidade Federal do Maranhão, São Luís, Maranhão 65080-805, Brasil

Ana Paula Barbosa Martins
Centre for Sustainable Tropical Fisheries and Aquaculture, College of Science and Engineering, James Cook University, Townsville, QLD 4811, Australia

Jorge Luiz Silva Nunes
Departamento de Oceanografia e Limnologia, Universidade Federal do Maranhão, Cidade Universitária do Bacanga, São Luís, Maranhão 65080-805, Brasil

Ibrahim M. Tobuni, Esmail A. Shakman and Ben-Abdallah R. Benabdallah
Zoology Department, Tripoli University, Tripoli, Libya

Fabrizio Serena
Environmental Protection Agency-Tuscany Region (ARPAT), Tripoli, Italy
Institute for the Coastal Marine Environment (IAMC)-Italian National Research Council (CNR), Mazara, Italy

Amber N. Reichert
Pacific Shark Research Center, Moss Landing Marine Laboratories, 8272 Moss Landing Road, Moss Landing, CA 95039, USA

Lonny Lundsten
Monterey Bay Aquarium Research Institute, 7700 Sandholdt Road, Moss Landing, CA 95039, USA.

David A. Ebert
Pacific Shark Research Center, Moss Landing Marine Laboratories, 8272 Moss Landing Road, Moss Landing, CA 95039, USA

Department of Ichthyology Research Associate, California Academy of Sciences, 55 Music Concourse Drive, San Francisco, CA 94118, USA

Wei-Jen Chen and Jhen-Nien Chen
Institute of Oceanography, National Taiwan University, No.1 Sec. 4 Roosevelt Road, Taipei 10617, Taiwan

Eve-Julie Pernet and Karine Olu
Département REM/EEP/Laboratoire Environnement Profond, IFREMER/Centre de Bretagne, Institut Carnot EDROME, 29280 Plouzané, France

R. E. Sherlock, K. R. Walz and B. H. Robison
Monterey Bay Aquarium Research Institute (MBARI), Moss Landing, CA 95039, USA

Anoukchika D. Ilangakoon
Member, Cetacean Specialist Group of IUCN, 215, Grandburg Place, Maharagama, Sri Lanka

Abigail K. Alling
Biosphere Foundation, P.O. Box 112636, Campbell, CA 95011-2636, USA

Esther D. Beukhof
Current address: Centre for Ocean Life, National Institute of Aquatic Resources (DTU Aqua), Technical University of Denmark, Jægersborg Allé 1, 2920 Charlottenlund, Denmark
Wageningen University & Research, Chair Group of Aquatic Ecology and Water Quality Management, PO Box 47, 6700 AA Wageningen, The Netherlands
Maritime Department, Wageningen Marine Reseach, PO Box 57, 1780 AB Den Helder, The Netherlands

Joop W. P. Coolen
Wageningen University & Research, Chair Group of Aquatic Ecology and Water Quality Management, PO Box 47, 6700 AA Wageningen, The Netherlands
Maritime Department, Wageningen Marine Reseach, PO Box 57, 1780 AB Den Helder, The Netherlands

Babeth E. van der Weide and Joël Cuperus
Maritime Department, Wageningen Marine Reseach, PO Box 57, 1780 AB Den Helder, The Netherlands

Hans de Blauwe
Department of Invertebrates, Royal Belgian Institute of Natural Sciences, Vautierstraat 29, 1000 Brussels, Belgium

Jerry Lust
Maritime Department, Wageningen Marine Reseach, PO Box 57, 1780 AB Den Helder, The Netherlands
Van Hall Larenstein, Integrated Coastal Zone Management, 8934 CJ Leeuwarden, The Netherlands

Pablo Covelo and Alfredo López
Coordinadora para o Estudo dos Mamíferos Mariños (CEMMA), P.O. Box 15. 36380, Pontevedra, Gondomar, Spain

Departamento de Biologia & CESAM, Campus de Santiago, Universidade de Aveiro, 3810-193 Aveiro, Portugal

Lidia Nicolau
Sociedade Portuguesa de Vida Selvagem (SPVS). Departamento de Biologia, Campus de Gualtar, Universidade do Minho, 4710-057 Braga, Portugal Departamento de Biologia & CESAM, Campus de Santiago, Universidade de Aveiro, 3810-193 Aveiro, Portugal

Rekha J. Nair and S. Dineshkumar
Central Marine Fisheries Research Institute, Ernakulam North PO, Kochi 682018, Kerala, India

Yves Letourneur
Laboratoire LIVE et LABEX « Corail », Université de la Nouvelle-Calédonie, BP R4, Nouméa 98851, New Caledonia

Thibaud Moleana
Laboratoire LIVE et LABEX « Corail », Université de la Nouvelle-Calédonie, BP R4, Nouméa 98851, New Caledonia
UMR BOREA 7208,MNHN/CNRS/UPMC/IRD, 61 rue Buffon, Paris Cedex 5 CP 53, 75231, France Aqualagon SARL, BP 2525 Mont-Dore, Nouméa 98800, New Caledonia

Luc Della Patrona
Ifremer LEAD-NC, 101 Promenade Roger Laroque, Nouméa BP 2059, 98846, New Caledonia

Tarik Meziane
UMR BOREA 7208, MNHN/CNRS/UPMC/IRD, 61 rue Buffon, Paris Cedex 5 CP 53, 75231, France

Firas Alshawy, Murhaf Lahlah and Chirine Hussein
Department of Marine Biology, High Institute of Marine Research, Tishreen University, Lattakia, Syria

Marco Loia and Barbara La Porta
Laboratory of Benthic Ecology - ISPRA, Italian National Institute for Environmental Protection and Research, Via di Castel Romano 100, Rome 00128, Italy

Luisa Nicoletti
ISPRA, Italian National Institute for Environmental Protection and Research, Via Vitaliano Brancati 60, Rome 00144, Italy

Francisco Polanco-Vásquez, Ana Hacohen-Domené, Thalya Méndez and Alerick Pacay
Fundación Mundo Azul, Blvd. Rafael Landivar 10-05 Paseo Cayala Zona 16, Edificio D1 Oficina 212, Guatemala City, Guatemala

Rachel T. Graham
MarAlliance, 32 Coconut Drive, Po Box 283, San Pedro, Belize

Karoline Magalhães Ferreira Lubiana and Mariana Cabral Oliveira
Laboratório de Algas Marinhas "Édson José de Paula", Departamento de Botânica, Instituto de Biociências, Universidade de São Paulo, Rua do Matão 277, São Paulo, SP CEP 05508-090, Brazil

Sônia Maria Flores Gianesella and Flávia Marisa Prado Saldanha-Corrêa
Departamento de Oceanografia Biológica, Instituto Oceanográfico, Universidade de São Paulo, Praça do Oceanográfico, 191, São Paulo, SP CEP 05508-120, Brazil

M. Sofía Dutto
Consejo Nacional de Investigaciones Científicas y Técnicas (CONICET), Av.
Rivadavia 1917, C1033AAJ Ciudad Autónoma de Buenos Aires, Argentina.
Instituto Argentino de Oceanografía (IADO-CONICET/UNS), Área Oceanografía Biológica, La Carrindanga km 7.5, B8000FWB Bahía Blanca, Argentina

Gabriel N. Genzano
Consejo Nacional de Investigaciones Científicas y Técnicas (CONICET), Av.
Rivadavia 1917, C1033AAJ Ciudad Autónoma de Buenos Aires, Argentina.
Instituto de Investigaciones Marinas y Costeras (IIMyC), Funes 3350, B7602AYL Mar del Plata, Argentina
Departamento de Ciencias Marinas, Facultad de Ciencias Exactas y Naturales, Universidad Nacional de Mar del Plata (UNMdP), Funes 3350, B7602AYL Mar del Plata, Argentina

Agustín Schiariti
Consejo Nacional de Investigaciones Científicas y Técnicas (CONICET), Av.
Rivadavia 1917, C1033AAJ Ciudad Autónoma de Buenos Aires, Argentina.
Instituto de Investigaciones Marinas y Costeras (IIMyC), Funes 3350, B7602AYL Mar del Plata, Argentina

Instituto Nacional de Investigación y Desarrollo Pesquero (INIDEP), Paseo Victoria Ocampo N° 1, Escollera Norte, B7602HSA Mar del Plata, Argentina.

Julieta Lecanda
Museo Municipal de Ciencias Naturales de Monte Hermoso, N. Fossatty (ex Rio Paraná) N° 250, 8153 Balneario Monte Hermoso, Buenos Aires, Argentina
Universidad Nacional del Sur (UNS), Bahía Blanca, Argentina

Mónica S. Hoffmeyer
Consejo Nacional de Investigaciones Científicas y Técnicas (CONICET), Av. Rivadavia 1917, C1033AAJ Ciudad Autónoma de Buenos Aires, Argentina
Instituto Argentino de Oceanografía (IADO-CONICET/UNS), Área Oceanografía Biológica, La Carrindanga km 7.5, B8000FWB Bahía Blanca, Argentina
Facultad Regional Bahía Blanca, Universidad Tecnológica Nacional (UTN), 11 de Abril 461, B8000LMI Bahía Blanca, Argentina

Paula D. Pratolongo
Consejo Nacional de Investigaciones Científicas y Técnicas (CONICET), Av. Rivadavia 1917, C1033AAJ Ciudad Autónoma de Buenos Aires, Argentina
Instituto Argentino de Oceanografía (IADO-CONICET/UNS), Área Oceanografía Biológica, La Carrindanga km 7.5, B8000FWB Bahía Blanca, Argentina
Departamento de Biología, Bioquímica y Farmacia, UNS, San Juan 670, B8000DIC Bahía Blanca, Argentina

Noel Vella, Adriana Vella and Sandra Agius Darmanin
Conservation Biology Research Group, Department of Biology, University of Malta, Msida MSD2080, Malta

Z. Belattmania, A. Chaouti, A. Reani and B. Sabour
Phycology Research Unit, Department of Biology, Faculty of Sciences, Chouaib Doukkali University, P.O. Box 20, 24000, El Jadida, Morocco

M. Machado, A. Engelen and E. A. Serrão
CCMAR– Centre of Marine Sciences, University of Algarve, Faro, Portugal

Debasish Mahapatro
Department of Marine Sciences, Berhampur University, Berhampur, Odisha, India

R. K. Mishra
National Centre for Antarctic and Ocean Research, Ministry of Earth Sciences, Government of India, Goa, India

S. Panda
Regional CCF, Angul, Odisha, India

Grisel Rodriguez-Ferrer and Craig Lilyestrom
Department of Natural and Environmental Resources, Recreational and Sport Fisheries Division, PO Box 366147, San Juan 00936, Puerto Rico

Bradley M. Wetherbee
Department of Biological Sciences, University of Rhode Island Kingston, 120 Flagg Road, Kingston, RI 02881, USA
Guy Harvey Research Institute and Save Our Seas Shark Research Center, Nova Southeastern University, 800 N Ocean Drive, Dania Beach, FL 33004, USA

Michelle Schärer
H.J.R Reefscaping, P. O. Box 1442, Boquerón 00622, Puerto Rico

Jan P. Zegarra
U.S. Fish & Wildlife Service, Caribbean Ecological Services Field Office, P. O. Box 491, Boquerón 00622, Puerto Rico

Mahmood Shivji
Guy Harvey Research Institute and Save Our Seas Shark Research Center, Nova Southeastern University, 800 N Ocean Drive, Dania Beach, FL 33004, USA

Masami Hamaguchi, Naoto Kajihara and Hiromori Shimabukuro
National Research Institute of Fisheries and Environment of Inland Sea, Fisheries Research Agency, 2-17-5 Maruishi, Hatsukaichi, Hiroshima 739-0452, Japan

Miyuki Manabe
Kagoshima Prefectural Fisheries Technology and Development Center, 160-10 Takada-ue, Ibusuki, Kagoshima 891-0315, Japan

Yuji Yamada
Kurashikiminami High School, 330 Yoshioka, Kurashiki, Okayama 710-0842, Japan

Eijiro Nishi
Yokohama National University, 79-1 Tokiwadai, Yokohama, Kanagawa 240-8501, Japan

William E. Yeomans
Clyde River Foundation, Graham Kerr Building, University of Glasgow, Glasgow G12 8QQ, Scotland, UK

John Clark
11 Kildary Avenue, Glasgow G44 3AX, Scotland, UK

Ana Hacohen-Domené and Francisco Polanco-Vásquez
Fundación Mundo Azul, Blvd. Rafael Landívar 10-05 Paseo Cayalá Zona 16, Edificio D1 Oficina 212, Guatemala City, Guatemala

Rachel T. Graham
MarAlliance, 32 Coconut Drive, PO Box 283, San Pedro, Belize

Alan A. Myers
School of Biological, Earth and Environmental Sciences, University College Cork, Cork Enterprise Centre, Distillery Fields, North Mall, Cork, Ireland

David McGrath
Department of Natural Sciences, Galway and Mayo Institute of Technology, Dublin Road, Galway, Ireland

Will Musk
Institute of Estuarine & Coastal Studies (IECS), University of Hull, Cottingham Road, Hull HU6 7RX, UK

Greg W. Rouse and Josefin Stiller
Scripps Institution of Oceanography, UCSD, La Jolla, CA 92037, USA

Nerida G. Wilson
Scripps Institution of Oceanography, UCSD, La Jolla, CA 92037, USA
Western Australian Museum, Locked Bag 49, Welshpool DC, WA 6986, Australia
University of Western Australia, Crawley, WA 6009, Australia

Benjamin Mueller
Carmabi Foundation, Willemstad, Curaçao

Ana Hacohen-Domené and Francisco Polanco-Vásquez
Fundación Mundo Azul, Blvd. Rafael Landivar 10-05 Paseo Cayala Zona 16, Edificio D1 Oficina 212, Guatemala City, Guatemala

Rachel T. Graham
MarAlliance, 32 Coconut Drive, PO Box 283, San Pedro, Belize

Gary A. Allport
BirdLife International, David Attenborough Building, Pembroke Street, Cambridge CB2 3QZ, UK

Christopher Curtis
Casa 31, Condominio Sommerschield II, Rua do Palmar No. 817, Maputo, Mozambique

Tiago Pampulim Simões
Number One Serviços Maritimos, Av. Karl Marx nr. 478, 6 andar, Maputo, Mozambique

Maria J. Rodrigues
WWF Mozambique, Av. Kenneth Kaunda 1174, Maputo, Mozambique

J. Tavera and S. Rojas-Vélez
Laboratorio de Ictiología, Grupo de Investigación SEyBA, Departamento de Biología, Universidad del Valle, Cali, Colombia

Asha de Vos
Oceanswell and The Sri Lankan Blue Whale Project, 131 W.A.D. Ramanayake Mawatha, Colombo 2, Sri Lanka

Takeshi Yuhara
Tohoku University Graduate School of Life Science, 6-3 Aoba, Aramaki, Aoba-ku, Sendai, Miyagi 980-8578, Japan

Hiroyuki Yokooka
IDEA Consultants, Inc. Institute of Environmental Ecology, 1334-5 Riemon, Yaizu, Shizuoka 421-0212, Japan

Masanori Taru
Faculty of Sciences, Tokyo Bay Ecosystem Research Center, Toho University, Miyama 2-2-1, Funabashi, Chiba 274-8510, Japan

Reshma Pitale and Deepak Apte
Bombay Natural History Society, Hornbill House, S.B. Singh Road, Mumbai, Maharashtra 400 001, India

Inês Maia Dias, Martijn Spierings and Babeth Van Der Weide
Wageningen Marine Research, P.O. Box 571780 AB Den Helder, The Netherlands

Joop W. P. Coolen
Wageningen Marine Research, P.O. Box 571780 AB Den Helder, The Netherlands
Wageningen University, Chair group Aquatic Ecology and Water Quality Management, Droevendaalsesteeg 3a, 6708 PD Wageningen, The Netherlands

Joël Cuperus
Rijkswaterstaat, Ministry of Infrastructure and the Environment, Zuiderwagenplein 2, 8224 AD Lelystad, The Netherlands

Erik C. Franklin
Hawaii Institute of Marine Biology, School of Ocean and Earth Science and Technology, University of Hawaii at Manoa, Kaneohe, HI, USA

Index

www.ingramcontent.com/pod-product-compliance
Lightning Source LLC
Chambersburg PA
CBHW080409190526
45161CB00003B/174